LESSONS FROM AMAZONIA

LESSONS FROM AMAZONIA

THE ECOLOGY AND

CONSERVATION OF A

FRAGMENTED FOREST

Edited by

Richard O. Bierregaard, Jr.,
Claude Gascon, Thomas E. Lovejoy,
and Rita C. G. Mesquita

Foreword by

Edward O. Wilson

Prologue by

Eneas Salati

Yale University Press   New Haven & London

Published with assistance from
the Smithsonian Institution

Designed by James J. Johnson and set in Melior and
Syntax types by Copperline Book Services, Hills-
borough, North Carolina.

Printed in the United States of America by
Sheridan Books, Ann Arbor, Michigan.

Library of Congress Cataloging-in-Publication Data
Lessons from Amazonia : the ecology and
conservation of a fragmented forest / edited by
Richard Bierregaard, Jr. . . . [et al.] ; foreword by
Edward O. Wilson ; prologue by Eneas Salati.
    p.   cm.
    Includes bibliographical references (p.).
    ISBN 0-300-08483-8 (alk. paper)

    1. Rain forest ecology—Amazon River Region.
2. Fragmented landscapes—Amazon River
Region.   3. Rain forest conservation—Amazon
River Region.   I. Bierregaard, Richard O.
QH112 .L48 2001
577.34'0981'1—dc21   2001026535

A catalogue record for this book is available from
the British Library.

The paper in this book meets the guidelines for
permanence and durability of the Committee on
Production Guidelines for Book Longevity of the
Council on Library Resources.

10 9 8 7 6 5 4 3 2 1

Contents

PART III   Fragmentation Effects

PART IV   Management Guidelines

PART V   Synthesis

Foreword

The Biological Dynamics of Forest Fragments Project began in 1979 with a single, easily understood focus. Its primary purpose was to assess the effect of reduction in rainforest area on biological diversity, and particularly on the number of species of plants and animals in remnant patches. Working off the elementary theory of island biogeography, it asked: What is the rate of species extinction in the forest fragment as the size of the fragment is varied? Would this local extinction rate eventually slow and halt, so that the number of species equilibrates? And finally, what processes occur in the demography and interaction of species following a reduction in habitat area?

In 1981 and again in 1985 it was my pleasant and immensely edifying experience to visit the experimental site north of Manaus as the guest of Tom Lovejoy and Rob Bierregaard. (The latter visit is described in the opening of my 1992 book *The Diversity of Life.*) In that period it was already evident to the designers of the project that in this—the largest experiment in history, and conducted in Earth's most complex ecosystem—answers to the key questions would come hard. *Everything* is complicated. Species respond to fragmentation in idiosyncratic and unexpected ways, some of the latter representing novel physical and biological phenomena. Every species, it turns out, really is connected to everything else, at least ultimately, the way the textbooks suggest. The researchers at the study site kept their big introductory questions of biodiversity in mind, but in approaching them they found it first necessary to explore relevant properties of many of the individual species taken in turn.

And so it came to pass that the experimental site plots, trimmed in size, stressed all around their edges, leaking biodiversity, became a window on a large proportion of the unsolved problems of basic and applied ecology. When the project began it was like a village, simple in structure, small in the population of its attending experts. Now it is more like a sizable town, with many activities and specialties pursued by a sizable community of scientists. The projects they pursue range across much of modern evolutionary and environmental biology. Two decades have seen the study of habitat fragmentation gathered into a sizable discipline. This classic and ongoing experiment addressed by this book is a key part of it.

Edward O. Wilson
Frank B. Baird, Jr., Professor of Science
Museum of Comparative Zoology
Harvard University

Prologue

There are very few projects that have developed long-term studies of the humid tropical forests of the Amazon. The Biological Dynamics of Forest Fragments Project is one of them, and certainly the one that most expanded our scientific knowledge of Amazonian tropical forest fragmentation. Like many pioneering initiatives of this kind, the BDFFP started as a simple scientific expedition and developed into a collaborative institutional project, in this case between the National Institute for Amazonian Research (INPA) and the World Wildlife Fund–US, during the years I was INPA's director.

Aside from the impressive amount of scientific data that the BDFFP has produced and summarized in this volume, the support given to graduate students (M.Sc. and Ph.D.), as well as fellowships for field research to scientists and students from different nationalities, is a most important result of this collaborative effort.

The literature published by researchers from the BDFFP is extensive and substantial but has been scattered throughout the scientific literature. This book is a long-awaited and convenient summary of two decades of research. We thank the book's editors for the hard work and dedication it took to produce this volume.

The chapters are organized as a survey of many of the aspects of forest fragmentation that have been studied throughout the years. In Part I, reviews of the fragmentation theory and the effects of deforestation on the region's ecosystems present us with the scientific basis on which this project is founded, and trace its institutional and scientific evolution.

Part II provides an introduction to the flora of the forest to set the stage for Part III, where forest fragmentation effects are demonstrated through studies of the different vertebrate, invertebrate, and plant communities. The results obtained from studies of the fragmentation effects on tree and bird communities, the first field studies initiated within the context of the BDFFP, are fundamental to our understanding of how parts of the Amazonian ecosystem react when isolated. The relation between forest isolation, peccaries, and tree frogs (when peccaries abandon small fragments, their wallows dry up, depriving some species of frogs of a place in which to reproduce) is becoming a classic example of how the vital ecological relations between different species can fall victim to the effects of habitat fragmentation.

Finally, in Part IV, the authors describe different ways that the data obtained by the BDFFP can be applied to biodiversity conservation in the Amazon. This demonstrates that the researchers of this project have gone beyond the confines of basic science and theory. They have become aware that their work should also respond to the Brazilian society's concerns and are providing much-needed guidelines about how to preserve, in areas that inevitably will be developed, the biodiversity of the largest humid tropical forest on the planet.

I hope that this project continues for many, many years and that answers to our questions concerning the Amazon forest can be based on the technical and scientific data generated by the BDFFP and other research projects all over the world, proving that science does not recognize national boundaries.

Eneas Salati
Technical Director
Brazilian Foundation for Sustainable Development
(Fundação Brasileira para o Desenvolvimento Sustentável)

Preface

The destruction of natural habitats and the fragmentation of what remains are the central issues in understanding and preserving functional, representative samples of our planet's naturally occurring ecosystems and biological resources. In tropical regions, this problem is especially pressing because human encroachment is threatening the last large tracts of tropical rainforest—the most diverse and complex of terrestrial ecosystems, where more than half of the world's species live on less than 7 percent of the planet's surface.

The inevitable results of human intrusion into forested areas are a reduction in native habitat and the creation of a mosaic of forest remnants in a sea of nonforest habitats. Very serious ecological consequences of such changes in the landscape can be expected and have been demonstrated.

Unlike other more devastated regions (such as the Atlantic forest along the southeastern coast of Brazil), less than 15 percent of the Amazon forest has been felled. This provides an opportunity to incorporate ecological considerations in the future process of economic development and natural resource use. To do this, however, relevant information on the impact of deforestation and forest fragmentation has to be translated into guidelines for the conservation and sustainable development of the region.

The Biological Dynamics of Forest Fragments Project was initiated in 1979 to study the impact of forest fragmentation on the integrity of the tropical rainforest in central Amazonia. Our understanding of the impact of forest fragmentation has increased by leaps and bounds over the years, though each new study commonly raises more questions than it answers. Since the start of the project, nearly 60 graduate theses have been defended and more than 300 scientific papers have been published, mostly in scientific journals, books, and symposium proceedings.

The immense data set collected by project researchers has direct relevance to the conservation and management of the region. Unfortunately, many natural-resource managers and decision makers do not have the time to read or even access to the scientific literature, where most of our results have been published. As scientists and conservation biologists, we must make an effort to transfer what we know from the scientific to the decision-making arena. This was one of our objectives in preparing this volume. To do so, however, means to rethink and translate much of the existing scientific literature into a format and language directly relevant to those who will use it to the benefit of the Amazon region.

In compiling the volume, we have attempted to satisfy two audiences. Field biologists will find exciting new results and models to stimulate their own research. We have also attempted, for the benefit of the wildlife managers and conservation planners with a less theoretical bent, to create conservation guidelines for the development of the central Amazon region. All of the chapters in the introductory and management sections, as well as several in the section on fragmentation effects, should appeal to this subset of our readers. We have directed our authors to distill management principles from their data sets and end each chapter with these "conservation lessons" in lieu of the more traditional opening abstract and have attempted to present the essence of the more technical papers in our introductions of each of the book's sections.

Obviously it is impossible to present all of the published and unpublished results of nearly two decades of research in one volume. We have included chapters reporting on both unpublished information that is relevant to the aim of the book as well as overviews of results from long-term, ongoing research projects. In our introductions to each of the book's sections, as well as our concluding chapter, we draw the reader's attention to the relevant publications in the large body of work carried out at the Forest Fragments Project that could not be included herein. Additionally, the entire list of project publications and theses are presented, sorted by topic, to provide a quick overview of areas that have been well studied and those needing much more work.

A great deal of research has been published on habitat fragmentation, but only recently have there appeared collected volumes summarizing the field. In 1996, J. Schelhas and R. Greenberg edited a series of essays, *Forest Patches in Tropical Landscapes,* that dealt with the value of forest remnants in fragmented landscapes. The central theme of that book was socioeconomic with a strong emphasis on the interaction between local human populations and the habitat remnants that they have created, and, indeed, the important role that such patches can play, for both the local economy as well as ecology.

In 1997, W. F. Laurance and R. O. Bierregaard, Jr., edited another volume, *Tropical Forest Remnants,* that brought together cutting-edge information on forest fragmentation in tropical systems worldwide. The studies presented an eclectic range of ecological systems and geographical areas, including strong coverage of our own Biological Dynamics of Forest Fragments Project. The resulting collection presented a pantropical overview of the state of the art in studies of tropical forest fragmentation.

The present volume differs from the previous two in that it presents a multidisciplinary, in-depth view of research results at a single forest site through nineteen years of coordinated and integrated research. The particular value of this book is that all of the research was conducted at the same site. The responses of different organisms and processes can be compared without confounding biogeographical factors. This book, therefore, provides a focused and intensive case history from a single study site and derives practical conservation and management guidelines for the Amazon forest.

The book is organized into five major sections: "Theory and Overview," "Forest Ecology and Genetics," "Fragmentation Effects," "Management Guidelines," and a concluding "Synthesis." In the first section we present a historical perspective of the BDFFP, followed by a review of the theoretical effects of fragmentation and a detailed description of the study site and the project's experimental design. Chapters on forest ecology and genetics deal with general aspects of the tropical rainforest in the Manaus area. The largest section of the book deals with fragmentation effects on plants, invertebrates, and vertebrates. Most of these chapters present new perspectives on how particular groups of organisms are affected by forest fragmentation. In the fourth section we introduce many of the issues involved in the creation of conservation and management guidelines regarding the Amazonian rainforest. The volume concludes with our synthesis of the findings of the project into a series of principles of forest fragmentation and their implications for conservation planning in the Amazon.

We were enormously heartened to learn, as our book began to take shape, of the Brazilian government's decision to protect a full 10 percent of its Amazonian rainforests—on the order of 50 million hectares. As the details of this most encouraging and ambitious project are worked out, it must be remembered that the plans made for development in the vast area *not* included in the government-protected land will have an enormous impact on the future of the region. Legislation and development guidelines, however, will only be as good as the information on which they are based. This book, which is being translated for a Portuguese edition, is an important step in making solid information available to the relevant people. It is our hope that this information is both useful and used as the pervasive influence of our species on Amazonia's immense natural resources and biodiversity increases.

A twenty-year effort such as this is the product of the dedication and perseverance of many people. We have tried to thank as many of them as possible in the chapters that follow. Here we offer our thanks to Heidi Downey, whose fine copyediting makes us look like better writers than we are, and to Jean Thomson Black and her colleagues at Yale University Press, who saw merit in our proposal to put this volume together and have helped us through the complicated process of getting our ideas between the covers of a book.

PART I
Theory and Overview

The research reviewed in this volume addresses habitat fragmentation, the most visible effect of our own species' incursion into the most diverse of all terrestrial ecosystems—tropical rainforests. Nearly all of the chapters report on research carried out at the Biological Dynamics of Forest Fragments Project study site, 80 km north of Manaus, Brazil. The four chapters in this section set the stage, logistically and theoretically, for many of the research reports that follow.

Project History

In the first chapter, Richard O. Bierregaard, Jr., who directed the project in Manaus from its inception in 1979 until 1988, presents an overview of the project's history. This chapter is a nuts-and-bolts look at how the project grew from a crew of five—a field director, a cook, and three trail cutters—to a team of more than fifty researchers and field technicians based at its own building on the campus of Brazil's National Institute for Research in Amazonia (INPA), the project's collaborating research institution. Bierregaard highlights the intellectual integration of the project into the Brazilian scientific community, both at INPA and throughout the rest of the country, as students and re-

searchers from the southern universities joined the research effort.

Conceived with a very directed focus, on the issue of the "minimum critical size" of a tropical rainforest ecosystem (Lovejoy and Oren 1981), Bierregaard relates how the project's researchers broadened their horizons from isolated forest patches to the whole landscape that surrounded them, while the project's directors, both at INPA and at World Wildlife Fund–US and later at the Smithsonian Institution, recognized the project's potential as a vehicle for education and training—both for graduate students heading for a career in research and education, as well as for important policy makers and the public at large.

The project grew from an idea that first tickled Thomas E. Lovejoy's fancy in 1976, when a heated debate rippled through a portion of academia and the conservation community interested in ecosystem preservation. The debate arose as Robert MacArthur and Edward O. Wilson's elegant model of island biogeography (MacArthur and Wilson 1963, 1967), which presented a theoretical and mechanistic explanation for the empirical observations that there tend to be more species of plants and animals on large islands than on small, was applied by conservationists in designing nature preserves. (We are delighted that Ed Wilson, one of the founders of the field of island biogeography, was willing to write a foreword for this book.)

Theoretical Background

In Chapter 2, Og DeSouza and colleagues begin by outlining MacArthur and Wilson's model, which predicts that the number of species on an island, or isolated in a nature preserve, will settle at an equilibrium determined by the balance of predictable rates of species arriving on the island and going extinct over time.

The ecology of habitat patches, while superficially similar to real islands surrounded by water, is fundamentally different in that the species found in the habitat "island" may also survive, if not even reproduce in (as opposed to just flying over or drifting through) the "sea" surrounding the patch. DeSouza and his coauthors present an overview of the ways that ecological processes can affect species richness after a fragmentation event, distinguishing between those that operate at the local and regional scales and a gray area between them known as "mesoscale."

Local processes are often biological (increasing or decreasing competitive interactions), whereas regional processes are often structural, involving physical factors that alter rates of immigration. Various feedback loops between local, regional, and mesoscale processes can cause the number of species in a fragment to increase, decrease, or remain the same. The authors provide a theoretical framework in which these different chains of events can be isolated at least conceptually—a first step in understanding and eventually managing the links between habitat fragmentation and ecosystem persistence.

Patterns and Causes of Deforestation

With this theoretical framework in mind, Claude Gascon and his coauthors in Chapter 3 discuss in more detail the impact that habitat fragmentation can have on a previously continuous forest habitat. In addition to looking at the effects of fragmentation on local ecosystems, the authors compare the extent of deforestation and its socioeconomic root causes, or "drivers," in three different regions in the Amazon basin: the states of Rondônia in the southwest; Pará and Maranhão in the southeast, where extensive deforestation has already occurred; and the area around Manaus in central Amazonia, where forest clearing has been kept to a minimum by the area's very remoteness.

Beyond the obvious loss of habitat and subsequent isolation of small populations, forest fragmentation also creates a great deal of new edge and "matrix" habitat. Effects associated with edges are proximately related to physical changes in the environment— usually increased heat and decreased humidity, which affect plants and animals in various ways. Studies at the BDFFP in Manaus as well as elsewhere during the 1980s and 1990s drew attention to the interactions that the flora and fauna of forest patches have with the modified ecosystem around them. Because the "sea" around isolated habitat fragments is often not completely inhospitable to many of the organisms that live in the primary habitat, understanding what happens in fragments depends on understanding how the native flora and fauna can use human-altered habitats as corridors for dispersal or even reproduction.

New roads, leases, and concessions to international timber companies looking for new areas to (over)exploit, and increasing pressure from urban centers (principally Manaus) in the central Amazon are sure to increase the pressures to develop the relatively undisturbed forests of this region. This impending accelerated deforestation will provide not an opportunity but rather an obligation to use the results of the BDFFP in regional planning and management.

Project Design, Logistics, and Research

The final chapter in this section is effectively the "material and methods" section for the whole volume. The project's first and penultimate scientific directors, Bierregaard and Gascon, briefly describe the undisturbed *terra firme* forest (upland forest, not subjected to seasonal inundation) in which the experiment is being carried out, and the

experimental design of the BDFFP. Their description includes a never-before-published description of the process of forest clearing and burning that isolated the experimental reserves. They provide a summary of the important baseline data sets and major topics that have been addressed over the two decades of the project's existence.

The history of isolation of the different fragments, or reserves, and the land use around them are documented. As will be seen in the chapters that follow, the land-use practices that the ranchers did (or did not) employ determined what sort of second-growth vegetation developed around the isolated reserves, and this in turn had a great effect on much of the fauna in the isolates.

CHAPTER 1

The Biological Dynamics of Forest Fragments Project

Overview and History of a Long-Term Conservation Project

RICHARD O. BIERREGAARD, JR., AND CLAUDE GASCON

The Biological Dynamics of Forest Fragments Project, in its twenty-third year as of this writing, is the world's longest-running and most comprehensive attempt to assess the effects of habitat fragmentation on a tropical rainforest ecosystem. Over its two decades of existence it has evolved from a somewhat narrowly focused experimental attempt to identify the "minimum critical size" (Lovejoy and Oren 1981) of an Amazonian rainforest ecosystem to a much more broadly focused investigation of the many effects that habitat fragmentation has on natural communities. Most significantly, perhaps, not only has the scope of our research broadened from the effects of fragment size, but the very scale at which we ask questions has changed.

The project has grown and evolved in some ways that we might—and others that we were unlikely to—have foretold in 1976, when the idea for the project began to take form in the mind of Thomas E. Lovejoy, then vice president for science at the World Wildlife Fund–US. The path from conception to reality was tortuous and both politically and financially challenging, again in ways that we had not envisioned. In this chapter we chronicle the logistical and scientific evolution of the project and may, in a general sense, provide lessons to anyone beginning a cooperative research project in Brazil. Readers will find a thorough description of the study area and of the project's experimental design, and an overview of the project's major research efforts (completed and in progress) in Chapter 4, and in the complete list of BDFFP publications and theses (p. 371).

Overview

The BDFFP grew from Lovejoy's realization that a great need for experimental data on the effects of habitat fragmentation could be filled easily, at least in concepts, by taking advantage of a Brazilian law requiring landowners in Amazonia to leave half their land in forest. Lovejoy saw that under the right circumstances cooperative ranchers would let us redirect their clear-cutting in such a way that a bit of the land that they had to leave in forest anyway would wind up as isolated patches of different sizes in the middle of pasture land. Because we would be arranging for the isolation of the forest remnants we would have the unprecedented opportunity to study them quantitatively before they were isolated from the surrounding forest.

This is publication number 268 in the BDFFP Technical Series.

Lovejoy knew that Brazil's National Institute for Research in Amazonia (INPA) was headquartered in Manaus and that just north of the city a series of large cattle ranches was being established under rigid control of the Manaus Free Trade Authority (SUFRAMA). This constellation of undisturbed primary forest being developed in a controlled situation, only 80 km from a major metropolitan area and the INPA campus, permitted experimental study of the effects of habitat fragmentation.

The results of the early years of study showed us that fragmented forest patches were not truly isolated from their surroundings. Rather, the two interacted in dramatic and often unexpected ways (although with time, the "unexpected" quickly becomes the expected in tropical ecosystems). And in turn, what happened in the area surrounding the forest remnants was dictated by a myriad of socioeconomic factors that we never dreamed, twenty years ago, would alter the very course of our experiment.

As the project grew and our basic scientific program evolved, we recruited Brazilian students from the universities in the south, initially as field interns and later as graduate students, several of whom earned M.Sc. and Ph.D. degrees under the auspices of the BDFFP. And moving a step further down the career path, several of these project alumni are now full-time project researchers, including the project's current scientific director, Heraldo Vasconcelos. As more energetic and promising Latin American students passed through the project, we realized that we had an opportunity to help train a sizable cohort of scientists who would no doubt assume influential positions on the Latin American conservation scene.

That the project would turn into an educational goldmine never crossed our minds as we set out hypotheses about fragment size

and species survival, but what better way to reach the ultimate goal of the project—rainforest conservation—than to emphasize the training and outreach aspects of our efforts? In television documentaries and newspaper and magazine articles, the project has been a flagship for tropical rainforest conservation. More than sixty master's and doctoral theses have been written, and twenty more are under way. A field course in tropical ecology, modeled after the successful Organization for Tropical Studies program in Costa Rica, established in 1993, has introduced more than one hundred students from all over Latin America to the beauties, mysteries, and rigors of the Amazonian rainforest and given them the basic skills required to ask and answer scientific questions in the forest. Dozens of Brazilian and foreign legislators and government agency officials have spent their first night in the rainforest at one of our camps, and a more structured program has been initiated to get our results into the hands of important policy makers in the region.

History of the Project

PRELUDE AND PLANNING

In the mid-1970s the impending threats to the world's tropical rainforests were recognized by only a handful of biologists, most of whom had worked in the tropics, drawn there more by their fascination for these exuberant ecosystems than any concern for their persistence. Conservation was for the most part the purview of "conservationists," while "biologists" sought ever-diminishing natural systems in which to ply their trade. Conservation was mostly a question of saving endangered species—most of them such charismatic vertebrates as the panda, African elephants, and peregrine falcons. The first issue of the journal *Conservation Biol-*

ogy was more than a decade from publication.

In 1974 the World Wildlife Fund–US, realizing that the programs and initiatives that it would undertake should be based on sound science, hired Lovejoy—the first Ph.D. to join its ranks—to mix science and conservation. Lovejoy, with experience in the Neotropics, saw tropical rainforest conservation as an issue looming ominously on the horizon. As tropical forests began to retreat from the onslaught of chainsaws, bulldozers, and hydroelectric projects, a key priority was clearly to save representative samples of tropical ecosystems.

So how big should protected areas be? This question was being passionately debated by biologists (see Chapter 2), but there were little data on which to base strategic planning decisions. In 1976, in fact, the question came up in a National Science Foundation office in Washington, D.C., where Lovejoy and a few of his colleagues were discussing an especially hot question in conservation biology: Would a single large reserve contain more species than a series of smaller reserves with an area equal to the single large one? When the assembled group could agree on only one point—that there simply were not enough data to answer the question—Lovejoy audaciously proposed that an experiment in the Amazon was both needed and feasible.

A Brazilian law required anyone developing rainforest land to leave half of their parcel in forest. Knowing that large tracts were going to be opened up near Manaus, the site of INPA, Lovejoy saw an opportunity to create an experiment at the landscape level. The only way to ever get before-and-after data on the effects of habitat fragmentation would be to have someone else underwrite the costs of the deforestation. The Brazilian government, through a program developed to spur agricultural development around the landlocked free-trade port of Manaus, was paying roughly 75 percent of the costs of establishing cattle ranches, in the misguided hope that the ranches would one day feed the growing population of Manaus. This gave us an opportunity to set up the experiment. INPA's presence provided an institutional, intellectual, and logistical base for the operation.

Late in 1976, Lovejoy flew to Manaus, where, somewhat to his surprise, his idea was enthusiastically received by both INPA's director, Warwick Kerr, and scientists at INPA and the managers of the Agricultural Research and Development District 80 km north of Manaus. It was on 15,000-hectare (ha) cattle ranches in this district that the experiment could be turned into a reality.

The plan was elegantly simple. We would establish a series of forest plots ranging in size from 0.1 to 10,000 ha in forests north of Manaus that were slated to be cleared for cattle pasture. By arrangement, the ranchers would leave our plots in the middle of their pastures, creating the world's largest manmade laboratory of island biogeography. By monitoring selected subsets of the flora and fauna before and after the isolation of these experimental plots, we planned to quantify species-area relationships in this particular ecosystem. With these numbers in hand, our intention was to address the issues inherent in the project's original title, the Minimum Critical Size of Ecosystems. Specifically, how big an area of undisturbed forest would be needed to preserve the full complement of species and, most important, their interactions in a given region?

The initial request, made in 1978, to the Brazilian government for authorization to conduct the experiment characterized the work as a foreign expedition. In 1979, as the proposal was nearing approval, we received the sage advice to reformulate the proposal as a joint effort between INPA and WWF-US.

Herbert Schubart, then head of the ecology department at INPA and eventually the institute's director, agreed to be Lovejoy's Brazilian counterpart and co-principal investigator. The new proposal was approved that year by the Brazilian National Research Council (CNPq).

With Brazilian colleagues and authorization on board, it remained to persuade the ranchers to give us access to their land and to leave our experimental forest plots in the middle of their pastures. It was our good fortune that all of the ranchers proved to be generous and tolerant of our presence, even if they did not not always fully understand what we were up to. The project has succeeded over the years in no small part because we have maintained cooperative relationships with the landowners. In fact, as we traipsed around in search of our experimental plots, we wound up doing most of the topographical surveys on which the ranchers planned their pastures.

The physical and temporal scale of the project was such that funding was unlikely to be forthcoming from the conventional sources, such as the National Science Foundation. So the seed money was obtained from the U.S. Fish and Wildlife Service, the Man and the Biosphere Program of US AID, and the Weyerhauser Family Foundation in exchange for a vague commitment to write inventory manuals. We were ready to launch the effort.

UP AND RUNNING, OR WELL-LAID PLANS
MEET REALITY: 1979 TO 1980

Richard O. Bierregaard, Jr., hired to direct the project in the field, moved to Manaus in mid-1979. In August he located the first reserve—still in pristine forest—and in October, data collection began. With no room on the INPA campus for us to set up offices, we established in-town headquarters in a residential neighborhood several kilometers from INPA. Our absence from the INPA campus was seen by some as symptomatic of our lack of integration with the scientific community at INPA, and more broadly in Brazil itself. This image problem has plagued us almost continuously (e.g., Gama 1997), but it eased somewhat in recent years when we moved to headquarters on INPA's campus.

In the field, executing what appeared to be a relatively simple research plan was everything but simple. On the ground, a 0.1 ha plot was impracticably small, and a 10,000 ha plot impossibly complex to see through to isolation. We had to select our study plots in areas for which no accurate maps existed and fit them into what we conceived the future landscape would look like so that the reserves would be fully isolated and the ranchers would not be inconvenienced. The carefully designed configuration of isolated forest patches that we had mapped out in Washington was unceremoniously discarded within weeks of cutting the first survey trails.

Because of our limited budget, we were not able to recruit and hire scientists to tackle specific lines of inquiry. Rather, we had to lure the research staff at INPA and other institutions into the project with nothing more than the offer of room and board, a field hand or two, a relatively dry spot in the forest in which to hang a hammock, and a chance to work in soon-to-be-perturbed primary rainforest.

As it turned out, in the early years we were able to cover at least representative groups of plants, vertebrates, and insects. Birds and canopy trees were studied by Bierregaard and Judy Rankin–de Mérona, respectively, monkeys by Márcio Ayres and Anthony Rylands, butterflies by the indefatigable Keith Brown, ants by Woody Benson, and frogs by Barbara Zimmerman.

Logistically, we were moving ahead at a

satisfactory pace. Two reserves (1 and 10 ha) were isolated in 1980 (plate 1) and seventeen reserves were demarcated and awaiting isolation by the latter half of 1981.

THE MIDDLE AGES: 1981 TO 1986

During the beginning of this period the ranchers' plans seemingly changed every week, frustrating the researchers, graduate students, and field interns who had come to Manaus to participate in the project (plate 10). Because our experimental perturbation (isolation) was based on the ranchers' continued development of their land, the outcome of the project hinged on factors outside of our control—including the Brazilian economy.

When the project's research permit was renewed, at the request of the Brazilian scientists who reviewed the renewal request, its name was officially changed to the Biological Dynamics of Forest Fragments Project. This change was indicative of a shift in the philosophy of the project. As we faced the reality that we would not see any 10,000 ha reserves isolated, we felt uncomfortable with the idea of extrapolating our results several orders of magnitude larger than our largest experimental plot. At the same time, we saw that throughout the areas of Amazonia under development a landscape mosaic of active and abandoned pasture and forest remnants of 100 to 1,000 ha in size was the seemingly inevitable result of colonization and development. The results of our studies, therefore, would have direct relevance to vast areas of Amazonia.

During this period, the Brazilian economy was staggering under huge international debt and rampant inflation. (During the project's first ten years there was a billionfold devaluation of the currency.) In the early 1980s, federal funds targeted for development projects in Amazonia started to dry up.

With cattle ranches unable to turn a profit, despite strong subsidies, the ever-decreasing federal funds were diverted from the agricultural sector. Without their subsidies the owners of the ranches where the BDFFP was being carried out were unable to afford the (real) cost of implementing pasture. They stopped clearing new forest, which meant that our experimental design of twenty-four isolated reserves would be not even half achieved. The financial situation was so bad that the ranchers could not keep open even the little they had already cleared, so we watched the "sea" around our experimental "islands" revert to scrubby second growth.

During this frustrating period we were sitting on some intriguing results from the two 1980 isolates; but with no replication they were nothing but intriguing anecdotes. Finally, in 1983 and 1984, eight more reserves were isolated, and we could get down to some real science (plates 2 and 6).

In the project's fifth year (1983) Luis Carlos de Miranda Joels was hired as assistant director. He took the initiative in increasing Brazilian involvement in the project, from field interns to the directorship.

TRANSITION: 1987 TO 1989

In 1987, Marina Wong, an ecologist with a strong ornithological background and much tropical experience, replaced Bierregaard as field director. In 1988, Roger Hutchings, a project participant with experience in the botanical ecology subproject, a member of the BDFFP Management Committee, and an M.Sc. candidate working on butterflies in the reserves, became the third field director.

Bierregaard's departure, along with that of Judy Rankin–de Mérona, longtime principal investigator of the botanical ecology, or phyto-demography, subproject, marked the end of the era of "mega-projects" run by principal investigators and staffed in the field by student interns, or *mateiros,* as the local woodsmen are known. There had been a growing number of graduate students working their own projects, but the winding down of the two core subprojects of the BDFFP heralded a paradigm shift in the structure of the project. From that point on, most of the research effort has revolved around graduate student research.

Lovejoy's 1987 move from WWF-US to the Smithsonian Institution shook the financial underpinnings of the project. While at WWF-US, Lovejoy had championed the fund raising for his brainchild. With his move to the SI, there was little incentive for WWF-US to raise money for what many saw as the Lovejoy Project, and Lovejoy, as an employee of the Smithsonian, could not raise funds for a WWF-US project. It became clear, as the budget began to dwindle, that the project would have to move to the Smithsonian.

The move was approved—thanks in no small part to the efforts of Frank Talbot, then director of the Smithsonian's National Museum of Natural History—and in 1989 the project moved to its new office under the umbrella of the museum's BIOLAT Program, which evolved into the Biodiversity Programs under Don Wilson's direction.

With Lovejoy at the foot of Capitol Hill, it was not long before the project became a showcase for tropical forest conservation among U.S. legislators. Over a two-year period, Lovejoy brought more than 7 percent of the U.S. Senate, including Al Gore, for an overnight stay in the rainforest. Although one distinguished senator, covered with mud after helping push a Toyota up a hill, was inspired to comment on the irony of the trip being called a "junket," these first-hand experiences with the ecosystem moved the issue of rainforest conservation from the distant and theoretical to the real world for these policy makers.

THE INPA AND SMITHSONIAN YEARS: 1989 TO THE PRESENT

In Manaus, the project underwent major changes in administration and logistics. Hutchings settled in as field director and oversaw a series of moves that led to the project's finally being installed on the INPA campus.

Space was limited at INPA, and we needed to finance the construction of our own offices and lab. With funds from the U.S. government, in 1992 we finished a building to house the project's administration and provide some research space.

With the project settled into the "house that Roger built," Hutchings handed over the reins to Claude Gascon, an alumnus of the project whose Ph.D. thesis focused on the ecology of tadpoles in BDFFP reserves. The move to INPA fostered stronger integration of the project with the student and research community there. BDFFP researchers dramatically increased their long-established but low-profile involvement with teaching and advising graduate students. Currently, the project is entrenched in the ecology department at INPA, has participation from INPA researchers and students from the ecology, forestry, entomology, and botany departments, and since 1998 has been under the scientific direction of Heraldo Vasconcelos, who did his M.Sc. and Ph.D. work under the project's aegis.

Since 1989 the project has moved strongly into the outreach arena. Most notably, it initiated a field course, modeled after the Organization for Tropical Studies operation in Costa Rica, which has proved to be a great success and has established a momentum of its own. Another noteworthy effort in outreach involved a collaboration with a research group in southeastern Amazonia that was studying the highly disturbed area of Paragominas. This collaboration culminated in a successful workshop and published proceedings (Gascon and Moutinho 1998).

Management and Oversight

From the beginning the project has been directed by a management committee composed of representatives of INPA and the project's directors, principal investigators, and researchers. The management committee deals with logistical, political, financial, and scientific issues. The annual review of project proposals is carried out by this committee. In recent years the annual review meeting has alternated between INPA and the Smithsonian in an effort to increase interchange between the two host institutions.

Careful documentation of the committee's activities has provided an important record of the project's development. Such a management structure is highly recommended for any international cooperative research effort of this scale. In fact, the BDFFP management model has been duplicated by other international collaborative projects at INPA.

We have periodically invited groups of Brazilian and international observers to visit the study area, speak with our students and researchers, and offer their criticisms and suggestions. Five such committees (aside from the project's management committee) have provided these important reviews of the BDFFP. The review committees were chaired by Robert May, Ângelo Machado, Michael Soulé, John Terborgh, and Stuart Pimm.

Funding

Until the late 1990s, the BDFFP's funding came almost solely from philanthropic foundations, corporations, the World Wildlife Fund, and the Smithsonian Institution, rather than from sources that more typically support basic research (such as the National Science Foundation). Recently, however, our long-term stability and substantial baseline data on the local ecosystem and land-use history have attracted support from more conventional basic-science sources, both within Brazil and internationally. We receive funds from the G-7 Pilot Program to Preserve the Tropical Rainforest, a competitive grants program directed at research into the conservation of the Amazon rainforest. We have also recently received competitive grants from the NASA-INPE-INPA Large-scale Biosphere Atmosphere Program to model the dynamics of biomass in the experimentally fragmented landscape and from the newly created Long-Term Ecological Research Program in Brazil. INPA has also provided logistic and financial support through scholarships for students and fellowships for the scientific director and other visiting scientists directly from the institution's budget or from the Brazilian Ministry of Science and Technology (MCT) and the National Research Council (CNPq).

Acknowledgments

In thanking the many people who have contributed to the success of the BDFFP over nearly twenty years we are sure to overlook many more people than we mention. With all due apologies to those not mentioned here, we highlight a number of people and organizations who have done so much for us over the years. Herbert Schubart, as the project's initial co-principal investigator; all the INPA directors who watched over us and helped us find our way through the Brazilian bureaucracy—Warwick Kerr, Eneas Salati, Herbert Schubart, Henrique Bergamin, José Seixas-Lourenço, and Ozório Fonseca; our first national assistant director,

Luis Carlos de Miranda Joels, who fixed so much of what had been done wrong and made major headway getting the project integrated into the Brazilian scientific community; INPA's international cooperation coordinators, George Nakamura, Ângelo dos Santos, Sérgio Fonseca, João Ferraz, Cláudio Ruy-Vasconcelos, and Vitor Py-Daniel; INPA's liaisons to the project's management committee, Niro Higuchi, Renato Cintra Soares, Heraldo Vasconcelos, Gil Vieira, and Roger Hutchings; our mateiros, who taught us so much about the forest and taught the gringos most of their Portuguese (*most* of which we could use in public); Ary Jorge Correia Ferreira, our indefatigable station manager, who rose from the ranks of our botanical field technicians to the highest position of responsibility on the staff; the ranch owners and managers, especially Aldo Busnello and Antônio Carlos Hoholben; and, of course, the more than two hundred field interns, graduate students, and researchers who collected and analyzed the data presented in this volume and in the project's Technical Series.

In Washington we must thank WWF-US for supporting the project from the outset and the Smithsonian for having seen fit to adopt us—especially Frank Talbot, who helped us get settled into the institution, and Marsha Sitnik, who may be the only person who *really* understands how the Smithsonian works; Peggy Rasmussen, Victor Bullen, and the late Pieter de Marez Oyens, stateside coordinators who packed innumerable "gringo pouches" and got us all back and forth; Judy Sansburry and more recently Nancy Shorey, who streamlined the finances and financial reporting; and Dalena Wright, who helped with funding our office and lab building.

Financial supporters include INPA, the Man and the Biosphere Program of US-AID, U.S. National Park Service, the Tinker Foundation, the McDonald's Corporation, Citibank, the A. W. Mellon Foundation, the Pew Charitable Trusts, the Weyerhauser Family Foundation, the John D. and Catherine T. MacArthur Foundation, US-AID Brazil, the Summit Foundation, Shell International, the Homeland Foundation, World Wildlife Fund–US, the Smithsonian Institution, and private contributors.

CHAPTER 2

A Theoretical Overview of the Processes Determining Species Richness in Forest Fragments

OG DESOUZA, JOSÉ H. SCHOEREDER, VALERIE BROWN, AND RICHARD O. BIERREGAARD, JR.

Ecosystem or habitat fragmentation refers to "any sort of process which results in reducing an original area's size, creating a new, smaller, or several split ones" (Lovejoy, Bierregaard, et al. 1986). The literature concerning the subject is growing fast and deep, promoting formerly overlooked issues to the status of main areas of research. Besides increasing in number, studies of habitat fragmentation have also gained complexity, as exemplified by the use of nontrivial analytical methods related to percolation theory and critical phenomena (e.g., Andrén 1994; Bascompte and Solé 1996, 1998). With this growth in the literature, the overall scenario related to habitat fragmentation may be easily misunderstood, especially by people new to the field.

In this chapter we present a brief synthesis of the predicted effects of ecosystem fragmentation on species richness to provide an overview of the ecological mechanisms thought to be involved. We do not intend to present an extensive review of the theme, nor any new theoretical insights. Readers interested in a more specific approach should refer to other chapters of this book (see, for example, Chapters 8, 10, 13, 19, and 21). Those interested in broader approaches, such as landscape mosaics, as well as a broader geographical perspective, will find good reviews in Forman (1995) and Laurance and Bierregaard (1997a).

Habitat Fragmentation

Ecosystem fragmentation can result from both natural and "non-natural" (human-induced) causes, such as rising ocean levels isolating tracts of vegetation or anthropogenic deforestation converting continuous habitats into scattered patches in a "sea" of disturbed habitat. Because of the structural similarity between such patches and true islands, as they themselves argued, MacArthur and Wilson's (1967) "theory of island biogeography," developed to explain patterns of species numbers on oceanic islands, should provide a valid first approximation to the biological dynamics in habitat patches left in man-made habitat mosaics. According to the model and its refinement by Brown and Kodric-Brown (1977), the number of species inhabiting an island or habitat fragment will be a dynamic equilibrium between opposing rates of extinction and immigration such that islands hold a relatively constant number of species.

Because the model treats all species equally, the identity of the species found on

This is publication number 269 in the BDFFP Technical Series.

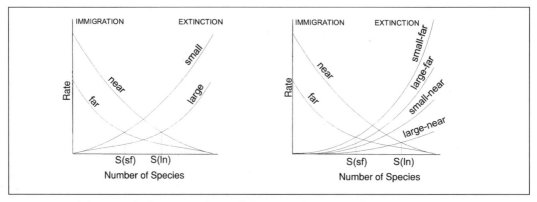

Fig. 2.1. Models of insular biogeography. At left, MacArthur and Wilson's (1967) model, representing the number of species in an island (S) being determined by a balance between extinction and immigration rates, which depend on the island's size and its distance from the source of colonizers. At right is Brown and Kodric-Brown's (1977) modification of the model, incorporating the "rescue effect" of immigration on extinction.

a given island would vary with time. On any island, the immigration rate (i.e., the number of new species arriving per unit time) falls as more species become established, because fewer immigrants will belong to new species. Simultaneously, the extinction curve rises, because the more species that are present, the more there are to become extinct, and the more likely any given one will become extinct because of ecological and genetic accidents related to small population sizes. The intersection of these curves represents the equilibrial species numbers and turnover rate (fig. 2.1 left).

Immigration rates depend on the proximity to a continent or other source of dispersing species, so that islands near the continent would present higher immigration curves than would isolated islands. Extinction rates depend on island size to the extent that the smaller the area, the stronger the detrimental effect promoted by the combination of diminishing population size and increasing probability of interference among species. Extinction rates, however, are also dependent on island isolation, because the same parameters that affect colonization by species new to the community also affect

the arrival of individuals belonging to the species already on the island. Providing that this immigration rate is sufficiently high, it will reduce the extinction rate because demographic and genetic contributions of immigrants tend to increase the size and fitness of insular populations, through what has been termed the "rescue effect" (Brown and Kodric-Brown 1977). High immigration rate also will have a statistical effect in reducing the apparent extinction rate simply by decreasing the probability that a given species will be absent during any census. Consequently, the extinction curves should be higher for islands that are small and distant, and lower for large and near ones. The equilibrial species number increases from small and distant islands to large and near ones (fig. 2.1 right).

These effects of the species-area relationship must be considered when analyzing the consequences of habitat fragmentation. One of the most universal relationships observed in biological communities is that between species and area—large areas will have more species than smaller ones. This relationship may be linked to effects of area per se or to an increased variety of habitats in larger areas. For a more complete discussion of species-area relationships, see, for example, Rosenzweig (1995).

Researchers using the theoretical framework derived for island situations in the study of ecosystem fragmentation must consider, for terrestrial ecosystems, that colonizers may also come *from,* rather than just pass over, the "sea of inhospitable habitats" surrounding the fragment. This changes how one should analyze immigration rates, because two types of sources of dispersing species must be considered: immigrants can come both from the "continent" (the nearest block of undisturbed habitat) and from the "sea." Moreover, ecosystem fragmentation promotes drastic and sudden changes in habitat, reducing its size and possibly, via sampling effects, the number of habitats in a given area. Some species, however, will survive this process, adding a confusing factor to the predictions of the theory of island biogeography, which assumes that an island is initially a species-empty environment.

Pursuing the island analogy, habitat fragments have been likened to "land-bridge" islands, which were once part of a more extensive habitat before becoming isolated by rising water levels and began with a full complement of the regional flora and fauna. In contrast, true "oceanic" islands, created by volcanos, for example, began as barren rocks devoid of life. In the latter case, immigration will initially be the dominant force, whereas in the case of land-bridge islands, extinction will predominate as ecosystems restricted to small areas "shed" species for the reasons outlined above. These extant species, present at the moment of isolation, may affect extinction and colonization rates in a way that may make these rates appear independent of fragment size and isolation. Therefore, although very useful, the equilibrium theory of island biogeography and associated species-area rela-

tions provide only a framework for the study of the biological dynamics of fragments (Forman 1995).

A number of factors in addition to fragment size and isolation must be considered in making predictions regarding the effects of fragmentation on terrestrial ecosystems. Most important, as Zimmerman and Bierregaard (1986) and others have pointed out, different species have different ecological requirements and therefore do not have equal probabilities of immigration or extinction on a real or habitat island.

In the three decades since MacArthur and Wilson developed their theory of island biogeography, ecologists and conservation biologists have adopted a much broader view and begun to look not just at individual fragments but at how the fragments interact with communities in the surrounding landscape. Indeed, an entire discipline, landscape ecology, has arisen as researchers began to question how populations interact with one another in fragmented landscape mosaics (e.g., Wiens and Moss 1999). Levins (1970) coined the term "metapopulations" to recognize that populations in a heterogeneous environment interact with each other, exchanging genetic material through emigration and immigration. Fahrig and Merriam (1994) argued that we should be focused on the spatial structure of fragmented habitats, because often the most important source of immigrants to a habitat fragment will be another fragment rather than a nearby "mainland" (continuous habitat) source with continuous habitat. To this end we have begun to measure our landscape mosaics from as far away as possible—via satellites—and have developed geographical information systems to quantify the structure and processes that are occurring at the landscape level (see Chapter 28).

Again, it is outside the scope of this chapter to synthesize this enormous and burgeoning discipline. Throughout the rest of this volume readers will encounter the reports of many studies dealing with landscape-level issues—from the genetics of rare species (Chapter 8) to metapopulations (Chapter 15) to spatial analyses of remotely sensed data describing the entire study area (Chapter 28). Here we shall focus on the processes in individual fragments that affect population levels.

The Biological Dynamics of a Fragmented Ecosystem

With habitat fragmentation, as opposed to oceanic island-based systems, there are three possible results following a fragmentation event: increment, diminishment, or maintenance of the species number of the community concerned (fig. 2.2). Once an ecosystem is fragmented, structural and biological changes are observed, with a consequent effect on the ecological processes operating in the fragment. Structural changes involve isolation of the remnant from its original matrix and increased proximity of its core to the bordering environment. Biological changes involve losses of individuals, species, and habitats. Structural changes are, therefore, linked to regional processes, such as immigration, whereas biological changes refer to local processes, such as species interactions. This dichotomy, however, is not as clear as we might like. Edge creation, for instance, is primarily a structural change, but it will affect habitat suitability (a local process) as well as immigration (a regional process). Thus, edge effects lie in a gray zone between local mechanisms that operate in the community living within the fragment, and large-scale processes related to the ensemble of local communities connected by dispersal. As such, edge effects deal with species interactions at local and regional

scales simultaneously. Mechanisms addressing such a broader geographical context of species interactions are referred to by some authors as operating at the "mesoscale" (Holt 1993).

The proximate consequence of these changes is local extinction, which is the most frequently reported effect of habitat fragmentation (e.g., DeSouza and Brown 1994; Bierregaard and Stouffer 1997; Stratford and Stouffer, 1999). Some researchers have reported that the number of species in the fragment relative to sites in the continuous forest may be maintained—or might even increase—after fragmentation (e.g., Malcolm 1997a; Tocher, Gascon, and Zimmerman 1997).

Interestingly enough, no matter whether species richness increases, decreases, or is kept constant after fragmentation, the ecological processes involved have many mechanisms in common (Gascon et al. 1999). What characterizes a particular process is, to a large extent, the number of mechanisms involved and the pathways linking such mechanisms. The dynamic equilibrium postulated by island biogeography theory is achieved by loops and feedbacks performed by these chains of mechanisms (see fig. 2.2).

Some immediate changes are promoted by ecosystem fragmentation (boxes in top row of fig. 2.2). These changes are the trigger mechanisms that will lead to the three possible outcomes: reduction, increase, or maintenance of species richness. As discussed above, the changes are related to local, mesoscale, and regional processes that, although they share several mechanisms, will be discussed separately.

LOCAL PROCESSES

Local processes are represented by losses of individuals and habitats (fig. 2.2, top boxes on the right). Losses of individuals may lead

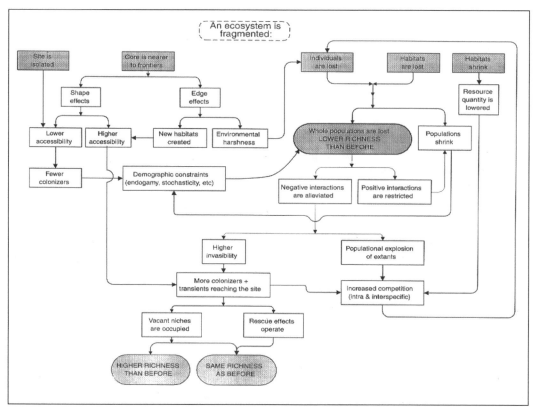

Fig. 2.2. Ecological processes determining species richness upon a fragmentation event. The boxes at top represent the immediate changes caused by habitat fragmentation. Rounded, shaded boxes represent the three possible outcomes of ecosystem fragmentation regarding species richness. Loops are indicated to make it clear that an observed value of species richness may not represent a static result but can be part of a dynamic equilibrium.

to local extinction if the individuals missed represent a whole population. This is likely to happen either to small populations or to species with high levels of spatial aggregation. Local extinction may be caused by other mechanisms, such as the loss of habitat, for species restricted to the lost habitat. It is often difficult to isolate the effects of each of the above mechanisms because they probably occur together. In areas with high beta (habitat-specific) and gamma (regional-turnover) diversity patterns, some species may be missed by the fragmentation event simply because of an incomplete sampling of all local habitats or of all regional subsets (see Schluter and Ricklefs [1993]). This is

particularly true for highly patchy environments, such as tropical forests, as shown by Zimmerman and Bierregaard (1986) for central Amazonian forest frogs.

Sampling effects also operate frequently in agricultural landscapes, where the preserved areas are normally spared because of their agricultural unsuitability (Usher 1987). That is, habitat remnants will represent only poor soils, steep topography, and so forth, thereby preserving only the community adapted to such situations. Local extinction may then be the final product of ecosystem fragmentation, specifically as a result of sampling effects, habitat loss, or any other factors. In such cases, the "species comprising a depauperate fauna should constitute a subset of those in richer faunas," the so-called nested subset hypothesis of Patterson (1987).

It is often difficult to isolate the importance of sampling effects (if, by chance, the fragment does not contain some of the original species) from habitat loss (the fragment does not contain some of the original species because they were restricted to one or more lost habitats not found in the fragment—which is itself a sampling effect at the habitat level). If one aims to verify the importance of these factors in the reduction of species richness, it is necessary to conduct rigorous sampling to estimate the heterogeneity of habitats in both pristine and fragmented ecosystems.

Missing individuals or habitats may not necessarily lead directly to local extinction but simply to reduction in population sizes ("Populations shrink" box in fig. 2.2) of some or all of the populations in the community. This reduction may in turn lead to demographic constraints, impairment of positive interactions, or alleviation of pressure from negative interactions. Small populations are more prone to extinction caused by chance events than are large populations (stochastic effects) and are likely to face important genetic constraints (Templeton et al. 1990). The smaller the population, the higher the chance that individuals will mate with close relatives, which may produce few offspring and offspring that are weak or sterile. This is called inbreeding depression, or endogamy. Outbreeding depression is the opposite trend: the paucity of mates may lead to hybridization with nearly genetically incompatible individuals (say, different species from the same genus, which often produces vigorous but sterile hybrids).

Small populations may also have their fitness severely affected by genetic drift (Nei, Maruyamo, and Chakraborty 1975), which is the chance increase or decrease of alleles. Stochastic and genetic constraints may lead to extinction of some populations in the fragmented ecosystem, lowering species richness of the community. It is important to notice that this extinction is secondary, induced by the decrease in population density and not directly by the loss of species or habitats. Such a decrease in species richness would not happen for some generations after a fragmentation event and would not be detected by sampling an area soon after its isolation.

Reductions in population sizes can adversely affect positive interactions if lost individuals belong to species playing mutualistic roles in the remnant (see Chapter 12). This would cause, ultimately, a decrease in the reproductive success of individuals in the remnants, reducing even further their population sizes.

A decrease in population densities may, on the other hand, lead to the reduction of competitive interactions, if resources have been unaffected by fragmentation. Similarly, predation pressures may be also diminished when predators' populations decline or disappear. This reduction of interactions may either ease the increment of population size of resident species through density compensation (Case 1975) or increase the invasibility of the remnant by immigrants (Tilman 1997). Density compensation occurs when extant populations grow in such a way that the community in the fragment will present the same total number of individuals as before fragmentation, but these individuals will be distributed over fewer species (see M. Williamson [1981] for a further discussion of density compensation and Terborgh, Lopez, et al. [1997] for a report on hyperabundant species in ecosystem fragments).

It may be difficult to identify the mechanisms leading to an increase in populations surviving in a habitat fragment, because, for example, two prey species whose populations are limited by a predator are, in essence, indistinguishable from two species competing for a resource. Populations may increase when predation pressure is reduced through an effect known as apparent competition (Holt 1977) or competition for enemy-free space (Jefferies and Lawton 1984), which may strongly affect the structure of ecological assemblages in fragments (Namba, Umemoto, and Minami 1999).

Density compensation may lead to a secondary increase in either intra- or interspecific competition, which, in turn, may lead to elimination of individuals from the remnant. If, on the other hand, reduction of interactions increases the invasibility of the remnant by immigrants, three possible outcomes may be observed, depending on the number and identity of immigrant species arriving in the fragment. First, the ecological niche of an invading species may overlap that of resident species, with the resulting competition leading to the elimination of individuals from the remnant, as discussed above. Second, an invading species may occupy niches that have been emptied totally by local extinction or partially by population shrinkage. Alternatively, if an immigrant belongs to a resident species, its arrival will increase population densities, thereby decreasing the strength of demographic constraints. The first event above is an adverse effect of higher invasibility, because it will feed back via the loop to local extinctions, leading to lower species richness. The second event may either maintain the same species richness as before fragmentation, when the invading species occupies niches freed by local extinction, or increase species richness, when the invader occupies niches partially freed by a reduction in population of residents. Species richness in the fragment may also increase when vacant niches are occupied by more than one invader species. The third event (the invader belongs to a resident species) can lead to restoration of original species richness, through "rescue effects" (Brown and Kodric-Brown 1977). It is important to realize that different processes may cause species richness to remain constant. Therefore, when species richness is deemed unchanged after fragmentation, it does not necessarily mean that the community is "immune" to the disturbance in question.

Whereas fragmentation often leads to a reduction in species richness through sampling effects for both species and habitat, fragmentation can preserve the original variety of habitats with nothing more than a reduction in their extent ("Habitats shrink" box). Residents would then be expected to experience resource shortages and a consequent increase in competition. Again the loop to local extinction is fed back as a secondary step induced by the increase in competition pressures and its consequential exclusion of individuals from the site.

MESOSCALE PROCESSES

Structural changes (top left boxes of fig. 2.2) can also trigger processes determining the species richness of fragments. By cutting down surrounding environments, one brings the core of a fragment closer to its borders, creating edges where they previously did not exist. Edges can be important determinants of habitat quality (Kapos 1989; Malcolm 1994), because the smaller the site, or the more linear (LaGro 1991), the greater the proportion of its area is exposed to winds, solar radiation, and the like. In such a scenario, more sensitive species would tend to disappear. Whereas such changes represent increased environmental harshness for some species, for others they may be seen as creating new habitats. These species would

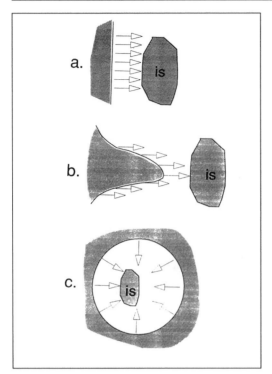

Fig. 2.3. Shape effects in colonization rates for passive dispersers.

distant margin (fig. 2.3c). The same can be true for the shape of the fragment: the number of propagules "hitting" the fragment should increase when the side of the site facing the source population expands and its position approaches a perpendicular to the main direction of colonists' travel (fig. 2.3a). Therefore, shape effects may promote both lower and higher accessibility of the site by immigrants, as compared to its original status within the forest. Higher accessibility implies the arrival of more immigrants, with consequences discussed above. Lower accessibility implies that fewer immigrants reach the site, which may lead to local extinction through inbreeding depression and demographic stochasticity (Jaenike 1978; Jennersten 1988; Klein 1989; Harrison 1991; Pease and Fowler 1997). A similar process can be triggered purely by isolating the remnant from its original matrix, which also reduces accessibility of the site.

gain easier access to the site, feeding the loops to maintenance, impoverishment, or enhancement of species richness in fragments.

REGIONAL PROCESSES

Immigration rates can also be affected by the shape of both the fragment and the potential source population, as proposed by Taylor (1987a, b) (fig. 2.3). According to this model, straight-line shores should "release" more colonizers than do pointed peninsulas (fig. 2.3b), at least for passive dispersers. Similarly, an island in the center of a circular lake should receive more immigrants than if it were located eccentrically, close to the shore. An island in the center of a lake can receive colonizers from the entire lake's perimeter, whereas an island near one shore might not receive colonizers from the more

Final Remarks

In this chapter we have presented an overview of the theoretical pathways by which ecological processes affect species richness after fragmentation. Using such an overview, the reader should be able to build a global picture of these intricate biological dynamics and be less likely to misinterpret results obtained in the field. Precise interpretation of field data is essential not only to the development of a sound theoretical framework with which to understand ecosystem fragmentation but also to the definition of sensible technical and political guidelines for conservation. In this regard, our flowchart (fig. 2.2) provides a checklist of the possible explanations for an empirically observed maintenance, increase, or decrease of species richness in a fragmented ecosystem. Possible explanations of the pattern may be tested individually, using sampling procedures or experiments designed for such aims.

Conservation Lessons

1. Ecosystem fragmentation may increase, diminish, or maintain the original number of species in the site concerned.
2. Maintenance of the original number of species does not necessarily mean that the community is immune to fragmentation effects.
3. Local extinction may be either a final product of ecosystem fragmentation or an intermediate step to an increase *or* decrease of species richness in the fragment.

4. Along with determining the best site to preserve, it is imperative to decide the best shape and geographical location of the fragment to be created, because species richness in the fragment depends strongly on these features.

Acknowledgments

We are indebted to Jay Malcolm, Athayde Tonhasca, Carlos Sperber, Júlio Louzada, and an anonymous referee, whose criticisms of previous versions of the manuscript greatly improved the final text.

CHAPTER 3

Deforestation and Forest Fragmentation in the Amazon

CLAUDE GASCON, RICHARD O. BIERREGAARD, JR.,
WILLIAM F. LAURANCE, AND JUDY RANKIN–DE MÉRONA

Habitat loss and fragmentation is the most important threat to forested ecosystems. Although fragmentation of natural habitats can have varied origins, deforestation in the Amazon is most commonly associated with human settlement or exploitation of the landscape. Historically, the effects of forest fragmentation were thought to be associated only with the loss of original habitat and the reduction in size of forest remnants. More recently, it has been shown that the ramifications of fragmentation are more numerous and diverse than previously assumed; the creation of forest edges and the appearance of a new matrix habitat (the modified landscape habitat surrounding habitat remnants) have wide-reaching effects at the local and landscape level. The importance and pervasiveness of each of these effects in a particular landscape depend in large part on the proximate causes of deforestation (e.g., timber harvesting versus family farms), which determine the resulting configuration of the landscape (Fearnside 1990a; Kahn and McDonald 1997).

In this chapter we describe the effects of forest fragmentation in the context of existing conservation biology theory (see also Chapter 2). We then examine the status of deforestation in the Amazon region, with emphasis on the central Amazon. Finally, we analyze modified landscapes from other Amazonian regions and discuss possible scenarios for the changes in the central Amazon region.

Components of Forest Fragmentation

Inevitably, tropical deforestation results in fragmentation of the forest into isolated patches embedded in a new matrix habitat. In an attempt to model and predict the consequences of forest fragmentation, conservation biology theory has relied on the framework provided by MacArthur and Wilson's (1967) island biogeography theory (e.g., Wilson and Willis 1975) and expanded by scientists in landscape ecology (e.g., Fahrig and Merriam 1994; Wiens and Moss 1999).

Recent studies, including this volume, have demonstrated that the effects of forest fragmentation on natural systems are more complex than can be predicted from simple surveys of species richness. For example, ecological processes (pollination, decomposition, and nutrient cycling) and forest functions (regulating hydrological and nutrient cycles, evapotranspiration, and the preven-

This is publication number 270 in the BDFFP Technical Series.

tion of fire) are directly linked to the integrity of the forest and are severely affected by human disturbance.

Fragmentation leads to a drastic increase in forest edge, which differs from natural ecotones (i.e., a natural gradient between two habitat types). The basic difference is one of degree of contrast between the two habitats. In a natural landscape, there is a relatively gradual transition between patches of adjacent habitat types, whereas in a fragmented landscape, at least for some time an abrupt contrast exists between forest and field.

Many biological consequences have been reported as a result of edge creation. In continuous tropical forests, sunlight usually is restricted to vertical penetration, but in a forest fragment sunlight can penetrate laterally along the fragment's margins. This single change seriously affects the microclimatic conditions of the forest near the edge (Kapos 1989; Murcia 1995). Such microclimatic changes will undoubtedly have substantial effects on the physiology and reproduction of plants and animals along forest edges. These effects will in turn affect forest structure and dynamics (Ferreira and Laurance 1997; Laurance, Bierregaard, et al. 1997) as well as having cascading effects on insects and animals that depend on plants for their life cycles (Didham 1997a; Chapter 18).

The loss of primary forest also results in the creation of a new matrix habitat. Only since the mid-1980s has the effect of the matrix on fragment biotas been seriously addressed (L. Harris 1984; Fahrig and Merriam 1994; Gascon et al., 1999). This has led to important advances in the appropriate scale for conservation actions (Soulé and Terborgh 1999). Matrix habitat appears to be important in the evolution of ecosystem dynamics in fragmented landscapes because it acts as a selective filter for movements of

species between fragments and other landscape features, facilitating movements of some species and impeding others; disturbance-adapted species will be present in the matrix and may invade forest patches and edge habitat; and the matrix type (pasture, degraded pasture, second-growth forest, etc.) can influence the severity of the edge effects in forest patches (Fahrig and Merriam 1994; Williamson, Mesquita, et al. 1998).

Although many forest species are able to use nonforested habitats (Bierregaard and Stouffer 1997; Malcolm 1997a; Tocher, Gascon, and Zimmerman 1997; Chapters 11, 19, 20, 24), more research is needed to determine how forest species react to different matrix habitats. Such data will allow us to understand which habitats represent high- or low-quality habitats for sensitive wildlife and may even help predict which species will be most vulnerable to extinction in isolated patches (Malcolm 1991; Laurance 1991b).

Finally, until recently, the importance of faunal corridors has also remained a largely theoretical debate because of the scarcity of empirical data (L. Harris 1984; Noss 1987; Simberloff and Cox 1987; Simberloff et al. 1992; Demers et al. 1995). The few studies that do exist suggest that corridors greatly enhance movements of animals among patches in a landscape (A. Bennett 1990; Merriam and Lanoue 1990; Saunders and Rebeira 1991; Dunning et al. 1995; Haas 1995; S. Laurance 1996; Lima and Gascon 1999), which in turn may decrease the probability of extinction of local populations.

Recent knowledge about how ecosystems are affected by landscape fragmentation suggests that much of the ecological degradation experienced can be accounted for by fragment size and the influences of edge effects and the surrounding matrix on the fragment. The magnitude of each of these components will greatly depend on the spatial configuration of the forest remnants and their connectivity in a modified landscape. For example, in an area with only few and scattered remnants, edge effects and the nature of the matrix habitat will degrade the integrity of forest patches. At the other extreme, in areas of selective timber extraction, subtle changes in the hydrological and microclimatic conditions of the forest will be more important (see Chapter 27 for a theoretical treatment of these effects).

Deforestation in the Amazon

Socioeconomic forces driving landscape change will ultimately determine the layout of the modified landscape, greatly influencing the qualitative effects of deforestation and the magnitude of those effects (Bierregaard and Dale 1996; Kahn and McDonald 1997). It is therefore important to examine both past examples of landscape changes in the Amazon and the present and future drivers of change in central Amazonia (see Chapter 26). This will enable us to predict the most serious threats to the central Amazon region and act to prevent them or diminish their effects (Laurance and Gascon 1997).

In the Amazon, as elsewhere, growing human populations and economic pressures are leading to widespread conversion of tropical rainforests into a mosaic of human-altered habitats and isolated remnants. Much of past forest clearing has taken place in a northeast to southwest arc around the heart of the Amazon forest. Deforestation in the states of Mato Grosso, Pará, and Rondônia combined account for more than 65 percent of all deforestation to date, and the state of Amazonas accounts for just over 5 percent of total deforestation (INPE 1998). Much of this deforestation comes as a result of major development projects, including: the expansion of road networks in the pre-viously inaccessible interior (Bélem–Brasília in 1960, paved in 1974; Transamazónica in 1968; Manaus–Porto Velho, Cuiabá–Porto Velho in 1967, paved in 1984; Manaus-Carcaraí in the mid-1970s; and Calha Norte, begun in the early 1980s and not yet completed); large-scale government colonization programs on the Transamazónica and in the state of Rondônia; giant hydroelectric projects (Tucuruí, Balbina, Samuel, Cachoeira Porteira); mining site development (Grande Carajás, Trombetas, Minerão do Norte); and other such extensive activities (Projeto Jarí).

Current estimates of deforestation show an increase in forest area being cleared yearly from 15,000 km^2 in 1994 to more than 29,000 km^2 in 1995 (INPE 1998). To date, more than 12 percent of the total Amazonian rainforest has been cleared for timber extraction and pasture creation (INPE 1998; updated from Skole and Tucker 1993). This does not take into account the large areas of forest that suffer selective logging each year (Whitmore 1997) and areas that suffer cryptic deforestation (Nepstad, Moreira, and Alencar 1999), for which the ecological impacts are poorly known.

Since 1965, details of the Brazilian Forest Code designed to protect the forest have contributed to a high degree of forest fragmentation by requiring 100-meter buffer zones to protect streams, limiting felling on steep slopes, and, most important, by requiring that 50 percent of each holding in the "Legal Amazon" (which includes the forested Amazon basin as well as drier savannas to the south) be conserved in tree cover (Brazilian Forestry Code: Decree Law 4771, Article 44). This has been interpreted as meaning anything from original forest to planted commercial tree species. Instituted as a mechanism of forest preservation, this law specifies neither the state of the forest to be preserved nor the minimum dimensions and geometry of each patch within a single holding or for a region. Despite modifications in 1984, 1986, and again in 1996, the code continues to be open to interpretations, which could result in anywhere from 0 to 80 percent of a landholding being left in native forest. This, combined with the regimented and dense geometries of official colonization projects and the spontaneous small holdings along most access roads, results in very small minimum diameters for most forest remnants.

The amount of deforestation varies from state to state within the region, with deforestation being greater in the more accessible and densely populated areas of older, western European occupation, as in the state of Maranhão, with 60.7 percent of its original primary forest cut, or where government incentives, combined with poor living conditions in more populous southern Brazil, have attracted huge numbers of migrants — for example, Rondônia is 14.2 percent deforested, mostly since 1980. Alarmingly, deforestation in Rondônia is increasing exponentially and giving no signs of slowing. Similar trends of lesser intensity can be seen in most other states in the region.

In the larger and less accessible state of Amazonas, deforestation is proceeding more slowly — only 0.8 percent of original forest has been cleared. However, because deforestation in this state is concentrated in a few major areas, such as around the capital of Manaus and along major roadways, the local impact is much greater than statewide figures would suggest. It is likely, however, that in the near future the rainforests in central Amazonia will be the focus of intense development pressure. This is probable for two reasons. First, tropical timber stocks worldwide are becoming scarce, and the Amazon now represents the largest natural source of such hardwood timber (see Chapter 27). Second, local and state governments

in the Amazon are already actively preparing the terrain for developers by investing in infrastructure in the region, especially the creation of new roads. As history has repeatedly shown — the state of Rondônia is a classic example — improved access to tropical forest is followed by increased deforestation.

The impending pressures on central Amazonia pose a serious threat to the ecological integrity of the region. Although serious and strict legislation exists in Brazil for the use of natural resources, enforcing these laws is more than problematic. For example, the Brazilian government agency responsible for natural resource management, IBAMA, has an average of one agent to patrol 18,000 km^2 of tropical rainforest — an impossible task. Environmental legislation passed in 1998 that severely punishes crimes against the environment with fines and prison sentences may prove helpful, although its efficacy will depend again on strict enforcement. These factors combine to form an alarming situation for the future of the Amazon rainforest.

While much of the central Amazon remains intact (Skole and Tucker 1993), we can learn much about the possible trajectories of future development based on the history of the two other regions substantially affected by development — eastern Amazonia, and Rondônia in the southwest.

The eastern Amazon basin has suffered heavy forest loss and land-use change. In the Zona Bragantina, it is estimated that less than 2 percent of the original forest cover remains. Other regions east and south of the city of Belém present a similar, although less dramatic, decrease in forest cover. More to the south, much of the area between the Xingú and Araguaia river basins has been extensively cleared of forest. The main driver behind deforestation in these areas has been logging of one hardwood species of timber — mahogany (*Swietenia macro-*

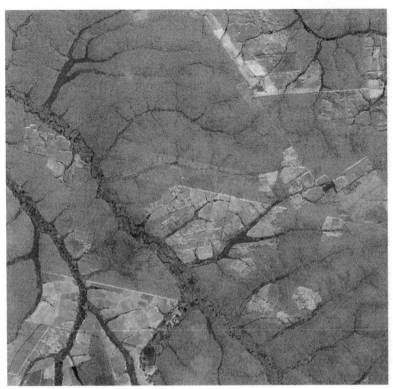

Fig. 3.1. Satellite image of forest fragmentation in southeastern Amazonia where timber extraction (principally for mahagony [*Swietenia macrophylla*]) has been the major driver for recent land clearing. Dark areas represent native forest, and lighter areas represent deforested land.

phylla). Huge areas of forest were cut down for the extraction of this valuable tropical timber species to supply the several hundred sawmills in the region. Cattle ranches and agricultural operations that subsequently occupied the land are but by-products of the initial land-clearing drive.

This deforestation scenario results in large cleared areas, once covered by tropical rainforest, dominating the landscape, with few, and small, forest remnants sprinkled about (fig. 3.1). There is little or no connectivity between these remnants. As cleared areas usually degrade in quality for pasture, many such areas are abandoned and left to natural regeneration. The extent of these areas is difficult to estimate, even with the best remote sensing imagery (Lucas et al. 1998).

In the state of Rondônia, at the other end

of the deforestation arc in the Amazon, forest clearing in the last several decades has also been severe (Skole and Tucker 1993). In contrast to eastern Amazonia, however, the prime driver of deforestation has not been timber extraction (although some extraction occurs in the initial process of forest removal). Rather, because of more productive soils, farms have been developed, prompting economic growth in the region (Leite and Furley 1985).

In a federally sponsored program of colonization, a template of roads was established, allowing access to otherwise unreachable areas (Dale and Pearson 1997) and leading to a massive wave of immigration from the south. At the peak of colonization in Rondônia, more than 2,000 families per month were moving into the state (Fearnside

Fig. 3.2. Satellite image of the "fishbone" deforestation pattern typical of areas where roads have been cut into the forest to facilitate colonization, as in Rondônia. Dark areas represent native forest, and lighter areas represent deforested land.

1986a). Small-scale farmers were the first to arrive and clear the forest. After crop yields fell, only three or four years after forest clearing, many families abandoned the land and sought new forest to fell. Subsequently, wealthy cattle ranchers amalgamated many small, abandoned clearings into large cattle pastures. This has produced a different trajectory of deforestation and resulting landscape. Roads are cut into the forest in a fishbone pattern, and most of the cleared areas are adjacent to the main roads and branch out as new roads are put in (fig. 3.2). Although a large percentage of the region is cleared of forest, the spatial pattern and connectivity of the remaining patches (Forman 1995) is clearly different from that in eastern Amazonia.

Future Landscape Configuration in Central Amazonia

Predicting what the central Amazon region will look like in ten or twenty years is a difficult and risky exercise, as is evidenced by widespread claims twenty years ago that much of the Amazon would be lost by the year 2000. As mentioned above, predictions rely on many factors, such as the potential incentives to deforestation and land-use change, human pressure from large urban centers, new legislation, which is being developed and adopted in Brazil, and outside forces, such as the state of the world economy. It is possible, however, to examine trends in some of the most important forces behind change and their potential for modifying the central Amazon landscape in the next decades. As described above, relatively little of the tropical forest in central Amazonia has been cleared to date; and most of the deforestation is centered around Manaus. However, there is intense and growing interest in this region in natural resource extraction, so this situation is unlikely to remain. How, then, can we predict and alter the trajectory of landscape and ecosystem change in the region?

We know that deforestation in Amazonia affects the regional hydrology, the global carbon cycle, evapotranspiration rates, biodiversity loss, the probability of fire, and a possible regional reduction in rainfall (Uhl, Kaufman, and Cummings 1988; Uhl and Kaufman 1990; Nobre, Sellers, and Shukla 1991; Bierregaard et al. 1992; Wright et al. 1992; Nepstad, Carvalho, et al. 1994; Vitousek 1994; Laurance, Bierregaard, et al. 1997). Because the spatial configuration of a modified landscape in the Amazon depends in large part on the economic and sociological forces behind deforestation (the Paragominas and Rondônia examples; Bierregaard and Dale 1996; Kahn and McDonald 1997; Whitmore 1997), and because many of the resulting ecological effects and their magnitude depend on levels of connectivity, and remnant size and degree of isolation, the ecological future of the region is inextricably linked with its economic future.

Roads and Waterways

Roads, which allow access to otherwise pristine areas of forest, have been the proximate cause of most of the deforestation in Amazonia. Since about 1995 the state of Amazonas and the federal government have invested a huge amount of resources in recuperating major highways through the heart of the Amazon. Previously, these highways were little more than mud roads with limited access. Now the main highway (BR-174) that links Manaus to the north (and ultimately to Venezuela) is totally paved and supporting much more traffic than before. The government plans to similarly recuperate the major highway that links Manaus to the south (BR-319).

The use and recuperation of major waterways for transportation of products will also create major new entryways into the Amazon rainforest. The federal government, through its Brazil em Ação development plan, has several such large-scale projects in the study phase, and the impact of these is essentially unknown. Although increased deforestation of primary forest along BR-174 has not yet occurred, human occupation on previously cleared land has increased significantly. The government actually proposed that the area along this highway become the next big agriculture cradle of Brazil, even though the soil is inadequate for the purpose (see Chapter 23).

Growing human presence in areas close to primary forest is almost inevitably the first step to increased deforestation. A new and improved network of main highways will undoubtedly lead to the appearance of a radiation of secondary roads cutting deeper into pristine areas, as occurred in Rondônia.

Timber Extraction

According to many sources, the heart of the Amazon will become the main focus of tropical hardwood timber extraction in the next decade (Federal House of Representatives Report 1997; Chapter 26). Since 1996, Malaysian and Chinese timber companies have invested heavily in land and sawmills in the Amazon. Many of these companies are now awaiting government authorization to initiate large-scale timber extraction.

Also, according to the federal report, unenforced authorization has led to a number of abuses and to serious damage to the forest ecosystem. Field surveys of timber extraction did not correspond to approved management plans; much more timber was extracted than planned, timber was bought from areas without approved management plans, and the same management plan was used for several different timber operations (see Chapter 26). Fortunately, many of the 1997 Federal Report's recommendations address these issues and are being implemented.

Even if timber extraction in the Amazon were performed according to the most modern, low-impact techniques (Uhl, Barreto, et al. 1998), the faunal impacts are still not completely understood. While low-impact logging is obviously far more desirable than clear-cutting, it does significantly open up the forest canopy, creating drier and warmer conditions (see Chapter 27). These changes may in turn render large areas of logged forest vulnerable to increased mortality and biomass loss (W. Laurance, S. Laurance, et al. 1997), as well as greatly increase the probability of fire in these areas (Nepstad, Carvalho, et al. 1994). The 1997 fires in many areas of Amazonia are witness to this increasing vulnerability. Finally, sustainable forest management models use rotation cycles of thirty years (BIONTE 1998), which may be completely unrealistic given recent estimates of tree ages in the Amazon (Chambers 1998) and the demand for an immediate return on investments. The tradeoff between conventional clear-cutting and low-impact logging must be evaluated carefully (see Chapter 26).

Urban Populations

Human pressures from large urban centers in the Amazon may also become important causes of landscape change in the near future. Since the establishment of a duty-free district (Zona Franca) in 1967, the city of Manaus, the only large metropolitan area in the central Amazon, has witnessed a settlement boom, with its population now surpassing 1.5 million. To date, much of the economic development of the city has relied on the free-trade zone, which has allowed international assembly industries (predominantly in the electronics sector) to import foreign-made components, assemble them in Manaus, and sell the finished goods at costs far below imported products. In its first decades the Zona Franca provided many jobs and probably diminished human pressure on the surrounding forests, in that its job market attracted people who otherwise would have been clearing the forest.

Manaus' tax-exempt status will last by law until 2013. Already, however, the lowering of import tariffs in Brazil outside the Zona Franca has undermined the attractiveness of the remote northern city to companies, and the number of jobs provided by the Zona Franca has already decreased substantially. As we approach the end of duty-free status, it is likely that companies will continue to divest of their activities in Manaus. A general collapse of the Zona Franca and the loss of its tens of thousands of direct jobs would perhaps spell disaster for large areas of forest that are now becoming accessible from Manaus thanks to the newly improved highways.

Fortunately, recent federal measures prevent the government agency responsible for human settlement and agrarian reform, INCRA, from using forested lands for colonization, as it used to. Settlement is now restricted to the several millions of hectares of land already deforested in the Amazon. Nevertheless, increased human presence in close proximity to forests can have indirect effects, as people may hunt and extract fruits, nuts, and other plant products.

Conclusions

The central Amazon region is on the verge of major changes. Several powerful drivers of land-use change are in place (timber extraction, soybean expansion, and roads) or may be forced onto the region (loss of jobs), with the potential for significantly altering the present-day landscape. It is probable that intense development will radiate out from the large urban centers simply because of the abundance of forest and the commodity of nearby markets and export infrastructure. This would result in the typical fishbone pattern of land use along major highway arteries. Without proper planning, this type of development will result in a series of small forest remnants with relatively little connectivity among patches.

Small but noteworthy measures, however, taken proactively, can greatly enhance the conservation value of such a landscape (Schelhas and Greenberg 1996; Laurance and Gascon 1997; Soulé and Terborgh 1999). In areas where timber extraction will occur, new and enforced legislation should avoid the repetition of past large-scale deforestation and fragmentation (the southern Pará example). The sustainable timber management plans required by law in Brazil prevent outright clear-cutting, permitting only selective extraction. Care should be taken, however, to prevent the insidious fragmentation that occurs from poking holes in the forest canopy over large areas (Nepstad, Moreira, and Alencar 1999; Chapter 27).

Conservation Lessons

1. Socioeconomic pressures will inevitably lead to substantial land-use change in central Amazonia, with associated fragmentation of the landscape.

2. Timber extraction and population expansion around Manaus are likely to be the dominant forces acting on the local landscape.

3. Past examples (Rondônia and Pará) can teach us how *not* to proceed with development in the region.

4. Technical knowledge of selective logging techniques and a growing understanding of the biological dynamics and ecology of fragmented landscapes should enable us to minimize many of the detrimental effects of developing the region.

5. Whereas existing Brazilian federal legislation indicates a desire to avoid the mistakes of the past, substantial difficulties in enforcement will have to be overcome.

6. Priority should be given to the careful creation of large protected areas (Ayres et al. 1998) in the Amazon, ahead of development frontiers and while the landscape allows for a proactive approach to conservation (Laurance and Gascon 1997; Soulé and Terborgh 1999).

The Biological Dynamics of Forest Fragments Project

The Study Site, Experimental Design, and Research Activity

CLAUDE GASCON AND RICHARD O. BIERREGAARD, JR.

In this chapter we present an overview of the Biological Dynamics of Forest Fragments Project's study site and experimental design—the underlying "materials and methods" section for most of the chapters that follow. We conclude by reviewing important research projects that have been carried out, as well as the major research themes that are currently being addressed in the field.

Experimental Design

The BDFFP began as the Minimum Critical Size of Ecosystems Project, with a research goal of identifying a minimum size of tropical forest habitat that would maintain most of the biotic diversity represented in an intact ecosystem (Lovejoy and Oren 1981). The research design entails studies of plant and animal communities in forest plots before and after their isolation by cattle ranchers opening new pastures (Lovejoy, Bierregaard, et al. 1983; Lovejoy, Rankin, et al. 1984). Pre-isolation data from the forest fragments under study makes this investigation nearly unique in this regard. Additionally, post-isolation communities in the plots are compared to control studies in adjacent, continuous forest through time. Although

the project and its study area have been described in detail elsewhere (Lovejoy and Bierregaard 1990), a brief summary will be given here, incorporating new information on the dynamics of the BDFFP landscape since those earlier reports.

As originally conceived, the BDFFP was to establish a replicated series of experimentally isolated forest reserves of 1, 10, 100, and 1,000 ha. The experimental design took advantage of a Brazilian law that required landowners in Amazonia to leave the forest standing on half of any land under development (now up to 80 percent is required by law). Reserves were delineated in primary forest on three cattle ranches, or *fazendas,* in the Distrito Agropecuário, a 500,000 ha agricultural research and development area, some 50 to 120 km north of the city of Manaus (fig. 4.1). The cattle ranches (Dimona, Porto Alegre, and Esteio) at the BDFFP site each have several thousand hectares of cleared land interspersed with patches of primary forest (including the project's experimental reserves). These ranches are themselves surrounded by primary forest, which extends unbroken for hundreds of kilometers to the north, east, and west (plate 3). The experimental reserves have been protected by presidential decree no. 91.884 (November 5, 1985).

This is publication number 271 in the BDFFP Technical Series.

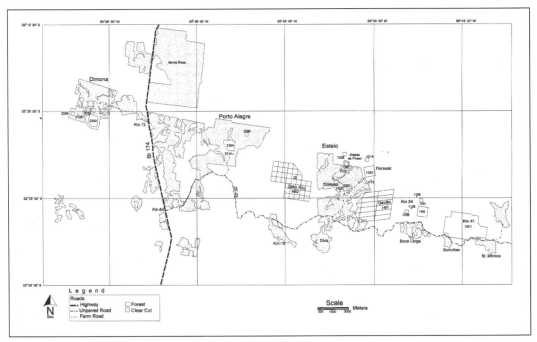

Fig. 4.1. Map showing the location of the study area of the Biological Dynamics of Forest Fragments Project, north of Manaus, Amazonas, Brazil. Source: INPE Landsat TM 5,4,3—RGB, 1995; prepared in June 188 by E. M. Venticinque and T. L. N. Fernandes.

The Study Site

FOREST DESCRIPTION AND CLASSIFICATION

The study area is classified as tropical moist forest and is remarkably diverse (table 4.1). Mean annual rainfall in the region is about 2,200 mm, with a range of 1,900 to 2,500 mm (see fig. 5.1). There is a pronounced dry season from June through October, with usually less than 100 mm of rain per month.

The canopy of the forest near Manaus is 30 to 37 m tall, with emergents as high as 55 m. The flora is remarkably diverse in tree species. As described in more detail by Rankin–de Mérona et al. (1992) and in Chapter 5, the project's phytodemographic inventory has identified more than 1,000 species or "morpho-species" of trees representing 61 families and 288 genera from its extensive inventory (more than 60,000) of stems with a diameter at breast height (DBH)

of at least 10 cm. (The term morpho-species is used for individuals apparently distinct from other species in the sample but not yet reliably identified.) Leguminosae is the most species-rich family, with 195 species and 33 genera. This family also represents 12.1 percent of all trees marked in the 66 one-hectare permanent plots. Other important families in the survey are Lecythidaceae (12.4 percent), Sapotaceae (11.9 percent), Burseraceae (9.3 percent), Euphorbiaceae (5.9 percent), Chrysobalanaceae (5.3 percent), Lauraceae (4.3 percent), Moraceae (4.1 percent), Annonaceae (3.5 percent), and Violaceae (2.5 percent). Tree diversity per hectare is also very high, reaching over 280 species in some 1 ha plots (Oliveira 1997). The understory (plate 5) is depauperate, in both number of species and density of flowering and fruiting plants, relative to other Amazonian forests (Gentry and Emmons

TABLE 4.1   Species Richness of Major
Biological Groups at BDFFP

Taxonomic group	BDFFP reserves	North America
Tree families	61	84
Tree species	>1,000	652
Butterfly species (excluding skippers)	>500	475
Frog species	>60	81
Lizard species	24	115
Snake species	63	115
Bird species	370	700
Mammal species (excluding bats)	52	234
Bat species	69	42

1987), except for palm species (Scariot 1996).

Although charcoal is ubiquitous in BDFFP soil samples, evidence of past human presence is scant (Bassini and Becker 1990). Rather, charcoal samples may represent evidence of recurrent ultra-dry periods in the Amazon in the past 1,000 years caused by repeated, strong El Niño events. The increased tree mortality in extremely dry years may render the tropical Amazon forest prone to large-scale fires, as were witnessed in 1997 around Manaus.

SOILS

The BDFFP landscape is located on Pleistocenic terraces of interglacial origin (Brazil, Projeto RADAMBRASIL 1978). Topography is undulating with mean altitudinal differences between plateaus and stream valleys of 40 to 50 m. Close to half of the land area is composed of sloping terrain, dissected by first- and second-order streams. Altitude above sea level varies between 80 and 100 m, common for Amazonian lowland forests (Brazil, Projeto RADAMBRASIL 1978). Soils are relatively nutrient-poor, sandy, or clayey xanthic ferralsols, called yellow latosols (Ranzani 1980; Chapter 23).

Isolation of Reserves

RESERVE SELECTION

Selection of reserve sites wound up being dictated by several logistical considerations. Because the ranchers often left 50 m buffers on each side of the many streams on their land (as required by law), we had to do extensive topographic surveys to find places where 1 and 10 ha reserves would "fit" into the landscape and be truly isolated after deforestation. We attempted to find areas in the forest that would be at least 200 m from any stream for the smaller reserves and set up clusters of reserves that could be serviced by one base camp. There was no place where we could place a 100 ha reserve that would not include some streams, so we chose to locate these larger reserves on stream headwater systems so that the only eventual connection from them would be along one stream leaving the reserve. This would also avoid potential contamination of aquatic ecosystems from runoff upstream of the reserve.

Twenty-four reserves, ranging from 1 to 1,000 ha, were demarcated on the three fazendas. However, due to economic downturns in the Brazilian economy and the perceived failure of cattle operations in the region, the ranchers cut far less land than they had planned and only 11 of our reserves were isolated (table 4.2).

HOW TO ISOLATE A FOREST FRAGMENT

Clearing of the forests by the ranchers was a time-consuming process that lasted throughout the dry season, usually from May through September. Clearing was done with ax, machete, and chainsaw, usually by crews of itinerant workers subcontracted by the ranchers.

The first step in the process was to clear the undergrowth of the area to be felled.

TABLE 4.2   Number and Status of Each BDFFP Fragment

Reserves	1 ha	10 ha	100 ha	1,000 ha	Control (Km 41)
Isolated	5	4	2	0	0
Nonisolated	3	4	2	2	1

Crews moved through the forest, clearing all stems and vines less than 10 cm in diameter with machetes and axes. This cleared the way for chainsawyers, who felled the canopy trees and turned the forest around the experimental plots in a parklike glen, devoid of any undergrowth. Species restricted to low-lying vegetation were effectively isolated at that time, despite the continued presence of a forest canopy around the reserves.

The felling of the canopy trees, an extremely dangerous operation, was accomplished as the sawyers cut halfway through about six trees and finally toppled them all like enormous dominoes, felling one into the rest. Wind gusts sometimes started the chain reaction before the sawyers were ready, with occasionally fatal results. Trees that were sawn through but were missed by their falling neighbors were potentially deadly booby traps amid the fallen chaos of the cleared forest, ready to topple without warning.

Once the forest was razed, the ranchers had to burn the fallen trees. The soil is too susceptible to compaction to permit mechanized clearing, and burning releases nutrients into the soil that are essential to establishing any sort of pasture (Chapter 23). The undergrowth that was cleared to give access to the chain sawyers also provided the fuel to get the trunks of the canopy trees burning. Burning itself could be attempted only after the fallen vegetation had dried for about two months winding up with at least 10 days with no rain (plate 6).

Once a sufficient drying period had elapsed, the ranchers torched the fallen for-est (plate 7). Even though the trees bake in the tropical sun, an enormous amount of moisture remains in the fallen forest; it boils off in steam as the drier wood burns. This moisture occasionally condensed and fell back on the fire in what the ranchers call "false rain," ruining the burn because not enough of the woody biomass was cleared to make the land practical for pasture installation.

To protect our reserves throughout this process we had to make sure that the trees felled immediately around the reserves were directed away from the reserves themselves. Then, prior to the burn, we removed as much of the tinder as possible from a strip 5 to 10 m wide around the reserve. On the day of the burn we coordinated with the ranch crew that was to start the fires around the edges of the clearing (usually hundreds of hectares in extent) so that we could set backfires burning away from the reserves. This technique enabled us to protect even the 1 ha reserves in the midst of raging bonfires hundreds of hectares in extent. By the time the fires from the edge of the clearing reached our reserves, our backfires had burned far enough away from the reserve edges that the larger fires sweeping downwind did not penetrate the isolates.

ISOLATION CHRONOLOGY

The isolation times and burning history for each of the areas have varied (table 4.3). As is seen in this volume, the history of land use on the ranches has strong influence on the dynamics of the fragment communities (e.g., Chaters. 19 and 20), as well as on the

TABLE 4.3. History of Surrounding Land Use and Isolation of BDFFP Reserves

Reserve	Size	Isolation distance	History
Fazenda Esteio: Colosso, Florestal, Cidade Powell, and Gavião camps			
1104	1 ha	150 m	Reserve isolated in 1980. The surrounding area was poorly burned. High second growth was present in 1983 (approx. 4 m). In August 1987 and August 1994, 100 m of second growth was cleared and burned around the reserve to maintain its isolation. Cattle grazed around reserve (plate 1).
1112	1 ha	400 m	Reserve isolated in 1983. Approximately 600 ha were cleared but never burned around this reserve. As of 1985 this whole area was high second growth dominated by cecropia. No grazing around this reserve (fig. 4.2).
1202	10 ha	700 m	Reserve isolated in 1980. The burn was successful and intense. High second growth was present in 1983 (approx. 4 m). When the ranchers burned the second growth around the reserve in November 1982 the fire got into the reserve and burned approx. 1 ha in the reserve's southeast corner. Grasses were planted in 1983 and cattle grazed around the reserve. The second-growth area between Colosso camp and the reserve was cut and burned in August 1985. Second growth was cleared again around the reserve in 1987. In 1989 and 1994, a 100 m band of second growth was again cleared and burned around the north, west, and south sides of the reserves (see plate 1).
1207	10 ha	100 m	Reserve isolated in 1983. Only a band of 100 to 150 m was cleared around the isolate on the east, west, and north sides. This area was never burned. Second-growth forest dominated by cecropia rapidly reconnected this reserve to the adjacent forest. No grazing around reserve (see fig. 4.2).
1301	100 ha	not isol.	A continuous forest reserve with one side exposed. East side of reserve isolated in 1983, but area was never burned. As of 1985 this whole area was high second growth dominated by cecropia.
1401	1,000 ha	not isol.	West side of reserve exposed in 1977–78 by the first clear-cut area on the farm. This area was pasture until 1989. Presently, high second growth (>15 m) delineates the west side of this continuous forest reserve.
Fazenda Porto Alegre: Porto Alegre camp			
3114	1 ha	300 m	Reserve isolated in 1983 by 1,000 ha clear-cut that was never burned. Rapidly surrounded by exuberant, cecropia-dominated second growth. A band of 100 m was cut and burned all around reserve in 1991 and 1994. Grass planted after cut and burn. No grazing around this reserve.
3209	10 ha	900 m	Reserve isolated in 1983 and rapidly surrounded by cecropia-dominated second growth (see plate 8). In 1991 and 1994, 100 m band around the reserve was cut and burned. No grazing.

TABLE 4.3. (continued) History of Surrounding Land Use and Isolation of BDFFP Reserves

Reserve	Size	Isolation distance	History
3304	100 ha	450 m	Approximately 700 m of the north and west margins were exposed in 1981 by pasture installation. Cattle grazed in this area. Reserve isolated in 1983 by 1,000 ha clearing that also left a connection with continuous forest along the stream about 2 km to the north (see plate 2). A 300 m break in the north end of the corridor was cut and burned in 1984. Cecropia-dominated regrowth on the east, south, and west sides was 1 to 1.5 m tall in late 1984 and nearly 10 m tall by 1988 (see plate 8). This reserve reisolated with the 1 and 10 ha reserves in 1991 and 1994.
Fazenda Dimona: Dimona camp			
2107	1 ha	150 m	Reserve isolated in 1984. Cut area was burned the same year. Pasture installed and cattle grazed. The area between camp and the reserve was cut in 1987 and burned two months later. In 1989, 1990, and 1994 a 100 m band was cleared of second growth around the reserve. *Vismia* spp. dominates second growth around reserve (see plate 4).
2108	1 ha	600 m	Reserve isolated in 1984 with Reserve 2107. Cut area was burned the same year. A 100 m strip around the reserve cleared in 1989 and 1994. Pasture planted and cattle grazed around reserve. *Vismia* spp. dominates second growth around reserve (see plate 9).
2206	10 ha	225 m	Reserve isolated in 1984. Cut area was burned the same year. A 100 m strip cleared in October 1989. In 1994, a 100 m strip around this reserve was again cut and burned. As with neighboring 2108, pasture installed and cattle grazed. *Vismia* spp. dominates second growth around reserve.
2303	100 ha	150 m	The north side of this reserve was exposed by pasture in 1980. In 1982, that area was solidly established pasture. Abandoned as pasture in 1983 and returned to high second growth (*Cecropia* spp. and *Vismia* spp.). The west side was cut and burned in 1984. In 1990 the reserve was finally isolated by a 200 m band cut and burned on the remaining east and south sides of the reserve. Currently, this reserve is surrounded by high second growth (*Cecropia* spp. and *Vismia* spp.).

nature and trajectories of forest recuperation on deforested lands (Chapters 24 and 25). The 1980 burn around reserves 1104 and 1202 was a successful one for the ranchers and enabled them to establish pasture and cattle in the area (plate 1).

The 1983 clearing was a disaster for the ranchers. One thousand hectares were cleared on the Porto Alegre fazenda, isolating reserves 3304, 3114, and 3209, and 600 ha were cleared at the Esteio fazenda, iso-

lating reserves 1112 and 1207 (fig 4.2). In September of that year a very early and intense rainy season began before the ranchers could burn. By the time the next dry season arrived, so much second growth and resprouting had occurred that it was impossible ever to clear the fallen forest from the 1,600 ha cleared on the two fazendas. The cleared areas then became, effectively, two enormous treefalls, which were soon dominated by a vast, nearly monocultural sea of

Fig. 4.2. Reserves 1207 (10 ha) and 1112 (1 ha) isolated in 1983. The surrounding forest was never burned and rapidly regenerated into a dense cecropia forest (see plate 8). Reserves 1202 and 1104, isolated in 1980, can be seen in the distance. Photo by Richard O. Bierregaard, Jr.

the pioneer species in the genus *Cecropia* (plate 8).

The 1984 season brought a more typical transition between dry and wet seasons and permitted the ranchers at the Dimona fazenda to clear and burn several hundred hectares around reserves 2107, 2108, and 2206. Pasture grasses were established and cattle grazed for several years around these reserves.

The final reserve isolation occurred in 1990 with the felling of a 200 m wide strip around the south and east sides of Dimona reserve 2303 (100 ha). A good burn was

achieved in the isolating strip and grasses planted, although cattle were not grazed. This area quickly reverted to second growth.

As pastures and the cattle operations themselves failed, the land surrounding our isolates reverted to second-growth forests dominated by *Cecropia sciadophylla,* which in some cases can reach 12 to 15 m in height. In contrast, second growth on fallow pasture is dominated by trees in the genera *Vismia* (Clusiaceae) and *Bellucia* (Melastomat-aceae), which can reach a height of 3 to 7 m around the 1984 isolates and can grow 5 to 10 m around the 1980 isolates (plate 9).

While confounding our analyses of fragmentation effects by reconnecting some of the isolates to continuous forest, the appearance of the two types of second-growth forest have provided ample opportunities to study forest regeneration and how it can be manipulated (Chapters 11, 24, and 25). For the smaller reserves (1 and 10 ha), we have periodically cleared 100 to 150 m strips around them to maintain their isolation. A well-kept log of the various uses of each of the felled areas surrounding the experimental isolates includes dates of forest felling, intensity of burn, time used as pasture, and number of times second growth was cut (table 4.3).

Logistics and Administration of the BDFFP

The BDFFP operates from a central office in Manaus on the campus of INPA (Instituto Nacional de Pesquisas da Amazônia), which handles administration and logistical support in Brazil. The headquarters includes administrative offices, a reading and study room, a computer room, a herbarium, and a small laboratory. Transport by four-wheel-drive vehicles takes project personnel and researchers between Manaus and any of the nine base camps set up 80 km north of the city. Each camp serves a suite of several reserves. Camps are rustic but comfortable and include a dormitory, a laboratory hut, and a separate kitchen area. Our continuous forest control camp (reserve 1501, also referred to as Km 41) has a generator providing 110V current and running water (plate 11).

Baseline Data, Major Surveys, and Ongoing Research at the BDFFP

In the process of studying habitat fragmentation, project researchers have accumulated a wealth of information about the functioning of the intact tropical rainforest ecosystem as control data for that collected in the experimental plots. The process of forest fragmentation, in this case caused by cattle ranchers clearing new pastures, creates extensive forest-edge ecotones and large areas of clear-cut pastures. Augmenting the studies of the effects of fragmentation on ecosystem functions at the species and community levels are important investigations using the study sites to address questions relating to the biology of extinction, edge effects, the processes of forest regeneration, and fragmentation effects on the genetic structure of tropical species. Rather than include exhaustive citations to published work in the text below, readers are referred to the BDFFP Publication Series, organized by topic, beginning on page 371.

Research carried out at the BDFFP is guided by the following research priorities: (1) studies of the effects of forest fragmentation on specific taxa, communities, ecological processes, species interactions, physical parameters, resource distribution, and the genetic structure of selected taxa; (2) studies investigating the process of forest regeneration; (3) taxonomic and systematic studies of poorly known or diverse taxa; (4) basic tropical ecology that can serve as the basis for future investigations of fragmentation effects; (5) studies on the recuperation of degraded areas.

An integral part of the BDFFP is the large-scale subprojects that were established at the project's inception to provide baseline data and help characterize the tropical forest that composed our research landscape. From physical characterization of soils to botanical and faunal surveys, these major subprojects have accumulated an immense amount of data through both time and space.

TOPOGRAPHY

Initial topographic surveys were performed over most of the projected reserves. This was in part to characterize the physical nature of the reserves, but the surveys also were used in selecting potential reserve sites based on similar topographical features. Topographic maps were hand drawn for each putative reserve. There were no maps even indicating stream locations, much less topography, when we began the process of reserve selection in 1979.

VEGETATION STRUCTURE

The physical structure of the reserves was characterized using a simple semi-quantitative habitat complexity index. One-hectare subsampling units were placed in 1 and 10 ha isolated reserves as well as in primary forest reserves. Within each 1 ha sampling unit, a 10 × 10 m grid was established and the intersection points were used to measure vegetation structure. A 2.5 m pole was used to make a vertical estimate of foliage thickness at each intersection point within the grid. Six height intervals were scored for foliage thickness: up to 2, 2−5, 5−10, 10−20, 20−30, and 30−40 m. At each intersection point and within each height interval, a semi-quantitative index of foliage thickness was given: 0 for less than 25 percent foliage cover; 1 for 25−50 percent; 2 for 50−75 percent; and 3 for more than 75 percent foliage cover (Malcolm 1991).

SOILS

A large-scale soil survey was performed over the entire set of reserves, including isolates and continuous forest reserves (Chapter 23). The sampling unit was composed of a 1 ha plot divided into twenty-five 20 × 20 meter subplots. Plots were located in both isolated and continuous forest reserves of all

sizes. Within the 1 ha reserves, the soil plots covered the entire reserve. In the larger reserves (10 and 100 ha), three to nine sampling plots were laid out within the reserves to cover within-reserve variation.

In each 1 ha soil sampling plot, 20 surface soil samples (up to 20 cm depth) were taken in 13 of the 25 subplots. The 20 subsamples were combined to homogenize microvariation in soil chemistry. Of this bulk sample, several replicates were taken and analyzed for physical attributes (clay, silt, fine sand, coarse sand, organic matter contents), soil tension, pH, as well as concentrations of sixteen different chemical elements. Concurrent with the surface soil samples, core samples were taken at regular intervals in all reserves (isolated and continuous) to provide detailed soil profiles analyses.

PHYTODEMOGRAPHIC DATA SET

This constitutes one of the largest subprojects and, with the soil and topography data sets, is the backbone for characterizing the tropical forest and related changes caused by isolation and fragmentation. In all, this subproject has long-term phytodemographic data for 66 one-hectare plots and 13 hectares of transects in isolated and continuous forest reserves (Rankin–de Mérona et al. 1992; Laurance, Ferreira, et al. 1998b; Laurance, Gascon, and Rankin–de Mérona 1999). All isolated reserves identified (fig. 4.1) have at least one to several 1 ha plots within their boundaries. In each 1 ha plot, all trees at least 10 cm DBH were tagged and mapped relative to their position within the 1 ha plot. Initial surveys were performed in the early 1980s, and the resulting data set includes more than 60,000 individual trees (fig. 4.3). About half of the trees have been identified to species or morpho-species level, and the remainder to at least family level (see Chapter 5).

Fig. 4.3. "Tio" Romeu Cardoso, a veteran *mateiro* who has collected many of the voucher specimens for the Phytodemography subproject, climbs toward the canopy. Photo by Scott Mori.

Identifications of the remaining taxa are proceeding. Each 1 ha plot has been resampled on two to five occasions over periods ranging from six to thirteen years, providing important information on forest dynamics (tree mortality, recruitment, and turnover rates; Chapter 9). The present research on this subproject includes studying tree mortality, recruitment, and turnover in fragmented and continuous forest reserves; assessing the effects of nonrandom deforestation patterns on remnant tree communities; contrasting the effects of square versus rectangular sampling plots for assessing tree community composition; assessing changes in forest biomass over time; and assessing soil-biomass relationships.

Large Tree Demography Plot

To complement the landscape-level phytodemography plots, a 100 ha contiguous forest plot was established in primary forest (reserve 1501, fig 4.1) to study the dynamics of tropical forests using a high-diversity, high-density family of tropical trees—the Lecythidaceae. The family is one of the most diverse plant families in the study area, is economically important (Brazil nuts), and the region is in the middle of the family's center of diversification (Mori and Lepsch-Cunha 1995; Chapter 6). The 100 ha plot was divided into 100 one-hectare units, each of which was subsequently divided into 20 × 20 m quadrats for mapping purposes. All individuals of the Lecythidaceae family of 10 cm DBH or more were located and tagged. Each of the 7,791 located individuals were mapped and identified to species level. Topography surveys were performed in the entire 100 ha plot as well as soil samples every 20 m.

Invertebrates

Many invertebrate groups have been sampled and surveyed in various components of the landscape. The most important and well sampled of these include butterflies (Hutchings 1991; Brown and Hutchings 1997), ants (Vasconcelos and Cherrett 1997; Chapter 16), dung and carrion beetles (Klein 1989), litter beetles (Chapter 18), bees and wasps (Chapter 17), fruit-flies (Chapter 14), termites (DeSouza and Brown 1994), and social spiders (Chapter 15). Although these subprojects and their corresponding data sets are less extensive than for other faunal groups, they nonetheless have provided important insights into fragmentation effects on the ecosystem.

Bird Surveys

Understory birds were censused in a mist-net based mark-recapture program from 1979 through 1992 (Bierregaard and Lovejoy

1988; Bierregaard and Stouffer 1997). Nets were strung in 100 m transects (8 nets) in 1 ha reserves, or 200 m transects (16 nets) in 10 and 100 ha reserves. Both isolated and continuous forest reserves were included in the survey. Matrix habitat (young second growth on abandoned pasture) has also been surveyed over the years, most intensively during the mid-1990s (Borges 1995; Chapter 20).

Throughout the course of the study, control data were collected in two ways—from mist-net lines in the isolated reserves prior to isolation and from net lines run in continuous forest located a few hundred meters to a few dozen kilometers from the experimental isolates. In many reserves, net lines were in continuous operation for several years prior to isolation. This subproject has accumulated more than 50,000 captures of about 25,000 individual birds in all components of the landscape.

In addition, observations of birds seen in and around the experimental and control plots (including many not caught in the mist-net samples) as well as the greater Manaus area have been compiled in an extensive regional checklist (Stotz and Bierregaard 1989; Cohn-Haft, Whittaker, and Stouffer 1997).

SMALL MAMMAL SURVEYS

Non-volant small mammals were surveyed in all isolated and many continuous forest reserves over a period of six years (1983–89) (Malcolm 1991, 1997). As well, the matrix habitat (cutover areas, maintained pasture, and abandoned pasture) was sampled for small mammal abundance and species composition. One-hectare plots were sampled in reserves of different sizes. Twelve terrestrial and arboreal trap stations were located within each 1 ha sampling unit. Each 1 ha plot was sampled at least three times during

the six years, and all small mammals were recorded, marked, and released. Over 2,500 individuals of eighteen species were recorded, marked, and released. This subproject is unique in that it includes systematic trapping of small mammals in both terrestrial and arboreal microhabitats.

BAT SURVEYS

This subproject was initiated to cover a major vertebrate group that had not been extensively surveyed as part of the BDFFP. One- and ten-hectare isolated reserves, as well as continuous forest plots, are being monitored using ground and arboreal mistnets, complemented by the use of acoustic monitoring and night-scope vision to locate net-shy bat species. To date, 69 species have been identified, including 39 species from 7,000 mist-net captures in the BDFFP landscape. Preliminary descriptions of community structure (Kalko 1998) and fragmentation effects are beginning to take shape (C. Handley, unpublished data).

FROG SURVEYS

Amphibian species were surveyed in all components of the landscape using different techniques to ensure that all species were recorded (Zimmerman and Bierregaard 1986; Zimmerman 1991; Gascon 1991; Chapter 19). All aquatic habitats (ponds and streams) were visited diurnally and nocturnally for species composition and abundance estimates. Further, 1 km transects were walked nocturnally in continuous forest, in isolated reserves of all sizes, and in all types of matrix habitat. Many of the now-isolated reserves were surveyed before isolation for amphibians (as is the case for birds) to provide before-and-after comparisons. Amphibian surveys have been conducted since the early years of the project and have identified to date over sixty species present in the BDFFP landscape.

Remote Sensing

Because we have a detailed history of land use in the study area, which includes a complete spectrum from undisturbed forest to vast, scorched-earth clear-cuts, the BDFFP study site provides an excellent testing ground for analyzing remotely sensed data from such satellites as LANDSAT. An ongoing collaboration with John Adams and his students at the University of Washington has provided us a remarkable tool for quantifying changes in landscape pattern and process, as described in Chapter 28.

The Future of the BDFFP

The BDFFP started off as a one-man operation and has successfully grown over twenty years into an efficient midsize research, training, and education project with a paid staff of twenty-five. This growth has allowed us to provide much-needed information to help mold development in the Amazon. Our main objectives over the next decades will continue to include our focal activities in research: the continuation of large-scale inventories of varied taxonomic groups and their response to forest fragmentation. There is no doubt, however, that we will explore new and unexpected avenues of pure and applied research, as has been the case with our studies of forest regeneration, which we never imagined would have become such an important part of our research effort. Further, we will continue the programs we have developed in recent years aimed at finding ways to transfer our accumulated knowledge to the decision makers' arena.

We are in an excellent position to continue to contribute significantly to the conservation debate in the Amazon through the dissemination of relevant information, all the more so because many of our Brazilian trainees and students can now influence the decisions that must be made if we are to find the path to sustainable development in the Amazon.

PART II

Forest Ecology and Genetics

lthough the adage says that one cannot see the forest for the trees, understanding a forest and how its flora and fauna respond to habitat fragmentation requires us to look carefully at the trees. Trees, after all, define the forest. Our descriptors for different forest types are often based on the trees found in them — evergreen, deciduous, dipterocarp, or coniferous forests, for example. Besides giving the forest its structure, trees are the primary photosynthetic engine that drives the whole ecosystem, and a myriad suite of animals have coevolved with the trees, depending on them for food and shelter while pollinating their flowers, dispersing their seeds, and even in some instances protecting them from attack by herbivores.

A Hyperdiverse Flora

Trees, then, are both players in and the ecological stage for the evolutionary play that is the rainforest ecosystem, and the Biological Dynamics of Forest Fragments Project in Manaus has invested heavily in studying plant communities in both continuous forest and forest fragments. William F. Laurance, in Chapter 5, sets the stage for this section with an overview of the diverse forests of the study site north of Manaus.

The backbone of the BDFFP studies of the local flora has been an extensive survey of the canopy trees in 66 one-hectare plots. The sample plots were spread across the 1,000 km² BDFFP study area to cover a range of continuous forest areas as well as all the experimental fragments (Rankin–de Mérona et al. 1992). This massive phytodemographic inventory includes nearly 60,000 trees and their voucher specimens, as well as data on growth, damage, and mortality of trees in continuous forest and in our experimental plots.

Plant communities in the rainforests of the central Amazon are extremely diverse (Prance 1990; Rankin–de Mérona et al. 1992; Oliveira 1997). In our 1 ha plots it is not uncommon to find more than 300 different tree species, counting only individuals with a DBH of 10 cm or more. Although our taxonomic work is not yet finished, we already have identified more than 1,000 species or morpho-species in our series of 69 one-hectare plots (individual trees that appear distinct from all others in our sample but are not readily assigned to a known species are termed morpho-species). It is also common to find less than 50 percent overlap in species composition from one 1 ha plot to another, even when the plots are separated by only a few hundred meters. Further complicating studies of these forests is the rudimentary state of our knowledge of their taxonomy. For example, in a single 25 ha plot,

seventy-one species of trees in the family
Sapotaceae were encountered. Despite this
being a family whose taxonomy across the
Amazon is fairly well known, twenty of the
species in the plot were new to science (Pennington, pers. comm.).

Although the bulk of our descriptive plant
community work has focused on canopy
tree species, baseline data on species richness and community structure have been
collected for understory palms (Scariot
1996; Chapter 10), *Heliconia,* and the Rubiaceae (Boom and Campos 1991), and we
have initiated work on the important guild
of woody lianas.

The Brazil Nut Family

In the Neotropical lowlands, the Brazil nut
tree family, Lecythidaceae, is one of the
abundant and ecologically important families in terra firme forests and the subject of
the second chapter in this section on forest
ecology and genetics.

In 1988, Scott A. Mori and Peter Becker
initiated an intensive study of the Lecythidaceae in a 100 ha plot in the continuous-
forest control reserve 1501, or Km 41. In
Chapter 6, Mori, Becker, and Dwight Kincaid summarize the years of work that they
and their students and colleagues have carried out in the Lecythidaceae Plot.

In their 100 ha study area the authors identified and mapped 7,791 individuals of
thirty-eight species of Lecythidaceae, three
of which were new to science. Their figures
suggest that the forests of central Amazonia
may have more individuals and a higher
species richness of this family than any other
forest in the world. The remarkable number
of species in the 100 ha plot agrees with
other analyses and suggests that reserves in
Amazonia will protect more tree species
than anywhere else in the Neotropics.

The complete (and unprecedented) inven-

tory of all the adult individuals of a tree family in a 100 ha plot enabled the authors to address important methodological issues about
how large an area needs to be inventoried to
answer different sorts of questions about the
forest. Using computer simulations the authors took random samples of different numbers of hectares and compared the results to
the known species richness for the plot.
They concluded that the number of species
encountered rises rapidly for small samples
(87 percent of the species are found on average with a sample of only 10 ha) and only
very slowly for larger samples. While their
exact numbers cannot be extrapolated to
other regions of the Neotropics, the principles should hold and be considered by anyone undertaking botanical or ecological studies in the region so as not to overtax their
own financial resources or those of the tax-
onomists who are often called on to identify
and archive unnecessarily large collections.

The Age of Tropical Rainforest Trees

Because the growth rings of many tropical
trees have not been reliably demonstrated to
be annual, the ages of individual trees and
the age structure of populations in rainforests have been merely the subject of more
or less educated speculation. In Chapter 7,
Jeffrey Q. Chambers and colleagues describe
the use of ^{14}C dating to provide the first accurate measurements of the age of large trees
in central Amazonia.

Understanding the age structure of the
rainforest has important management and
conservation implications: models of global
carbon cycling will have more accurate estimates of the residence time of carbon in
wood; foresters should be better able to harvest at sustainable levels; and conservation
biologists will be able to assess the interaction of age structure with genetic diversity
and population stability. Chambers dis-

cusses in detail how our knowledge of the age of trees in the forest can improve our understanding in all three of these areas.

By ^{14}C dating very large trees at a local sawmill, Chambers found trees up to 1,400 years old, and two trees of the same diameter whose ages differed by 750 years, suggesting trees can get to the canopy after extended periods of growth suppression. Because these ancient trees will be reproducing with trees many generations younger, maintaining obviously successful genes in the population, Chambers argues that these canopy giants have a value far greater than that calculated on the basis of their round-wood volume.

For many of the people for whom ^{14}C dating would be helpful, the method is expensive or unavailable. It has been shown that some tropical trees do indeed have annual rings, so the information available in tree rings about demography and climate history may indeed be available. Because the generation time of trees may exceed the research lifespan of most biologists and research projects by an order of magnitude, the information available from tree rings or ^{14}C dating opens an exciting opportunity to understanding the ecological dynamics of the forest.

The Genetics of Rare Species

One of the characteristics of the rainforest ecosystem across all taxa is a preponderance of rare species. For example, if we define rarity as any species with a relative abundance of 2 percent or less, 90 percent of the understory birds sampled in the BDFFP reserves are rare (Bierregaard 1990). With trees, 30 to 40 percent of the species are commonly found to occur at densities of less than one tree per hectare. Species may be rare in a given environment for a number of ecologically varied reasons: the species may

be rare across its entire geographic range or may be rare only at the periphery of its range; they may be specialized on rare microhabitats related to soil type or drainage, leading to a distribution with widely scattered pockets of high density; they may be species that have just colonized an area; or they might be in population decline.

Nadja Lepsch-Cunha, Claude Gascon, and Paulo Kageyama explain, in Chapter 8, how rarity is a broad concept that can cover very different demographic and genetic population structures. Based on genetic and demographic data from two tree species in the BDFFP study area, they discuss the genetic consequences of small populations and compare contrasting hypotheses concerning the mechanisms, evolutionary and ecological, that might lead to and maintain rare species. They show that the effects of human disturbance on species with seemingly similar demographic characteristics can be very different if the species have different genetic characteristics.

Conservation biologists and wildlife managers must be concerned about the potential risk of inbreeding in small (rare) populations. As the authors point out, rare species might be self-compatible breeders that have eliminated many of the recessive deleterious genes in the population. In contrast, obligatorily outcrossing species are more likely to suffer inbreeding depression when they become rare. Thus one of the effects of habitat fragmentation—a reduction in population size—is likely to affect species differently, depending on their evolutionary history. These same traits will determine whether *ex situ* conservation efforts, such as maintaining species in botanical gardens, might be expected to succeed or whether conservation efforts will require very large reserves, where large populations of outcrossing species can be maintained.

CHAPTER 5

The Hyper-Diverse Flora of the Central Amazon

An Overview

WILLIAM F. LAURANCE

This chapter provides an overview of the diverse central Amazonian flora for the nonspecialist reader. In it I describe the forests of the Biological Dynamics of Forest Fragments Project study area, place the local flora in a regional context, and discuss ecological and historical factors that may have contributed to the remarkably high richness of tree species. Those wishing further details can consult the works of Ghillean Prance (Prance, Rodrigues, and da Silva 1976; Prance 1982, 1990), the late Alwyn Gentry (Gentry 1986, 1988, 1990; Gentry and Dodson 1987; Gentry and Emmons 1987), and other researchers (e.g., Rankin–de Mérona, Hutchings, and Lovejoy 1990; Rankin–de Mérona et al. 1992; Mori and Lepsch-Cunha 1995; Nee 1995; Scariot 1998; Oliveira and Daly 1999; Oliveira and Mori 1999; Ribeiro et al. 1999).

Climate

Forests of the BDFFP's 1,000 km² study area are broadly typical of the terra firme (non-flooded) forests throughout the central Amazon region. These are lowland (50–100 m elevation) tropical moist forests. Temperatures are warm to hot (monthly minima and maxima usually range from 19° to 21° and 35° to 39° C, respectively), and remain quite constant throughout the year (monthly means vary only 2° C). There is little seasonal variation in day length (Oliveira and Mori 1999).

Annual rainfall in the study area ranges from about 1,900 to 3,500 mm, with a distinctive dry season from June to October (mean annual rainfall from 1988–98 was 2,651 ± 402 mm [mean ± SD] at BDFFP camp Km 41). Some earlier accounts (based on long-term rainfall data for the city of Manaus) have probably overemphasized the severity of the study area's dry season, suggesting that monthly dry-season rainfall averages less than 100 mm (e.g., Anon. 1978). In fact, the BDFFP area appears consistently wetter than Manaus, especially during the drier months. From 1988–98, for example, June–October rainfall was 69 percent higher in the study area than in Manaus, while total annual rainfall was 9 percent higher). Mean rainfall for all months exceeds 100 mm in the study area, although the driest months average under 150 mm (fig. 5.1). Nevertheless, the dry season can become considerably more severe during periodic droughts. During the 1997 drought, one of the worst this century (Hammond

This is publication number 272 in the BDFFP Technical Series.

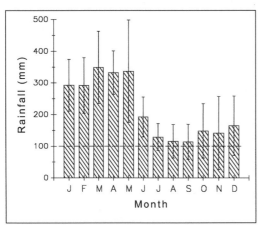

Fig. 5.1. Mean monthly rainfall (± SD) in the BDFFP study area from September 1987 to September 1999 (data collected at Camp Km 41). This period includes the major El Niño drought of 1997 and a lesser El Niño drought in 1992. The dotted line demonstrates that even the driest months average more than 100 mm of rainfall.

and ter Steege 1998), dry season rainfall was less than a third of normal, and the number of days without rain nearly doubled (Km 41 records).

Soils

As discussed in Chapter 23, the BDFFP study area overlies poor to extremely poor soils, as do large expanses of the Amazon basin (Brown 1987). The dominant soils are sandy or clayey latosols (Beinroth 1975) derived from heavily weathered alluvial deposits from the Tertiary. Clay particles in these soils tend to be aggregated, giving the soil poor water-holding characteristics, even with high clay contents (Richter and Babbar 1991). The soils are typically acidic, high in toxic aluminum, and very poor in crucial plant nutrients such as phosphorus, calcium, and potassium (Chauvel, Lucas, and Boulet 1987).

Because the soils are so poor, plants in the

region have evolved remarkably efficient mechanisms to conserve, scavenge, and recycle scarce nutrients (Herrera et al. 1978; Stark and Jordan 1978). As a result, the area supports dense evergreen rainforests, despite being rooted in soils that barely sustain even short-term slash-and-burn agriculture (Chapter 23).

Nevertheless, soil nutrients and texture do affect the vegetation. Within the BDFFP's network of 69 permanent 1 ha plots, soil features account for a third of the total variation in above-ground tree biomass, which rises along a gradient of increasing clay, nitrogen, organic carbon, and total exchangeable bases, and declines at sites with high sand and aluminum concentration (Laurance, Fearnside, et al. 1999). Biomass declines even further on white-sand soils, which are usually associated with larger stream gullies. In these areas, a sparse *campina* shrubland or scrubby *campinarana* forest develops (Guillaumet 1987; Duivenvoorden 1996). Although data are limited, it is clear that soils and topography also strongly influence the distribution and composition of plant species in terra firme forests (Guillaumet 1987; Kahn 1987; Gentry 1990; Prance 1990; Rankin–de Mérona et al. 1992; see also Tuomisto et al. 1995; Ruokolainen, Linna, and Tuomisto 1997; Sabatier et al. 1997 for studies in other parts of the Amazon).

The low-nutrient status of central Amazonian soils has important implications for forest ecology. Soil infertility could account for the apparent scarcity of flowers, fruits, and seeds among understory (Gentry and Emmons 1987) and canopy (M. J. G. Hopkins, pers. comm.) plants in the area, which can be seen as yet another adaptation for conserving nutrients (Van Shaik and Mirmanto 1985; Leigh 1999). Plants growing on poor soils also tend to have low rates of leaf production and high concentrations of defensive compounds, such as foliar tannins and phenolics, evolved to deter herbivores (Waterman 1983; Nichols-Orians 1991). Such defensive compounds may be responsible for the low levels of herbivory observed on leaves of central Amazonian trees (Vasconcelos 1999). The scarcity of fruits, seeds, flowers, and high-quality foliage means that food resources for animals are probably severely limited, and this may well account for the very low population densities of most fauna in the region (e.g., Powell 1985; Chapter 22).

Forest Dynamics

Relative to other Neotropical terra firme forests, natural forests in the study area exhibit rather low dynamics. Turnover rates of trees (the average of annualized mortality and recruitment rates for stems of at least 10 cm DBH) from the early 1980s to 1999 was only 1.20 ± 0.37 percent per year (X ± SD, for plots more than 300 m from forest edge, n = 28), compared to 1.66 ± 0.46 percent per year for other nonflooded Neotropical forests (see Phillips and Gentry 1994; Laurance, Ferreira, et al. 1998a). This situation changes dramatically if the forest is fragmented (see Chapters 9 and 13).

Several factors contribute to the low dynamics of central Amazonian forests. Growth rates of trees, and hence the intensity of competition for light among individuals, is likely to be somewhat reduced on poor soils (Leigh 1999), and this may lower rates of tree mortality. Large-scale disturbances are also rare. Downpours from convectional thunderstorms can cause intense local disturbances, but these events are uncommon, affecting only a tiny fraction of the basin (less than 0.05 percent) each year (Nelson 1994; Nelson, Kapos, et al. 1994). Charcoal fragments are common in soils of

the study area, indicating past fires (Bassini and Becker 1990), but there is no evidence of agriculture, and the Km 41 vicinity appears to have been continuously forested for at least 4,500 years (Piperno and Becker 1996). Finally, small (usually less than 0.2 ha) ponds created by wet-season rains often kill some trees (Mori and Becker 1991), but these affect only a small area of the forest each year.

Forest Architecture

Forests in the BDFFP study area generally range from 35 to 40 m in height, with emergent trees sometimes exceeding 50 m. The density of trees is high (averaging 613 ± 90 stems/ha [mean ± SD] for at least 10 cm DBH trees, based on data from 69 one-hectare plots). There are relatively few large trees (10.8 ± 3.9 stems/ha of more than 60 cm DBH), as is typical of sites on poor soils, and this probably accounts for the abundance of smaller trees (densities of small trees tend to be reduced in areas with many large trees, which sequester much light, water, and nutrients; Leigh 1999). Giant trees (more than 150 cm DBH) are present at very low densities. Some of these massive trees are ancient, up to 1,400 years old (Chambers, Higuchi, and Schimel 1998; Chapter 7).

The forest understory is often quite dense, with many saplings and stemless palms (especially *Astrocaryum* spp. [Arecaceae]), but is depauperate in flowering and fruiting shrubs (Gentry and Emmons 1987). Herbs, epiphytes, and climbing vines are present but in lower densities than in many other Neotropical forests (Guillaumet 1987; Gentry 1990). About half of all trees (≥ 10 cm DBH) bear at least one liana (≥ 2 cm DBH), and often several trees per hectare are heavily infested with lianas (Laurance et al., un-

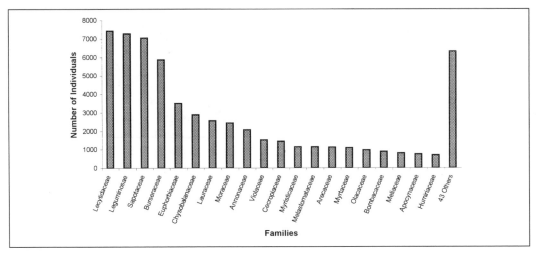

Fig. 5.2. Abundances of the twenty most common tree families from permanent 1 ha plots in the BDFFP study area (abundance = number of at least 10 cm diameter-at-breast-height (DBH) stems in 66 ha of plots).

published data). Hemi-epiphytic stranglers such as *Ficus* spp. (Moraceae) and *Clusia* spp. (Clusiaceae) are common. Local variation in topography, soils, and the ages and sizes of treefall gaps creates considerable heterogeneity in forest structure.

Floristic Composition

Knowledge of the central Amazonian flora has improved markedly in recent years, partly through work associated with the BDFFP (e.g., Rankin–de Mérona et al. 1992; Mori and Lepsch-Cunha 1995; Nee 1995; Scariot 1998; Oliveira and Daly 1999; Oliveira and Mori 1999; Chapter 6), and partly from extensive botanical collections and the production of a high-quality plant guide from Ducke Forest Reserve near Manaus (Ribeiro et al. 1999). This has helped to remedy serious deficiencies in understanding the local flora, especially for poorly studied groups like herbs, vines, and epiphytes.

Gentry (1990) provided an invaluable comparison of the Manaus-area flora to those of several other well-studied sites in the Neotropics (particularly La Selva in Costa Rica, Barro Colorado Island in Panama, and Manu Park in Peru). Although there are broad similarities among these sites in terms of dominant plant families, each has unique characteristics, and the Manaus flora is the most distinctive overall (table 5.1). Five major tree families, Chrysobalanaceae, Lecythidaceae, Myristicaceae, Burseraceae, and Bombaceae, are far more species-rich in Manaus than at the other Neotropical sites, and several key families at the other sites are poorly represented in Manaus. The Manaus flora is further distinguished by having a number of families (e.g., Duckeodendraceae, Rapateaceae, Rhabdodendraceae, Peridiscaceae) that appear specialized for the notoriously poor soils of the central and northern Amazon (Gentry 1990). Among trees, a few key families predominate; although sixty-three families have been identified in the BDFFP's permanent plots, the four largest, Lecythidaceae, Leguminosae, Sapotaceae, and Burseraceae, account for more than half of all individuals (fig. 5.2).

All of the well-studied Neotropical floras

TABLE 5.1. Distinctive Features of the Central Amazonian Flora Relative to Other Tropical and Neotropical Floras

1. Tree diversity in the central Amazon is among the highest in the world, comparable to that in such hyper-diverse areas as western Amazonia.

2. In the central Amazon, non-tree taxa (herbs, shrubs, stemless palms, climbing vines, and epiphytes) have only moderate diversity and are often less abundant than in other well-studied neotropical areas.

3. Weedy taxa (e.g., composites, sedges, grasses) are poorly represented in Amazonian floras, relative to those of Central America.

4. Compared to other Neotropical sites, the central Amazonian flora has many tree species (and some tree families) specialized for living on poor to very poor soils.

5. At the family level, the Manaus flora is quite distinctive; five tree families— Chrysobalanaceae, Lecythidaceae, Myristicaceae, Burseraceae, and Bombaceae—are far more prevalent and diverse than in other well-studied Neotropical floras.

6. At the generic level, the Manaus flora is highly distinctive from other well-studied Neotropical sites; genera such as *Piper*, *Ficus*, *Psychotria*, and *Philodendron* are very poorly represented, while others, including *Licania*, *Protium*, *Eschweilera*, and *Swartzia* are exceptionally diverse.

7. At the species level, the Manaus flora apparently overlaps little with that of other Neotropical sites; in fact, most well-studied sites in South and Central America tend to have unique floras at the species level.

8. Beta diversity (across microhabitats) in the Amazonian tree flora is extremely high, and even nearby plots can have remarkably low overlap in species composition.

Sources: Oliveira and Mori 1999; Ferreira and Rankin-de Mérona 1997; Gentry 1990; Rankin–de Mérona et al. 1992; Oliveira 1997.

tend to be rather different at the generic level, but the Manaus flora is especially distinctive (Gentry 1990). It is unusual in having an overwhelming predominance of trees belonging to a few relatively large genera

(table 5.1; Gentry 1990; Oliveira and Daly 1999). In addition, many major genera at other sites are poorly represented in Manaus, while several genera of limited importance elsewhere are extremely diverse at Manaus.

Floristic Diversity

The high tree species diversity of the Manaus flora is striking. To date, roughly 1,300 tree species (including specimens identified to genus and morpho-species level—i.e., individuals apparently distinct from other species in the sample but not reliably identified) have been documented in the BDFFP plots (S. G. Laurance, pers. comm.). The number of known taxa may well rise as identifications continue to improve, although an unknown number of morpho-species will eventually be subsumed into previously identified taxa. Some groups are especially well represented; for example, the BDFFP study area supports the highest density and species richness of Lecythidaceae ever observed (Chapter 6). Indeed, Oliveira and Mori (1999) demonstrated that the local (alpha diversity) richness of trees near Manaus is as high as that in western Amazonia (280–285 species/ha for at least 10 cm DBH stems), which was previously thought to be the major center for Neotropical tree diversity (Gentry 1988; Valencia, Balslev, and Paz y Miño 1994).

The exceptional richness of the Manaus tree flora is even more remarkable given that the area overlies poor soils, receives much less rainfall than western Amazonia, and has limited dynamics (see Phillips and Gentry 1994; Clinebell et al. 1995; Oliveira and Mori 1999). Mori (1990) and Oliveira and Daly (1999) have proposed that the Manaus area represents a "crossroads" where many species with varying geographic distributions within the Amazon meet. They based their case on a large-scale analysis of South American tree distributions, and argued that the number of taxa whose ranges ended at Manaus was far higher than expected by chance, and that this pattern was not simply caused by sampling artifacts. The observed pattern is consistent with the notion that Manaus was a key center of reconvergence of populations isolated within different Pleistocene forest refuges, with incipient speciation presumably occurring within the different reserves (Oliveira and Daly 1999). If their hypothesis is correct, the historical dynamics of the forest are partly responsible for the remarkably high tree diversity in the Manaus area.

Unlike trees, several other plant groups are not particularly well represented in Manaus (see table 5.1). Relative to other Neotropical sites, there are relatively few herbs and shrubs, and only moderately diverse assemblages of lianas and epiphytes (Guillaumet 1987; Gentry 1990). The limited diversity of epiphytic and understory taxa appears related to the area's poor soil and relatively strong dry season (Gentry and Emmons 1987). Lianas, which favor forest disturbances, may also be constrained by low rates of turnover in these forests (Laurance, Perez-Salicrup, et al., in press). Other deficiencies could simply arise from collecting artifacts (Nelson, Ferreira, et al. 1990). Orchids, for example, were thought to be poorly represented in the Manaus area (Gentry 1990), but recent collections at Ducke Reserve have revealed many epiphytic taxa that were previously unknown in the region (Ribeiro et al. 1999). Palms can achieve high local densities (Kahn 1987; Scariot 1998) and are quite diverse in the Manaus area if one includes *varzea* forests (B. W. Nelson, pers. comm).

Rarity and Other Features

The forests of the central Amazon, like those of many other tropical forests, are notable for having many rare species (Ackerly, Rankin–de Mérona, and Rodrigues 1990; Rankin–de Mérona et al. 1992). Among the roughly 1,300 tree species in the BDFFP plots, for example, only about 12 percent (160 species) would not be considered rare (average density of at least 1 stem/ha; S. G. Laurance, pers. comm.). Moreover, in addition to high alpha diversity, these forests exhibit striking beta and gamma diversity; even plots separated by only a few hundred meters can have surprisingly different plant assemblages (Hubbell and Foster 1986; Oliveira 1997; Laurance, Ferreira, et al. 1999), and there is often little overlap in the floras of different habitats or regions (Gentry 1990). Finally, the bulk of tropical plant species are obligately outcrossing, and most depend on coevolved pollinators and seed-dispersers for long-term survival. All of these attributes have important implications for conservation (Chapter 13).

Conservation Lessons

1. Central Amazonian forests in the vicinity of Manaus have extremely high tree diversity and may be an important "crossroads" where species with varying geographic distributions meet.

2. Neotropical floras are highly differentiated; among well-studied sites, there apparently is little overlap in species composition. The Manaus flora is distinctive among that in well-studied sites, with many species adapted for the very poor soils of central and northern Amazonia.

3. Networks of numerous nature reserves will clearly be needed to capture a high proportion of the floral diversity of the Amazon. The high levels of local, across-

habitat, and across-regional diversity mean that nature reserves must be stratified across major biogeographic, climatic, soil, and topographic gradients.

4. Combinations of large (e.g., 10^5–10^7 ha) and small (e.g., 10^3–10^4 ha) reserves will be needed to preserve large and functionally intact ecosystems, as well as the multitudes of locally endemic species in the Amazon.

Acknowledgments

I thank Bruce Nelson, Michael Nee, Susan Laurance, Leandro Ferreira, and Heraldo Vasconcelos for commenting on drafts of the manuscript, and Susan Laurance for providing useful data from the BDFFP Herbarium. This work was supported by World Wildlife Fund-US, National Institute for Research in the Amazon, Smithsonian Institution, MacArthur Foundation, Andrew W. Mellon Foundation, Conservation, Food, and Health Foundation, and the NASA Long-Term Biosphere-Atmosphere Experiment in the Amazon.

CHAPTER 6

Lecythidaceae of a Central Amazonian Lowland Forest

Implications for Conservation

SCOTT A. MORI, PETER BECKER, AND DWIGHT KINCAID

The Lecythidaceae are often ecologically important members of Neotropical forests where they are especially abundant in lowland, nonflooded forests. In various ecological studies of lowland forests, especially in Amazonia and in the Guyana floristic province, the family often ranks among the ten most important families of trees (Black, Dobzhansky, and Pavan 1950; Cain et al. 1956; Prance, Rodrigues, and da Silva 1976; Prance and Mori 1979; Balslev et al. 1987; Mori and Boom 1987; Mori, Becker, et al. 1989). These studies, however, provide little information on sample variability—mostly because of the time and expense involved in replicating samples in the species-rich lowland forests of the tropics.

The purpose of this study was to provide information on the frequency, density, dominance, and species richness of Lecythidaceae based on a 100 ha sample from central Amazonian Brazil. The information provided herein will furnish the baseline data needed for comparing changes in the composition of Lecythidaceae as a result of forest fragmentation or climate change. Moreover, our experience with sampling diversity of Lecythidaceae provides insight into the number of hectares that should be sampled in order to reach an understanding of the species richness of this ecologically important family of tropical trees.

Study Site

The Lecythidaceae study plot is located in Reserve 1501, a 1,000 ha control reserve of the BDFFP located within more or less continuous forest. Reserve 1501 is also called Km 41 because it is situated 41 kilometers along state highway ZF-3 from federal highway BR-174 (the Manaus-Boa Vista highway) (see fig. 4.1).

The mean annual temperature for Manaus, some 80 km south of Reserve 1501, is 26.7° C with monthly means fluctuating only by about 2° C. Maximum temperatures range between 35° and 39° C and minimum temperatures between 19° and 21° C. Cool air masses, often occurring at the transition between the rainy and dry seasons, can drop temperatures to 17° C. There is a distinct dry season between July and September, and these months normally receive less than 100 mm of rain (BDFFP 1990; Lovejoy and Bierregaard 1990). Local winds may on occasion be strong enough to topple trees (Nelson 1994; Nelson, Kapos, et al. 1994), although no large blowdowns are known to

This is publication number 273 in the BDFFP Technical Series.

have occurred since 1987 at Reserve 1501. The difference in day length between the longest and shortest days of the year at Manaus is about eighteen minutes (List 1950).

The plot is dissected by several small streams, especially one that flows north to south in the middle of the eastern half of the plot, and there is a plateau in the northeastern corner (Mori and Becker 1991). Although there are wetter areas along small streams and periodic small ponds form for unusually long periods during years of excessive rainfall (Mori and Becker 1991), this plot is typical of the terra firme habitat found throughout central Amazonia. As in any other tropical forest, the formation of small and mid-sized gaps is common (Denslow 1980).

The reserves of the BDFFP are located on extensive Tertiary sediments within the ancient meander plain of the Amazon. However, Reserve 1501 is not situated near any major river, and as a result there are no recent alluvial deposits in the plot. The soils in the plot are sandy or clayey latosols that have been subjected to long periods of leaching and are therefore generally poor in nutrients (Lovejoy and Bierregaard 1990). In the Lecythidaceae plot, soils dominated by clay are prevalent, but soils richer in sand occur in the northwestern and southwestern corners (P. Becker et al., unpublished data). Charcoal is ubiquitous in the soil of the plot (Bassini and Becker 1990), but there is no evidence, based on a study of plant remnants in the soil (phytoliths), that crops were ever grown there (Piperno and Becker 1996).

Reserve 1501 is situated in terra firme forest at between 80 and 110 meters altitude. The forest is dominated mostly by species of Sapotaceae, Lecythidaceae, and Burseraceae (Oliveira 1997). Palms, especially spiny ones of the genus *Astrocaryum,* are abundant in the understory. Within all of the re-serves of the BDFFP, there are at least 57 families (Oliveira 1997) and over 800 species of trees (BDFFP 1990). Nee (1995) has published a preliminary vascular plant flora of the BDFFP reserves, and Oliveira (1997) has reported on total tree species richness per hectare based on a sample of all trees of at least 10 cm DBH (diameter at breast height, or 1.3 m from the ground on the uphill side of the trunk) in three hectares of the Lecythidaceae plot.

Methods

The area for the 100 ha plot was selected by Becker in consultation with Marc van Roosmalen to include the major types of topography. A factor in plot selection was proximity to the camp at Reserve 1501. The plot was professionally surveyed, and care was taken to ensure that each 20 × 20 m quadrat contained 400 m² in horizontal projection, as was subsequently recommended by Dallmeier (1992). A stake was placed at the corners of all of the 20 × 20 m quadrats, and a plaque marked with the x and y coordinates was affixed to the stake. Each of the 100 hectares is identified by the x,y coordinates of its southwesternmost stake.

All individuals of Lecythidaceae at least 10 cm DBH were located by experienced woodsmen under the supervision of a Brazilian student intern and confirmed by Mori. Individuals of Lecythidaceae are relatively easy to identify because of the fibrous nature of their bark (Mori, Becker, et al. 1987), but mistakes, especially in distinguishing between species of Annonaceae and species of Lecythidaceae, are sometimes made. The position of each individual was then recorded by measuring the distance and azimuth to the individual from one of the corner stakes.

After the trees were marked and meas-

TABLE 6.1. Density, Dominance, and Diversity of Lecythidaceae per Hectare in a 100 ha Plot at Reserve 1501

Hectare Coordinates	Density (trees/ha)	Dominance (m²/ha)	Diversity (spp./ha)	Hectare Coordinates	Density (trees/ha)	Dominance (m²/ha)	Diversity (spp./ha)
0,0	142	5.26	12	25,5	51	2.13	11
0,5	89	4.24	15	25,10	71	2.92	14
0,10	107	3.57	13	25,15	70	4.15	18
0,15	113	5.06	22	25,20	58	4.59	20
0,20	109	4.05	18	25,25	65	3.48	16
0,25	92	3.50	18	25,30	55	3.24	23
0,30	122	5.65	20	25,35	87	5.00	18
0,35	125	4.38	15	25,40	82	5.35	18
0,40	87	2.54	16	25,45	70	3.51	19
0,45	93	4.60	19	30,0	65	2.76	15
5,0	90	4.04	20	30,5	51	4.84	18
5,5	86	3.16	19	30,10	49	3.11	13
5,10	73	3.34	15	30,15	90	5.01	13
5,15	99	4.01	18	30,20	77	3.78	18
5,20	149	5.71	18	30,25	90	4.83	16
5,25	117	4.12	22	30,30	84	4.09	17
5,30	82	4.00	17	30,35	90	2.92	19
5,35	86	3.02	19	30,40	100	4.64	21
5,40	75	2.65	15	30,45	85	4.12	18
5,45	79	3.33	24	35,0	61	2.62	16
10,0	50	2.26	17	35,5	52	2.85	13
10,5	65	2.64	16	35,10	75	3.34	15
10,10	73	3.85	21	35,15	85	4.13	16
10,15	103	3.69	18	35,20	54	2.23	14
10,20	77	3.63	20	35,25	79	4.59	20
10,25	116	4.34	18	35,30	83	2.80	17
10,30	69	3.33	16	35,35	95	3.49	20
10,35	89	3.58	18	35,40	85	3.72	16
10,40	75	5.11	20	35,45	102	2.83	21
10,45	84	3.42	18	40,0	45	3.04	13
15,0	45	3.19	18	40,5	73	3.50	17
15,5	62	2.40	15	40,10	64	3.33	15
15,10	68	2.67	12	40,15	55	2.69	13
15,15	64	2.89	13	40,20	78	4.25	16
15,20	80	3.67	20	40,25	70	2.98	14
15,25	82	3.01	20	40,30	58	3.91	19
15,30	91	4.08	18	40,35	65	5.00	21
15,35	74	3.74	15	40,40	74	3.11	18
15,40	60	2.64	18	40,45	65	3.11	19
15,45	72	4.50	16	45,0	89	3.65	17
20,0	100	4.25	21	45,5	73	4.85	19
20,5	78	3.51	18	45,10	79	4.64	21
20,10	72	4.49	13	45,15	52	2.77	18
20,15	74	4.34	18	45,20	53	4.03	17
20,20	68	3.12	16	45,25	57	3.84	16
20,25	76	3.85	17	45,30	56	3.01	15
20,30	81	6.56	18	45,35	59	3.93	17
20,35	69	4.31	20	45,40	63	3.74	15
20,40	64	4.55	15	45,45	60	3.43	18
20,45	74	4.77	20	Totals	7,791	376.34	
25,0	72	3.84	17				

TABLE 6.2.   Frequency, Density, Dominance, and Within Family Importance Values for Species of Lecythidaceae in a 100 ha Plot at Reserve 1501

Species[1]	Absolute frequency[2]	Absolute density[3]	Absolute dominance[4]	Relative frequency[5]	Relative density[6]	Relative dominance[7]	WFIV[8]
ALLI	5	5	.46	0.3	0.1	0.1	0.5
BEEX	1	1	1.04	0.1	<0.1	0.3	0.3
CADE	29	40	4.35	1.7	0.5	1.2	3.4
CAMI	27	29	9.93	1.6	0.4	2.6	4.6
COAL	75	192	15.85	4.3	2.5	4.2	11.0
CORI	63	208	13.81	3.6	2.7	3.7	10.0
CUGU	18	21	1.79	1.0	0.3	0.5	1.8
CULO	14	14	.36	0.8	0.2	0.1	1.1
CUMU	20	24	1.29	1.2	0.3	0.3	1.8
CUST	41	77	5.31	2.4	1.0	1.4	4.8
CUTA	2	3	.09	0.1	<0.1	<0.1	0.1
ESAF	51	190	11.26	2.9	2.4	3.0	8.3
ESAT	96	571	24.50	5.5	7.3	6.5	19.3
ESBR	61	105	1.51	3.5	1.3	0.4	5.2
ESCA	8	10	1.01	0.5	0.1	0.3	0.9
ESCO	28	55	2.35	1.6	0.7	0.6	2.9
ESCR	95	1,539	64.85	5.5	19.8	17.2	42.5
ESCY	75	275	31.47	4.3	3.5	8.4	16.2
ESGR	93	335	8.63	5.4	4.3	2.3	12.0
ESLA	41	57	3.89	2.4	0.7	1.0	4.1
ESMI	73	180	12.25	4.2	2.3	3.3	9.8
ESPE	61	100	1.43	3.5	1.3	0.4	5.2
ESPS	48	118	8.18	2.8	1.5	2.2	6.5
ESRC	68	309	8.91	3.9	4.0	2.4	10.3
ESRN	13	16	1.48	0.8	0.2	0.4	1.4
ESTE	84	250	6.49	4.8	3.2	1.7	9.7
ESTR	95	1,321	53.52	5.5	17.0	14.2	36.7
ESWA	99	926	22.90	5.7	11.9	6.1	23.7
GUEL	30	111	2.64	1.7	1.4	0.7	3.8
LEBA	53	105	2.22	3.1	1.3	0.6	5.0
LEGR	31	40	3.64	1.8	0.5	1.0	3.3
LEPA	22	30	4.72	1.3	0.4	1.3	3.0
LEPI	36	46	4.96	2.1	0.6	1.3	4.0
LEPR	90	362	23.23	5.2	4.6	6.2	16.0
LERE	7	11	.90	0.4	0.1	0.2	0.7
LEZA	29	51	10.59	1.7	0.7	2.8	5.2
LE01	7	8	.23	0.4	0.1	0.1	0.6
LE05	40	51	3.39	2.3	0.7	0.9	3.9
INDETS	4	4	.92	0.2	0.1	0.2	0.5
TOTAL	1,733	7,791	376.35	≈100.0	≈100.0	≈100.0	≈300.0

Notes: [1]Species codes: ALLI, *Allantoma lineata* (Martius ex Berg) Miers; BEEX, *Bertholletia excelsa* Humboldt & Bonpland; CADE, *Cariniana decandra* Ducke; CAMI, *C. micrantha* Ducke; COAL, *Corythorphora alta* R. Knuth; CORI, *C. rimosa* W. Rodrigues subsp. *rimosa;* CUGU, *Couratari guianensis* Aublet emend. Prance; CULO, *C. longipedicellata* W. Rodrigues; CUMU, *C. multiflora* (J. E. Smith) Eyma; CUST, *C. stellata* A. C. Smith; CUTA, *C. tauari* Berg; ESAF, *Eschweilera amazoniciformis* Mori; ESAT, *E. atropetiolata* Mori; ESBR, *E. bracteosa* (Poeppig ex Berg) Miers; ESCA, *E. carinata* Mori; ESCO, *E. collina* Eyma; ESCR, *E. coriacea* (A. P. de Candolle) Mori; ESCY, *E. cyathiformis* Mori; ESGR, *E. grandiflora* (Aublet) Sandwith; ESLA, *E. laevicarpa* Mori; ESMI, *E. micrantha* (Berg) Miers; ESPE, *E. pedicellata* (Richard) Mori; ESPS, *E. pseudodecolorans* Mori; ESRN, *E. rankiniae* Mori; ESRC, *E. romeu-cardosoi* Mori; ESTE, *E. tessmannii* R. Knuth; ESTR, *E. truncata* A. C. Smith; ESWA, *E. wachenheimii* (R. Benoist) Sandwith; GUEL, *Gustavia elliptica* Mori; LEBA, *Lecythis barnebyi* Mori; LEGR, *L. gracieana* Mori; LEPA, *L. parvifructa* Mori; LEPI, *L. pisonis* Cambessèdes; LEPR, *L. prancei* Mori; LERE, *L. retusa* Spruce ex Berg; LEZA, *L. zabuajo* Aublet; LE01, *Lecythis* sp. 01; LE05, *Lecythis,* sp. 05. [2]Number of hectares in which species is found. [3]Number of trees of a species per 100 ha. [4]Basal area of a species per 100 ha. [5]Total number of occurrences of a species of Lecythidaceae/total number of occurrences of all species of Lecythidaceae. [6]Total number of trees of a species of Lecythidaceae/total number of trees of all species of Lecythidaceae. [7]Total basal area of a species of Lecythidaceae/total basal area of all species of Lecythidaceae. [8]Relative frequency + relative density + relative dominance.

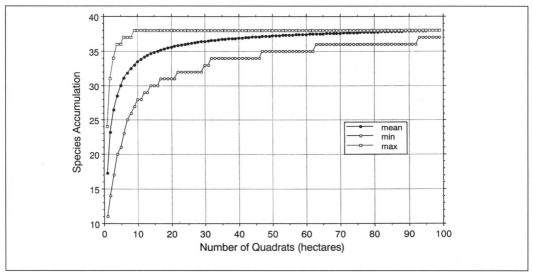

Fig. 6.1. Species-area curves generated from sampling hectare combinations without replacement. N = 10,000 except for the single and 2 ha samples which are exact solutions. The lower curve represents the minimum number of species, the middle curve the mean number of species, and the upper curve the maximum number of species for each combination of quadrats.

ured, Mori began the long process of identification to species. Each tree was visited, and the bark, leaves, and flowers and fruits, when available, were examined. When there was any doubt about the determination, a herbarium specimen was collected as a voucher. The vouchers are mostly deposited in the herbaria of the Instituto Nacional de Pesquisas da Amazonia (INPA), the New York Botanical Garden, and the Smithsonian Institution. A list of the vouchers is available from Mori. All identifications, locations of the trees, DBH, and other data are kept in a database (dBase III) under the supervision of Becker.

The field crew, which lived at Reserve 1501 for five years, was asked to make collections of all species that came into flower or fruit. The fertile material sometimes revealed mistakes in determination that had to be resolved on subsequent visits by the senior author to all of the trees with dubious determinations. This process was made possible by the lists of species and their locations

generated from the database. Species concepts follow Mori and Lepsch-Cunha's floristic treatment of all species of Lecythidaceae known to occur in the 100 ha plot (Mori and Lepsch-Cunha 1995). In addition, Mori and Prance (1990) and Prance and Mori (1979) present taxonomic treatments, including distribution maps, of the 198 species of Neotropical Lecythidaceae known at the time of the publication of these monographs. An additional seven species have since been described (Mori 1992a, 1995; Mori and Lepsch-Cunha 1995).

Notes on the habitat preferences of the species of Lecythidaceae found at Reserve 1501 can be found in Mori and Lepsch-Cunha (1995). With the exception of *Allantoma lineata,* a species found along streams, all of the species in the Lecythidaceae plot prefer terra firme forest.

Absolute density, absolute dominance, and species richness of Lecythidaceae were tabulated for each of the hectares sampled (table 6.1), and absolute frequency, absolute

TABLE 6.3.   Samples of 1, 2, 5, 10, 20, 25 and 50 Hectares

Sample size and combinations	Species richness	Frequency	Probability	Mean species richness per sample
1 ha sample				17.3
(all 100 possible	11	1	.0200	
combinations of	12	2	.0100	
1 taken out of 100)	13	8	.0800	
	14	3	.0300	
	15	12	.1200	
	16	12	.1200	
	17	10	.1000	
	18	23	.2300	
	19	8	.0800	
	20	11	.1100	
	21	6	.0600	
	22	2	.0200	
	23	1	.0100	
	24	1	.0100	
2 ha sample				23.2
(all 4,950 possible	14	1	.0002	
combinations of	15	10	.0020	
2 taken out of 100)	16	24	.0048	
	17	42	.0085	
	18	93	.0188	
	19	181	.0366	
	20	330	.0667	
	21	575	.1162	
	22	708	.1430	
	23	762	.1539	
	24	733	.1481	
	25	611	.1234	
	26	449	.0907	
	27	216	.0436	
	28	135	.0273	
	29	52	.0105	
	30	24	.0048	
	31	4	.0008	
5 ha sample				30.0
(10,000 combinations	21	1	.0001	
of 5 taken out of 100)	23	4	.0004	
	24	28	.0028	
	25	96	.0096	
	26	262	.0262	
	27	635	.0635	
	28	1,201	.1201	
	29	1,737	.1737	
	30	2,040	.2040	
	31	1,824	.1824	
	32	1,253	.1253	
	33	658	.0658	
	34	220	.0220	
	35	36	.0036	
	36	5	.0005	

TABLE 6.3. (continued) Samples of 1, 2, 5, 10, 20, 25 and 50 Hectares

Sample size and combinations	Species richness	Frequency	Probability	Mean species richness per sample
10 ha sample				33.4
(10,000 combinations	28	2	.0002	
of 10 taken out of 100)	29	23	.0023	
	30	179	.0179	
	31	659	.0659	
	32	1,567	.1567	
	33	2,607	.2607	
	34	2,671	.2671	
	35	1,697	.1697	
	36	517	.0517	
	37	69	.0069	
	38	9	.0009	
20 ha sample				35.6
(10,000 combinations	31	2	.0002	
of 20 taken out of 100)	32	13	.0013	
	33	230	.0230	
	34	1,140	.1140	
	35	3,169	.3169	
	36	3,601	.3601	
	37	1,631	.1631	
	38	214	.0214	
25 ha sample				36.1
(10,000 combinations	32	1	.0001	
of 25 taken out of 100)	33	52	.0052	
	34	432	.0432	
	35	2,107	.2107	
	36	4,043	.4043	
	37	2,810	.2810	
	38	555	.0555	
50 ha sample				37.2
(10,000 combinations	34	1	.0001	
of 50 taken out of 100)	35	61	.0061	
	36	1,383	.1383	
	37	5,011	.5011	
	38	3,544	.3544	

Note: The samples were run without replacement of the hectares, and each sample run, except those of 1 and 2 ha, was performed 10,000 times.

density, and absolute dominance (Curtis and Cottam 1962) were calculated for each of the species of Lecythidaceae present in the 100 ha plot (table 6.2). Finally, relative frequency, relative density, and relative dominance of each species of Lecythidaceae in relation to all other species of the family were calculated and summed to give the Within Family Importance Value (WFIV) in the manner of Boom and Campos (1991) for the Rubiaceae (table 6.2).

Known species richness of Lecythidaceae in the Neotropics was obtained from the *Flora Neotropica* monographs of Prance and

Mori (1979) and Mori and Prance (1990), as well as from papers with new species published subsequent to these monographs (Mori 1992a; Mori and Lepsch-Cunha 1995).

Species-area curves were prepared based on exact and approximate randomization analyses (performed by Kincaid) of the number of species added to the total with each additional hectare sampled. For each sample run, the hectares were sampled randomly without replacement. Analyses were based on 10,000 sample runs for sample combinations greater than two (fig. 6.1 and table 6.3).

Results

The 100 ha plot at Km 41 harbors thirty-eight species of Lecythidaceae (table 6.2). Only one other species, *Lecythis poiteaui* Berg, is known to occur elsewhere at Km 41. The species in the Lecythidaceae plot are represented by 7,791 individuals at least 10 cm DBH, which are found throughout the entire plot.

FREQUENCY

Individuals of Lecythidaceae are found in all of the 100 hectares (table 6.1). Five species (*Eschweilera atropetiolata, E. coriacea, E. grandiflora, E. truncata,* and *E. wachenheimii*) occur in more than 90 percent and sixteen species occur in at least half of the hectares (table 6.2). In contrast, the least frequent species is *Bertholletia excelsa* (Brazil nut), which is represented by only one individual in the entire plot. Ten species were found in fewer than 25 percent of the hectares sampled.

DENSITY

A total of 7,791 individuals of Lecythidaceae at least 10 cm DBH were recorded in the 100 ha plot (see tables 5.1, 5.2). As few

as 45 individuals were found in hectare 15,0, and as many as 149 in hectare 5,20 (table 6.1). The mean density of Lecythidaceae per hectare is 77.9 ± 19.9 (mean ± SD). Some species, such as *Eschweilera coriacea* and *E. truncata,* are abundant, with 1,539 and 1,321 individuals, respectively, in the 100 ha plot (table 6.2).

DOMINANCE

Total basal area for all Lecythidaceae in the 100 ha plot is 376.34 m² (table 6.1), and average basal area per hectare is 3.76 ± 0.86 m² (mean ± SD). Hectare 20,30, with 6.56 m², possesses the greatest basal area, and hectare 25,5, with 2.13 m², the least basal area of Lecythidaceae. *Eschweilera coriacea* (64.85 m²/100 ha) and *E. truncata* (53.52 m²/100 ha), because of their high densities and relatively large sizes, are the most dominant species of Lecythidaceae in the 100 ha plot (table 6.2).

WITHIN FAMILY IMPORTANCE VALUE

The ecologically most important species of Lecythidaceae in the 100 ha plot are: *Eschweilera coriacea* (WFIV = 42.5), *E. truncata* (36.7), *E. atropetiolata* (19.3), *E. cyathiformis* (16.2), *Lecythis prancei* (16.0), *Eschweilera grandiflora* (12.0), *Corythophora alta* (11.0), *Eschweilera romeu-cardosoi* (10.3), and *Corythophora rimosa* (10.0). These nine species account for 58 percent of the possible 300-point WFIV index.

SPECIES RICHNESS

The total number of species in the 100 ha plot is thirty-eight (table 6.2). A single other species, *Lecythis poiteaui,* is also known from the reserve, but it does not occur in the plot (Mori and Lepsch-Cunha 1995). Hectare 25,5, with only eleven species, and hectare

5,45, with twenty-four species of Lecythidaceae, represent the least and most species-rich hectares, respectively. The mean number of species per hectare in the plot is 17.3 ± 2.6 (mean ± SD).

We generated three simulated species-area curves for the 100 ha Lecythidaceae plot (fig. 6.1). The lower curve represents the fewest and the upper curve the greatest number of species sampled per specified number of hectares. The middle curve gives the mean number of species sampled. Species richness was calculated for 1, 2, 5, 10, 20, 25, and 50 ha samples—each, except for the 1 and 2 ha samples, based on a draw of 10,000 combinations of the number of hectares sampled from the 100 ha plot. For example, a sample of 10 hectares yields as few as 28 (P = 0.0002), as many as 38 (P = 0.0009), and a mean of 33.4 species (table 6.3). A 20 ha sample yields as few as 31 (P = 0.0002), as many as 38 (P = 0.0214), and a mean of 35.6 species.

Discussion

The high frequency, density, and species richness of Lecythidaceae in the 100 ha plot at Reserve 1501 suggest that central Amazonia may harbor the world's greatest number of individuals and species richness of Lecythidaceae.

At least 45 and as many as 149 individuals of Lecythidaceae were found in 100 percent of the 100 ha sampled (table 6.1), demonstrating that this family is found everywhere in the terra firme habitat of this area. A mean density of 77.9 ± 19.9 (mean ± SD) individuals of Lecythidaceae per hectare is considerably higher than that found in any other study. The next highest reported density of Lecythidaceae is the 50.3 individuals/ha found in central French Guiana by Mori and Boom (1987).

The mean basal area of Lecythidaceae in the 100 ha plot of 3.76 ± 0.86 m² (mean ± SD) per hectare is considerably lower than the 5.47 m² reported for the family in central French Guiana (Mori and Boom 1987). The average basal area per tree of Lecythidaceae in the 100 ha plot is 483.0 cm² (calculated from data provided in tables 5.1 and 5.2) in contrast to 1,086.7 cm² per tree in central French Guiana (Mori and Boom 1987) indicating that, at least for Lecythidaceae, the French Guianan forest possesses considerably larger trees. Although this has not yet been tested, the forests of central Amazonia appear to be relatively small in stature in contrast to some other areas, such as central French Guiana (Mori, pers. obs.). Small tree size in this part of central Amazonia may be attributed to some combination of the relatively poor soils, low quantity, and seasonality of rainfall found in the area.

Oliveira (1997), in his study of overall tree diversity among individuals of trees at least 10 cm DBH in hectares 10,10; 25,30; and 5,35 of the Lecythidaceae plot, found 285 species among 618 individuals, 280 species among 654 individuals, and 280 species among 644 individuals in these hectares, respectively. His data demonstrate that central Amazonia possesses overall tree species richness in the at least 10 cm size class comparable to that found by Gentry (1988) of about 300 species/hectare in Amazonian Peru and by Valencia, Balslev, and Paz y Miño (1994) of 307 species/hectare in Amazonian Ecuador.

In the hectares studied by Oliveira, the Lecythidaceae comprise 21 species among 73 individuals, 23 species among 55 individuals, and 19 species among 86 individuals, respectively (table 6.1). In these hectares, Lecythidaceae comprise 6.8 to 8.2 percent of the total species and 8.4 to 13.4 percent of the total individual trees at least 10 cm DBH.

The presence of thirty-eight species of

Lecythidaceae in such a limited area is far higher than that heretofore reported for the family. Thirty-one species of Lecythidaceae have been found in central French Guiana (Mori and collaborators 1987; Mori, unpublished data), but that figure is from an unspecified and much larger area. Mori (1990) has pointed out that slightly more than 50 percent of Neotropical Lecythidaceae are found in Amazonia, a finding that agrees with Gentry's claim (1982) that Amazonia is a center for overall species richness of trees and lianas. Therefore, reserves in Amazonia, more than anywhere else in the Neotropics, will protect disproportionately high numbers of trees and lianas. Moreover, because trees are the structural species that provide the habitat for interstitial species (e.g., herbs, shrubs, epiphytes) (Huston 1994), conservation programs designed to protect tree species richness also conserve numerous other associated species of plants and animals.

The per-hectare species richness of Lecythidaceae in the 100 ha plot is the highest heretofore reported for this family. The mean species richness for the 100 ha of 17.3 ± 2.6 (mean ± SD) species/ha is similar to the eighteen species that Prance, Rodrigues, and da Silva (1976) found in a single hectare of comparable forest in central Amazonia. Such high species diversity of trees in a family per hectare, however, is not unusual in Amazonian forests. For example, Valencia, Balslev, and Paz y Miño (1994) found 20 species of *Pouteria,* 19 species of *Inga,* and 14 species of *Protium* in a hectare surveyed in Amazonian Ecuador.

The high species richness of central Amazonian Lecythidaceae is partially the result of a mixture of species from western Amazonia (e.g., *Eschweilera bracteosa* and *E. tessmannii*), species from the Guyana floristic province (e.g., *Eschweilera collina* and *Lecythis poiteaui*), and widespread species (e.g., *Couratari guianensis* and *Es-*

chweilera coriacea) that may have migrated into central Amazonia after the recession of a lake (Lago Amazonas) that covered large expanses of central Amazonia in the late Pleistocene into the early Holocene (Frailey et al. 1988; Mori 1991). Others (Tuomisto, Ruokolainen, and Salo 1992), however, argue that the presence of lacustrine sediments in Amazonia can be explained by mechanisms other than the presence of a Pleistocene Lago Amazonas.

The occurrence of possible endemic taxa of Lecythidaceae in central Amazonia (e.g., *Corythophora rimosa* subsp. *rimosa, Couratari longipedicellata, Eschweilera amazoniciformis, E. cyathiformis, E. seudodecolorans, E. rankiniae, E. romeu-cardosoi, Gustavia elliptica, Lecythis barnebyi, L. gracieana, L. parvifructa,* and *L. retusa*) may be an artifact of incomplete botanical exploration as noted by Nelson, Ferreira, et al. (1990) for other taxa. These species, however, may also represent peripheral populations of forest species cut off from more widespread ancestral species by savanna habitats caused by climatic fluctuations or by the appearance and disappearance of water barriers, occurring in the Pleistocene and Holocene. It is unlikely that these endemics are the result of adaptations to habitat heterogeneity (Tuomisto et al. 1995) because they are found today in habitats similar to those of their relatives. For example, the morphologically most similar species to *L. barnebyi* and *E. amazoniciformis* are the Guianan species *L. poiteaui* and the eastern to central Amazonian species *E. amazonica* R. Knuth (Mori 1990). In both cases, the ranges of the two species pairs overlap in central Amazonia and all four species are found in terra firme forest. If this apparent endemism is real, then species richness of Lecythidaceae in central Amazonia is considerably enhanced by mechanisms not yet understood.

At the present time there are 205 known

species of Neotropical Lecythidaceae (Prance and Mori 1979; Mori and Prance 1990; Mori 1992a; Mori and Lepsch-Cunha 1995). Consequently, protection of the 100 ha Lecythidaceae plot at Reserve 1501 would include 17.6 percent of all known species of Neotropical Lecythidaceae. This figure, however, will decrease as other areas become better known and as undescribed species of Lecythidaceae from other areas are published.

To protect the thirty-eight species of Lecythidaceae in the plot, much larger tracts of forest are needed to ensure that the minimal population sizes essential for maintaining each of the species are included in proposed reserves. For example, it is evident that *Bertholletia excelsa,* with only one individual in the 100 ha, could not be protected in reserves as small as the Lecythidaceae plot at Reserve 1501. Nonetheless, well-placed reserves of adequate size (thousands of hectares) in central Amazonia based on knowledge of overall species distributions and minimal population sizes of the species of Lecythidaceae can go a long way toward protecting the species richness of this family. It is difficult to extrapolate to other families, however, because other families may differ as to where they reach their greatest species richness. The Lecythidaceae, for example, have high species richness in central Amazonia, whereas the Moraceae may be more species rich in western Amazonia (Ducke and Black 1954)— that is, a reserve designed to protect the greatest number of species of Lecythidaceae will not always protect the greatest number of species of Moraceae. Moreover, reserves should not be designated based on species richness alone without also considering higher taxa and ecosystem diversities as well, so that a reserve of some thousands of hectares that would be effective for trees in the Lecythidaceae would be far too small to

preserve large, territory-demanding carnivores.

To be able to address the causes of change in forest stand characteristics as a result of forest fragmentation and climate change, it is necessary to possess baseline data from undisturbed forests. Although we know that fire has had an impact within the Lecythidaceae plot (Bassini and Becker 1990; Piperno and Becker 1996), these fires probably took place 6,000-400 yr before the Pleistocene, and therefore fire's impact on present frequency, density, dominance, and species richness of Lecythidaceae may no longer be obvious. Piperno and Becker (1996), using data from soil phytoliths, provided evidence that the Lecythidaceae plot has been under continuous forest cover since at least 4,500 BP and has never been cleared for swidden agriculture.

The largest individuals of Lecythidaceae in the plot are probably less than 400 years old. Trees of *Bertholletia excelsa* in the 140−150 cm DBH class have been estimated by radiocarbon dating to be only 270 years old (P. Camargo et al. 1994). The lone *B. excelsa* in our plot is 115 cm DBH and therefore probably became established after the last significant fire swept the plot. Lecythidaceae are exceedingly vulnerable to fire such that forests regenerated from old slash-and-burn fields usually have few individuals and species of the family (Prance 1975). Smaller, more localized, disturbances such as normal gap formation (Denslow 1980), tree mortality resulting from periodic flooding (Mori and Becker 1991), and blowdowns caused by excessive winds (Nelson 1994; Nelson, Kapos, et al. 1994) frequently occur, but their influence is similar in all central Amazonian terra firme forests. We therefore conclude that the values of frequency, density, dominance, and species richness of Lecythidaceae presented here are typical for central Amazonian terra firme forest and therefore can be used as baseline data to monitor changes in forest composition resulting from forest fragmentation or climatic change in this area.

Species-area curves have attracted the attention of botanists for many years because they indicate how well a sample represents overall species richness in a given area (e.g., Arrenhius 1921; Gleason 1922, 1925; Preston 1948) and have been the focus of a relatively recent review (McGuinness 1984). The species-area curves we present (fig. 6.1) have the advantage of being based on a known universe of the organisms being sampled—i.e., the number of species of Lecythidaceae at least 10 cm DBH in a 100 ha plot on terra firme in central Amazonian Brazil. Because all species of Lecythidaceae in this part of the world attain diameters greater than 10 cm DBH, the species we have tabulated probably represent all of the species of Lecythidaceae actually found in the plot. Exceptions would be the presence of different species that have not yet reached 10 cm DBH or our failure to recognize a species as belonging to the Lecythidaceae. We suggest that similar, contiguous 100 ha plots in this part of Amazonia will yield similar species-areas curves for Lecythidaceae and perhaps even be representative of the species-area curves of other, ecologically similar, families of trees.

The most striking lesson to be relearned from our species-area curves of Lecythidaceae is that the species richness sampled increases rapidly for small sample sizes and slowly for large sample sizes. Therefore, it is relatively easy to sample most common species of Lecythidaceae in a given area in central Amazonia with a small sample size, but very difficult to sample all species, even with a large sample. A sample of a single hectare yields a mean species richness of 17.3 spp./ha, a two-hectare sample 23.2 spp./ha, a five-hectare sample 30.0 spp./ha,

a twenty-hectare sample 35.6 spp./ha, and a fifty-hectare sample 37.2 spp./ha (table 6.3). All thirty-eight species, obviously, are inventoried with 100 percent probability only when all 100 ha are sampled. Since botanists began doing quantitative inventories in Amazonia (Black, Dobzhansky, and Pavan 1950) it has been apparent that species continue to accumulate with increased sample size. This is especially true if the samples are not taken from contiguous plots (Black, Dobzhansky, and C. Pavan 1950)—a phenomenon that probably reflects the complicated geological history of Amazonia as well as its extreme habitat richness (Tuomisto 1995).

In designing ecological and biodiversity studies, it is extremely important to establish what questions are to be asked and then determine how large a sample size is needed to answer those questions. For example, because we wanted to know exactly what species of Lecythidaceae were present in our 100 ha plot, as well as establish their ecological relationships, the entire plot was sampled. In our 10,000-sample experiment, the probability of sampling all thirty-eight species of Lecythidaceae with a 50 ha sample is only 0.3544 (table 6.3), indicating that considerably more time, effort, and money are needed to sample all species of Lecythidaceae present even in an area as small as 100 ha.

If the purpose of an ecological study is to determine density, dominance, and total number of species present without defining what all of the species are, then much smaller sample sizes can be used. For example, suppose that the goal of a study is to determine the number of species of Lecythidaceae present in a 100 ha plot of central Amazonian terra firme forest without providing the names of all of the species. One can then extrapolate from a much smaller sample size of 10 ha because, based on our data, 87 percent of all species present have already been sampled. Use of our data, however, will not be valid for forests outside central Amazonia, for forests on different soils, or for forests under markedly different disturbance regimes.

Because there are limited resources available for doing systematic and ecological studies in Amazonia, it is important that "oversampling" be eliminated from projects at the proposal stage. Oversampling not only taxes the resources of those undertaking a project, it also affects the systematics community asked to identify and archive the voucher collections needed to document ecological studies. Therefore, before undertaking ecological and biodiversity studies in Amazonia, the costs involving specialist determination and specimen handling and storage must be considered (Mori 1992b).

Conservation Lessons

1. The Lecythidaceae is one of the dominant tree families of undisturbed terra firme forests in central Amazonian Brazil, accounting for 6.8 to 8.2 percent of the species and 8.4 to 13.4 percent of the individual trees at least 10 cm DBH. Therefore this family, along with other dominant families, such as Sapotaceae and Burseraceae, is useful for monitoring changes in forest composition resulting from human disturbance and climatic change.

2. Species of Lecythidaceae are found in all 100 hectares of the 100 ha plot. An average hectare harbors 77.9 ± 19.9 (mean ± SD) individuals, a basal area of 3.76 ± 0.86 m^2 (mean ± SD), and 17.3 ± 2.6 (mean ± SD) species of Lecythidaceae. Marked deviations from these values indicate major disturbance—for example, fire—in the relatively recent past.

3. Species richness is so high in central Amazonia that it is difficult to sample all species present in hectare plots. Because of the expense of sampling in such species-rich areas, the number of hectares surveyed should be designed to answer the ecological and conservation questions addressed. Excessive sampling drains financial and human resources without adding information.

4. The current composition of Lecythidaceae in the 100 ha plot indicates that it has not undergone major disturbance within the past 400 years, and therefore large-scale fires, massive blowdowns, extreme flooding, and so forth have probably not had a direct influence on current stand composition. Nevertheless, this plot has experienced large-scale fires in the distant past and is subject to the continual smaller-scale disturbances common to Amazonian forests. Because this plot is representative of central Amazonian forests, it and the surrounding forest at Reserve 1501 merit protection as a biological reserve.

5. Central Amazonia is home to a greater number of species of Lecythidaceae than anywhere else in the world, and, in general, central Amazonia is a center of diversity of woody plant families. The presence of high species richness of trees and lianas justifies the establishment of large biological reserves throughout central Amazonia.

Acknowledgments

We are grateful to the following agencies for making our study of Lecythidaceae in central Amazonia possible: the Companhia do Vale do Rio Doce for providing the funding needed to establish the 100 ha plot; the World Wildlife Fund and the Smithsonian

Institution for their financial support; and the Mellon Foundation for providing Mori with a fellowship to work on this project at the Smithsonian Institution from December 1, 1990 to June 1, 1991. Additional financial support was provided by the Fund for Neotropical Plant Research of the New York Botanical Garden through generous contributions from the Beneficia Foundation. We thank the Brazilian field interns, N. M. L. da Cunha, M. A. de Freitas, M. Mustragini, and A. A. de Oliveira, who supervised field operations at various stages of the project, and our field crew, P. A. C. Lima Assunção, R. Cardoso, J. F. Menezes, E. da Costa Pereira, and C. F. da Silva, for their indispensable help in gathering data. We thank INPA for facilitating our work, and the principal investigators, Richard O. Bierregaard, Jr., and Thomas Lovejoy for establishing and directing this project. We are grateful for the help provided by the on-site field directors of the BDFFP, Marina Wong, Roger Hutchings, and Claude Gascon.

CHAPTER 7

Tree Age Structure in Tropical Forests of Central Amazonia

JEFFREY Q. CHAMBERS, TIM VAN ELDIK,
JOHN SOUTHON, AND NIRO HIGUCHI

"How old are the trees?" is one of the first questions that come to mind while walking through a tropical rainforest. In fact, little is known about the ages of tropical trees, and even less is known about the distribution of ages across the forest landscape. The structure of tree ages provides some of the basic data needed to model population dynamics and is important for managers interested in calculating sustainable timber yields. The distribution of ages within populations also influences genetic diversity and population stability. Old trees can reproduce with trees many years younger and prevent the loss of genetic traits with the passing of generations. At an entirely different scale, the global carbon cycle is the focus of considerable international attention because potentially disastrous climatic changes are predicted to result from increases in atmospheric CO_2 and other trace gases (Kattenberg et al. 1996). For tropical rainforests, the residence time of carbon in wood is an important and yet essentially unknown parameter of climate change models.

The lack of information on tropical tree ages persists in large part because dating tropical trees is difficult. In extratropical regions, dating is facilitated by the existence of annual growth rings, but in tropical forests the lack of consistent annual variation in cambial activity results in either no ring structure or a ring structure whose annual formation is uncertain (Fahn et al. 1981). To take advantage of rings to age trees, annual rings must first be validated by other studies. In the absence of such validation, the only way to directly age a tree, in lieu of tracking a tree for its entire life, is by radiocarbon (^{14}C) dating. Here we present radiocarbon dates for forty-four large central Amazon trees and discuss the implications for forest management and conservation.

Methods

Trees were selected from a logging operation near the city of Itacoatiara, about 250 km east of Manaus, Brazil. Here, the Swiss-owned Precious Woods has established a sustainable timber harvesting operation called Mil Madeireira in 80,000 hectares of primary forest, where they have identified sixty-fix species as commercially valuable. Our samples were either selected at the log yard, where boles were stacked prior to processing, or through use of a geographic information system that precisely located stumps from all harvested trees in 2,000 ha

This is publication number 274 in the BDFFP Technical Series.

Fig. 7.1. Collecting tree-ring data and wood samples from the Mil Madeireira log yard. Photo by Christopher Dick.

blocks (fig. 7.1). The vegetation is dense tropical evergreen forest (terra firme, or not seasonally inundated) similar to that of the Biological Dynamics of Forest Fragments Project. The climate is similar to that in Manaus, where mean annual rainfall is about 2,200 mm and mean average temperature is 26.7° C, with a distinct dry season from June to September (see fig. 5.1). Soil type varies gradually with elevation, comprising Oxisols on plateaus (about 80 percent clay), Utisols on slopes, and Spodosols (2 to 5 percent clay) in valleys associated with small streams (Bravard and Righi 1989). Nearly 1,200 tree species have been identified in nearby Ducke reserve (Ribeiro et al. 1999), and the total for the BDFFP sites is expected to at least equal this (Chapter 5). Each hectare contains about 350 different tree species (more than 10 cm DBH) many of which occur at densities of less than one tree per hectare.

The centermost wood (first ring) from the base of the tree provides a date commensurate with the sapling stage of a tree. In all species sampled, a ring structure was evident, and this made identifying the center relatively easy. However, some boles were hollow, and samples had to be taken either at the upper end of the base section (typically 5–6 meters above the ground), or if the hollow was continuous along the stem, from the periphery of the hollow (typically less than 5 cm diameter) at the base. Trees were chosen according to size with approximately 100 cm the minimum base diameter. From a survey of 18 ha from BDFFP permanent inventory plots, trees larger than 100 cm DBH represent just 0.08 percent of stems larger than 10 cm DBH, or about one meter-sized tree every two hectares. The samples collected at Mil Madeireira are representative of the largest and presumably oldest trees from 4,000 ha that were selectively logged.

Forty-four trees from fifteen species were radiocarbon dated by Accelerator Mass Spectrometry (AMS) at Lawrence Livermore National Laboratory (see Taylor, Long, and Kra 1992 for a review of AMS techniques). Living wood is in equilibrium with atmospheric ^{14}C, but when the wood dies, the ^{14}C content gradually declines by radioactive decay. By comparing the current radiocarbon content from the centermost wood with the atmospheric value at the time of death (Stuiver and Becker 1986), the age of the tree can be determined. The accuracy of radiocarbon dating varies with the age of the organic material because atmospheric ^{14}C concentrations respond to production rate changes caused by natural variation in cosmic radiation. As a result, there are fluctuations in the calibration curve of radiocarbon age versus true (calendar) age (Stuiver and Becker 1986). Because of these fluctuations, radiocarbon analysis can only unambiguously date organic material that is more than about 350 years old. Radiocarbon dates on younger material typically correspond to several possible calibrated age ranges between 1650 and 1950 A.D., unless other evidence allows discrimination between these possibilities. Standard errors of calendar dates for trees more than 350 years old are approximately ±50 years, and for trees less than 350 years old are approximately ±100–150 years.

Results

Radiocarbon ages, partially presented in Chambers (1998), demonstrated that large trees of the central Amazon can be very old, with some trees living more than 1,000 years (table 7.1). The oldest tree was a *Cariniana micrantha* with a calendar age of almost 1,400 years. Tree age was positively correlated with tree diameter, but the relationship was weak ($r^2 = 0.25$). The weak correlation between size and age was not simply due to differences between species. For example, two individuals of *Cariniana micrantha* with a base diameter of 180 cm differed in age by about 900 years. There were also differences among species. The largest trees from *Dinizia excelsa,* for example, were never more than about 300 years old, and the distribution of ages was much less variable. Age variation between species was related to life history strategy. Shade-tolerant, slow-growing species were older and exhibited more variability in age than did shade-intolerant fast-growing species.

Discussion

The largest trees in the central Amazon are not necessarily the oldest, conforming with temperate studies that also show poor correlations between size and age (White 1980).

The growth rates of tropical tree species, including canopy emergents, vary considerably (Clark and Clark 1992), and a weak correlation between age and size, and a strong correlation between age and average growth rate age is not surprising. The oldest trees in any size class are those that grew the slowest. Previous maximum age estimates for rainforest trees centered around 500–700 years (e.g., Lieberman and Lieberman 1987; Clark and Clark 1992), although Condit, Hubbell, and Foster (1995) predicted trees older than 1,000 years based on mortality rate extrapolations.

These results suggests that for shade-tolerant, slow-growing species, there are a variety of pathways to reach the canopy. This pathway may comprise periods of fast growth in illuminated gaps, and periods of protracted slow growth beneath the canopy, comprising a number of growth and suppression events as transitory gaps periodically appear and close, with relatively few individuals attaining a canopy position. Demonstrating the prevalence of growth suppression, Clark and Clark (1992) studied six species representing a range of life history strategies in a Costa Rican rainforest and showed that about 20 percent of all individuals showed no measurable growth. Ashton (1981) suggested that suppressed trees are destined to die and that only individuals experiencing optimum growth rates will reach the canopy. For shade-intolerant species this is may be true, but what constitutes optimum growth for shade-tolerant species is quite variable.

POPULATION DYNAMICS AND CONSERVATION

Because trees are long lived, it is difficult to study the long-term dynamics of populations. Given an estimated minimum average life span of 150 years (see below) for a central Amazon tree in dense terra firme forest, population studies that last for years, or perhaps for decades, capture only a small fraction of the life span of most trees. Some commercial tree species are being exploited at high rates, and understanding age structure for these species is important for making sound management decisions.

With "roundwood" selling for about US$700 per cubic meter (Veríssmo et al. 1995), big-leaved mahogany (*Swietenia macrophylla*) is the most valuable tree species in the Amazon. As reviewed by Snook (1996), mahogany regenerates after large-scale disturbance, which results in essentially even-aged stands with few juveniles. Supporting this theory, Gullison et al. (1996) in the Bolivian Amazon found that mahogany was regenerating only in one out of five stands studied, and regeneration was due to hydrological disturbances. In the eastern Amazon, disturbances differ in scale from those discussed by Snook (1996), and regeneration in large gaps would not be inconsistent with published studies. Also, fire may play a more important role in the long-term disturbance history in eastern Amazonia than the hurricanes and flooding reported by Gullison et al. (1996, e.g.) in the western reaches of the basin.

Meggers (1994b) has compiled evidence suggesting that extensive drought, and perhaps widespread fires, linked to mega-ENSO (El Niño Southern Oscillation) events occurred in Amazonia 450, 700, 1,000, and 1,500 years ago. Existing populations of mahogany in terra firme forests of the Brazilian Amazon could be relics of such a catastrophic historical disturbance. However, regeneration of mahogany in large treefall gaps, or blowdowns (e.g., Nelson et al. 1994), would not be inconsistent with existing studies. If mahogany is regenerating in large gaps, the largest and presumably oldest trees act as an important source of seeds (Gullison et al. 1996), and their complete removal would halt regeneration. Mapping out age distributions for populations of mahogany can shed light on its regeneration strategy and distribution history and help make informed management decisions possible.

Regardless of the regeneration dynamics of mahogany, because of its high value, loggers will often explore large tracts of forest to find and harvest this single species (see Chapter 26). Mahogany can thus act as catalyst, opening up the remaining forest to logging and agricultural uses that have a much greater impact than the removal of mahogany alone (Fearnside 1997e). Because mahogany appears to be a shade-intolerant, fast-growing species (Nelson et al. 1994; Gullison et al. 1996), suggesting that the largest trees are probably relatively young (less than 300 years), enrichment planting of mahogany in selectively logged forests may be a rational alternative to exploiting large tracts of primary forests.

Tropical timber markets need to become less fixated on a few species and to realize that many Amazon tree species produce valuable timber. The Mil Madeireira logging operation, for example, has identified sixty-six species with valuable properties, although few have commercial markets. Understanding age distributions will help determine which of these species are most amenable to sustainable exploitation once they have become established on national and international markets.

Genetic Diversity

"The life of a tree is long compared to the evolutionary time scale of its major enemies" (Harper 1977, p. 635), and in a tropical forests pests, pathogens, and decomposers constitute a large, active, and diverse community of organisms. Such natural events as large-scale blowdowns (Nelson,

TABLE 7.1. Ages for Large Central Amazon Trees as Determined by Radiocarbon Dating at Lawrence Livermore National Laboratory's Center for Accelerator Mass Spectroscopy

Species	Common name	Family	Radiocarbon date	1997 age	Base diam. (cm)	Ave. growth (cm/year)
Bagassa guiananensis	Tatajuba	Moraceae	300	350	110	0.31
Brosimum parinarioides	Amapá	Moraceae	280	350	100	0.29
Cariniana micrantha	Tauari vermelha	Lecythidaceae	1440	1370	180	0.13
Cariniana micrantha	Tauari vermelha	Lecythidaceae	780	720	140	0.19
Cariniana micrantha	Tauari vermelha	Lecythidaceae	650	660	140	0.21
Cariniana micrantha	Tauari vermelha	Lecythidaceae	470	560	150	0.27
Cariniana micrantha	Tauari vermelha	Lecythidaceae	440	550	150	0.27
Cariniana micrantha	Tauari vermelha	Lecythidaceae	360	460	180	0.39
Caryocar glabrum	Piquiarana	Caryocaraceae	180	250	120	0.48
Caryocar villosum	Piquiá	Caryocaraceae	90	230	130	0.57
Caryocar villosum	Piquiá	Caryocaraceae	390	510	100	0.20
Dinizia excelsa	Angelim vermelha	Leguminosae	180	250	120	0.48
Dinizia excelsa	Angelim vermelha	Leguminosae	140	250	102	0.41
Dinizia excelsa	Angelim vermelha	Leguminosae	110	170	100	0.59
Dinizia excelsa	Angelim vermelha	Leguminosae	70	100	150	1.50
Dinizia excelsa	Angelim vermelha	Leguminosae	−20	80	100	1.25
Dipteryx odorata	Cumarú	Leguminosae	1220	1170	120	0.10
Dipteryx odorata	Cumarú	Leguminosae	730	710	100	0.14
Dipteryx odorata	Cumarú	Leguminosae	290	360	110	0.31
Hymenolobium spp (2)	Angelim pedra	Leguminosae	380	510	120	0.24
Hymenolobium spp (2)	Angelim pedra	Leguminosae	540	580	140	0.24
Hymenolobium spp (2)	Angelim pedra	Leguminosae	900	840	230	0.27
Hymenolobium spp (2)	Angelim pedra	Leguminosae	490	560	110	0.20
Hymenolobium spp (2)	Angelim pedra	Leguminosae	430	550	100	0.18
Hymenolobium spp (2)	Angelim pedra	Leguminosae	420	550	100	0.18
Hymenolobium spp (2)	Angelim pedra	Leguminosae	450	550	100	0.18
Hymenolobium spp (2)	Angelim pedra	Leguminosae	320	400	100	0.25
Iryanthera grandis	Arurá vermelha	Myristicaceae	480	560	110	0.20
Lecythis poiteaui	Jaraná	Lecythidaceae	940	900	100	0.11
Manilkara huberi	Massaranduba	Sapotaceae	380	510	140	0.27
Manilkara huberi	Massaranduba	Sapotaceae	220	270	110	0.41

TABLE 7.1. (continued) Ages for Large Central Amazon Trees as Determined by Radiocarbon Dating at Lawrence Livermore National Laboratory's Center for Accelerator Mass Spectroscopy

Species	Common name	Family	Radiocarbon date	1997 age	Base diam. (cm)	Ave. growth (cm/year)
Ocotea rubra	Louro gamela	Lauraceae	410	530	120	0.23
Ocotea rubra	Louro gamela	Lauraceae	390	510	100	0.20
Ocotea rubra	Louro gamela	Lauraceae	380	500	100	0.20
Ocotea rubra	Louro gamela	Lauraceae	280	350	110	0.31
Ocotea rubra	Louro gamela	Lauraceae	270	340	110	0.32
Parkia pendula	Angelim fava	Leguminosae	60	170	110	0.65
Peltogyne catingae	Violeta	Leguminosae	180	250	80	0.32
Schlerolobium spp (2)	Tachí	Leguminosae	100	230	100	0.43
Tabebuia serratifolia	Ipê	Bignoniaceae	650	660	80	0.12
Tabebuia serratifolia	Ipê	Bignoniaceae	650	660	80	0.12
Tabebuia serratifolia	Ipê	Bignoniaceae	540	580	90	0.16
Tabebuia serratifolia	Ipê	Bignoniaceae	440	550	80	0.15
Tabebuia serratifolia	Ipê	Bignoniaceae	300	350	90	0.26

Note: Average growth rate was calculated by dividing the diameter of the tree by its age; 1997 age is the measured age of the tree when it was harvested.

Kapos, et al. 1994) and perhaps widespread fires (Meggers 1994b) provide other agents of mortality in Amazonia. Given such an environment, a thousand-year-old tree has demonstrated impressive persistence, some of which may be the result of genetic factors. Functional differences within populations often reflect genetic differences and are not simply the result of environmental variation (Harper 1977, p. 601). Thus, ancient trees may harbor valuable genetic traits useful for developing commercial silvicultural strains (e.g., fast growth and resistance to pests), and their harvest would represent a loss well beyond their function as an individual.

Carbon Cycling

Wood represents a long-term storage for atmospheric carbon whose residence time can exceed a millennium for wood stored in the center of a large canopy-emergent tree. A more important parameter for global carbon cycling models is the mean residence time of carbon in wood, which is a function of stand mass and tree mortality rates. Global models, such as the Frankfurt Biosphere Model (Kohlmaier et al. 1997), use a number of untested assumptions to estimate the residence time for woody tissues in tropical evergreen forests at about forty years. Based on BDFFP forest inventory data for an 18 ha plot, the standing stock of above-ground woody tissues in central Amazon terra firme forests is about 330 Mg ha^{-1} (SE = 10); according to Chambers et al. (in press) the production of woody detritus is about 4.2 Mg ha^{-1} yr^{-1}. Assuming that stand mass is not changing dramatically over time (Phillips et al. 1998), the mean residence time of carbon in woody tissues is given by standing stocks divided by production (Olson 1963), or about eighty years, twice the estimate of Kohlmaier et al. (1997). In forest fragments, because mortality rates are considerably

higher (Laurance, Ferreira, et al. 1998a; Chapter 9 and 13), the mean residence time of carbon will be substantially less, and average tree age lower, than the primary forest.

The mean residence time is not the same as the average age of a tree because only the centermost wood is the same age as the tree. An object-oriented model was developed using the BDFFP permanent inventory data for trees with a DBH greater than 10 cm, which simulates forest stand and carbon cycling dynamics based on tree growth and mortality rates (Chambers 1998). The model predicted that the median tree age in a 100 ha plot was 165 years, and on average, approximately every 40 ha harbored a tree more than 1,000 years old. The model assumed that growth rates up to 10 cm DBH were equivalent to the growth rate of a 10 cm tree. Because growth rates for larger trees are often higher than for small, often suppressed, trees (Hartshorn 1980; Clark and Clark 1992), the average age at death is probably even older. Based on median growth rates through all size classes, Clark and Clark (1992) estimated that the average time required for five emergent species to reach 30 cm DBH was 260 years in Costa Rican forests.

Another important component of global carbon cycling models, and hence climate change predictions, is recovery rate for deforested areas. Because growth rates in secondary and logged forests are measurably higher than in the primary forests (N. Higuchi, unpublished data), recovery times should be shorter than suggested using long-term average growth rates from primary forest. The dynamics of regrowing forests are complex, however, and cannot be based on growth rates alone. Nepstad, Uhl, and Serrão (1991), for example, found that accumulation of stand mass and species in secondary forests of the eastern Amazon is inversely related to the intensity of use prior to abandonment. Moreover, Fearnside and Guimarães (1996) calculate that clearing and burning of secondary forests almost completely offset carbon taken up by regrowing forests in the Brazilian Amazon.

The distribution of tree size classes can also influence recovery times, because large trees, although rare, represent a large portion of forest mass (Brown, Mehlman, and Stevens 1995). At the BDFFP permanent plots, 50 percent of the above-ground stand mass is contained in the largest 8 percent of the trees. Saldarriaga et al. (1988), using a chronosequence of secondary forests recovering for up to 80 years (the oldest reported study), estimated that it takes about 200 years for secondary forests to attain mature forest mass values. Some species, however, may not yet have reached maximum size distributions even after 200 years. The average age for the largest trees of the species *Cariniana micrantha,* for example, is 720 years (table 7.1), suggesting that some emergent species will not be represented in the largest size classes for many centuries. To test whether the absence of some species in the largest size classes affects stand mass recovery times, studies in secondary forests older than 200 years need to be carried out. Considering that Neotropical evergreen forests account for about 25 percent of total global terrestrial net primary production (Mellilo et al. 1993; Potter et al. 1993), age and size distributions in these forests can have a large impact on global carbon cycling rates.

SUSTAINABLE FORESTRY

Sustainable timber production is not usually considered when logging tropical forests (see Chapter 26). Even for logging companies that do consider sustainability a criterion, allowable volume extractions do not account for different life history strategies among commercially valuable species. This is not surprising because very little practical information is available to managers of high-diversity forests. The average growth rate is not even available for most commercial tree species.

Clark and Clark (1996) found that the mean growth rate among a number of emergent tree species differed, although in most cases these differences were not statistically significant. Ages and average growth rates of commercially valuable trees from the central Amazon also vary considerably (table 7.1). However, there are some distinct differences among commercially valuable species. *Dinizia excelsa,* for example, was the fastest-growing species radiocarbon dated, with a mean age and 95 percent confidence interval of 170 ± 35 years. In contrast, mean ages and 95 percent confidence intervals for *Cariniana micrantha* and *Tabebuia serratifolia* were 720 ± 135 years and 560 ± 55 years, respectively. If commercial-sized trees of some species are typically much older than other species, sustainable harvest cycles should reflect these differences.

Once a forest is selectively logged, the increase in light and other resources allows for increased growth for trees left in the gaps. The successional sequence in these gaps is a function not simply of the size structure of populations but also of the growth-rate response. Because the spatial extent and size of canopy gaps is larger in a partially logged forest, by the second harvest the floristic and size structure will differ from the original forest (see also Chapter 27). Selective logging is likely to increase the abundance of fast-growing light-demanding species and reduce the abundance of more shade-tolerant species. Fast-growing light-demanding emergents (e.g., *Dinizia excelsa,* and perhaps *Swietenia macrophylla*) may thrive in a landscape modified by selective logging (see also Chapter 13). If the extraction of timber is deemed more valuable than maintaining an intact forest, the overall strategy should be to develop low-cost management practices that take into account differences in life history strategies and tilt competitive interactions in favor of the most valuable species. Strategies should also focus on ways of increasing the efficiency of land use to limit conversion of old-growth forest for commercial interests.

TREE RINGS IN TROPICAL FORESTS

Radiocarbon dating is expensive (more than US$500 per sample) and not readily available to most ecologists and forest managers. Contrary to conventional beliefs, many tropical rainforest trees exhibit a ring structure, and if these rings are annual, dating tropical trees would be facilitated. Rings are formed due to a reduction in cambial activity related to environmental stress, and tropical trees that grow in regions that experience flooding, or drought, for example, invariably produce rings. If stress occurs more than once a year, however, or if the stress is mild, rings may not be annual, or may vary from year to year (Fahn et al. 1981). Phenological events, such as leaf flush or flowering, can also cause ring formation.

Although tree rings are common in many tropical regions, in most cases it is not known whether the rings are formed on a consistently annual basis. One method used to confirm annual ring formation exploits the rapid increase in atmospheric ^{14}C caused by atmospheric nuclear testing in the late 1950s and early 1960s (Nydal and Lövseth 1970). Radiocarbon concentrations reached twice normal atmospheric levels in about 1964 and have gradually declined because the above-ground test ban went into effect. This "bomb spike" can be used as a tracer to date biotic carbon of recent (past forty years) origin. By measuring the ^{14}C content of rings

produced since the late 1950s, and then comparing those values to the known atmospheric content, it can be precisely determined whether rings are annual (Worbes and Junk 1989).

Using this method Worbes and Junk have demonstrated that the rings of trees in flooded forests of the Amazon are annual. Rings in these forests are caused by the yearly stress induced by inundation during the high-water season. Preliminary results using the same method show that the rings from a pioneer species (*Pourouma* sp.) at a BDFFP upland site are also annual. Contrary to these results, Vetter (1995), using a cambial scarring method, found that *Scleronema micranthum* can produce more than one ring per year. Overall, the mechanisms of ring formation for tropical forests are complex and may result from a combination of processes. By employing a variety of methods to study ring structures for a number of species, those that consistently produce annual rings can be identified, and ages can be determined from cores of living trees. This would allow age structures to be determined at a regional scale, which is important for evolutionary and other historical studies.

Annual rings would not only promote demographic studies but would also provide a means to trace climate, or other environmental conditions, for hundreds to thousands of years. D'Arrigo, Jacoby, and Krusic (1994), for example, have generated a tree ring width chronology using *Tectonis grandis* ("teak") from Indonesia that dates back to 1565. The data were compared with ring chronologies from the United States and Mexico, and with ENSO records, demonstrating that ring chronologies from rainforests are capable of recording ENSO events. The Amazon basin is subject to prolonged drought during ENSO years, and reconstructing a climate history from tree rings can demonstrate how forests have responded historically. To obtain the longest climate records, the oldest individuals are typically selected, and, fortunately, these oldest trees often have the highest quality records (Fritts 1991). If it can be determined that the ring structures found in the trees from table 7.1 are annual, promising studies into the environmental history of the Amazon would be possible.

By measuring how ages are distributed within tree populations, a new dimension can be given to relatively short-term population studies, shedding light on historical events, as well as on the future trajectory of a community. A reconstruction of past events based on ring chronologies and other ancillary data for an old-growth forest in New Hampshire, for example, demonstrated that the forest was destroyed by a fire in 1665 and a hurricane in 1938 (Henry and Swan 1974). If the annual rings can be verified for a number of tropical forest species in the Amazon, tree cores can be used to reconstruct community history.

Even if Amazon terra firme trees with annual rings are discovered, it is a time-consuming and difficult process to measure the age structure of an entire tree community. It is important to note that detailed information on tree age structure is not critical for making many informed management decisions. Information on size distributions, a much more manageable task, may be more useful for making predictions about future population structure (Harper 1977, p. 603). However, age structure is the only way to trace the history of the forest, and this information may be critical for making evolutionary studies, understanding forest disturbances, and making some important management decisions about the sustainable harvest of timber.

Conservation Lessons

1. The central Amazon harbors giant trees, some of which are more than 1,000 years old.

2. Fast-growing, light-demanding species (e.g., *Dinizia excelsa, Swietenia macrophylla*) may thrive in a selectively logged landscape and be ideal candidates for enrichment planting.

3. Thousand-year-old trees can reproduce with trees that are many generations younger and prevent the loss of genetic diversity. The oldest individuals may also harbor genetic traits like fast growth rates or resistance to pests that may be useful for selective breeding programs.

4. Sustainable harvesting programs should consider the life history strategies and average growth rates of commercial species when determining allowable volume schedules and harvest cycles. Fast-growing, light-demanding species are probably more amenable to sustainable harvesting than are slow-growing, shade-tolerant species.

5. Growth rings, where consistently annual, will be a valuable tool for determining long-term growth rates of commercial species and for elucidating historical patterns of disturbance in tropical forests.

6. It will take hundreds of years for a deforested area to accumulate the carbon that was lost to the atmosphere after burning. When calculating forest recovery times, the size, age, and species composition of the recovering forests are important factors.

7. The residence time for carbon in woody tissues in central Amazon terra firme forests is about eighty years, and the average tree with a DBH of more than 10 cm is at least 150 years old.

8. Forest managers and ecologists should collaborate toward understanding forest function and structure in an applied context.

Acknowledgments

We would like to thank INPA and the BDFFP for logistical support while in Manaus; workers at Mil Madeireira and the CAMS laboratory for assistance; and Chris Dick and Simon Lewis for help taking tree cores. This work was supported in part by Lawrence Livermore National Laboratory's Laboratory Directed Research and Development program (95-DI-005) under the auspices of the U.S. Department of Energy subordinate to Contract No. W-7405-Eng-48 with the University of California.

CHAPTER 8

The Genetics of Rare Tropical Trees

Implications for Conservation of a Demographically Heterogeneous Group

NADJA LEPSCH-CUNHA, CLAUDE GASCON, AND PAULO KAGEYAMA

Recently, much discussion on the nature of rare species in tropical communities has appeared in the literature. This discussion has focused on the different geographical distributions, life history patterns, and evolutionary pathways that could lead to a species either being or appearing to be rare. The identification of types of rarity has direct repercussions for our understanding of species diversity patterns and for the conservation and management of tropical communities.

This discussion originated from common frequency distribution patterns of species abundance in ecological systems in which there are only a few common species and a preponderance of rare species (Fisher, Corbet, and Williams 1943; Preston 1948; MacArthur 1957; Whittaker 1965; May 1975). Existing studies show that rarity is not peculiar to tropical communities but rather a characteristic of many mature communities at various latitudes.

It is important to understand the causes of rarity and the patterns of variation in population abundance within a species' geographic distribution. A number of scenarios is possible. Is a locally rare species rare over its entire geographic distribution, or is it rare only at the margins of its geographic distri-

bution? Is its population decreasing or has it recently colonized the locality (Hubbell and Foster 1986)? Is rarity a property of the species, or is the species only temporarily rare or common? Are population fluctuations stochastic, or are populations in equilibrium (Connell, Tracey, and Webb 1984)?

Little is known about geographic patterns of species abundance. Until recently, few ecologists included comparative geographic abundance data in their studies and rarely studied the abundance and geographic distribution of populations (Brown, Mehlman, and Stevens 1995). The few studies that quantified spatial variation in population abundance within a species' entire geographic distribution suggested that high population densities, in general, are restricted to a few locations (Rabinowitz 1981; Brown 1984; Rabinowitz, Cairns, and Dillon 1986; Schoener 1987; Lawton 1993; Kunin and Gaston 1993; Brown, Mehlman, and Stevens 1995). In most of a species' range, populations occur at low densities, and as a consequence the majority of species are locally rare (Rabinowitz, Cairns, and Dillon 1986; Brown, Mehlman, and Stevens 1995).

Independent of spatial variation in population abundance patterns, 30–40 percent of the tree species in tropical forests are rare,

This is publication number 275 of the BDFFP Technical Series.

generally defined as populations with one individual or less per hectare (e.g., Prance, Rodrigues, and da Silva 1976; Ashton 1984; Hubbell and Foster 1986; Rankin–de Mérona and Ackerly 1987; Gentry 1990). The recognition of rarity and the level of significance given to rare species will greatly influence decisions regarding conservation and management. Existing knowledge of the demographic status of species in the tropics is limited and hampers our attempts to develop effective conservation guidelines.

Using demographic and genetic data from two tree species found in the study areas of the Biological Dynamics of Forest Fragments Project (see fig. 4.1), this study set out to: (1) review the different definitions of rarity and discuss their significance with regard to tropical tree communities; (2) discuss the genetic characteristics of small or reduced population sizes; (3) discuss two contrasting hypotheses regarding the evolutionary and ecological mechanisms responsible for the origin and maintenance of rare species; and (4) show how different demographic patterns are related to different genetic structures of ecologically similar species with different population abundance patterns within their geographic range.

Rarity: A Broad Concept for Different Demographic and Genetic Structures

For conservation purposes, the International Union for the Conservation of Nature (IUCN 1984, 1988) established five categories of rarity: extinct, threatened, vulnerable, rare, and insufficiently known species. These categories combine the intrinsic natural aspects of rarity (patterns of geographic distribution and population abundance), as well as the species' response to habitat changes. In this sense, the categories extinct, threatened, and vulnerable, in general, reflect a species' response to anthropogenic disturbance,

whereas a rare species refers to its natural abundance.

Rare species may include, for example, species found in restricted geographic locations (endemics), as well as species found in habitats that are isolated and geographically widespread, even though in both cases the populations may be locally abundant (see examples in Primack 1993). Also considered rare are species with low population densities but wide geographic range. In this chapter we focus on these rare species with widespread, low population densities.

Rabinowitz, Cairns, and Dillon (1986) pointed out that the lack of precision in defining rare species results in grouping together biologically diverse species; a clear understanding of the types of rarity is necessary before we can think of managing and protecting rare species. They show that rarity is influenced by three distinct factors, two of which have already been commented upon: geographic extension (widespread or restricted), geographic variation in population density (always low or high in some places), and ecological specificity (high or low specificity for the habitat).

Using plant data from the British Isles, Rabinowitz, Cairns, and Dillon (1986) defined eight categories from a combination of these three factors. Seven of the categories represent different types of rarity, and four of them represent species whose populations are always in low abundance. For example, 6 percent of the British Isles plant species possess populations with densities that are always low, and 14 percent of the species are restricted geographically (endemics).

How would tropical tree species be distributed within these different categories? Characteristics of the tropical forest (large expanses of continuous forest, high proportion of endemism, and high species diversity) differ substantially from the flora of the British Isles, and a different frequency distribution of species within these categories would be expected for the tropics. No similar data exist for the tropical forests that would allow large-scale analysis (Bawa and Ashton 1991). On a local scale, the 50 ha plot of Barro Colorado Island, Panama, for example, included 21 of 300 species with one individual with a diameter at breast height (DBH) of at least 1 cm, and 15 percent of the species comprised 80 percent of all stems (Hubbell and Foster 1986). Among 1,300 tree species in the BDFFP plots, only about 12 percent (160 species) would not be considered rare (less than 1 stem/ha; Chapter 5). It is possible, however, to hypothesize the distribution of tropical tree species within these categories of rarity based on existing ecological knowledge. Bawa and Ashton (1991) distinguished four generalized types of rarity: (1) species that are uniformly rare, with population densities always low, including emergent and pioneer genera and species (*Bombax, Ceiba, Couratari multiflora, Durio*); (2) species that are common in some places but rare in others—a group that can be represented by many species in a given site (the same as the "diffusive" rarity of Schoener [1987]); (3) species that are locally endemic and, in general, abundant where they occur (many epiphytes, ground herbs and shrubs); and (4) species with spatially grouped populations but with the total population density always low (*Cecropia* spp.).

Kageyama et al. (1997) distinguished five groups of tropical arboreal species, using ecological characteristics and demographic responses of the species to anthropogenic disturbance: (1) low-density species in primary forest that become abundant in secondary forest—in general, these are gap emergents and opportunists (*Schysolobium amazonicum, Myracrodruom urundeuva*); (2) species that are rare in primary forests and absent in other habitats that are affected negatively by anthropogenic disturbance (*Theobroma grandiflorum, Hevea brasiliensis*); (3) habitat specialists in forest environments with restrictive edaphic characteristics, such as hilltops, and which may become very common in secondary forests (*Rapanea ferruginea, Dodonea viscosa*); (4) pioneer species that are common in the first stages of succession in secondary forests (*Cecropia* spp., *Vismia* spp.; plates 8 and 9; see Chapters 24 and 25); and (5) shade-tolerant species that are common in primary forests and have varied responses in secondary forests (*Euterpe oleracea, Escheweilera coriaceae*).

In general, both classifications converge to common categories, and the recognition of rare species is implicit, as well as is variation in population abundance. Harper (1981) specifically included population variation through evolutionary time in his categories: plants that remain rare for long periods; plants that had been rare and are now common; and plants that were common and have become rare.

The variation in population abundance through time is directly linked to the genetic structure and evolution of a species. Population size is a critical parameter in determining the amount of genetic variation in a population (Lande and Barrowclough 1987). The genetic consequences of population bottlenecks, such as the loss of the heterozygosity, can be neutralized when a population quickly recovers in numbers (Crow and Kimura 1970; Nei, Maruyamo, and Chakraborty 1975).

Habitat fragmentation has considerable potential for rapidly reducing population size and increasing population isolation (Young, Boyke, and Brown 1996). Conservation theory suggests that small and isolated populations can, for a number of reasons, be characterized by reduced population viability and reduced time of persistence (Simberloff 1988; Durant and Mace 1994; Chapter 2).

One of the genetic effects of population reduction that can influence extinction probability is the increase of inbreeding and genetic drift. The increase of inbreeding causes inbreeding depression — that is the reduction of fitness in self-fertilized progeny relative to outcrossed progeny through the loss of heterozygosity itself and the segregation of deleterious recessive alleles in homozygote condition (Charlesworth and Charlesworth 1987; Simberloff 1988; Ellstrand and Ellam 1993). Genetic drift causes the loss of genetic variation which will reduce the potential of a species to adapt to biotic and abiotic habitat changes (Lande 1988, 1998; Barret and Kohn 1991; Ellstrand and Ellan 1993). More recent models incorporate the effect of mildly deleterious new mutations that accumulate and can become fixed by random genetic drift with an associated decrease in fitness (Lande and Barrowclough 1987; Alvarez-Buylla et al. 1996).

Initially, it was expected that inbred species would suffer lower levels of inbreeding depression due to the elimination of most deleterious and semi-deleterious alleles (genetic load) with time (purging), while outcrossing species would be more susceptible due the permanence of these alleles. However, empirical data have shown that the occurrence of purging is low and that it is an inconsistent power within natural populations (review in Byers and Waller 1999). Data from natural populations are now accumulating and show that in outcrossing populations mildly and highly deleterious mutations and polymorphisms for overdominant alleles may all contribute to inbreeding depression, while in highly selfing populations, mildly deleterious recessive mutations are expected to be the main cause of inbreeding depression, because highly deleterious alleles would have

already been purged (Koelewijn 1998, Charlesworth et. al 1990; Husband and Schemske 1996; Byers and Waller 1999).

Studies have shown that most self-fertilizing species express the majority of their inbreeding depression late in the life cycle, at the stage of growth and reproduction, whereas outcrossing species express much of their inbreeding depression either early, at seed production, or late (Husband and Schemske 1996). The purging of genetic load and its extent will depend on genetic and population factors such as dominance variance, mutation effects, and selection history, rendering universal generalizations difficult (Byers and Waller 1999).

Thus, the effects of genetic drift and inbreeding, within and among populations, on the genetic structure of rare plants probably depend on whether a species has always been rare or has recently become rare as a result of the anthropogenic influence or natural stochasticity (Lande 1988). Species differ in their sensitivity to inbreeding depression (Raulls, Ballou, and Templeton 1988; Lacy, Petric, and Warnecke 1993, Husband and Schemske 1996; Byers and Waller 1999) as a result of different genetic structures and rates of population growth (see Nei, Maruyamo, and Chakraborty 1975). The strength and timing of inbreeding depression thus depends on demographic behavior, mating and reproductive systems, genetic diversity, and the potential for pollen transfer across long distances. Depending on these characteristics, genetic and demographic population responses to stress and disturbance will also vary.

Test of Hypotheses

Many evolutionary and ecological hypotheses exist to explain the origin and maintenance of tropical diversity (Connell 1978; Denslow 1987; Hamrick and Loveless 1989).

Other researchers have attempted to recognize general patterns of genetic and ecological structure in tropical species (Hamrick and Loveless 1989; Eguiarte et al. 1992; Stacy et al. 1996). Early studies of tropical forests suggested that high species diversity was the result of nonadaptive speciation resulting from inbreeding and genetic drift as a consequence of low density of conspecifics (Corner 1954; Baker 1959; Fedorov 1966). It was thought that pollinators would be unlikely to move among widespread conspecifics with asynchronous flowering. In this way, opportunities for outcrossing among tropical trees, particularly over long distances, would be limited. This "nonadaptive" hypothesis assumes that the genetic structure of tropical trees is characterized by high inbreeding, small population sizes, and spatial genetic differentiation (Eguiarte et al. 1992).

More recently, experimental studies on reproductive systems and on the behavior of pollinators of tropical tree species have shown high levels of self-incompatibility in hermaphroditic species and have shown that certain animal pollinators are capable of flying long distances to visit conspecific trees (Ashton 1969; Janzen 1971; Bawa 1974; Janzen 1975; Bawa and Opler 1975). These results have led to an alternate hypothesis, the "microniche equilibrium," which assumes that tropical arboreal diversity results from the adaptation of trees to specific niches (Dobzhansky 1950; Janzen 1970; Connell 1971; Denslow 1987). Thus, populations would have efficient mechanisms of pollen and seed dispersal, large effective population sizes, low inbreeding levels, and little spatial genetic differentiation.

The conservation and management implications of these two hypotheses are different. In small inbred populations, as the nonadaptive hypothesis predicts, a species can be preserved with only a few individuals coming from many populations, and *ex situ* preservation would be practical and possible (Eguiarte et al. 1992). Alternatively, if tropical tree populations have large effective population sizes and low inbreeding levels, as the microniche hypothesis suggests, the deleterious effects of maintaining small populations should be severe in the short run because of inbreeding depression, and severe in the long run because of genetic drift. In this latter case, viable conservation proposals should contemplate the maintenance of large ecological reserves, and conservation *ex situ* becomes impossible (Franklin 1980; Eguiarte et al. 1992).

Using published data on demography and genetics of rare tropical arboreal species, the nonadaptive and microniche hypotheses can be tested. The abundance criterion used for defining low-density species was less than 0.5 individual per hectare based on work from Barro Colorado Island (Hubbell and Foster 1986). The following characteristics were compiled from existing studies: the breeding system through the multiloci outcrossing rate, gametic compatibility, effective movement of pollen, fixation index (F) of progenies and adults, and gene diversity or expected heterozygosity (table 8.1).

The multiloci outcrossing rate (t_m) varies between 0 and 1.0 for complete autogamy and exogamy, respectively; species with mixed mating systems will have intermediate values (Ritland 1983). This rate is always close to 1.0 for common species and is variable for low-density species. Most of the low-density species have t_m values equal or close to 1.0, but three Bombacacea, *Ceiba pentandra, Cavanillesia platanifolia,* and *Dipteryx oleifera,* and two species of *Shorea* (Dipterocarpaceae), *S. megistophylla* and *S. trapezifolia,* show values lower than 0.8.

Within these five species, the two *Shorea* spp. are distinct because they are endemic species with relatively large populations

Table 8.1. Multilocus Outcrossing Rates (t_m) and Within-Population Genetic Diversity Estimates of Rare and Common Tropical Tree Species

Species and reference number	Adult densities	No. of trees in BCI[a]	Pop./year[b]	No. of trees sampled for Rs/He[c]	No. of loci sampled for Ms/He	Outcrossing rate t_m (±SE)	F Adult[d]	F Progeny[d]	Mean expected heterozygosity He[e]
Rare species									
A. Low-density, emergent, early or secondary successional species, animal-pollinated and wind-dispersed									
Cavanillesia platanifolia[1,2,3,4]	14/50ha[e]	21	1.1987	16/13	4/42	0.57 (0.02)	−0.198	0.275	0.123 A[i]
	"	"	1.1988	11/—	5/—	0.35 (0.03)	"	0.475	"
	"	"	1.1989	3/—	4/—	0.21 (0.05)	"	0.649	"
	common		2.1989	18/—	6/—	0.66 (0.07)			
Ceiba pentandra[2,5,6,7]	43/84ha[e]	40		11/~44	6/26	0.69 (0.03)	−0.297	0.184	0.106 A
Tachigali versicolor[2,6]	8/50ha[e]	82		23/~44	2/27	0.93 (0.04)			0.023 A
Pithecellobium pedicellare/elegans[8,9]	<1/ha to 2/ha			38/—	4/4	0.95 (0.02)	0.04	0.02	0.16 to 0.56 Pg,i
			1.1991	19/31−90	5/6	0.97 (0.03)			0.140 A
			1.1992	47/31−90	5/6	0.91 (0.01)			"
Schizolobium parahybum[3]	1?/50ha	1		—/11	—/32				0.091 A
Cedrella fissilis[10]	34/270ha			5/34	7/11	0.92 (0.05)	0.033	0.099	0.243 A
Couratari guianensis[11]	27/400ha			—/27	—/8		0.518		0.429 A
Courari multiflora[11]	41/400ha			4/41	7/8	0.95 (0.04)	0.166	0.114	0.436 A
Platypodium elegans[2,6]	36/50ha[e]	49	1.1988	5/~44	10/28	0.92 (0.04)			0.183 A
			1.1989	5/—	10/—	0.90 (0.04)			
B. Low-density, canopy, or emergent species, pollinated and dispersed by animals									
Enterolobium schomburgkii[3]	2?/50ha	2		—/13	—/29				0.122 A
Bertholletia excelsa[12]	1/6ha to 15-20/ha			29/29	2/2	0.85 (0.03)	−0.105[h]	0.002	0.499−0.580 A
							0.309[h]	0.041[h]	
Myracrodruom urundeuva[14]	PF[h]		1	—/—	—/3			0.495	0.243 P
	SF[h]		2	—/—	—/3				0.356 P
Calophyllum longifolium[15]	25/84ha	5	1.1992	8	7	1.03 (0.09)			
		4	1.1993	8	7	1.03 (0.04)			
Spondias mombin[15]	14/84ha	2	1.1992	10/—	5/—	0.99 (0.16)			
		3	1.1993	11/—	5/—	1.30 (0.11)			
Dipteryx oleifera[18]	25/50ha					0.74 (0.07)			

Species and reference number	Adult densities	No. of trees in BCI[a]	Pop./year[b]	No. of trees sampled for Rs/He[c]	No. of loci sampled for Ms/He	Outcrossing rate t_m (±SE)	F Adult[d]	F Progeny[d]	Mean expected heterozygosity He[e]
C. Low-density, understory, and animal dispersed species									
Ficus maxima[3]		6		—/6	—/20				0.136 A
Inga minutula[3]		3		—/11	—/25				0.240 A
Tocoyena pittieri[3]		5		—/6	—/26				0.186 A
Trichanthera gigantea[3]		3		—/7	—/28				0.026 A
D. Common, endemic, canopy, or emergent species, animal dispersed									
Shorea congestiflora[16]	6.4/ha		1.1990	17/—	6/—	0.87 (0.02)	0.088	0.067	
Shorea trapezifolia[16]	24/ha		1.1990	27/—	2/—	0.54 (0.04)	−0.060	0.295	
			1.1991	26/—	4/—	0.62 (0.03)	−0.300	0.237	
Shorea megistophylla[17]	10/ha PF		1	8/8	4/4	0.87 (0.06)	−0.247 (PF+LF)	0.151 (PF+LF)	0.348 (PF+LF)
	<2/ha LF[h]		2	22/22	4/4	0.71 (0.03)			
Stemonoporus oblongifolius[18]			4 pop.	24/35−43	5/9	0.84 (0.02)	−0.101	0.085	0.273−0.334A
E. Other rare species									
Acacia melanoceros[3]		1		—/13	—/33				0.107 A
Ficus costaricana[3]		7		—/8	—/19				0.248 A
Ficus obtusifolia[3]		8		—/6	—/19				0.254 A
Ficus popenoei[3]		6		—/6	—/21				0.158 A
Koanophyllon wetmorei[3]		0		—/7	—/24				0.095 A
Lozania pittieri[3]		3		—/8	—/14				0.062 A
Myrospermum fructescens[3]		9		—/22	—/37				0.257 A
Pseudobombax septenatum[3]		9		—/10	—/24				0.121 A
Tetrathylacium johansenii[3]		7		—/7	—/20				0.050 A
Common species									
A. Common, canopy, or understory species, pollinated and dispersed by animals									
Hevea braziliensis[20]			1	—/26	—/4			0.210	0.290 P
			2	—/27	—/4			0.223	0.319 P
Carapa guianensis[21]	10/ha		2 pop.	20/~17	5/6	0.97 (0.02)			0.280 to 0.340 P
	15/ha		9 pop.			0.99 (0.03)			

Table 8.1.   (continued) Multilocus Outcrossing Rates (t_m) and Within-Population Genetic Diversity Estimates of Rare and Common Tropical Tree Species

Species and reference number	Adult densities	No. of trees in BCI[a]	Pop./ year[b]	No. of trees sampled for Rs/He[c]	No. of loci sampled for Ms/He	Outcrossing rate t_m (±SE)	F Adult[d]	F Progeny[d]	Mean expected heterozygosity He[e]
Trichilia tuberculata[2,6,7]	10/ha			17/~44	3/27	1.08 (0.03)	0.074	0.034	0.089 A
Euterpe edulis[23]	54–200/ha		8 pop.	4-16/~23	7/7	0.94 (0.02) to 1.04 (0.05)	−0.190 to 0.082	0.005 to 0.117	0.405 to 0.493 A
Faramea occidentalis[6]	high			—/~44	—/14				0.000
Astrocaryum mexicanum[22]	60–168/ha		4 pop.	—/20–77	5/5	0.93 (0.17) to 1.05 (0.07)	−0.515 to −0.34	−0.235 to 0.048	
Swartzia symplex[19]			3 pop.	—/48	—/36				0.272 A
Brosimum alicastrum[2,6,7]	22/50ha[e]			9/~44	7/21	0.88 (0.04)	−0.271	0.067	0.125 A
Sorocea affinis var. ochnaceae[2,19]	15.4/ha		1 2	8/48 7/—	5/36 5/—	1.09 (0.05) 0.97 (0.02)			0.239 A
B. Common, canopy, or understory species, animal pollinated, wind or explosive mechanism of dispersion									
Alseis blackiana[19]	high		3 pop.	—/48	—/26				0.374 A
Acalypha diversifolia[19]	high		3 pop.	—/48	—/29				0.273 A
Hybanthus prunifolius[19]	high		3 pop.	—/48	—/42				0.247 A
Rinorea sylvatica[19]	high		3 pop.	—/48	—/35				0.106 A
C. Common, canopy, or understory, early or secondary successional species									
Cordia alliodora[28,24]	218/~25ha		11 pop.	49/17–30	4/8	0.97 (0.03)			0.084–0.193 P
Cecropia obtusifolia[25]	14 PF		9 pop.	40/68	8/8	0.97 (0.02)	−0.128 to		0.380–0.420 A
	800–1200 SF[h]							0.182	
Beilschmedia pendula[2,7]	2.36/ha[e]			14/—	3/—	0.92 (0.06)	−0.300	0.043	
Bauhinia forficata[26]					4/	0.98 (0.03)		0.109	0.503 A
Aeschynomene sensitiva var. sensitiva[27]			11 pop.	—/5	—/4				0.184–0.364 P
Aeschynomene sensitiva var. amazonica[27]			10 pop.	—/5	—/4				0.100–425 P

Notes: [a]Number of trees (10 cm DBH in the 50 ha plot on BCI [Condit et al. 1996]). [b]Pop.: population(s); different populations sampled are indicated by numbers; number of populations are indicated by a number followed by "pop." Year: different years of sampling. [c]Rs: reproductive system; He: mean expected heterozygosity. [d]F adult and F progeny: fixation indices of adults and progenies. [e]Based pm DBH of estimated adult size from the 50 ha plot on BCI (Murawski & Hamrick 1991). [f]Average of sampled adult. [g]Range of individual loci. [h]PF: primary forest; SF: secondary forest; LF: selectively logged forest. [i]A: parameter estimated from adults; P: parameter estimated from progenies.

Sources

1. Murawski et al. 1990. BCI—Panama.
2. Murawski and Hamrick 1991. BCI—Panama.
3. Hamrick and Murawski 1991. BCI—Panama.
4. Murawski and Hamrick 1992a. BCI and Darien—Panama.
5. Murawski and Hamrick 1992b. BCI—Panama.
6. Hamrick and Loveless 1986. BCI—Panama.

7. Murawski and Hamrick, unpublished data cited in Murawski 1995.
8. O'Malley and Bawa 1987. La Selva—Costa Rica.
9. Hall, Walker, and Bawa 1996. La Selva—Costa Rica.
10. Gandara 1996. Atlantic Forest—Brazil.
11. Lepsch-Cunha 1996. Amazonas—Brazil.
12. O'Malley et al 1988. Amazonas—Brazil.
13. Buckley et al 1988. Amazonas—Brazil.
14. Moraes 1992. Atlantic Forest—Brazil.

15. Stacy et al. 1996. BCI—Panama.
16. Murawski, Dayanandan, and Bawa 1994. Sri Lanka.
17. Murawski, Nimal Gunatilleke, and Bawa 1994. Sri Lanka.
18. Murawski and Bawa 1994. Sri Lanka.
19. Loveless and Hamrick 1987. BCI—Panama.
20. Paiva, Kageyama, and Vencovsky 1994; Paiva, Kageyama, et al. 1994. Amazonas, Acre—Brazil.
21. Hall, Orrell, and Bawa 1994. Costa Rica.
22. Eguiarte, Perez-Nasser, and Piñero 1992. Mexico.
23. Reis 1996. Atlantic Forest—Brazil.
24. Chase, Boshier, and Bawa 1995. Costa Rica.
25. Alvarez-Buylla and Garay 1994. Mexico.
26. Santos 1994. Atlantic Forest—Brazil.
27. Hill et al. 1978. Amazonas—Brazil.
28. Boshier, Chase, and Bawa 1995. Costa Rica.

(high densities) in nonexploited areas (Murawski, Dayanandan, and Bawa 1994). Furthermore, the estimates were made based on individuals collected in sites where selective harvesting (population thinning) had already occurred. With the exception of *Cavanillesia platanifolia,* other rare species in table 8.1 had low densities in the undisturbed sites where they were studied. *C. platanifolia* had fluctuating population densities, and even though it presented relatively low outcrossing rates in both sites where it was studied ($0.21 < t_m < 0.66$), it presented higher rates in Darien (Panama), where it is common, and lower rates on Barro Colorado Island, where its density is low (Murawski and Hamrick 1992a). These results suggest that larger outcrossing rates occur in sites with large populations. The main differences in breeding systems between these species with low outcrossing rates are: *S. megistophylla,* although primarily outcrossing, showed apomixis and self-fertilization confirmed in an isolated tree (Murawski, Dayanandan, and Bawa 1994); *C. platanifolia, C. pentandra,* and *S. trapezifolia* showed outcrossing rates between 0 and 1.0, indicating a possible self-compatibility. Tests could not detect apomixis in any of these three species (Murawski, Hamrick, et al. 1990; Murawski and Hamrick 1992b; Murawski, Dayanandan, and Bawa 1994). These latter three species plus *D. oleifera* possess common ecological characteristics, such as the demand for light in early life stages, rapid growth, and persistence above the canopy (emergent) in the mature forest (Murawski, Dayanandan, and Bawa 1994).

At the community level, Bawa, Perry, and Beach (1985) found that two of three rare hermaphroditic tree species in La Selva (Costa Rica) were self-compatible, contrasting with the majority of the most common species, which were self-incompatible. It should be remembered that only three rare species were studied, and generalizations about rarity and self-compatibility should be made with caution. *Hymenolobium pulcherrimum,* one of the two rare self-compatible species, also possesses characteristics typical of pioneer species, and like the Bombacaceae, their seeds are wind dispersed.

Murawski and Hamrick (1991) speculate that outcrossing or incompatibility systems can be relaxed in some very low-density species that otherwise would have impaired reproductive capacity (Ashton 1984). However, the study of the mating system and the pollen dispersal of three low-density tree species, *Spondias mombin, Calophyllum longifolium,* and *Turpinia occidentalis,* showed outcrossing rates equal to 1.0 (table 8.1) for these species (Stacy et al. 1996). Although this does not necessarily mean self-incompatibility, previous studies have shown *S. mombin* to be self-incompatible (Bawa 1974; Bawa and Opler 1975; Stacy et al. 1996), and *C. longifolium* suffered a high rate of fruit abortion during the fruiting period. These results, in reference to the mating system discussed here, show that the five rare species that present t_m values smaller than 0.8 seem to have peculiar characteristics. They either are endemics with large primary forest populations (Dipterocarpaceae) or are species with colonizing or initial succession characteristics, and likely to be in the Bombacaceae.

The association between low density and self-compatibility may not be as strong when the dioecious tree species in the tropical forest are considered. Murcia (1996) and Bawa, Perry, and Beach (1985) comment and cite various studies in which many dioecious species are rare or uncommon in tropical forests, but it is not known to what extent these species could have alternate systems of reproduction like apomixis.

As there is little possibility that 100 percent of an adult population would flower synchronously, as shown in phenological studies in tropical forests (Newstrom et al. 1994; Alencar 1998), long-distance pollen movement would be expected in various low-density species with outcrossing rates close to 1.0. Nevertheless, the spatial distribution of reproductive individuals must be investigated.

Murawski and Hamrick (1991) have studied the effects of flowering density on the outcrossing rate of nine tropical species, including *C. platanifolia* and *C. pentandra*. These two species plus *Quararibea asterolepis* (common), all of the Bombacaceae family, have shown elevated levels of inbreeding with a decrease in flowering density of trees. The other six species studied, four common (*Beilschmedia pendula, Trichilia tuberculata, Brosimum alicastrum,* and *Sorocea affinis*) and two low density (*Platypodium elegans* and *Tachigali versicolor*), did not show this relationship. *P. elegans* and *T. versicolor,* the two rare ones, had outcrossing rates close to 1.0. *Tachigali versicolor* showed a clumped distribution of reproductive individuals and long periods of flowering time and substantial evidence of pollen flux over long distances (more than 500 m).

A population of *Couratari multiflora,* a species of very low density with an outcrossing rate close to 1.0 (see table 8.1), was studied in a continuous forest reserve at the BDFFP, where it showed an extended flowering period (Lepsch-Cunha and Mori 1999). Nevertheless, despite the clumped spatial distribution of adults, the flowering individuals did not show a clumped distribution. The median distance between flowering individuals was between 600 and 1,000 m. This suggests a very low possibility of mating among very close individuals.

Through genetic markers and paternity exclusion techniques, which directly infer the patterns of effective pollen movement, Hamrick and Murawski (1990) found for *P. elegans* and *T. versicolor*—both with hermaphroditic flowers pollinated by bees that foraged in large areas—25 percent and 20 percent of the actual pollen movement occurred beyond 500 m and 750 m, respectively. Stacy et al. (1996) showed pollen movement of up to 300 m or more for *C. longifolium* and *S. mombin* in consecutive years. In these two species as well as in *T. occidentalis,* a large fraction of pollen movement occurred at distances of some hundreds of meters, well beyond the distance to the closest neighbor. Nason and Hamrick (1997) showed effective pollen movement through fragments extending from 80 to 1,000 m and 6 to 14 km for the low-density species *S. mombin* and five monoecious species of *Ficus,* respectively. C. Dick (1999; see Chapter 12) showed that pollen movement in the BDFFP reserves reached 3.5 km among trees of the emergent *Dinizia excelsa* located in pasture and surrounding fragment and continuous forests. Spatially isolated trees in pasture of *Symphonia globulifera* formed links among isolated populations (Aldrich and Hamrick 1998). Gandara (1996) showed, by following rare alleles, that in *Cedrela fissilis,* a species with low density and a multiloci outcrossing rate close to 1.0 (table 8.1), that the trees received pollen from a distance of about 950 meters. The population phenology of very rare *C. multiflora* showed long-distance movement by pollinators, which was confirmed by outcrossing rates equal to 1.0 for the trees examined (Lepsch-Cunha 1996; Lepsch-Cunha and Mori 1999). Therefore, long-distance pollen dispersal is not uncommon among tropical trees.

Another method of estimating how many matings within a population are random or preferential is through the fixation index (F) of Wright (1965). This index, defined as (H_e

$- H_o)/H_e$, where H_o is the observed and H_e is the expected frequency of heterozygotes, measures the fraction of reduction of heterozygosity in relation to that expected in a randomly mating population, according to the expectations of the Hardy-Weinberg Equilibrium (Brown and Weir 1983). Values of F equal to 0.0 indicate that a population is in Hardy-Weinberg equilibrium, and values significantly greater or less than zero indicate, respectively, an excess of homozygotes or heterozygotes. The comparison between the fixation indices of adults and progenies (seedlings originating from a single reproductive event) indicate which evolutionary forces are most likely to be important in populations.

The adult fixation index values (F_{adult}) were close to zero (F is less than 0.2) or negative in the common and rare species, with the exception of the rare *Couratari guianensis* (table 8.1). These results clearly suggest that the tropical forest tree species present low levels of inbreeding and a tendency to possess an excess of heterozygotes in the adult population. This excess can be explained by the variable distances of pollen movement and seed dispersal, by negative preferential outcrossing ("dissassortative mating," or preferential outcrossing of genetically different individuals) or by selection favoring heterozygotes.

For example, Eguiarte et al. (1992) found a positive relation between individual heterozygosity and growth rates in *Astrocaryum mexicanum*. Theoretically, species with low or negative fixation indices may suffer inbreeding depression when greater levels of inbreeding occur. The effect of the inbreeding depression can be seen in *Pithecellobium pedicellare* through the presence of albino and chorotic seedlings in progenies (O'Malley and Bawa 1987), indicating that outcrossings between relatives can be detrimental to survival.

The F_{adults} when compared to $F_{progeny}$ reflect the change in heterozygosity through the phase of seedlings to adults in a population. This change points to the most probable evolutionary forces having acted between the seedling phase and the adult phase, such as selection, drift, and migration. The $F_{progeny}$ better reflects the mating system, despite the possibility of selection acting between the fertilization phase and the emergence of the seedlings.

The fixation indices of progeny ($F_{progeny}$) are similar to those for adults within a species—with the exception of *Hevea brasiliensis,* which has a higher value (greater than 0.2). The $F_{progeny}$ for rare species tended to be larger than for common species. This indicates a greater tendency for random matings or some type of negative differential selection (for example, gametic self-incompatibility) in common species. The $F_{progeny}$ of the rare species show a tendency of more inbreeding in these populations. Among the rare species, *C. platanifolia, Bertholletia excelsa* (in one population), *Myracrodruom urundeuva,* and *S. trapezifolia* had $F_{progeny}$ greater than 0.2, which can be considered high. It is interesting to note that *M. urundeuva* is dioecious, typically of low density in primary forest, but becomes common in disturbed areas (Moraes 1992). This species regenerates naturally in pasture, and the high fixation index (F) suggests a possible bottleneck and the subsequent founding of the population by a few individuals, or the presence of apomixis (Moraes 1992). Many other rare species have $F_{progeny}$ values between 0.1 and 0.2. Similar values were found in only two common species, adults of *C. obtusifolia* and progenies of *Bauhinia forficata*. These last two species are pioneers, have high outcrossing rates (Alvarez-Buylla and Garay 1994; Santos 1994), occur in high density in large clearings, and are considered rare in relation to the total density of a local population (see above). *Bauhinia forficata* is the most common species in a small disturbed fragment in a semi-deciduous forest in the state of São Paulo, Brazil (Nascimento, Tabanez, and Viana 1996).

The expected heterozygosity (H_e) is estimated from allelic frequencies and represents the expected proportion of heterozygotes in an ideal population with random mating, absence of natural selection, and random genetic drift (Brown and Weir 1983; Nei 1987). These values are extremely variable in the two groups (table 8.1) but in general are high when compared to the average for all plant species (Hamrick and Godt 1990). However, more low-density species have heterozygosity values close to zero than do common species with the exception of *T. tuberculata* and *Faramea occidentalis*. Interestingly, these understory species are two of the most common species on Barro Colorado Island (Condit, Hubbell, and Foster 1992, 1994). All the low-density species sampled in the larger areas (*P. pedicellare, C. fissilis, C. guianensis,* and *C. multiflora*) showed high values of heterozygosity (0.243 to 0.436).

Hamrick and Murawski (1991) compared genetic diversity on Barro Colorado in sixteen low-density and sixteen common species (Hamrick and Loveless 1989). On average, the diversity values were greater for common species. The rare species were selected on the basis of low density on the entire island beyond the 50 ha study plot, and for having five to twenty-five individuals larger than 1.5 m tall in the study plot. This minimum size and the chosen densities used by the authors may represent extreme cases of rarity. This could mean that the populations are naturally dying out or are recently colonizing the site. For example, two of these species with heterozygosities close to zero—*Trichantera gigantea* and *Schizolo-*

bium parahyba ("guapuruvu")—have, respectively, one and three individuals at least 10 cm DBH in the plot, and *Enterolobium schomburgkii* has two (Condit, Hubbell, and Foster 1996). These last two species are represented by emergent individuals. *Tachigali versicolor,* the species with low heterozygosity and very low density of adults, is semiparous (a peculiar characteristic of species wherein individuals have only one reproductive event in their lives), which can result in differentiated genetic patterns. Thus, the data of Hamrick and Murawski (1991) should be viewed with caution with respect to generalizations about rare species.

Through these few examples, some generalizations are suggested. Genetic diversity estimates of rare species were more variable than for common species, which suggests different types of rarity. First are species with high outcrossing rates, high values of gene diversity, low fixation indices in adults and progenies, and long-distance pollination, such as *C. multifora, P. pedicellare, P. elegans,* and *S. mombin.* For these species, the distance between individuals does not impede gene flow, and species are exogamics, adapted to mating with widely spaced individuals. Second are pioneer species that fluctuate in density, such as *C. platanifolia, D. oleifera, C. pentandra,* and *C. guianensis*). They tend to have intermediate outcrossing rates (mixed mating system), presence of self-compatibility or probably apomixis in some cases, and greater fixation indices in progeny. Third are dioecious species (*M. urundeuva*) or hermaphrodites with an outcrossing mating (*C. fissilis*) and low density in primary forest that can benefit from regeneration in perturbed areas (*S. mombin*). Fourth are species that may be suffering from inbreeding due to a decrease in population size (*B. excelsa, C. guianensis*). Finally, one last group (e.g., *Shorea* spp.) is composed of common endemic species that are becoming rare through exploitation, and at times are confused with some pioneer species of the second group (*C. platanifolia*).

As seen here, most tropical forest tree species studied to date show high genetic diversity and low or negative inbreeding coefficients, which tends to refute the "non-adaptive" hypothesis. As mentioned previously, this hypothesis predicts that species with widely spaced individuals would tend toward a more autogamic mating system (self-fertilization) and inbreeding through kin mating, promoting small neighborhood sizes (number of individuals that exchange genetic material). This would promote high speciation locally. The occurrence of different types of rarity suggests different patterns of genetic diversity and lends support to the microniche equilibrium hypothesis.

Hamilton's (1999) use of the chloroplast DNA genome (cpDNA) to study of gene flow and seed dispersal in the *Corythophora alta,* a common Lecythidaceae canopy tree in the BDFFP fragments, provides an exciting new avenue to understanding tropical tree gene flow and the effects of fragmentation on genetics. CpDNA is maternally inherited in most angiosperms and dispersed only through seeds (Reboud and Zeyl 1994, cited by Hamilton 1999). Therefore, the study of cpDNA is a direct estimate of previous seed dispersal that led to trees being established. Hamilton found that forest fragments of 1 and 10 hectares all contain trees with a single cpDNA haplotype, indicating that they had a single maternal founder and that seed dispersal has been limited to small areas of continuous tropical forest.

Couratari multiflora and *Couratari guianensis*

Couratari multiflora and *C. guianensis* (Lecythidaceae) are arboreal species widely dis-

tributed in the Neotropics. *C. guianensis* extends from Costa Rica through all of the Amazon, and *C. multiflora* is found from Venezuela and Guiana through the Brazilian Amazon (Mori and Prance 1990). They are similar ecologically; both typically occur in low density, are emergents with asymmetrical flowers pollinated by large, long-tongued bees, and their seeds are wind dispersed (Nelson, Absey, et al. 1985; Mori and Boeke 1987; Mori and Prance 1990). However, these two species have different levels of variation in population abundance over their geographic range.

From available information (Mori and collaborators 1987; Mitchell and Mori 1987), *C. multiflora* is a uniformly low-density species (Bawa and Ashton 1991). This species is expected to show a preferentially exogamic mating system (high rates of outcrossing), as well as efficient long-distance mechanisms of pollen and seed dispersal (Bawa and Ashton 1991). On the other hand, *C. guianensis* shows varying population density—low in some sites and common in others (Mitchell and Mori 1987). The genetic consequences of population size reduction will greatly depend on the predominant population density of the species; it is, however, expected that it may cause allelic fixation and a decrease in the level of heterozygosity.

In this section we present data on the within-population genetic diversity of these two species to test predictions based on their type of rarity. *Couratari multiflora* is expected to show a very low (close to 0) inbreeding coefficient and a random mating system. *C. guianensis,* on the other hand, should show some level of inbreeding and a more mixed mating system as a result of adaptations to population fluctuations. The specialization of the flowers, pollination by large bees that probably promote long-distance pollen movement, wind dispersal

TABLE 8.2.   Measures of Population Genetic Structure for Adults, Juveniles, and Progeny of *Couratari multiflora* and *Couratari guianensis*

	Density and spatial pattern			
	Couratari multiflora		*Couratari guianensis*	
	1 individual/ 10 ha	Aggregated	1 individual/ 14 ha	Random or aggregated
Age class	Adult	Progeny	Adult	Juvenile
Number of individuals	41	103	27	26
Total alleles	26	25	22	24
Total loci	8	8	8	8
A[1]	3.25 (0.45)	3.57 (0.30)	2.75 (0.25)	3.00
A_e[2]	1.92 (0.69)	1.93 (0.65)	1.70 (0.50	1.76
P (0.95%)[3]	100.0	85.71	100.0	100.0
P (0.99%)[3]	100.0	100.0	100.0	100.0
H_{om}[4]	0.359	0.380	0.203	0.258
H_{em}[5]	0.431	0.429	0.420	0.415
F_m[6]	0.176	0.114	0.518	0.377
Percentage loca in HWE[7]	50.0	72.4	25.0	63.5

Notes: [1]Mean number of alleles per locus. [2]Mean expected number of alleles per locus. $A_e = 1/p_i^2$, where p_i equals the allelic frequency of the ith locus. [3]Proportion of loci analyzed that are polymorphic. [4]Mean observed heterozygosity. [5]Mean expected heterozygosity. $H_e = 1 - p_i^2$ for each locus, where p_i is the allelic frequency esitmated from allele i. [6]The mean of $F = 1 - (H_o/H_e)$ for each locus. [7]Hardy-Weinberg Equilibrium.

of seeds, and the widespread geographic distribution suggest high genetic variability for both species.

Genetic data were gathered from a study performed in a 10,000 ha nonisolated reserve (Reserve 1501; fig 4.1) of the BDFFP (Lepsch-Cunha 1996; Lepsch-Cunha, Kageyama, and Vencovsky 1999). Selected species with known densities and ecological characteristics were chosen from a taxonomic and demographic database for trees in the Lecythidaceae family in a 100 ha forest plot (Mori and Lepsch-Cunha 1995; Chapter 6).

A census of trees considered adults (at least 20 cm DBH) of the two species was performed using the 100 ha Lecythidaceae forest plot as a starting point, and more trees were included from an adjacent area to increase sample size. The final census included a 400 ha area in which twenty-seven *C. guianensis* and forty-one *C. multiflora* adult trees, respectively, were located and mapped. Also, because they were easily recognized, twenty-six juveniles (1 to 20 cm

DBH) of *C. guianensis* were sampled within this area (table 8.2). Although no juveniles of *C. multiflora* were included, this species produced enough fruits to permit analyses of the genetic diversity of progenies (families composed of daughter-plants). The analysis of *C. multiflora* progenies was based on only four experimentally germinated families (cohorts of seedlings from fruit from the same mother tree), because high predation rate by animals (Lepsch-Cunha 1996) made collecting viable, mature seeds difficult.

Standard electrophoretic procedures of isoenzyme analysis were used (Soltis and Soltis 1990; Alfenas et al. 1991). Estimates of genetic parameters for *C. multiflora* progenies and adults, as well as for the *C. guianensis* young and adults, were quantified based on Brown and Weir (1983) and Nei (1987). Conformity tests to the Hardy-Weinberg Equilibrium were made using the BIOSYS program (Swofford and Selander 1989).

Estimates of genetic diversity were based

on percentages of polymorphic loci (P), mean and effective number of alleles (A, A_e), and mean expected heterozygosity (H_{em}). All parameters were similar between progenies and adults of *C. multiflora* (table 8.1). The rejection of the Hardy-Weinberg Equilibrium null hypothesis for the distribution of the genotypes was greater in adults (50 percent of the loci) than in progenies (37.6 percent), and most of the deviations occurred because of an excess of homozygotes in both categories.

Adult and young individuals of *C. guianensis* showed similar values of genetic diversity. Although the results were not significant, young plants showed higher values for the number of alleles and for the observed mean heterozygosity. Significant deviations from the Hardy-Weinberg Equilibrium were found in 36.5 percent and 75.0 percent of the juvenile and adult loci, respectively, all due to an excess of homozygotes. The means of the fixation indices (F_m) were 0.377 and 0.518, respectively, for young plants and adults. These results suggest that the juvenile population has lower levels of inbreeding than that of adults.

The two species showed similar percentages of polymorphic loci, mean and effective number of alleles, and mean expected heterozygosity, and these estimates were high when compared to other tropical tree species (table 8.1). However, *C. guianensis* had two loci represented only by homozygotic individuals (despite non-allelic fixation), and the fixation indices for most of the loci were high and positive. Moreover, observed mean heterozygosities (H_{om}) for juveniles as well as for adults of *C. guianensis* (0.258, 0.203) were much smaller than those of progenies and adults (0.380, 0.359) of *C. multiflora*. Nevertheless, both species have similar evolutionary potential for generating gene diversity (H_{em}; table 8.2). This probably reflects intense gene flow via seeds

or some other form of maintenance of allelic diversity in the *C. guianensis* population.

We interpreted the *C. guianensis* results as supporting two explanations or hypotheses. The first hypothesis is that genetic diversity may indicate some adaptation to fluctuating population density through substantial levels of long-distance gene flow. Five adults in our study population were observed biweekly over a five-year period, and they produced small quantities of fruit (S. Mori, pers. comm.), in contrast to other, denser populations, close to the study area, which were observed yielding many fruits (P. A. Assunção, INPA, and C. Dick, Harvard University, pers. comm.). Based on this information, the second hypothesis is that the high F(s) may indicate a negative association between population density and reproductive success, broadly defined as both quality and quantity of offspring. In this case, when the population is reduced in size, inbreeding increases and the fruit set drops. However, even if *C. guianensis* is adapted to reduction in population size or if it is suffering the combined effects of small effective population size and reduced outcrossing rate, this species will need immigration from "source" areas outside the low-density population areas. In the first hypothesis, the main conservation problem related to fragmentation will be the maintenance of genetic diversity in the next generations. In the second, the low production of fruits coupled with demographic stochasticity probably will lead to population decline or eventual extinction in isolated populations (Bawa and Ashton 1991).

Couratari multiflora is an allogamous species with a multiloci outcrossing rate equal to 1.0, relatively low inbreeding coefficients, with half or most of the loci in Hardy-Weinberg Equilibrium for both adults and progenies, and with high levels of genetic diversity. These results suggest a large neighborhood area and high gene flow. The phenology data discussed in the previous section, along with the mating system, show that a distance of 1,000 m between trees does not impede pollen flow. In this sense, this species appears to be adapted to out-crossing in low-density populations. This species would, therefore, need large areas, on the order of thousands of hectares, to maintain an effective population size and not suffer from inbreeding depression. These results emphasize the need to distinguish between different types of rarity based on genetic characteristics of populations, as the response of species with seemingly similar demographic characteristics to human disturbance can be very different.

Conservation Lessons

1. There is a need to distinguish between different types of genetic and demographic structures of tropical tree species populations, as many of them are represented locally with populations of widely spaced individuals. By not recognizing these differences, effective conservation and management guidelines that maintain genetic variability for rare species may be impossible.

2. Studies investigating inbreeding depression of tree populations in fragments are extremely important, because most of the conservation and management actions depend on the knowledge of the importance of inbreeding in populations. Indi-rect measures of inbreeding, such as flower and fruit production, percentage of germination, and changes in the distribution curve of size and age classes in disturbed and undisturbed sites, can be revealing. Correlations among quantitative characteristics (growth, height, fruit production) and heterozygosity levels should be tested.

3. Studies investigating the effects of different-sized fragments on the pollinator community (Aizen and Feinsinger 1994a, b; Murcia 1995, 1996) are essential.

4. Results from this chapter suggest that large reserves (thousands of hectares) are necessary for maintaining effective population size of two rare tree species, although each species will show a different response to anthropogenic disturbance.

5. In the face of these results, the *ex situ* conservation of tropical tree species would be much more costly than it is for most crop species capable of self-fertilization. Preferable alternatives for conserving these species include mixed plantings in an agrosilviculture system, where corridors of native trees are intermingled with agricultural crops, providing both a conservation as well as an economic benefit.

6. For native trees with economic potential (e.g., lumber, fruit, extracts), seeds and seedlings with high genetic variability could be produced in or harvested from large conservation land tracts and distributed to agrosilvicultural operations.

PART III
Fragmentation Effects

Fragmentation Effects on Plant Communities

Habitat loss and fragmentation have turned into the most important threat to biological diversity for all terrestrial ecosystems. In such regions as the Amazon rainforest, which is still largely intact, ever-increasing social and economic pressures are compelling developments of a vast and great wildernesses. Whether the drivers of development and deforestation are timber or mineral extraction, urban expansion, or agricultural operations, the result will inevitably be land-cover change and forest fragmentation. Although the scale of change in the landscape may depend on the root causes of deforestation, we now have sufficient information to begin to understand the major components of forest fragmentation and their ecological impacts on the local biota. Much needs to be learned.

Until recently, most of our understanding of habitat fragmentation has come from studies in temperate regions, where landscapes have, in some cases, been modified for centuries. However, much new information has been forthcoming from research carried out in tropical regions during the past two decades (e.g., Laurance and Bierregaard 1997a), especially from the BDFFP. In our introduction to this section of the book we outline the components of forest fragmentation, highlighting the major results from the BDFFP and other research efforts.

In particular, we draw the reader's attention to published results from the BDFFP that are not reviewed in the rest of this volume (see also Chapter 4) and the BDFFP Technical Series, page 371).

Ecological Consequences of Forest Fragmentation

AREA AND INSULARIZATION

Habitat fragmentation, by definition, involves a reduction in original area and isolation of remaining patches of forest. Theory predicts that fewer species will be present in smaller areas than in large ones. Many causes have been suggested to explain this relationship between species richness and area. The simplest explanation is that many species are lost in the resulting landscape due to a decrease in habitat heterogeneity (Williams 1943; Simberloff and Abele 1982; Haila, Hanski, and Raivio 1987; Schwarzkopf and Rylands 1989; Norton, Hobbs, and Atkins 1995), but other factors certainly are important (Chapter 2).

Because farmers and settlers tend to clear accessible areas and those on the most productive soils first (Wiens 1976; Forman and Godron 1986), forest fragments are often a nonrandom subset of the original habitats in a landscape. Consequently, many specialized species may be excluded from the forest patches because of their strong association with a particular habitat type, suggesting that fragment location will be important in the initial inclusion or exclusion of species (Zimmerman and Bierregaard 1986; Chapter 2). This problem is compounded in tropical regions where many species have small ranges, and small areas can have a high percentage of endemic species (Terborgh and Winter 1983; Gentry 1986). Species with large home ranges can also be excluded from patches that do not provide the minimum area for survival (Lovejoy, Bierregaard, et al. 1986; Chapter 21; Chapter 22; Rylands and Keuroghlian 1988).

Not surprisingly, results from the BDFFP have shown a decrease in species richness for many faunal groups in central Amazonia (Powell and Powell 1987; Rylands and Keuroghlian 1988; Klein 1989; Schwarzkopf and Rylands 1989; Bierregaard et al. 1992; Morato 1993; DeSouza and Brown 1994; Stouffer and Bierregaard 1995a; Tocher, Gascon, and Zimmerman 1997; Bierregaard and Stouffer 1997). These results obviously suggest that bigger is better and that reserve design should maximize size to ensure that as many of the original species present in the forest will be conserved after disturbance. However, as pointed out elsewhere, forest remnants of all sizes serve varied functions, and even small fragments have some conservation value (Nepstad, Uhl, et al. 1996; Schelhas and Greenberg 1996; Turner and Corlett 1996; Chapters 13 and 16).

Isolation does not necessarily result in immediate local extinctions. In the BDFFP experimental forest isolates, an initial increase in capture rates of birds in fragments was detected immediately following forest isolation, probably as a result of birds taking refuge in the fragments (Bierregaard et al. 1992). However, capture rates and species richness in fragments eventually fell to below pre-isolation levels (Stouffer and Bierregaard 1995a). A somewhat different situation was found for frogs; Tocher, Gascon, and Zimmerman (1997) showed that after ten years of isolation, some frog species were more abundant in the fragments compared to populations in continuous forest.

In many cases, populations can persist in patches at low population density. Small populations are, however, much more vulnerable to random demographic and genetic effects that often lead to local extinction (Gilpin and Soulé 1986; Menges 1992). In addition, independent life history stages of a given species may be affected differentially as a result of isolation (Chapter 10).

Although declines in species richness were observed for all broadly defined faunal groups in fragments compared to continuous forest, in some cases fragments actually harbor more species after isolation than in the same physical area before deforestation. This has been shown for amphibians (Tocher, Gascon, and Zimmerman 1997) and butterflies (Hutchings 1991; Brown and Hutchings 1997), although the nature of the post-isolation community varied among these groups. Amphibians, for example, lost few of the pre-isolation complement of species, and most of the increase in species richness was due to the appearance of species associated with open areas. Conversely, for butterflies, more than 40 percent of the original complement of species in a patch of forest is lost after isolation, presumably displaced by light-loving specialists (Hutchings 1991).

We have witnessed changes in community composition after habitat isolation, as well as in the structure of the community as Malcolm has shown, for instance, for the vertical distribution of small mammals (Malcolm 1997a). Whereas the small mammal community shows higher total biomass in the canopy stratum in continuous forests, this dominance shifts to the understory in small isolated fragments. Malcolm (1991) explained this as a response to changing insect biomass, an important food source, which follows the same trend.

It is important to remember that many of the changes witnessed in fragments are neither immediate nor static in nature. In some cases, the natural temporal variation in species abundance in primary forest is accentuated in forest fragments (Malcolm 1991). In some cases, species composition changes though time in the fragments and in the surrounding areas (Tocher, Gascon, and Zimmerman 1997; Chapter 19). This is clearly the case also for euglossine bees that have been sampled at various times since the fragments were established in the early 1980s. Powell and Powell (1987), based on data collected in 1983, were the first to show a decrease in visitation rates of euglossine bees at chemical baits in the fragments after isolation. In 1988–89, no difference in abundance of bees was found in the same general area by Becker, Moure, and Peralta (1991) (see also Chapter 17).

ISLAND BIOGEOGRAPHY THEORY AND ECOLOGICAL PROCESSES

In the Amazon, regardless of whether a particular taxonomic group showed an increase or decrease in species richness after isolation, larger isolates invariably maintained more species than smaller ones. These results demonstrate a clear species-area relationship, which makes the types of post-isolation increase in species richness mentioned above somewhat unexpected. For example, according to the model of island biogeography proposed by MacArthur and Wilson (1963, 1967), fragments can be expected to contain fewer species after isolation because more extinctions and less immigration should occur in isolated patches (see Chapter 2).

MacArthur and Wilson's island biogeog-raphy model spurred a massive amount of theoretical and field research and could certainly be considered one of the most important research paradigms in ecology during the 1970s and '80s. In the real world, however, the theory has proven limited in its capacity to predict the range of biological and ecological changes associated with habitat fragmentation (Bierregaard et al. 1992) and of questionable usefulness in providing guidelines for reserve design (Zimmerman and Bierregaard 1986).

While the theory of island biogeography relies on one response variable (the number of species) to measure ecosystem change, available data suggest that a much more complex suite of modifications occur as a result of habitat fragmentation. Not only are populations affected, but ecological processes such as pollination and decomposition can be altered by forest isolation (Murcia 1996). Decreased decomposition rates in the Manaus fragments were shown by Klein (1989) as a result of a decrease in species richness of dung and carrion beetles in those habitats. Malcolm (1991) showed that in continuous forest at the BDFFP study site, insect biomass was concentrated in the forest canopy, whereas in fragments it was concentrated in the understory. This shift will almost certainly alter patterns of insect-plant interactions (pollination, herbivory, etc.). Extensive work on bees at BDFFP (Powell and Powell 1987; Becker, Moure, and Peralta 1991; Morato 1993, 1994; Chapter 17) strongly suggests that there will be modified reproductive success of associated plant species (Aizen and Feinsinger 1994a, 1994b). In most cases we expect decreased pollination rates or efficiencies, but as Chris Dick has shown (Chapter 12), in some instances isolated trees in a landscape mosaic of forest patches and disturbed land may experience increased fecundity as a result of changes in pollinator populations.

Tightly coevolved species associations, which are common in tropical systems, may lead to cascading extinctions. Harper (1987, 1989) demonstrated that the absence of army ants in small fragments also led to the disappearance of obligate ant-following birds. Facultative ant-following birds, however, were less severely affected. Zimmerman and Bierregaard (1986) hypothesized that the absence of peccaries in fragments would reduce the survival of certain frog species that depend on peccary wallows for reproduction. These sorts of second-order extinctions are not addressed in the graphical immigration and extinction models of island biogeography theory, which treats all species as both identical and independent.

EDGE CREATION AND EDGE EFFECTS

Fragmentation leads to a drastic increase in forest edge, which differs from natural ecotones, or gradients between two habitat types. The basic difference is one of degree of contrast between the two habitats. In a natural landscape, much less contrast usually exists between patches of adjacent habitat types, whereas in a fragmented landscape, an abrupt contrast occurs (e.g., forest vs. field), especially immediately after isolation of a forest fragment (see fig. 13.5a).

Many biological and physical consequences have been reported as a result of edge creation, penetrating as much as 250 m into the forest (see fig. 29.1), and we now believe that edges may be the most important component of forest fragmentation, at least for relatively small reserves (hundreds of hectares or less). In continuous tropical forests, sunlight usually is restricted to vertical penetration, but in a forest fragment it can penetrate laterally along the fragment's margins. This single change dramatically affects the microclimatic conditions of the forest near the edge (Kapos 1989, Murcia 1995).

In Manaus, increases in temperature and decreases in relative humidity and soil moisture have been detected up to 80 m from the edge inside the forest (Kapos 1989; Camargo and Kapos 1995). These changes, however, are not permanent but evolve with time as the edge is closed up by growth of vegetation (Camargo and Kapos 1995).

Microclimatic changes associated with edge creation probably are one of the causative factors, along with increased wind turbulence (Laurance 1997), that explain the striking increases in tree mortality and forest turnover and changes in forest structure near edges as described in Chapters 9 and 13 (see also Wandelli 1991; Malcolm 1994;. Ferreira and Laurance 1997). Although the BDFFP study was not designed to investigate edge effects per se, edge effects are necessarily associated with fragmentation, particularly in smaller reserves. Statistical analyses can dissect out these two factors and determine their relative importance. In fact, when doing so, most of the change in tree mortality and damage found in forest fragments can be attributed to edge effects as shown in Chapters 9 and 13. This translates into substantial loss of standing biomass and release of carbon from forest stocks (W. Laurance, Laurance, et al. 1997). Also, because of increased sunlight along the edge, leaf fall increases and the recruitment pattern of seedlings is modified (Sizer 1992).

As edge effects drive many of the changes in the vegetation in forest fragments (Malcolm 1994; Laurance, Ferreira et al. 1998a, 1998b), these edge habitats become more similar to open areas and second-growth forests. These changes may explain some faunal changes in the fragments. The butterfly community in fragments shows a large shift in community composition; light-loving specialist species displace forest-interior species soon after forest isolation (Hutchings 1991; Brown and Hutchings

1997). These light-loving specialists are better suited to the modified conditions that exist in small fragments due to edge effects, and they often find abundant food sources in the surrounding second-growth vegetation (see below).

MATRIX HABITAT AND LANDSCAPE CONFIGURATION

As habitats are fragmented, primary forest is converted into new land cover surrounding the remnant patches of the original habitat. This modified land has been termed the "matrix" in which the remnants are imbedded. Until recently, the effect of the matrix on fragmented biotas has been seriously underestimated. However, the attention of BDFFP researchers, and of the rapidly growing school of landscape ecology worldwide, is now focusing considerable attention on this issue (L. Harris 1984; Fahrig and Merriam 1994; Offerman et al. 1995). As discussed in Chapter 3, the matrix habitat will be important in the evolution of ecosystem dynamics in fragmented landscapes for several reasons. Most important, it acts as a filter for movements of species between fragments and other landscape features, facilitating movements of some species and impeding others. Also, disturbance-adapted species will be present in the matrix and may invade forest patches and edge habitat (e.g., Brown and Hutchings 1997; Chapters 10 and 14). And finally, the matrix type (pasture, degraded pasture, second-growth forest, etc.) can influence the severity of the edge effects in forest patches (Williamson, Mesquita, et al. 1998).

A number of recent studies have highlighted the importance of the matrix. Laurance (1991b) found that the best ecological correlate of mammal vulnerability to extinction in rainforest patches of Australia was a species' abundance in the matrix habitat. Likewise, Malcolm (1991) found a significant positive correlation between the abundance of small mammal species in fragments and their abundance in the matrix habitat in central Amazonia. In Sweden, Aberg et al. (1995) also demonstrated that the type of matrix habitat influenced the occurrence of hazel grouse (*Bonasa bonasia*) in isolated fragments. Angermeier (1995) showed that impeded movement (as measured by degree of isolation of aquatic habitats) was one of the main factors predicting extinction proneness in fish. Finally, the degree of patch isolation was important in predicting the abundance of euros (*Macropus robustus*) in a fragmented landscape in Australia (Arnold, Weeldenburg, and Ng 1995).

As forest clearing proceeds, the loss of primary habitat may lead to the disappearance of many forest-associated species in the transformed matrix (Borges 1995; Tocher, Gascon, and Zimmerman 1997). This is shown by Vasconcelos for ants (Chapter 16), Tocher, Gascon, and Meyer for frogs (Chapter 19), and for birds by Stouffer and Borges (Chapter 20). All of these studies demonstrate that clearing forest will result in impoverished faunas in the disturbed habitats. However, the level of disturbance and recovery will be important as far as community composition is concerned. After several years, communities in the less disturbed habitats (those that have been less intensively used or burned), and in areas that have recovered some second-growth vegetation, will have recovered many of the primary forest species compared to more severely disturbed habitats.

The appearance of a new matrix habitat will allow species associated with these modified habitats to colonize the area and invade the fragments (Tocher, Gascon, and Zimmerman 1997; Hutchings 1991; Malcolm 1991; Chapter 14). Some of these invading species will represent potential competitors or predators for native species in fragments (Janzen 1986a). Overall, it is likely that the composition and dynamics of the matrix will strongly influence communities in isolated fragments (Malcolm 1991).

Students of habitat fragmentation have recognized that the populations in individual fragments, or islands, are not distinct entities but are interconnected through immigration to the fragments from adjacent tracts of continuous forest (the mainland), as well as other fragments. Such immigration links the small populations into a "metapopulation" (Levins 1970), or "population of populations," which exchanges genetic material to a degree determined by the extent to which the matrix, through either its nature or physical extent, affects immigration to the fragments in it. The sudden appearance of barriers in the modified landscape can significantly alter the metapopulation dynamics of surviving species. The presence of a new matrix habitat (i.e., pasture) can limit dispersal, movement, and colonization, as shown for small mammals in Australia (Laurance 1991b), Canada (Oxley, Fenton, and Carmody 1974), and Amazonia (Malcolm 1991).

For species that are important in ecological process (pollination and decomposition), barriers to movement between fragments may have long-term effects on ecological functions in those fragments. In a series of Chaco dry forest fragments in Argentina, for example, most of the sixteen plant species investigated by Aizen and Feinsinger (1994a) showed a decrease in the number of pollen tubes, fruit set, and seed set, quite possibly because important plant pollinators disappeared from fragments and were unable or unwilling to traverse the matrix to recolonize them. Species such as primates and frugivorous or nectar-feeding birds that depend on keystone resources during parts of the year, such as fig trees,

may be negatively affected by fragmentation if they cannot move to areas where that resource is present (Howe 1977, 1984; Terborgh 1986; Mills, Soulé, and Doak 1993).

The configuration of fragments within the matrix is also a key factor determining long-term population survival and large-scale abundance patterns of many taxonomic groups (Askins, Philbrick, and Sugeno 1987; Burkey 1989; Potter 1990; Rolstad 1991; Fahrig and Merriam 1994; Offerman et al. 1995; S. Robinson, Thompson, et al. 1995; Villard, Merriam, and Maurer 1995; Flather and Sauer 1996). Lawton and Woodroffe (1991) showed that breeding water voles (*Arvicola terrestrius*) are less likely to be in patches of forest isolated by large expanses of matrix habitat. Potter (1990) demonstrated the importance of patch configuration within the matrix for population viability of brown kiwis (*Apteryx australis*) in New Zealand. If large and small patches exist, the small ones may serve as stepping-stones for movement between larger patches. As fragmentation of the landscape increased, the distance moved by three small mammal species increased, but fewer individuals moved altogether (Diffendorfer, Gaines, and Holt 1995).

Until recently, the importance of faunal corridors has also remained a largely theoretical debate because of the scarcity of empirical data (L. Harris 1984; Noss 1987; Simberloff and Cox 1987; Simberloff et al. 1992; Demers et al. 1995). The few studies that do exist suggest that corridors greatly enhance movements of animals among patches in a landscape (Bennett 1990; Merriam and Lanoue 1990; Saunders and Rebeira 1991; Dunning et al. 1995; Haas 1995; S. Laurance 1996), which in turn may decrease the probability of extinction of local populations. In one 100 ha fragment at the BDFFP, Bierregaard (1990) reported that obligate army-ant following birds disappeared immediately after a 2 km corridor of forest that had linked the reserve to continuous forest was severed. Research also has shown that communities of small mammals and litter frogs do not differ in composition or abundance between linear remnants and adjacent continuous forest areas, suggesting that corridors can have great potential for increasing connectivity of populations in isolated fragments (Lima 1998).

Outstanding Issues

Although students of habitat fragmentation at the BDFFP and around the world have made substantial strides toward understanding the processes of ecosystem change and degradation that accompany fragmentation, there is much to be learned. How general are the results to date? No process that has been demonstrated in the field has been validated for all taxonomic groups. Replication of published studies with different taxa will be extremely valuable. We need to refine our abilities to measure many aspects of edge effects. Detailed studies of population demographics in fragments (e.g., Sieving and Karr 1997) will help us determine whether fragments are serving as sources of colonists for other fragments or "sinks," where emigrants arrive but cannot reproduce at replacement levels (Pulliam 1988). This would provide a much-needed refinement of current species-area estimations based solely on presence-or-absence data. As the ratio of matrix to remnant habitat increases, is there a threshold below which fragmentation effects are insignificant relative to the more obvious loss of habitat (Fahrig 1997; Bender, Contreras, and Fahrig 1998; Monkkonen and Reunanen 1999)? Can we discover clever ways to ask and answer fragmentation-related questions about species with generation times that exceed our own lifespans? And perhaps most important, as Margules (1999) asked, how do

we transfer knowledge gained at our reductionist, local scale to the landscape scale of planning and decision making?

Fragmentation Effects on Plant Communities

In mid-October 1979, Judy Rankin–de Mérona participated in the BDFFP's first foray into the field to collect data. That trip launched the Botanical Ecology subproject, now known as the Phytodemography Project. This began a research effort to map, collect voucher specimens, and identify every tree with a DBH of at least 10 cm in sixty-nine individual hectares scattered throughout the BDFFP study area—the largest vouchered botanical inventory ever carried out in the Amazon.

In this first chapter on the effects of habitat fragmentation, Rankin–de Mérona and longtime collaborator Roger Hutchings present a comparison of tree mortality, damage, and recruitment at the edge and in the core of a 100 ha forest fragment with control data from nearby continuous forest. Not surprisingly, windthrows took their toll along the edge of the reserve, and damage seemed to penetrate the reserve in dominolike fashion. Their data show that more than 20 percent of the reserve has been structurally damaged. Even in the apparently intact core of the reserve, 500 m from its edge, there are measurable changes, with the number of dead and damaged trees surpassing the number of small trees growing into, or being "recruited into," the smallest size class of trees in the study. In their control plots, mortality and recruitment were similar, suggesting that the forest is indeed in equilibrium.

Rankin–de Mérona and Hutchings use these results to argue that buffers for protected areas will need to be on the order of kilometers rather than the 100 m currently prescribed by law.

Amazonian Palms and Conservation Strategies

Beneath the canopy trees studied by Rankin–de Mérona, Hutchings, and their colleagues, the understory of the BDFFP forests is characterized by abundant palms. Aldicir Scariot, in Chapter 10, summarizes his work on the thirty-six species that make up the palm community in the BDFFP study area.

Working in isolated 1, 10, and 100 ha fragments, Scariot sampled more than 23,000 individual palms in 110 sample plots. His censuses were made ten to fifteen years after fragment isolation and compared to control plots in nearby continuous forest. Despite the short period of time since isolation (relative to the generation time of palms), he was already able to confirm that nonforest species are replacing forest species in the isolates and that population densities are declining. This confirms the work presented by Benitez-Malvido in Chapter 11, who also shows that seedling density of forest interior tree species has decreased in fragments of all sizes compared to continuous forest interior. These results suggest that delayed responses of fragmentation will appear in the phytodemography of plant communities, compounding the immediate effects on these same communities, as shown in Chapters 9 and 13. The replacement of forest species by nonforest species argues for careful monitoring of not only species number, but also species identity, in a situation similar to that found by Hutchings (1991) and Brown and Hutchings (1997) with butterflies.

Comparisons of the samples at different life stages (seedling and adult) showed higher similarities for seedlings than for adult forms, indicating that the various species have different patterns of age-specific survivorship. Because seedlings are the life stage most strongly affected by fragmentation, the long-term viability of isolated populations remains in doubt.

Because palms play an important role in the ecology of the rainforest, as an important food source for a diverse group of animals, and are also important in the local economy in areas under settlement, regional conservation and management plans should not neglect this valuable part of the local flora. Scariot argues for a network of many large reserves, with special attention at the regional level to the geographical distribution of palms across the Amazon basin, while at the landscape level reserves should adequately sample local habitat heterogeneity. Most important, the design of reserves should focus on the preservation of ecological processes rather than simply the species themselves.

Seedling Ecology in Forest Fragments

Knowing that fragmentation has dramatic effects on the adult trees of the forest canopy, it is logical to expect that the growth or survivorship of seedlings will also be altered in isolated forest patches. Surprisingly little research had been carried out on seedlings in forest fragments until Julieta Benitez-Malvido began her studies in the BDFFP experimental reserves.

As she reports in Chapter 11, Benitez-Malvido studied cohorts of naturally occurring seedlings as well as transplants of greenhouse-germinated seedlings of three species in the Sapotaceae—the dominant family of canopy trees in the Amazonian forests around Manaus. Her objective was to determine whether fragment size and position within a fragment would affect seedling abundance, levels of herbivory, growth, and survivorship. Understanding these relationships would be as a first step in determining how fragmentation will affect recruitment of canopy trees in a frag-mented landscape.

Benitez-Malvido found lower densities of naturally occurring seedlings in fragments than in continuous forest, but surprisingly no effect of fragment size on seedling den-sity. She found higher levels of damage by herbivores in continuous forest and larger fragments than in smaller fragments, demon-strating that fragmentation is affecting plant-animal interactions. This result is somewhat surprising given Malcolm's (1991) result that insect biomass tends to shift down from the canopy in fragments until one remembers that insects, the most diverse group of or-ganisms in the forest, will certainly not all respond to fragmentation in the same man-ner. Benitez-Malvido also found that the three tree species studied had species-specific responses to fragmentation effects, probably indicative of adaptations to slightly different micro-environments in the undisturbed forest.

In concert, her findings show that ecolog-ical processes associated with seedling es-tablishment have been altered in the exper-imental forest fragments and point the way to future studies in this area.

Forest Fragmentation and Bee Pollination

The remarkable diversity of tree species in rainforests makes wind-borne pollination inefficient. Because the nearest potential mate for a tree may be hundreds of meters away, there has been strong selective pres-sure for plants to rely on animals to transfer pollen from one individual plant to another. In the Amazon, birds and bats are important pollinators, but bees are the predominant pollen carrier, or vector.

Besides the "usual" effects of habitat frag-mentation—reductions in population size, changes in microclimate, etc.—disturbance

of the forest has brought another player onto the pollination scene. Africanized honey bees, introduced in Brazil in 1956, have pro-liferated across the Amazon in the wake of colonization and agricultural clearings and are commonly found in the types of land-scape mosaics typified by the BDFFP study area.

In Chapter 12, Chris Dick discusses the complex interactions between plants, polli-nators, and fecundity for a large timber tree species in the forests and clearings of the BDFFP study area. He begins with a review of the history of Africanized honey bees in Latin America and describes the role they might play in competing with native species. He makes the interesting point that most plant species rely on a fairly broad suite of pollinators. We may infer that what has become fairly standard dogma when dis-cussing rainforest conservation—that co-evolutionary interdependencies between plants and pollinators make these systems particularly vulnerable to habitat fragmen-tation—may be overstated.

His work is focused on one species of tree, *Dinizia excelsa*, one of the largest in the Amazon and one that, in part because of its timber value, is often left standing in the middle of otherwise clear-cut forests. Dick climbed trees in open pasture, isolated re-serves, and continuous forest to record the species that were visiting their flowers and later followed up with estimates of fecun-dity of the study trees. The Africanized honey bees were found in the pasture and forest fragment trees but not in continuous forest, while native species were found in continuous forest and fragments but not in the pasture. The fecundity of the trees in pasture and fragmented forests was higher than in continuous forest, which, Dick ar-gues, is most likely the result of the efficient pollination by Africanized honey bees.

Because Africanized honey bees forage

over kilometers, remnant trees may serve as genetic links to apparently isolated populations and serve as catalysts for forest regrowth by attracting pollinators and seed dispersers. The presence of the Africanized honey bees is not likely to be entirely positive, as a number of native plants with highly specialized pollination requirements may suffer reduced pollination rates if they are visited frequently by the more generalist Africanized honey bees.

A Synthesis of Habitat Fragmentation and Plant Communities

In Chapter 13, William Laurance provides an overview and synthesis of the effects of fragmentation on plant communities and discusses the implications that these effects have for landscape management. Laurance, who moved to Manaus in 1996 to coordinate the botanical ecology work at the BDFFP, brings to his analyses a pantropical perspective derived from his work with Australian rainforest fragments.

Laurance reviews the basic structure of the Phytodemography Project and reviews the evidence suggesting that trees in forest fragments suffer increased mortality and damage not just in the immediate aftermath of the isolation event, but perhaps indefinitely. Such changes are likely to exacerbate the microclimatic changes that may be one of the initial causes for the increased mortality and can lead to a shift in the very structure of the forest with an increase in disturbance-adapted plants, such as vines and lianas.

Because tropical forest tree species are mostly rare and often involved in tightly co-evolved, mutualistic relationships, forest fragmentation is expected to lead through a number of pathways to a loss in species richness of the local flora. Accompanying this erosion of biodiversity is a recently documented and unanticipated collapse of biomass within 100 m of a forest fragment's edge—some plots have lost as much as 36 percent of their total biomass. Most of the biomass that is lost from the fragments will wind up as greenhouse gases in the atmosphere. Laurance and his colleagues have estimated that this biomass collapse associated with worldwide habitat fragmentation may be the equivalent of burning more than 1 million hectares of forest per year.

Laurance concludes with a discussion of landscape management guidelines relating to the minimum size of forest reserves and the minimum width of "faunal corridors," or belts of forest left connecting otherwise isolated fragments to each other or to adjacent expanses of continuous forest. He proposes that networks of both large and small reserves are needed across the landscape, with the smallest of the "small" being on the order of a few thousand hectares. Corridors, he argues, should be uninterrupted by any breaks, made up of primary forest, and no narrower than 300 m.

CHAPTER 9

Deforestation Effects at the Edge of an Amazonian Forest Fragment

Tree Mortality, Damage, and Recruitment

JUDY M. RANKIN–DE MÉRONA AND ROGER W. HUTCHINGS H.

Since the mid-1960s and the launching of the Transamazon Highway and its associated government colonization plans, the rainforest landscape of the Amazonian terra firme in Brazil has undergone a radical transformation caused by human intervention. These changes have important implications for the overall structure and ecological processes of the forest communities, as well as for the species that compose them (Chapter 3).

The immediate pattern and rate of deforestation on a site are linked to the intended land-use pattern, topography, and the deforestation techniques employed, as well as more generally to population pressure, government incentives, and other social and economic factors (Fearnside 1989a; Kahn and McDonald 1997; Chapter 23). In areas of planned or spontaneous agricultural colonization, as well as larger-scale agricultural projects, a mosaic of different-sized forest fragments is created, intermingled with cattle pasture or annual crops. In general, forest fragments are fewer and more isolated from continuous forest and each other the larger the scale of the enterprise, the more uniform the topography, and the longer the time since initial deforestation.

Despite their arbitrary establishment and varying size, the importance of remnant forest patches grows with the overall area of the Amazon forest affected by deforestation and fragmentation. Until recently, little quantitative data existed on the behavior of animal species of the tropical rainforest subject to such forest fragmentation now so commonly encountered in Rondônia and Pará, and none existed at all on the demography of rainforest trees.

Forest fragments offer the opportunity to study the ecological organization and maintenance of the tropical rainforest community, both as an intact entity and under conditions of direct and indirect disturbance. Studies of forest fragments in other tropical regions have concluded that the number of species in a community decreases with time after isolation (Terborgh 1974; Laurance and Bierregaard 1997a). Results from studies of rainforests of the central Amazon show changes ranging from loss of woody plant biomass (Laurance 1997; Laurance et al. 1997) to modifications in forest-edge microclimate and plant-soil water relations (Kapos 1989).

Eighty kilometers north of Manaus, Brazil, in the isolated forest fragments of the Biological Dynamics of Forest Fragments Project, the effects of forest fragmentation on

This is publication number 276 in the BDFFP Technical Series.

trees can be seen in the form of increased mortality rates and in the different sources of mortality in the years that follow isolation in small fragments and at the edges of larger ones (Lovejoy, Rankin, et al. 1984; Laurance, Ferreira et al. 1998a). Given the low population densities—less than one tree per hectare for many tree species (Rankin–de Mérona and Ackerly 1987; Rankin–de Mérona et al. 1992)—increased mortality may soon lead to local species loss in isolated patches. Whether the same results apply and the same mechanisms are at work in replicate samples within the same community, within broad taxonomic groups, and over a range of different forest fragment sizes is under investigation in research projects of the BDFFP (Chapter 4).

The objective of our study was to investigate the effects of deforestation on the population structure and dynamics of rainforest trees in isolated forest patches formed by large-scale deforestation. This investigation is being conducted as part of the Botanical Ecology Subproject of the BDFFP, under the auspices of the bilateral international agreement between Brazil's National Institute for Research in Amazonia (INPA) and the Smithsonian Institution in Washington, D.C. (see also Chapter 13).

Study Location

This study was conducted in the terra firme rainforest of the Agricultural Research and Development District (Distrito Agropecuário da SUFRAMA) outside Manaus (fig. 9.1) in the experimental and control reserves of the BDFFP. A detailed description of the experimental design of the BDFFP can be found in Lovejoy and Bierregaard (1990) and in Chapter 4.

The data were collected in 1 ha inventory plots of an isolated 100 ha forest patch (Reserve 3304) of the Porto Alegre cluster, one

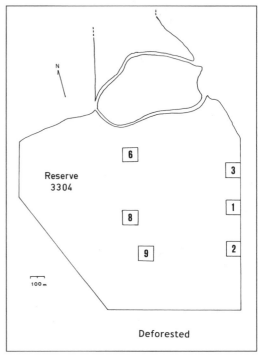

Fig. 9.1. Map of Reserve 3304 on the Porto Alegre ranch, Distrito Agropecuário da SUFRAMA, showing the inventory hectares analyzed for the current study.

of the many replicate forest fragments under observation (see fig. 4.1; table 4.3; and plate 2). This rainforest is characterized by a large number of trees at least 10 cm DBH per hectare (647 ± 50.5 trees/ha for a 15 ha sample), with most trees less than 60 cm DBH (Rankin–de Mérona et al. 1992). Exceptional individuals with diameters up to 160 cm DBH do occur but at very low densities. Species diversity is high, with a minimum estimate of more than 800 tree species in 55 families occurring in the project study area —the final count may surpass 1,200 as the problematic (sterile botanical material) trees in the inventory are identified (see also Chapters 6 and 13). The dominant families are Burseraceae, Sapotaceae, Lecythidaceae and Leguminosae, in that order (Rankin–de Mérona et al. 1992). The number of species per hectare is estimated at 350.

Methods

Reserve 3304 was demarcated during 1983 in a tract of continuous rainforest slated for clear-cutting and conversion to *Brachyaria* sp. pasture by the property owners. An initial forest inventory was conducted during and immediately after patch isolation in mid-1983 through 1984 to record as much as possible the original forest composition and structure. Ranch crews who felled the trees were accompanied by project personnel, permitting the formation of clean, straight patch edges; felled trees were thrown away from the stand inside the patch boundaries. The very low number of vines in the canopy and the care with which felling proceeded resulted in remarkably little damage to the remaining trees. After isolation, the patch was surrounded by early secondary vegetation and, to a smaller extent, by *Brachyaria* sp. cattle pasture, separating it from the nearest continuous forest by a band of about 500 m or more except for a narrow (75–150 m) corridor of forest connecting the reserve to continuous forest about 2 km to the north (plate 2).

The initial forest inventory was conducted in nine noncontiguous 1 ha blocks and a single 1 ha transect. The data reported here are for the three hectares on the eastern edge of the patch (hectare numbers 1, 2, and 3) and three hectares in the patch core (numbers 6, 8, and 9), approximately 500 m from the eastern edge (fig. 9.1). (The eastern edge is usually the windward edge in the region.) The 1 ha plots were divided into contiguous 20 m by 20 m blocks, which were processed one at a time, with all data and trees collected before proceeding to the next block. Block corners were marked with 1 m PVC pipe labeled with the block coordinates. All trees of at least 10 cm DBH in the designated plots were measured for DBH and bole length (trunk to first major branch). Each

tree received a uniquely numbered aluminum tag that was affixed to the trunk with a noncorrosive nail, and a botanical voucher sample, sterile when no fertile material was available, was collected and registered in the field under the same tree number. Representative voucher specimens are preserved in the INPA herbarium and the laboratories of the BDFFP.

In 1986 and again through 1988 we returned to the surveyed hectares to record diameter growth, condition, and mortality of trees already surveyed plus ingrowth of trees previously below the 10 cm DBH limit. All original survey trees were revisited for DBH remeasurement, but botanical samples were taken only from new or flowering trees. In the field we made a detailed description of the "Status" (well, damaged, or dead), "Condition" (trunk or crown broken, uprooted, standing dead, partial trunk, or crown death), and "Circumstances" (caused by wind, another tree, fire, animals; location of damage; and other contributing factors or indicative symptoms) of damaged, dying, or dead trees. We remapped areas where treefalls or new trees had appeared.

Results

OVERALL EDGE-CORE COMPARISONS

At the time of the initial forest inventory there was no significant difference in the number of trees per hectare in the three core and three edge hectares (edge mean = 677.7 ± 56.5, core mean = 648 ± 34.0, t = 0.779, P = 0.234, df = 4). Comparisons of the mean number of dead and damaged trees in the core and edge hectares and the number of trees recruited into the 10 cm DBH class over the observation interval show significant differences for all three groups with the means being consistently greater for the edge hectares (dead: t = 2.738, P = 0.026;

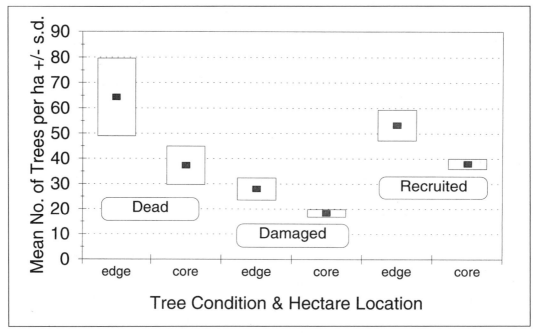

Fig. 9.2. Mean number of trees dead, damaged, and recruited in three edge and three core hectares of Reserve 3304.

damaged: t = 4.182, P = 0.007; new trees: t = 3.625, P = 0.011; all with df = 4; fig. 9.2). Recruitment or ingrowth into the smallest diameter class (10−15 cm) over the six hectares was correlated with the number of dead and damaged trees but not with the original number of trees in the inventory, suggesting that tree damage and death created growth opportunities for trees already present in the understory with diameters just under 10 cm. Growth differences for larger trees have yet to be analyzed.

Hectare 2 on the eastern edge of the patch, and hectare 8 in the core, 550 m from the eastern edge, were selected for more detailed analysis of the types and sources of damage suffered by the inventory trees. An examination of the stand's diameter distributions from the original 1983−84 inventory showed them not to differ significantly (Kolmogorov-Smirnov two-group test, D_{max} observed = 0.035, P = 0.05).

EDGE HECTARE 2

In edge hectare 2 there was an overall decline in the number of trees in the plot over the observation period, with deaths being three times greater than recruitment (76 versus 25; table 9.1). In all, 76 trees, or 10.2 percent of the original stand, died. However, there was no significant difference in the frequency distribution of live trees by DBH class in 1984 and late 1988 (fig. 9.3; Kolmogorov-Smirnov two-group test D_{max} = 0.019, P = 0.05). Of all dead trees, 15 (21 percent) were found intact, with no outward signs of damage (table 9.1). The greatest damage was breakage caused by other trees, whereas deaths by breakage from wind damage were slightly greater than by tree damage. Dead and damaged trees were disproportionately concentrated in the 10−15 cm diameter class and to a lesser degree in the 20, 25, and 30 cm classes (fig. 9.4). These distributions are significantly different (Kolmogorov-Smirnov two-group test D_{max} = 0.245, P =

TABLE 9.1.   Inventory Population Summary in Representative Core and Edge Sample Hectares, 1984–88

	Number of Individuals	
Status	Reserve 3304, Hectare 2 (Reserve edge)	Reserve 3304, Hectare 8 (Reserve core)
Original inventory population in 1984	742	618
Still alive in 1988	691	601
New trees	25	16
Damaged	59	44
Dead	76	31
Lost	2	1
Status uncertain	1	

0.05); however, a global comparison of the mean number of dead versus damaged trees by quadrat for the hectare showed no statistically significant difference (mean damaged = 2.80 ± 2.5, mean dead = 2.84 ± 2.3, t = −0.822, P = 0.208).

The twenty-five sample quadrats of the hectare were grouped into five subsamples forming bands parallel to and at different distances from the forest edge. The occurrence of dead and damaged trees was then examined as a function of distance from the edge. For all except the standing dead trees the numbers were greater near the edge than away, with a slight tendency for the number of dead trees to peak between 20 and 40 m, while damaged but living trees peaked at 0 to 20 m (fig. 9.5). One-way analyses of variance (ANOVA) for damaged and dead trees other than standing dead showed significant differences for the quadrat subgroups at the different distances from the edge (damaged trees: F = 4.024, P = 0.015; other dead trees: F = 6.507, P = 0.002; df = 4, 20, 24). Standing dead trees were few and of similar abundance regardless of proximity to the edge, with no significant difference found between the quadrat subsample bands (F = 1.250, P = 0.322, df = 4, 20, 24).

The direct and indirect role of wind can be seen in the pattern of dead and damaged trees as a function of distance from the windward forest edge. The greatest amount of wind damage and number of deaths oc-

curred along the open edge in the 0 to 20 m band of quadrats, and the greatest amount of damage and death caused by other trees occurred in the 20 to 40 m zone; in a domino-like pattern, the windthrown trees of the edge caused damage in the forest behind as they fell (fig. 9.6; see also fig. 13.5b). These trends were seen when the data for damaged or dead trees were examined separately. At the present time the number of dead and damaged trees has dropped to levels resembling the core or native forest at 60 m from the forest edge. This agrees with data on the distance of penetration of edge effects found for this reserve based on a study of forest and edge microclimate and plant-water relations (Kapos 1989; Laurance, Bierregaard, et al. 1997; fig. 29.1).

CORE HECTARE 8

In the reserve core, trends were observed that more closely, although not completely, resembled sample hectares in continuous native forest in the same region. In general, fewer tree deaths occurred than at the edge, representing only 5 percent of the original stand, but the number of damaged trees continued to be large (table 9.2). Damage by wind and other trees was approximately equal (table 9.3). In contrast to the edge hectare, where several uprooted trees survived damage when they lodged on those behind them rather than toppling com-

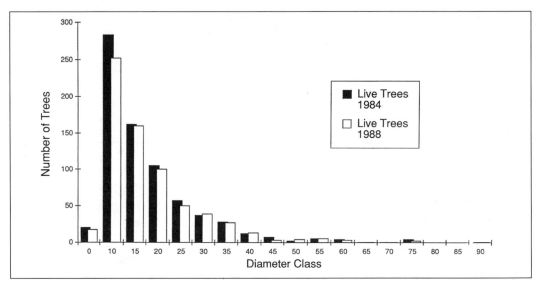

Fig. 9.3. Live trees present in edge hectare 2 in 1984 and 1988 by diameter class.

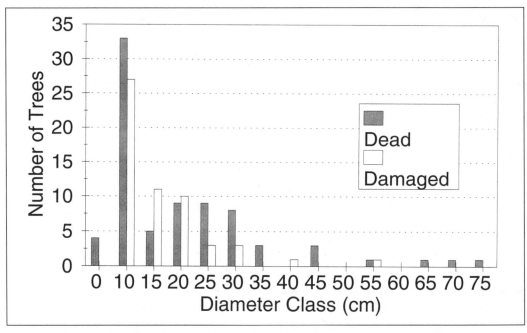

Fig. 9.4. Diameter class distribution of damaged and dead trees in edge hectare 2 in 1988.

pletely, none of the four uprooted trees in this hectare survived. Deaths again outnumbered recruitment but by only twice as much. The number of standing dead trees (12) was similar to that for the edge hectare (14). The hectare thus suffered a small de-cline in the number of live trees during the study period, but the diameter-class structure was not significantly different from that of the original inventory (fig. 9.7; Kolmogorov-Smirnov two-group test $D_{max} =$ 0.009, P = 0.05).

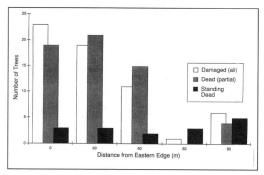

Fig. 9.5. Damaged, standing dead, and other dead trees in edge hectare 2 as a function of distance from the forest edge based on 20 m wide bands parallel to the edge and composed of 5 sample quadrats each.

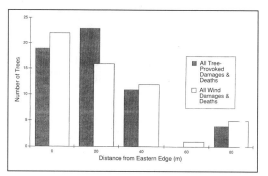

Fig. 9.6. Wind and tree-provoked damage and tree deaths in edge hectare 2 as a function of distance from the forest edge.

A comparison of the mean number of dead versus damaged trees by quadrat showed no significant difference within the hectare (mean damaged = 1.56 ± 1.6, mean of all dead trees = 1.24 ± 1.1, t = 0.832, P = 0.205, df = 48). These means are approximately half as large as those for edge hectare 2 but are significantly different only for the comparison between all dead trees for the two hectares (t = 3.139, P = 0.001, df = 48) and not for damaged trees. The diameter class distributions of dead and damaged trees within core hectare 8 were also not significantly different (fig. 9.8; Kolmogorov-Smirnov two-group test D_{max} = 0.155, P = 0.05).

An analysis of the twenty-five quadrats of core hectare 8 in five 20 m wide transects (similar to that done for edge hectare 2) was performed to detect any differences in damage and deaths as a function of distance from the patch edge. The same categories (damaged, standing dead, and all other dead trees) were examined. ANOVAs for each category showed no significant differences between the distances for any of the categories (damaged trees: F = 1.322, P = 0.296; stand-

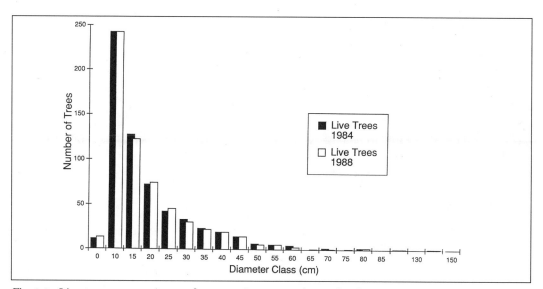

Fig. 9.7. Live trees present in core hectare 8 in 1984 and 1988 by diameter class.

TABLE 9.2.   Sources of Damage and Mortality: Reserve 3304, Hectare 2 (Edge)

	Trunk broken	Uprooted	Standing dead	Unknown
Trees damaged				
By another tree	28	1		
By wind	23	3		
Unknown cause	3	1		
SUBTOTALS	54	5		
Trees killed				
By another tree	19	9		
By wind	26	4		
By termites?			1	1
No damage			15	
Unknown cause	1	1		
SUBTOTALS	46	14	16	1

TABLE 9.3.   Sources of Damage and Mortality: Reserve 3304, Hectare 8 (Core)

	Trunk broken	Uprooted	Crown damaged	Trunk partially dead	Other	Standing dead
Trees damaged						
By another tree	17	0	1	1	0	
By wind	17	0	0	0	2	
Unknown cause	6	0	0	0	0	
SUBTOTALS	40	0	1	1	2	
Trees killed						
By another tree	8	2				
By wind	6	2				
By termites?						
No damage					1	
Unknown cause						
SUBTOTALS	14	4	0	0	1	12

ing dead trees: F = 0.565, P = 0.691; other dead trees: F = 0.575, P = 0.684, df = 4, 20, 24), despite a substantially large number of damaged trees observed in the 550 to 570 m quadrat band (fig. 9.9). Examination of the individual data shows this to be largely the result of a single windthrown tree that directly felled eight other trees, four of which were in the same quadrat. No apparent pattern of direct or indirect wind action could be observed for this hectare (fig. 9.10).

When the diameter-class distributions of the live trees in edge hectare 2 and core hectare 8 of Reserve 3304 were compared at the end of the observation period they con-tinued to be statistically indistinguishable (Kolmogorov-Smirnov two-group test, D_{max} = 0.054, P = 0.05) despite significant differences found in the mean number of dead trees. This suggests that growth by surviving trees and recruits has at least in part compensated for this loss in the individual diameters classes.

COMPARISONS WITH UNDISTURBED
CONTINUOUS FOREST AND OTHER
FOREST FRAGMENTS

Five-year results for 5 ha of continuous native forest in the project study area show low

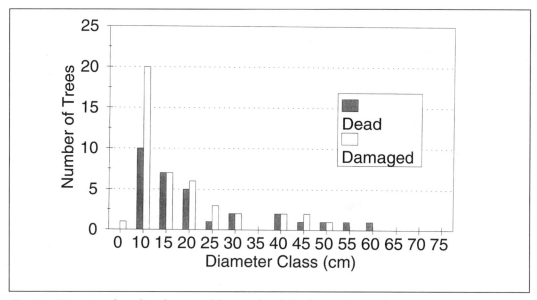

Fig. 9.8. Diameter class distribution of damaged and dead trees in core hectare 8 in 1988.

rates of recruitment by trees less than 10 cm DBH, very low growth rates for trees of at least 10 cm DBH (only 12 percent of the surveyed trees grew enough to advance from one size class to another based on a 5 cm interval width), and low mortality and slow forest turnover rate in comparison to other Neotropical rainforests (Rankin–de Mérona et al. 1990, and unpublished data). The present native forest appears extremely stable demographically; no significant differences were found in overall population structure over five years (Kolmogorov-Smirnov two-group test, D_{max} = 0.010, P = 0.05), with the number of deaths being almost exactly compensated for in all size classes by growth plus recruitment. This is in great contrast to edge hectare 2 of Reserve 3304, where deaths were three times as great as recruitment, and core hectare 8, which, while showing fewer deaths than the edge hectare 2, still has twice as many deaths as recruits. When compared on a per hectare, per year basis, edge hectare 2 showed almost twice as many deaths as native forest, while recruitment was about the same. In contrast, core

hectare 8 showed fewer recruits than any of the native forest hectares and deaths in the middle range for native forest.

Previous observations in other isolated forest fragments of the BDFFP have shown a large number of standing dead trees in the years following isolation (Lovejoy, Rankin, et al. 1984; Rankin–de Mérona, unpublished data). In an isolated 10 ha forest patch (Reserve 1202), seventy-four standing dead trees (3.7 trees/ha/year) were encountered within two years of patch isolation. When partitioned into nine contiguous hectares, the greatest number of standing dead trees were found to occur in corner hectares, which have two exposed edges, on the side of the reserve that had been the windward side on the day the surrounding felled forest was burned (see Malcolm 1994). The core and other edge hectares had lower or similar numbers.

The number of standing dead trees in edge and core hectares in Reserve 3304 were examined in this light. Values ranged from 1.4 to 2.8 trees/ha/year for the three core hectares and 1.4 to 3.1 trees/ha/year for the

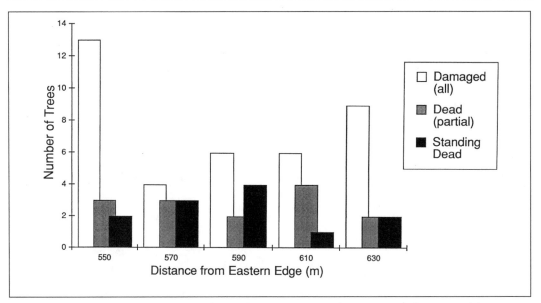

Fig. 9.9. Damaged, standing dead, and other dead trees in core hectare 8 as a function of distance from the forest edge based on 20 m wide bands parallel to the edge and composed of 5 sample quadrats each.

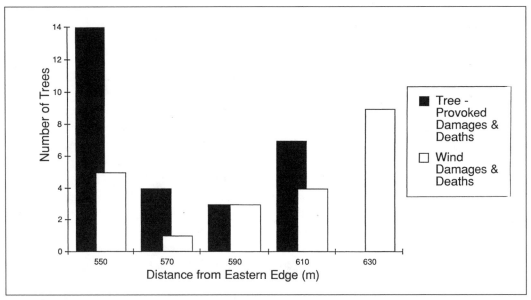

Fig. 9.10. Wind and tree-provoked damages and tree deaths in core hectare 8 as a function of distance from the forest edge.

three edge hectares. The mean numbers of standing dead trees calculated on a per quadrat basis for edge hectare 2 and core hectare 8 were found to be not significantly different (mean for ha 2 = 0.600 ± 0.764, mean for ha 8 = 0.44 ± 0.651, t = 0.797, P = 0.2146, df = 48). The numbers of standing dead trees in Reserve 3304 are in the upper end and overlapping the range for native forest on a per hectare, per year basis (2.8 to 3.1

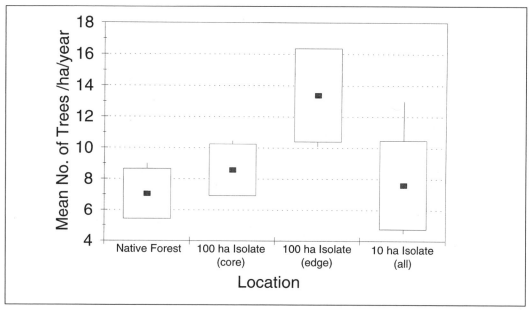

Fig. 9.11. Standing dead trees per hectare per year in continuous undisturbed forest (1), the core hectares of Reserve 3304 (2), the edge hectares of Reserve 3304 (3), and isolated 10 hectare Reserve 1202 (4).

trees/ha/year versus 0.3 to 3.0 trees/ha/year for undisturbed; fig. 9.11).

Discussion

Edge hectare 2 of the 100 ha reserve showed heavy and well-defined effects of wind on the forest edge. These data are similar to those from other forest patches in the study area where great changes due to wind damage have also taken place. The most dramatic case seen to date on the project was at the Dimona cluster in late October 1985, when the western edge of Reserve 2303 (100 ha) was heavily damaged by windthrown trees and almost the entire 1 ha patch Reserve 2108 was similarly affected. As a result, Reserve 2108 can no longer be considered a closed forest stand. Significant wind damage has also been sustained on the eastern (normally the windward) edges of most of the other isolated reserves, as has been graphically illustrated for Reserve 1202

where fallen trees on the forest edge occurred almost exclusively on the eastern edge during the first two years after isolation (Lovejoy, Rankin, et al. 1984; see also fig. 13.5b).

At the present, damage and deaths caused directly or indirectly by wind extend up to 60 m from the forest edge. Whether the domino effect of trees toppling back on those behind continues will depend on how fast the margin can reform a semblance of closed vegetation and whether conditions outside the fragment remain the same. In the case of this and other BDFFP forest fragments, the significant losses in aboveground biomass caused by tree mortality have not been offset by either tree growth and recruitment or the increase by other plants, such as lianas, in the decade following isolation (W. Laurance, Laurance, et al. 1997). Reconstitution of vegetation similar in stature to the original forest stand and capable of buffering the fragment at the fragment edge thus seems improbable.

Even if stand degradation stops at the current point, Reserve 3304 has been significantly affected. Considering only the more overt damage at the margin and assuming that it is similar over the whole periphery, then in less than five years, 22.5 percent of the patch has been significantly affected by isolation and the reserve reduced to an effective size of 77.5 ha. Data are still being analyzed on the condition of live trees that may indicate low survival potential for otherwise undamaged trees on the edge.

We suspect that tree deaths not caused by apparent mechanical damage may be due to stress from changes in forest microclimate and soil hydric conditions. A significant decrease in soil water potential has been detected at the margins of project reserves, although understory plants did not show signs of water stress, and it has been suggested that plants may adapt to conditions at the edge by closing stomata (Kapos 1989). Data from Reserve 3304 on the penetration of microclimate changes due to edge effects extend an estimated 60 m in from the edge, the point at which the present study shows a return to core levels for damaged and dead trees, and may be due in large part to the degree of opening of the overhead canopy rather than the vertical profile of the edge itself (Camargo and Kapos 1995). Results from the present study suggest that edge conditions contribute to the number of standing dead trees.

Several factors complicate the interpretation of these data. In particular, the number of standing dead trees may be underestimated because when large-scale wind damage takes place, standing dead trees are easily brought down and thus will be counted as felled dead trees rather than standing dead. Recensuses have been made every two years in most reserves, and every year in a selected few, thereby decreasing the chance that standing dead trees will be missed.

Nonetheless, damage on some edges is so severe that an underestimate is inevitable.

Another complicating factor is the timing of intense storms, which cause damage such as was seen at Dimona in 1985. If such a storm occurred in the first year after isolation, among the many trees felled by the wind might be some affected by physiological stress (and thus indeed be weaker than healthy trees) before the threshold for death to a standing individual was reached.

Finally, it is now becoming evident that the specific conditions at the time of isolation and those that follow may play an important part in the behavior of trees as well as other organisms, such as frogs, birds, and monkeys, which react to changes much more quickly than do trees (see Chapters 19, 20, and 21). The high numbers of standing dead trees in the corner hectares of 1202 (Lovejoy, Rankin, et al. 1984) are most likely related to stress at the time of the burn to clear the felled trees around the reserve. Flames scorched the leaves of some trees, and leaf litter inside the forest margin burned up to 25 m inside the edge (see fig. 13.5a). The fire was hot and brief enough to melt PVC stakes marking the quadrats within 2 cm of ground level while the length of the stake that slumped over to the ground was untouched. In contrast, the forest around Reserve 3304 was cleared but never burned, thus eliminating the fire-related stress placed on trees at the edge (see table 4.3 and plate 8).

Observations on the many project reserves over the years since the first isolation have shown that events may take place that, even when occurring on a regional scale, might differentially affect forest fragments. The variability observed in data for both isolated and undisturbed forest shows that our ability to answer the question of whether forest fragments of the same size and age will behave in a similar fashion over time or whether they will diverge in some or all of their characteristics will depend on sampling from a large number of replicates.

Conclusions

Inventory hectares at the margin of Reserve 3304 differed from those in the patch core in most of their demographic aspects within five years of patch isolation, with the edge stands showing a significant net decline in live trees. These data in turn differ from those for native forest where tree populations are extremely stable—recruitment equals mortality. A large portion (22.5 percent) of the 100 ha reserve has already been visibly affected by its isolation from continuous forest. A domino pattern of windthrown and broken trees falling back from the forest edge and subsequently damaging others suggests that this phenomenon could continue to propagate damage farther into the forest interior if edge conditions remain the same.

Comparisons of data from core hectares and native forest suggest that the effects of isolation may have already reached as far as the apparently intact core area, 500 m from the edge. While the mechanisms of these effects have not yet been elucidated at our site, Laurance (1997) has shown how wind turbulence generated by the edge may affect forest structure hundreds of meters from the edge. These results suggest that substantial buffer zones will have to be included in any management plans whose objectives include the maintenance of pre-isolation tree population structure in rainforest fragments. Effective buffers will be on the order of one or more kilometers, in function of fragment geometry, rather than the tens or hundreds of meters currently employed, for example, in the protection of creek headwaters and stream courses in the Brazilian Amazon.

Conservation Lessons

1. The forest area immediately affected by clear-cutting is larger than the cleared area itself because of immediate direct and indirect effects on the trees in the adjacent but untouched native forest stands. These effects, the absolute extent of which will depend on total deforestation surface and geometry, are rapidly visible up to tens of meters from the actual cutting limit and result in significant local tree demographic and habitat changes.

2. The mechanisms responsible for the observed changes in forest structure and dynamics after fragmentation are such that, in the case of transformation to permanently maintained nonforest cover, it is reasonable to expect further propagation of these effects into the remnant native forest stand, thus further enlarging the total affected area.

3. Although the results presented here are based on observations too recent to confirm long-term trends, subsequent analyses of other forest fragments under study at the BDFFP site (Laurance, Ferreira, et al. 1998a) also suggest that indirect effects of deforestation may act at greater distances than those immediately observed here. Thus further research is needed to determine to what extent the survival, the effective tree cover, and the natural species composition can be permanently maintained in the face of medium or large-scale deforestation. Until such results are available, it is imperative that measures designed to protect rainforest stands in areas experiencing extensive deforestation include buffer zones on the order of one or more kilometers rather than tens or hundreds of meters, as are currently in effect.

4. Regional development plans should be conceived in such a way as to avoid those spatial occupation patterns creating multiple narrow, and thus fragile, forest corridors. Not only is the true extent of the indirect effects of rainforest deforestation still unknown, but even once established, the enforcement of any protection measures to maintain the minimum necessary tract dimensions becomes more difficult in areas with complex landscape geometries. A classic example is the fishbone disposition of the secondary access roads in the government colonization areas in Rondônia (fig. 3.2). Here, deforestation works inward from the many parallel access road margins, creating long, narrow forest corridors. Under such conditions, violations of protected areas having important consequences for large but linear tracts may easily be committed before any detection is possible. Under such circumstances, only a single violation is necessary to create an isolated and almost certainly nonviable rainforest fragment.

Acknowledgments

The authors wish to thank all INPA and project staff, field crew, and interns who contributed to the collection of the data set and the production of this chapter. Special thanks are due Manuel Jonas Pereira, subproject field supervisor. Financial support was received from the World Wildlife Fund–US, the Weyerhauser Family Foundation, the Sequoia Foundation, and the Smithsonian Institution.

CHAPTER 10

Effects of Landscape Fragmentation on Palm Communities

ALDICIR SCARIOT

The high levels of rarity of vascular plants and high number of species per unit area are major features of tropical rainforests (Gentry 1992). About half of the described plant and animal species occur in these forests, which cover less than 7 percent of the world's land surface (Wilson 1988). Despite its unique characteristics, more than 1 percent of the tropical forest biome is being deforested per year and more than 1 percent is being degraded (Myers 1988a).

The recent and rapid reduction of tropical rainforest cover since the turn of the century implies an alarming decrease of biodiversity. Simberloff (1986) estimated that at current rates of deforestation, the 92,000 plant species residing in Neotropical forests would be reduced by 13,594 at the end of the twentieth century. However, with potentially higher deforestation rates, he estimated that up to 60,000 species would disappear. Alternatively, Myers (1988b) estimated that 50 percent of the 34,000 endemic species in ten "hot spots" (areas with an exceptional concentration of species and endemism, and rainforest sites suffering rapid rates of degradation) would become extinct within twenty years. These represent unprecedented extinction rates (Myers 1981).

The primary cause of the decline of rainforest species diversity is habitat loss (Ehrlich 1988). Habitat destruction results in habitat fragmentation, which increases the loss of original habitat, reduces the size of habitat patches, and increases the isolation of habitat patches (Andrén 1994). Forest fragments may differ with respect to their shape, size, microclimate, light regime, soil, degree of isolation, and type of ownership (Saunders, Hobbs, and Margules 1991). Consequently, forest fragmentation may influence local and regional patterns of biological diversity because of the loss of unique microhabitats, habitat insularization and associated changes in dispersal and migration patterns, and small- and large-scale soil erosion (Soulé and Kohm 1989). Moreover, edge effects, which can alter the distribution, behavior, and survival of plant and animal species, will be magnified in areas of high forest fragmentation (Kapos 1989; Murcia 1995; Ferreira and Laurance 1997).

The planning and management of nature reserves must consider the effects of forest fragmentation on the persistence of species. If the area of a nature reserve is below the minimum required to maintain a species' population, the species will be at risk of extinction in the reserve. Two major points

This is publication number 277 of the BDFFP Technical Series.

must be considered when evaluating the ability of native species to survive in a particular reserve. First, each species has a minimum viable population below which it becomes difficult to find mates, to produce genetically variable offspring (for outcrossing species), to survive random fluctuations in size, and to produce new colonizing populations over the long term. Isolated populations below this minimum size are unlikely to persist (Boecklen and Simberloff 1986). Second, small isolated populations tend to have lower levels of heterozygosity than do large, wide-ranging populations (Shaffer 1981; Soulé 1983; Ellstrand and Elam 1993). Populations that have experienced genetic bottlenecks are likely to suffer a reduction in heterozygosity, which may in turn lower the capacity of the population to adapt to changing conditions (Barrett and Kohn 1991). Consequently, estimates of the minimum area necessary for the long-term persistence of a species must be large enough to minimize the probability that severe genetic bottlenecks will occur. This minimum area will depend on the population density of the target species and differ greatly among taxa. For example, Ackerly, Rankin–de Mérona, and Rodrigues (1990) estimated that a minimum of 200 ha of Amazonian rainforest would be necessary to maintain the most common plants in the Myristicaceae, suggesting that a much larger area would be required to maintain rarer species of this family.

Despite the debate concerning the optimum size and spatial arrangements of nature reserves (Quinn and Hastings 1987, 1988; Gilpin 1988), it is clear that the effects of forest fragmentation will be magnified in relatively small reserves, which are likely to contain far fewer forest species than larger ones. This observation would seem to favor the design of large reserves.

Detailed community and demographic studies that document the effects of forest fragmentation on species abundance, diversity, and survivorship are scarce in the Amazonian rainforest (Laurance, Bierregaard, et al. 1997; Offerman et al. 1995; Scariot 1996, 1999). Despite the scarce data available, it is important to understand the dynamics of small versus large reserves and continuous forest to provide guidance for the design and management of nature reserves and for conservation policy.

In this chapter I describe aspects of diversity and floristic composition of a central Amazonian palm community in a set of forest fragments and continuous forest. I also discuss the effects of short-term isolation of forest fragments on the palm community persistence and implications for its conservation.

The palm family (Palmae) was chosen for this study for four major reasons. First, they are a dominant vascular plant family in tropical forests. Second, palms are important components of the forest structure in Amazonia. Third, palms serve as key resources for animals during periods of food scarcity (Terborgh 1986). And finally they are used by humans. Hence, negative effects of forest fragmentation on palms will influence both the perceived and real value of forest fragments managed for conservation.

Methods

The reserves used for the current study are located on three cattle ranches (Dimona, Porto Alegre, and Esteio) located in terra firme forest 80 km north of Manaus, Brazil (see fig. 4.1; table 4.3). At Dimona and Porto Alegre, four reserves were sampled at each site: 1, 10, and 100 ha, and continuous adjacent forest. At Esteio, no 100 ha reserve exists, consequently I sampled in one 1 ha, one 10 ha, and one continuous forest reserve. The 1, 10, and 100 ha reserves were isolated from continuous forest, between 1980 and 1984, by a distance of 100–350 m. All forest fragments are approximately square in shape. (For a more detailed description of the reserves and surrounding forest, see Chapter 4.)

A field survey of the palm species was carried out from August 1993 to April 1994 in the eleven reserves described above. In each reserve, ten 20 m × 20 m plots were randomly positioned but restricted to topographically flat areas, resulting in a total of 110 plots totaling 4.4 hectares. By restricting plots only to topographically flat areas, I avoided sampling biases that would have resulted if I had included the specialist palm species that occupy the low-lying and steep areas of this region.

Every individual palm occurring inside each plot was classified according to its life history stage (seedling, juvenile, or adult). For arborescent palms (those with an aerial stem), seedlings were identified as those plants that did not present a well-defined aerial stem at the time of census; juveniles were identified as those that had a defined aerial stem but had no signs of upcoming or previous reproduction; and adults were identified as those individuals that either were reproducing at the time of the census or had previously reproduced. For nonarborescent palms (those with a subterranean stem), seedlings included those individuals that still had undivided leaves; juveniles were identified as individuals that had either leaves divided in pinnae or leaves that were in the process of dividing; and adults were identified as those individuals that either had reproduced or were reproducing for the first time during the census. These criteria, based on functional life stage, avoided any misclassification.

SPECIES IDENTIFICATION AND DATA ANALYSIS

In each 20 m × 20 m plot, I recorded and identified all palm individuals to the subspecies level when possible, following Henderson (1995). For the analyses described below, each taxon was considered as a species. Unidentified species were classified as morpho-species (individuals apparently distinct from other species in the sample but not reliably identified), and voucher specimens were prepared to allow subsequent comparisons with material deposited at the Brazilian National Institute for Research in Amazonia (INPA) and Brazil's National Center for Research in Genetic Resources and Biotechnology (CENARGEN) herbaria.

I estimated species richness (S), the Shannon index of diversity (H'), and the equitability index (J) in all sampled reserves for each life stage and for all three life stages pooled. All ten plots from each reserve were pooled. Species richness (S) is defined as the number of species present in a reserve or in a site. The Shannon index was calculated as described in Magurran (1988):

$$H' = -\Sigma\, p_i \ln(p_i)$$

where p_i is the proportion of the sample represented by the ith species. Equitability (J') was estimated by $J' = H'/\ln(S)$. Using a randomized incomplete blocks design analysis of variance (ANOVA), I tested for possible differences among reserve sizes using sites as blocks on the above indices. When ANOVA results showed differences among means, pairwise Tukey HSD tests were conducted to detect differences at the 0.05 significance level.

To describe changes in species composition across reserve sizes and across sites, I recorded presence-absence data for all species for each reserve size class and for each site. The similarity in species composition between two reserves or between two sites (beta diversity) was evaluated using the Sorensen's coefficient (S_S) for each life stage. All ten plots in each reserve were pooled and treated as a datum. The number of plots observed was constant across reserve size classes, so the differences among them in the floristic composition of the palm community should be realistically characterized by the Sorensen coefficient. The coefficient of similarity of Sorensen, S_S, varies from 0 (completely dissimilar) to 1.0 (completely similar) and is defined for two samples, A and B, in Krebs (1989) as:

$$S_S = \frac{2a}{2a + b + c}$$

where a is the number of species present in samples A and B, b is the number of species present in sample B but not in sample A, and c is the number of species present in sample A but not in sample B.

To determine whether reserve size affected the floristic composition of palms, Canonical Discriminant Analyses (CDA) were conducted using the SAS procedure CANDISC (SAS Institute 1989). For these analyses, reserve size was considered the classification variable and species abundance per plot the quantitative variable. A separate CDA was conducted for each life stage. This analysis seeks to exhibit differences among reserve sizes based on differences in species abundances.

Results

PALM COMMUNITY COMPOSITION

Eleven genera and thirty-six species were found in the study area (table 10.1). I did not find any species in the unsampled area of any reserve that were not recorded in the sampled plots. Of the eleven species that were not present in all reserve size classes, only three species (*Astrocaryum acaule, Bac-*

tris killipii, and *B. oligocarpa*) did not occur in 1 ha reserves (which had thirty-three palm species). Five species (*Attalea maripa, Bactris balanophora, Bactris* sp2, *Geonoma maxima* var. *chelidonura,* and *Geonoma* sp.) did not occur in 10 ha reserves (which had thirty-one palm species); seven species (*Astrocaryum acaule, Attalea maripa, Bactris maraja* var. *maraja, Bactris oligocarpa, Geonoma* sp., *G. stricta* var. *stricta,* and *Lepidocaryum tenue*) were not present in the 100 ha reserves (which had twenty-nine species); and five species (*Astrocaryum acaule, Bactris* sp2, *B. maraja* var. *maraja, Bactris oligocarpa,* and *Lepidocaryum tenue*) did not occur in the continuous forest (which had thirty-one species).

Although few species were restricted to a single site, none of the three sites included all the thirty-six species observed in this study (table 10.1). Of the thirty-four species occurring at Dimona, two were found only at this site: *Bactris balanophora* (five individuals) and *Lepidocaryum tenue* (in large colonies). Of the thirty-one species occurring in Porto Alegre, only one species (*Astrocaryum acaule* [one individual]) was found only at this site. None of the thirty-two species observed at Esteio was restricted to this site. Species absent from Esteio were *Astrocaryum acaule, Bactris balanophora, Bactris* sp2., and *Lepidocaryum tenue;* species absent from Dimona were *A. acaule* and *Geonoma* sp.; species absent from Porto Alegre were *Attalea maripa, Bactris balanophora, B. maraja* var. *maraja, Geonoma stricta* var. *stricta,* and *Lepidocaryum tenue.* Five species did not occur in continuous forest and were thus classified as nonclosed-forest species.

Six palm species had subterranean trunks, twenty-nine had aerial trunks, and only one (*Desmoncus phoenicocarpus*) was lianescent. There was no association between the number of species of each habit (i.e., aerial

TABLE 10.1. Palm Species Occurring in Topographically Flat Areas in the PDBFF Reserves and Sites in Central Amazonia

Species	Presence by reserve size				Presence by fazenda		
	1 ha	10 ha	100 ha	CF	Esteio	Dimona	Porto Alegre
Astrocaryum acaule		x					x
Astrocaryum gynacanthum	x	x	x	x	x	x	x
Astrocaryum sciophilum	x	x	x	x	x	x	x
Attalea attaleoides	x	x	x	x	x	x	x
Attalea maripa	x			x	x	x	
Bactris acanthocarpa var. *acanthocarpa*	x	x	x	x	x	x	x
Bactris acanthocarpa var. *intermedia*	x	x	x	x	x	x	x
Bactris balanophora	x		x	x		x	
Bactris constanciae	x	x	x	x	x	x	x
Bactris elegans	x	x	x	x	x	x	x
Bactris gastoniana	x	x	x	x	x	x	x
Bactris hirta var. *hirta*	x	x	x	x	x	x	x
Bactris hirta var. *pulchra*	x	x	x	x	x	x	x
Bactris killipii		x	x	x	x	x	x
Bactris maraja var. *maraja*	x	x			x	x	
Bactris oligocarpa		x			x	x	x
Bactris simplicifrons	x	x	x	x	x	x	x
Bactris sp. 1	x	x	x	x	x	x	x
Bactris sp. 2	x		x		x	x	x
Bactris tomentosa var. *tomentosa*	x	x	x	x	x	x	x
Desmoncus phoenicocarpus	x	x	x	x	x	x	x
Euterpe precatoria var. *precatoria*	x	x	x	x	x	x	x
Geonoma aspidiifolia	x	x	x	x	x	x	x
Geonoma deversa	x	x	x	x	x	x	x
Geonoma maxima var. *chelidonura*	x		x	x	x	x	x
Geonoma maxima var. *maxima*	x	x	x	x	x	x	x
Geonoma maxima var. *spixiana*	x	x	x	x	x	x	x
Geonoma sp.	x			x	x		x
Geonoma stricta var. *stricta*	x	x		x	x	x	
Iriartella setigera	x	x	x	x	x	x	x
Lepidocaryum tenue	x	x				x	
Oenocarpus bacaba	x	x	x	x	x	x	x
Oenocarpus bataua var. *bataua*	x	x	x	x	x	x	x
Oenocarpus minor	x	x	x	x	x	x	x
Socratea exorrhiza	x	x	x	x	x	x	x
Syagrus inajai	x	x	x	x	x	x	x

Note: Nonclosed-forest species are those not occurring in continuous forests.

or subterranean trunk) and reserve size (X^2 = 0.067, P = 0.87, df = 3) or the number of species of each habit and site (X^2 = 0.21, P = 0.89, df = 2). Palms of both types of habit were similarly represented in all reserve sizes and at all sites. Most of the species had aerial trunks, but only five (*Attalea maripa, Euterpe precatoria* var. *precatoria, Oenocarpus bacaba, O. bataua* var. *Bataua,* and *Socratea exorrhiza*) were large arborescent species (up to 20 m tall). Six species were midstory palms: *Astrocaryum gynacanthum, Bactris balanophora, B. constanciae, Iriartella setigera, Oenocarpus minor,* and *Syagrus inajai,* reaching no more than 12–15 m tall. All other twenty-five species were confined to the understory. The species composition of this palm community was dominated by stemmed, understory palms.

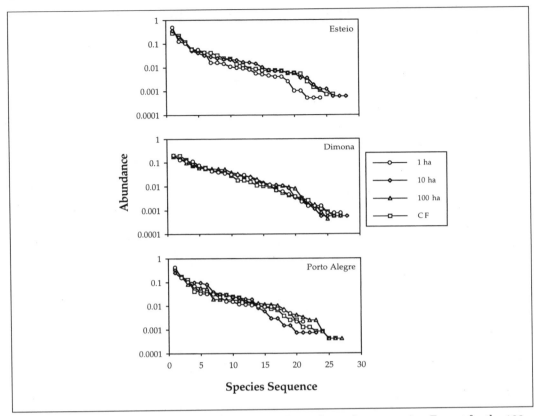

Fig. 10.1. Sequence of palm species ranked by abundance for each reserve size. Except for the 100 ha reserve, all other reserves have three sites sampled. Note logarithmic scale on vertical axis.

DIVERSITY CHANGES ASSOCIATED WITH FRAGMENTATION

Dominance-diversity curves were quite similar among sites and reserve sizes (figs. 10.1 and 10.2). Curves exhibited the log-normal species-abundance distribution that is characteristic of species-rich communities (Preston 1962; May 1975). These curves describe tropical plant communities in which there are typically few highly abundant species, with most species occurring at low densities.

No significant differences were detected for species richness (S) among reserve sizes (F = 0.10, P = 0.96, df = 3 for adults; F = 0.79, P = 0.55, df = 3 for juveniles; F = 2.91, P = 0.14, df = 3 for seedlings; and F = 0.49, P= 0.71, df = 3 for total). This suggests that palm

species richness has not been affected by reserve size since the fragments were isolated (tables 10.2 and 10.3). However, the elimination of nonclosed-forest taxa from the analysis of species richness may unveil patterns of species replacement in forest fragments compared with continuous forest. In fact, when nonclosed-forest species were removed from the analysis, corrected species richness revealed significant differences among reserve sizes in the pooled life stages (F = 7.86, P = 0.02, df = 3). The 1 ha reserves had significantly lower total forest-species richness than did continuous forest (P = 0.02, Tukey test; fig. 10.3), and the 10 ha reserves seem to already show signs of decreasing total species richness compared with the continuous forest (P = 0.07, Tukey

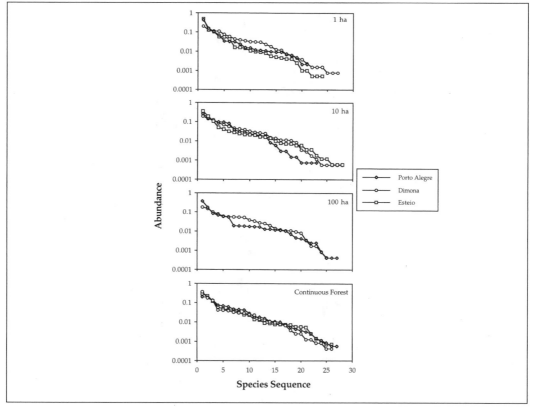

Fig. 10.2. Sequence of palm species ranked by abundance for each site. Except for the Esteio site, all other sites have four reserves sampled. Note logarithmic scale on vertical axis.

test). The ANOVA F-test for the seedling life stage showed a tendency to detect significant differences (F = 5.24, P = 0.05, df = 3); the continuous forest and 100 ha reserves showed a tendency to have significantly higher seedling richness of forest species than did 1 ha reserves (P = 0.07 for both comparisons, Tukey test). Although the reserves have been isolated for only a short time (10–15 years), it is already possible to detect that species composition in the forest fragments is changing.

If edge effect influences the whole area of forest fragments, then it may be the predominant factor determining species richness. However, if edge effects do not cover the whole area of the forest fragment, then the lower number of forest species in some forest fragments compared with the number

in continuous forest may be because the species were not present when the fragment was isolated or because their populations did not survive the isolation.

For the Shannon diversity index, no significant differences were detected among reserve size categories (adults, F = 0.09, P = 0.96, df = 3; juveniles, F = 0.75, P = 0.57, df = 2; seedlings, F = 0.83, P = 0.53; all stages pooled, F = 2.04, P = 0.23). The Equitability index also did not differ significantly among reserve sizes (adults, F = 0.10, P = 0.95, df = 3; juveniles, F = 0.35, P = 0.79, df = 3; seedlings, F = 0.53, P = 0.68, df = 3, all life stages pooled, F = 1.02, P = 0.46, df = 3). Shannon and Equitability indexes were affected by the number of species and the number of individuals sampled.

Estimated coefficients of similarity of

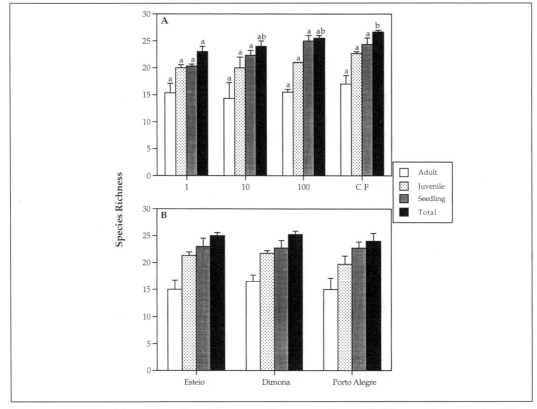

Fig. 10.3. Species richness (nonclosed forest species removed) at each life stage (mean and 1 SE) for the reserve sizes (A) and sites (B) for all life stages. The sum of all individuals in all life stages is represented by the total. Bars with different superscript letters for the same life stage differ significantly at P > 0.05.

species composition (Sorensen index) among sites and reserves showed greater variation for adult stage than did comparisons of younger age classes (tables 10.4, 10.5). Adults also had the lowest Sorensen coefficient means (S_s = 0.809 for reserves and 0.824 for sites). The highest means for the Sorensen coefficient (indicating the highest level of similarity) were found for the pooled life stage class (S_s = 0.897 for reserves and 0.917 for sites), although this result is highly influenced by the large numbers of seedlings in the class. Accordingly, the seedling class is also characterized by the second highest level of similarity among reserves (S_s = 0.888) and among sites (S_s = 0.901). The higher similarity among sites

and reserves for seedlings than for adults indicates that patterns of survivorship may differ among sites and reserves. Alternatively, rare species may be not present at some sites as adults but present as seedlings because many more seedlings were sampled (because they are more abundant) and also because some species do disperse and germinate in habitats unsuitable to them as adults. Surprisingly, the highest similarity for the pooled life stage classes (total) is between 1 ha and continuous forest (S_s = 0.937) and between the 100 ha and continuous forest (S_s = 0.933). In general, the Esteio and Porto Alegre sites are quite similar in all life stage classes except for the adults. Excluding the adult life stage, Dimona and

TABLE 10.2. Comparison of Density, Diversity, and Equitability in Palm Communities in Central Amazonia

	1 ha	10 ha	100 ha	CF
Density/ha	4248.33 ± 510.81	3984.16 ± 311.78	5901.25 ± 91.25	7187.50 ± 810.64
Species richness	24.00 ± 1.73	26.00 ± 1.53	26.00 ± 1.00	26.67 ± 0.33
Species richness (without fugitives)	23.00 ± 1.00	24.00 ± 1.00	25.00 ± 0.50	26.67 ± 0.33
Diversity (H')	2.15 ± 0.22	2.39 ± 0.11	2.45 ± 0.22	2.32 ± 0.09
Equitability (J')	0.67 ± 0.05	0.73 ± 0.03	0.75 ± 0.07	0.70 ± 0.02

Note: Means are of the three sites for each reserve size, except the 100 ha reserves, which are represented by only two sites.

TABLE 10.3. Comparison of Density, Species Richness, Diversity, and Equitability of the Palm Community by Sites in Central Amazonia

	Esteio	Dimona	Porto Alegre
Density/ha	5327.50 ± 825.12	5535.62 ± 1179.62	4984.37 ± 620.85
Species richness	25.66 ± 0.88	26.75 ± 0.63	24.5 ± 1.50
Species richness (without fugitives)	25.00 ± 0.58	25.25 ± 0.63	24.00 ± 1.47
Diversity (H')	2.127 ± 0.14	2.59 ± 0.03	2.19 ± 0.06
Equitability (J')	0.65 ± 0.03	0.78 ± 0.01	0.68 ± 0.01

Note: Means are estimated from four reserves for each site, except for the Esteio site, which has only three reserves.

TABLE 10.4. Community Similarity (Sorensen Index) Estimated for Pair-Wise Comparisons of Reserve Size Classes

		Reserve size		
		1 ha	10 ha	100 ha
Adult	10 ha	0.77		
	100 ha	0.80	0.77	
	CF	0.83	0.89	0.80
Juvenile	10 ha	0.88		
	100 ha	0.89	0.85	
	CF	0.89	0.91	0.90
Seedling	10 ha	0.86		
	100 ha	0.91	0.88	
	CF	0.88	0.88	0.93
TOTAL	10 ha	0.87		
	100 ha	0.90	0.87	
	CF	0.94	0.87	0.93

Note: A separate analysis is presented for each life stage. Total represents all three life stages pooled.

TABLE 10.5. Community Similarity (Sorensen Index) Estimated for Pair-Wise Comparisons of Sites

		Site	
		Esteio	Dimona
Adult	Dimona	0.77	
	Porto Alegre	0.81	0.90
Juvenile	Dimona	0.89	
	Porto Alegre	0.90	0.87
Seedling	Dimona	0.90	
	Porto Alegre	0.93	0.88
TOTAL	Dimona	0.94	
	Porto Alegre	0.92	0.89

Note: A separate analysis is presented for each life stage. Total represents all three life stages pooled.

TABLE 10.6.   Canonical Correlations, Proportion of Total Variance Explained by the Two First Canonical Variables, and the Significance Level

	Canonical variable	Squared canonical correlation	Proportion	F-test significance
Adult	1	0.40	0.41	0.0112
	2	0.36	0.34	0.0653
Juvenile	1	0.66	0.61	0.0001
	2	0.48	0.28	0.0257
Seedling	1	0.75	0.69	0.0001
	2	0.42	0.17	0.0150
TOTAL	1	0.77	0.68	0.0001
	2	0.48	0.18	0.0028

Note: A separate analysis is presented for each life stage. Total represents all life stages pooled.

Porto Alegre sites have the lowest similarity among all pairwise comparisons.

ORDINATION OF RESERVES

CDA from all life stages yielded statistically significant canonical variables (table 10.6). Two canonical variables were sufficient to account for most of the multivariate dispersion. The two variables explained similar amounts of variation in the adult life stage; however, for the younger life stages and for all life stages pooled the first canonical variable explained a higher proportion than the second one (table 10.6). The differences in floristic composition among reserve sizes increase in early life stages (fig. 10.4). In fact, in the adult life stage there is not much dispersion among plots of different size reserves (fig. 10.4a), but the dispersion increases toward the early life stages. For juveniles, seedlings and total, in the 1 ha, 10 ha, and continuous reserves, most of the multivariate dispersion is explained by the first canonical variable and in the 100 ha reserves mainly by the second canonical variable (fig. 10.4b). Higher differences in floristic composition occur in the seedling and total life stages (figs. 10.4c, 10.4d).

Discussion

COMMUNITY COMPOSITION AND
SPECIES DISTRIBUTION

The palm family is one of the eleven plant families that most contribute to species richness in any lowland Neotropical forest (Gentry 1988). Henderson, Galeano, and Bernal (1995) cite 67 genera and 550 species of palms in the Americas. Of the 34 genera and 151 species (189 taxa) occurring in Amazonia, there are eight (24 percent) genera and 140 (75 percent) endemic taxa (Henderson 1995). The number of genera (11) and species (36) found in this study (in topographically flat areas—plateaus only) represents 32 percent of the genera and 20 percent of the palm species and varieties of the whole Amazon region. Adding species that do not occur in topographically flat areas but occur only in low wet and swampy areas (*Mauritia flexuosa, Mauritiella aculeata,* and *Hyospathe elegans*) increases the figures to 41 percent of the genera and 22 percent of the species and varieties. This area therefore has high genera and taxa richness of palms.

Detailed comparisons with other studies are difficult because the taxonomic status of most of the palm species studied here has

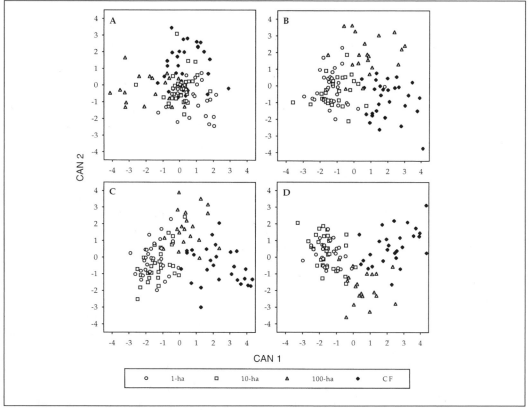

Fig. 10.4. Canonical discriminant analysis (CDA) ordination of palm densities in reserves of 1 ha, 10 ha, 100 ha, and continuous forest (CF) for each life stage. (A) adults, (B) juveniles, (C) seedlings, and (D) total (all life stages pooled).

been neglected in systematic works. Botanical inventories conducted in tropical forest generally record individuals with a DBH of at least 10 cm (e.g., Rankin–de Mérona et al. 1992). Consequently, these studies miss most of the palm species and other species occurring in the midstory and understory. As a result, these inventories do not correctly represent the palm community or the plant community in general, missing one very important component of the structure and dynamics of tropical forests. The only comparable work is that of Kahn and Castro (1985), who, in a site near my study area, found thirty-two palm species in 1.2 ha; I found thirty species in the 1.2 ha sampled in the three continuous forests. Except for the

two most diversified genera, *Bactris* and *Geonoma,* the species of which were not reported by the authors (thus preventing any taxonomic comparison), all but three species are the same. Two of these species (*Manicaria* sp. and *Attalea spectabilis*) occurred in their study area but not mine, while one species (*Desmoncus phoenicocarpus*) occurred in mine but not theirs.

Based on herbarium collections, Henderson (1995) estimated that an area of 1 degree latitude by 1 degree longitude (about 110 × 110 km) had an average of thirty taxa (range: 17–58) with the highest value near Iquitos, Peru. Such values must be interpreted with caution, as they are not exhaustive field collections but rather estimates based on available material in herbaria. Because of the lack of inventories and intensive specimen collections it is difficult to assess the regional diversity (diversity of a range of habitats in a geographical area) among areas of gamma diversity (overall diversity of a group of areas of alpha diversity) in the Amazon basin.

Besides the high richness of the sampled area, palms dominate the landscape; they are one of the most common vascular plant families in tropical forests. Among all arborescent species with DBH of at least 10 cm in the upper Urucu River drainage, only nine arborescent palm species were found by Peres (1994c). They made up the sixth family in abundance, representing 6.76 percent of all tree species. The 23,225 palm individuals occurring in the 4.4 ha sampled in this study represents a density of 5,278 palms per ha, considering all reserves. However, the mean number of palms per ha in the continuous forest in the BDFFP area is 7,188 palms per ha, much higher than the 2,917 per ha found in topographically flat areas by Kahn and Castro (1985) and even much more than the 900 per ha found by Kahn, Meija, and Castro (1988) in eastern Amazonia. This number is lower than the 9,860 palms per ha found by Kahn et al. (1988) in western Amazonia. Palm abundance increases from the east to the west of the Amazon Basin. On a local scale the abundance and species richness of palms also varies according to topography, soil, and level of habitat disturbance (e.g., Clark et al. 1995). The importance of palms in the Amazonian landscape is due more to their abundance than to their species richness.

INVASIVE SPECIES

The replacement of forest species by nonclosed-forest ones may decrease the biological value of a reserve unit, considering that among the thirty-six species sampled in this study five are characteristic of nonclosed-forest. The level of disturbance in smaller forest fragments may be the direct cause of invasion of these areas by nonclosed-forest species. In disturbed areas, microclimatic conditions change and some species are adversely affected; unaffected competitors may establish and increase their densities (Tilman 1996). The lower total palm densities in the 1 and 10 ha reserves may favor invasive species. Resources available or conditions amenable to forest species decrease with forest fragmentation, and the consequent density decrease of forest species may allow for the invasion of palm species (Scariot 1996). Nonclosed-forest species such as *Lepidocaryum tenue,* which occurs only in disturbed areas (pers. obs.), succeeded in becoming established in small forest fragments, later expanding its population through vegetative propagation and eventually establishing large colonies. Nonclosed-forest species may confound results of studies that do not take into account the identity of the species. Nonclosed-forest species should be conserved in disturbed areas or on the edges of reserves designed to preserve forest species.

COMMUNITY SIMILARITY

In my study areas of the BDFFP, large reserves more closely resemble the continuous forest in palm community structure and floristic composition. As expected, the palm seedling species composition of the 100 ha reserves is more similar to that in continuous forest (S_S = 0.93) than that in any other fragment size. This indicates that at this critical life stage, fewer species are recruiting seedlings in the small reserves (pairwise comparisons: 1-10, 1-100, 1-CF; 10-100-ha, and 10-CF) compared to the continuous forest. Eventually, species composition in small reserves may diverge even more from continuous forest because of this lower recruitment.

The Dimona–Porto Alegre pair of sites had higher similarity for adults. If there has been no differential adult mortality among sites since the isolation of the reserves, then the observed differences in similarity may be intrinsic to the sites, although sampling error or sampling design cannot be disregarded. Esteio and Porto Alegre are the most similar sites regarding juveniles and seedlings, while Esteio and Dimona are even more similar regarding all life stages pooled. There is no clear trend of similarity between sites found for all life stages. I observed more similarity in palm communities among sites than did Rankin–de Mérona et al. (1992), who compared tree communities among continuous forest sites in this region.

If the requirements for species to successfully establish, survive, and reproduce differ among species, then the number of species present in a given area will be determined more by habitat diversity than by the size of the reserve. However, increasing the geographic area sampled or preserved generally also increases the number of habitats; consequently, area should have a positive influence on species diversity. Any set of small reserves may eventually contain more species than one continuous reserve of the same area as a consequence of the habitat heterogeneity sampled by the archipelago of reserves, but many of the forest species may be absent from small reserves as shown here. Also, because none of the sites in this study had all thirty-six species of palms, it is clear that an extensive series of reserves covering a wide range of the habitat heterogeneity is needed to preserve most of the species.

SPECIES ABUNDANCE

Populations of species with long-lived individuals are less likely to go extinct because of demographic accidents (Pimm 1991). Due to the short-term isolation of the reserves and the long life span of palms, one expects that the most sensitive life stage is seedling and the least sensitive stage is adult. Little is known about palm life spans, but they have long life cycles. For example, *Astrocaryum mexicanum* reproduces only after thirty-nine years of age (Sarukhán 1980), *Euterpe globosa* (*E. precatoria*) reaches maturity at fifty-one years (Van Valen 1975); and *Balaka microcarpa* flowers between forty and sixty years of age (Ash 1988). In the current study, adult survivorship had apparently not been affected by the 10–15 years since forest fragmentation. Thus, the CDA analyses show that the floristic composition at the adult life stage is not yet affected by forest fragmentation. The long lifespan of palms and the short isolation time of reserves buffer against changes in the florist composition of adults. For the earlier life stages, dispersion of the sites in the CDA is more evident, suggesting effects of isolation on the earlier stages. Isolation seems to be affecting the floristic composition of seedlings in the 1 and 10 ha forest fragments compared to the continuous forest. Analyses of all life stages pooled (total) present results similar to those for the seedlings because there are so many more seedlings than adults or juveniles. Of the life stages examined here, seedlings may be the first life stage to be affected by forest fragmentation.

PALMS AS KEYSTONE RESOURCES

Palms are keystone species in the tropical rainforest, representing a major vertebrate food source during periods of food scarcity (Terborgh 1986). During the dry season when food availability is low in the BDFFP reserves, capuchin monkeys rely heavily on *O. bataua* fruits, and also on *A. maripa* in years when this less abundant species produces fruits (Chapter 22). A decrease in palm density in fragments may affect animal species that depend on their fruits and seeds during times of food scarcity. However, the direct and indirect effects of a decrease in palm population density on other species will depend on the strength of such relationships; the more specific the relationship, the higher the probability that the dependent species will be affected. If only a few palm species are part of such specialized interactions, protection of these taxa would be critical for the persistence of the overall community. However, if many or most of the palm species are of similar importance, efforts to save a only a few keystone species will fail to protect the rest of the community (Mills, Soulé, and Doak 1993).

Our inability to detect statistically significant effects may be a result of small sample (population) sizes rather than the absence of a biologically meaningful effect, and even large populations may go extinct following catastrophes (Mangel and Tier 1994). When previously undetected effects become apparent, species that depend on these palm species for survival may have already decreased in density or been forced to migrate; they may even be doomed to extinction.

Concluding Remarks

Because of their importance in forest structure and composition, their high taxa richness, and their use by humans and wildlife, the palm family must be a primary target of any conservation plan in the Amazon basin. This study has shown that even a short time after isolation, the palm community is already showing the consequences of habitat fragmentation, such as species invasion and reduction in population size. These effects

may be magnified with time. It is important to note that studies that rely only on species numbers and do not take into account the identity of the species and habitat preferences may fail to detect such species invasions as the ones shown here, where non-closed-forest species are invading isolated reserves. Additionally, the high heterogeneity of Amazonian forests even at small spatial scales demands careful planning to establish reserves in this region. Small reserves are better than no reserves at all. Despite their value, however, small reserves are not substitutes for large ones because the former can not maintain viable populations of all key taxa.

The demography of the palm community (survival and reproduction) has already been affected by site and fragmentation effects as evidenced by the CDA analysis. In fact, the seedling life stage is the most affected, in turn affecting recruitment to later life stages (Scariot 1996). The decreased densities in small reserves may also initiate unpredictable changes in forest structure and in food resources for animal species, decreasing the biological value of these reserves. Despite these constraints, small reserves may be important in a metapopulation scenario, where apparently fragmented populations are connected via seed dispersal and consequent gene flow (Levins 1970). However, the best approach to maintaining the integrity of this group and other species directly and indirectly associated with it is establishing many large reserves taking in account the geographical distribution of the species to encompass site heterogeneity. Indeed, any long-term plan of preserving biodiversity must be designed to preserve the patterns and the ecological processes of the target species.

Conservation Lessons

1. Due to its importance in forest structure and composition, its high taxonomic richness, and its use by humans and wildlife, the palm family must be a primary target of any conservation plan in the Amazon basin.

2. Palms respond to forest fragmentation through reduction in the population densities of forest species, invasion of forest fragments by an invasive species, and density increases of secondary forest species. These effects may be magnified with time.

3. The seedling life stage is the most strongly affected by forest fragmentation, which in turn may affect recruitment to later life stages and compromise the long-term persistence of forest palm populations.

4. The decreased densities in small reserves may also initiate unpredictable changes in forest structure and in food resources for animal species, decreasing the biological value of these reserves.

5. The best approach to maintaining the integrity of this group and other species directly and indirectly associated with it is establishing many large reserves. Small reserves may also be important in a metapopulation scenario provided gene flow is occurring.

6. Any long-term plan for preserving biodiversity must be designed to save not only patterns but also ecological processes.

Acknowledgments

I thank G. R. Colli and R. Henriques for their valuable help with the ordination analysis. S. J. Mazer, J. Endler, S. Cooper, H. Paz, and M. Pacheco provided valuable encouragement, discussions, and critique to my research. G. Colli, R. Henriques, C. Gascon, and two anonymous reviewers made comments on earlier versions of the manuscript. Many thanks to the staff of the BDFFP in Manaus. Sebastião S. de Souza and Manoel Jonas Pereira were tireless assistants in the field. My research was supported by the World Wildlife Fund, National Geographic Society, Fundação Botânica Margaret Mee, US-AID Brazil, the Andrew W. Mellon Foundation, and INPA/Smithsonian Institution (BDFFP). During the period of this work I received a fellowship from EMBRAPA.

CHAPTER 11

Regeneration in Tropical Rainforest Fragments

JULIETA BENITEZ-MALVIDO

Forest fragmentation affects the rates of mortality, growth, and recruitment of trees (Laurance et al. 1997; Chapter 9) and seedlings (Sizer 1992; Benitez-Malvido 1995, 1998). Fragmentation reduces the population size of tropical trees (Alvarez-Buylla et al. 1996) and of their animal vectors, such as pollinators (Powell and Powell 1987; Aizen and Feinsinger 1994a, 1994b; Didham, Ghazoul, et al. 1996; Chapter 12) and seed dispersers (Rylands and Keuroghlian 1988; Klein 1989; Bierregaard and Stouffer 1997; Chapters 20 and 21). Reduced animal dispersal may limit the potential of plant species to colonize available areas in isolated fragments (Primack 1992, 1993; Chapman and Onderdonk 1998).

Most studies of the effects of forest fragmentation on tree communities describe vegetation alterations at the edge of the fragments (Willson and Crome 1989; Williams-Linera 1990a, 1990b; Brothers and Spingarn 1992; Sizer 1992; Matlack 1994). However, little is known about how the environmental modifications produced by fragmentation affect the seedling community of mature forest (Whitmore 1989), not just considering the edge of the fragments but also their core, their size, and the edge shape (Benitez-Malvido 1998; Chapter 9).

In this chapter I present results of an ongoing study, started in 1991, on seedling ecology in tropical forest fragments (Benitez-Malvido 1995; Benitez-Malvido 1998). To evaluate the impact of fragmentation on forest regeneration I used the natural tree seedling community of mature forest and seedling transplants of three native species of Sapotaceae (Benitez-Malvido and Kossmann-Ferraz 1999; Benitez-Malvido et al. 1999). I tested whether fragmentation affects seedling abundance, seedling survival, growth, and damage by herbivores in different sized fragments, at different positions within the fragments (center, edge, and corner), and at different distances from the edge (Benitez-Malvido 1998). My study considered only shade-tolerant species and mature phase forest as they constitute the majority of species and the area, respectively, in tropical rainforests (Whitmore 1989).

Methods

The study was carried out as part of the Biological Dynamics of Forest Fragments Project in the project's study area (see fig. 4.1; table 4.3). The study area is a mosaic of extensive forest, forest fragments (including the experimental fragments of the BDFFP),

This is publication number 278 of the BDFFP Technical Series.

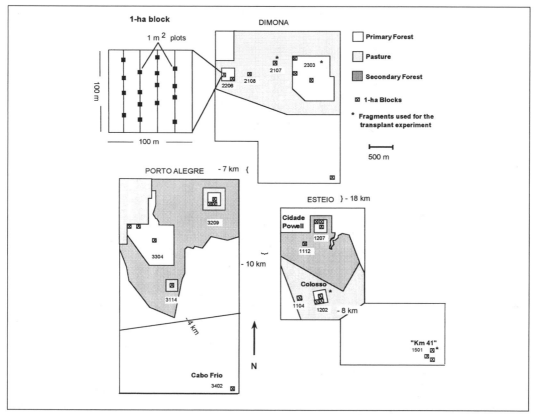

Fig. 11.1. Maps of the study sites at the three *fazendas* (cattle ranches): Dimona, Porto Alegre, and Esteio. The fragments are separated by distances of 0.3 to 28 km (after Benitez-Malvido [1998], reprinted by permission of Blackwell Science).

and large anthropogenic clearings associated with three 15,000 ha cattle ranches (see Chapter 4). The mature forest is all terra firme and is described in more detail in Chapter 5, as well as in Lovejoy and Bierregaard (1990).

Throughout this chapter, the four-digit numbers identifying the forest fragments (see fig. 11.1) are those used by Lovejoy, Bierregaard, et al. (1986) (see also table 4.3).

SEEDLING DENSITY

All woody trees 5–100 cm tall were considered seedlings (no vine or palm seedlings were considered). From 1991 through 1993 the density of naturally occurring shade-tolerant seedlings was measured within iso-lated forest fragments at the Dimona, Esteio, and Porto Alegre *fazendas* (cattle ranches). I used four fragment types: continuous forest (control); 100 ha fragments; 10 ha fragments; and 1 ha fragments (Benitez-Malvido 1998). Fragments were sampled using 1 ha blocks as experimental units, each of which contained twenty plots arranged in a stratified random manner. Groups of four 1 m² plots were located along five equidistant transects, 20 m apart (fig. 11.1).

To assess the variation between forest edges and interior I sampled three 1 ha blocks within each 100 and 10 ha fragment. The three 1 ha blocks were located in the center, edge, and corner of the fragment. In the blocks at the edge, samples were taken at

20, 40, 60, 80, and 100 m from the edge using the equidistant transects (fig. 11.1).

Experimental Seedlings

Seedlings of three native tree species of Sapotaceae were grown from seeds in a shade house at the Department of Tropical Silviculture of INPA (Benitez-Malvido 1995). I used seedlings from the Sapotaceae because it is the family with the most species in the study area (Rankin–de Mérona et al. 1990). The species were *Pouteria caimito* (Ruiz and Pavón) Radlkofer, *Chrysophyllum pomiferum* (Eyma) Pennington, and *Micropholis venulosa* (Martius and Eichler). Although little is known about the ecology of these species, the three species are defined as large-seeded, long-lived canopy tree species, are native to the study area (Benitez-Malvido 1995; Benitez-Malvido and Kossmann-Ferraz, 1999; T. Pennington, unpublished data), and are considered rare, as is typical of many tropical trees.

For the transplant experiment I selected a single fragment of each type. The fragments selected were: continuous forest at Km 41 (Reserve 1501); a 100 ha fragment at Dimona (2303); a 10 ha fragment at the Colosso reserve cluster (1207; see fig. 4.2); and a 1 ha fragment at Dimona (2107; see fig. 4.1 and plate 4). For future research the replication of each fragment type (continuous forest, 1, 10, and 100 ha fragments) is desirable to guarantee that responses of seedlings are due to fragmentation effects rather than to unexplored site effects.

Between January and May 1992, seedlings of the three species were transplanted to random positions in 1 m² plots within the four different fragment types, at different distances from the 100 and 10 ha fragment edge, and at different positions within 100 and 10 ha fragments (fig. 11.1). A total of 2,080 seedlings were planted at a density of 13 seedlings/plot as follows: *Pouteria cai-*

mito, 5 seedlings/plot (n = 800); *Chrysophyllum pomiferum*, 3 seedlings/plot (n = 480); and *Micropholis venulosa*, 5 seedlings/plot (n = 800). The number of replicates for each species depended on the availability of seedlings. Seedlings were about three months old when transplanted. The density of 13 seedlings/m² is similar to that found in the continuous forest control block (Benitez-Malvido 1998).

Seedling survival, relative growth rate (Hunt 1990), and herbivore damage have been recorded for more than five years. Herbivore damage has been estimated by evaluating the standing percentage of leaf area removed. Each leaf from every seedling has been assigned to one of the following categories of damage: 0, intact; 1, 1–6 percent; 2, 6–12 percent; 3, 12–25 percent; 4, 25–50 percent; 5, 50–100 percent. The score of each leaf was used to define an index of herbivory (IH) per plant and species (Domínguez, Dirzo, and Bullock 1989) as:

$$IH = \epsilon n_i i / N$$

where *i* is the category of damage, n^i is the number of leaves in the ith category of damage, and *N* is the total number of leaves on the plant.

To avoid any confounding influence introduced by differences in the vegetation of the surrounding matrix (Milne and Forman 1986; Stouffer and Bierregaard 1995a, 1995b; Mesquita, Delamônica, and Laurance 1999), I used edges and corners that were at least five years old and had similar vegetation around them at the time the seedling density sampling and the transplantation experiment took place (see Benitez-Malvido 1998).

Statistical Analysis

The effect of fragmentation on seedling density and performance (growth and survival) was investigated using three comparisons (see Benitez-Malvido 1998). First, the effect of fragment type (fragment versus continuous forest) on seedling density and performance was assessed by comparing continuous forest with fragments of different sizes. I used the central blocks of 100 and 10 ha fragments for the comparisons. Second, the effect of position in the fragment (center, edge, and corner) was evaluated by comparing different positions within the 100 and 10 ha fragments. And, finally, I analyzed the effect of distance from edge by comparing five different distances from the edge within 100 and 10 ha fragments, and regressed the data against distance from edge.

For the seedling density measurements, the data were analyzed using generalized linear models (GLMs) for nested analysis of count data (Crawley 1993; see Benitez-Malvido 1998) using the GLIM statistical package (Green and Payne 1994). Because fragments were not evenly distributed among the three fazendas, not all the 1 ha blocks were used for the comparisons. Where there was more than one 1 ha block per fragment type at any fazenda, I randomly selected a block for that fragment type.

Because seedling species were not transplanted at the same time, each species was analyzed separately. For each comparison the seedling parameters were analyzed through GLMs. I used a one-way analysis of variance (ANOVA) to test for the effect of fragment type, and a nested ANOVA for the effects of position in the fragment and distance from the edge. Comparisons of herbivory indices were carried out with the Kruskal-Wallis test. Data transformation was applied when required (see Benitez-Malvido 1995).

Results

Findings from the experimental seedlings on the impact of fragmentation on seedling abundance and performance are based on more than five years of post-transplantation data. Results showed that fragment size, position within the fragment, and distance from edge had an effect on the abundance and performance of seedlings.

EFFECT OF FRAGMENT SIZE

The mean density of seedlings/m^2 (mean ± SD) for all fragments declined from continuous forest (14.4 ± 3.1); to 100 ha fragment center (12.9 ± 2.4); to 10 ha fragment center (11.8 ± 3.0); to 1 ha fragments (9.6 ± 2.7). Forest fragments consistently had lower seedling densities than areas of continuous forest, but fragment area unexpectedly had few effects on seedling density (fig. 11.2; Benitez-Malvido 1998).

Results suggest that fragmentation reduces the levels of herbivory on seedlings. Seedlings of *P. caimito* transplanted into continuous forest and large fragments were more severely attacked by herbivores than those in smaller fragments (fig. 11.3), with a significant decline from continuous forest to 100 ha, to 10 ha, to 1 ha fragment.

EFFECT OF POSITION AND DISTANCE FROM EDGE

The corners of the 100 ha fragments had significantly lower seedling density than the centers and edges. There were no significant differences at different positions within 10 ha fragments, but densities within the three positions were similar to those found in 100 ha fragment corners (fig. 11.4; Benitez-Malvido 1998). In the seedling transplants, survival for *C. pomiferum,* and herbivore damage for *M. venulosa,* were higher at the centers of 100 and 10 ha fragments than in their edges and corners (fig. 11.5).

Seedling density increased significantly with distance from the edge within 100 and

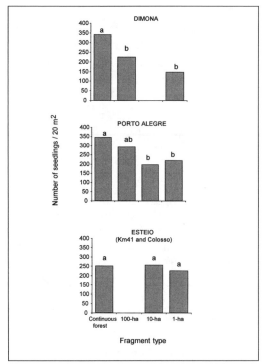

Fig. 11.2. Seedling density per 20 m² for tree seedlings in different fragment types within three *fazendas* north of Manaus. Different letters at the top of the bars denote significant differences among fragments. Data were unavailable for 10 ha fragment at Dimona and for 100 ha fragment at Esteio (Benitez-Malvido 1998).

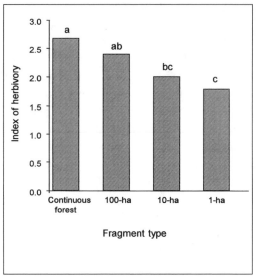

Fig. 11.3. Index of herbivory (medians) for *Pouteria caimito* seedlings over fourteen months of being transplanted into different fragment types. Significant differences are indicated with different letters at the top of the bars. Herbivory indices (following Domínguez et al. 1989), in percentage of leaf area eaten: 0, intact; 1, 1–6%; 2, 6–12%; 3, 12–25%; 4, 25–50%; 5, 50–100%.

10 ha fragments (fig. 11.6; Benitez-Malvido 1998), with higher increases for the 100 ha fragments. For the species examined and variables considered (growth, survival, and herbivory), some seedlings improved their performance, others were negatively affected, and still others were unaffected by edge exposure (fig. 11.7). For example, seedlings of *C. pomiferum* showed increased growth toward the forest interior, whereas survival was unaffected by distance from the edge. In contrast, *M. venulosa* seedlings had lower survival away from the forest edge. There was no edge-related pattern in damage by herbivores to seedlings for any species.

Discussion

Forest fragmentation at this Amazonian site decreased the number of tree seedlings and influenced their performance. Reduced seedling density within the fragments may be the result of an interplay of factors of two kinds: those that reduce the probability of seedling establishment and those factors that increased seedling mortality.

FRAGMENTATION EFFECTS ON SEEDLING ESTABLISHMENT

Successful establishment of seedlings within fragments could be hampered by three ef-

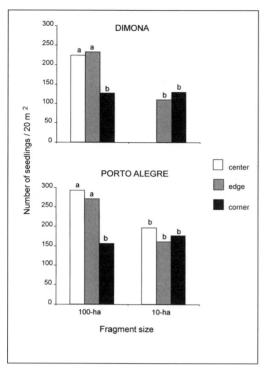

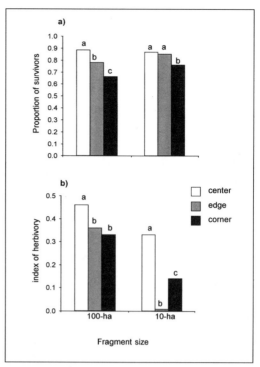

Fig. 11.4. Seedling density per 20 m² for tree seedlings in three different positions within 100 and 10 ha forest fragments at *fazendas* Dimona and Esteio. Different letters, at the top of the bars, denote significant differences between positions. Data were unavailable for "center" position at Dimona (after Benitez-Malvido [1998], reprinted by permission of Blackwell Science).

Fig. 11.5. Patterns of seedling performance at different positions within 100 and 10 ha fragments. (a) Survival of *C. pomiferum* seedlings after forty-two months of transplantation. (b) Index of herbivory of *M. venulosa* seedlings forty months after transplantation.

fects. First, seed rain may be reduced. Most tropical trees (about 80 percent) are pollinated by animals, which also help to disperse the seeds. Fewer animal vectors in fragments (primates, birds, bats, bees, etc.), mean that there is less seed rain and hence reduced seedling recruitment (Powell and Powell 1987; Aizen and Feinsinger 1994a, 1994b; Bierregaard and Stouffer 1997; Rylands and Keuroghlian 1988; Chapman and Onderdonk 1998; Chapters 17, 20, and 21). Other factors affecting seed rain are changes in tree phenology—that is, a ceasing of flowering or setting of fruits and seeds (Lovejoy, Bierregaard, et al. 1983) and an in-

creased mortality of reproductive trees within the fragments (Laurance et al. 1997; Chapters 9 and 13). Second, increased seed predation could result from an increased number of frugivorous mammals within small forest fragments (Malcolm 1991; Laurance 1994). Finally, drastic changes in relative humidity and temperature in the soil along forest edges (Kapos 1989; Bierregaard et al. 1992; Chen, Franklin, and Spies 1992; Kapos et al. 1993; Camargo and Kapos 1995) can prevent germination, especially of seeds that do not tolerate dehydration and are produced mostly by shade-tolerant tree species (Vázquez-Yánes and Orozco-Segovia 1984).

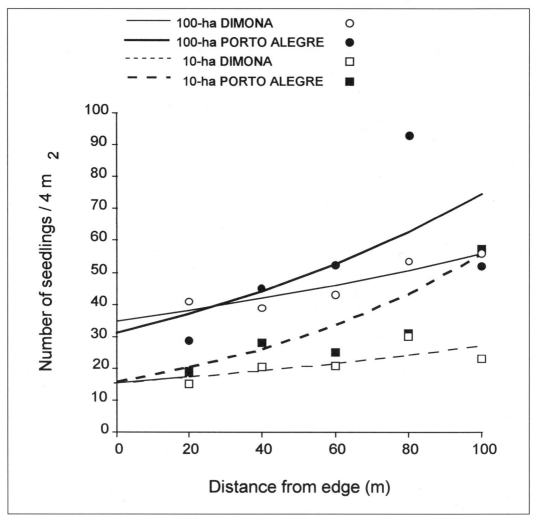

Fig. 11.6. Seedling density at different distances from the forest edge within 100 and 10 ha forest fragments. The number of seedlings increased significantly toward the forest interior (after Benitez-Malvido [1998], reprinted by permission of Blackwell Science).

FRAGMENTATION EFFECTS ON SEEDLING MORTALITY

Increased seedling mortality within the fragments might be a consequence of drastic modifications in the physical environment, such as wind, relative humidity, and temperature (Kapos 1989; Bierregaard et al. 1992; Chen, Franklin, and Spies 1992; Kapos et al. 1993; Camargo and Kapos 1995). Along edges, causes of seedling mortality might also be related to increased litter fall (Clark and Clark 1989; Sizer 1992; Benitez-Malvido 1995) and diminished light penetration (Matlack 1994; Camargo and Kapos 1995) caused by the thick foliage that develops after edge creation (Malcolm 1991; Benitez-Malvido 1998). Moreover, shade-tolerant seedlings along edges can be also hindered by having to compete with pioneer species that have higher growth rates. Higher seedling mortality within the fragments may also be related to harsh external

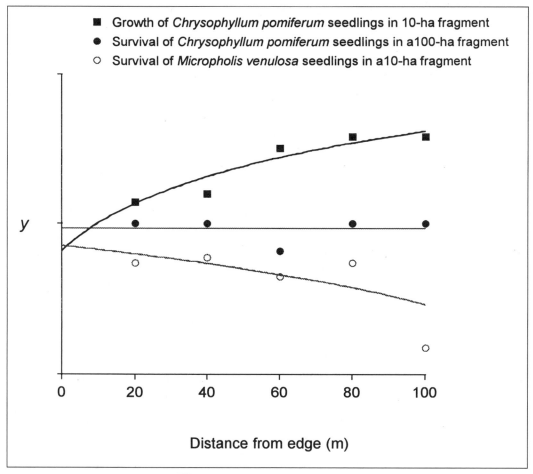

Fig. 11.7. Patterns of edge effects on transplanted seedlings. The *y* axis represents growth or survival.

events, such as fire, hot dry winds (Janzen 1986a), and cattle overgrazing and trampling (Aizen and Feinsinger 1994a).

EFFECT OF FRAGMENT SIZE AND EDGE SHAPE

Larger fragments generally hold greater habitat diversity and are more likely to retain, when compared with smaller fragments, larger populations of the original species and more intact interiors (which may not be found at all in 1 ha fragments), so plant interactions with animals (i.e., pollination, seed dispersal, herbivore damage)

are likely to remain much less changed from continuous forest (Saunders, Hobbs, and Margules 1991). In my experiments, modifications of plant-animal interactions within the fragments were evident through differences in herbivore damage to seedlings. The higher levels of herbivore damage found in continuous forest, in larger fragments, and in central blocks within fragments may be the result of higher moisture levels, which have been shown to be correlated with greater insect abundance and diversity in tropical ecosystems (Strong 1977; Klein 1989). It is likely that some herbivores that normally feed on the seedlings in continuous forest are absent from forest fragments (Didham, Hammond, et al. 1998; Didham, Lawton, et al. 1998; Chapter 18). However, I found no edge-related patterns on herbivory. It is likely that marked changes on herbivory rates along edges take place immediately after edge creation, and at the reserve margins and not more than 20 m from it. The flush of plant growth near reserve margins soon after edge formation may attract insects by providing new food resources (Lovejoy, Bierregaard, et al. 1986).

The response of transplanted seedlings to edge exposure was species-specific. Variation in plant responses to edge exposure may be caused by the particular micro-habitat requirements, such as shade tolerance, of each species (Bazzaz 1984). Although the three study species are defined as shade-tolerant tree species (Benitez-Malvido and Kossmann-Ferraz, 1999), I have observed different responses by the three species to increments in light levels. Seedlings of the three species grew more than four times faster in a secondary regrowth patch with high light levels than they did in the shaded forest understory (J. Benitez-Malvido, unpub. data). In the regrowth patch, seedlings of *M. venulosa* are, on average, 2.4 and 2 times taller than those of *C. pomiferum* and *P. caimito,* respectively (J. Benitez-Malvido, unpub.data). Variations in light availability could be intensified by the accelerated forest dynamics within the fragments (Laurance, Ferreira, et al. 1998a).

The different positions (corner, edge, and center) within the rectangular fragments appeared to have important effects on seedling abundance and performance. In general, the corners of the larger fragments had lower seedling density, survival, and herbivory than their centers. Fragment corners display a more damaged overstory and understory than edges, probably because of wind and fire exposure, and they seem to be poorer habitats for seedling establishment and for the insects that feed on seedlings. Differences in seedling abundance and performance among positions within fragments of 100 and 10 ha could be explained by Malcolm's (1994) model for edge-related effects. The model predicts that a point near a corner of a forest fragment would be more severely affected than one near a linear edge because a point near a reserve corner is in proximity to more edge.

Conclusions

Several findings suggest that forest fragmentation influences seedling regeneration (Benitez-Malvido 1998; Chapman and Onderdonk 1998; Jules 1998; Scariot 1999). Reduced seedling density and reduced levels of herbivory found in fragments may indicate changes of diversity patterns in the understory plant community (Dirzo and Miranda 1991; Chapman and Onderdonk 1998; Scariot 1999). Because the identification of much of the seedling flora in the BDFFP area is still difficult, we are unable in many instances to distinguish between invasive and native species, and to know whether some seedling species tend to dominate or disappear (see Chapter 10 for examples with palms, whose seedlings are readily distinguished).

My study showed that edge effects were more important than fragment area in affecting seedling abundance, and also that edge shape influences seedling abundance, performance, and levels of attack by herbivores. Fragment corners appeared to be more exposed to deleterious edge effects than centers and linear edges.

The results of this study as well as others (Laurance et al. 1997; Laurance, Ferreira, et al. 1998b; Jules 1998; Chapter 10) indicate that descriptions of ecological processes (e.g., herbivore damage to seedlings, recruitment, vegetation dynamics) in tropical forest fragments smaller than 100 ha seem not to represent functional ecosystems in the central Amazon and should be interpreted with caution. Although ecological processes in small forest fragments may not occur at rates typical of undisturbed forest, they nonetheless continue; surviving trees within the fragments and the surrounding vegetation can become important propagule sources through seed production or as perching sites for many animals that can carry and deposit seeds to the soil (Guevara and Laborde 1993).

Long-term studies are needed to evaluate the consequences of forest fragmentation on the seedling community. The biological factors and environmental conditions that influence seedling recruitment, growth, and survival at the time of edge creation might not be the same as time elapses and the conditions at the edge change—for example, when a new vegetation layer, with different structure and composition, develops (Williams-Linera 1990a, 1990b; Chen, Franklin, and Spies 1993; Matlack 1994; Jules 1998; Laurance, Ferreira et al. 1998b).

Tree seedlings are probably the most important regenerative pool of shade-tolerant, long-lived canopy tree species and are the major component of diversity, structure, and function of the rainforest tree community. Fragmentation strongly affects seedling populations and hence the regenerative potential of the forest.

Future research priorities in the forest fragments should include: (1) studies on phenological phenomena, such as flower, fruit, and seed set, as well as seed rain, seed dispersal, and seedling recruitment to understand the patterns that regeneration follows in fragmented habitats; (2) studies of how the modifications in forest structure and light regimes affect seedling establishment and performance; (3) studies on the possible changes in seedling species diver-

sity and composition within fragments; and (4) studies of ecologically and economically important tree families (e.g., Lecythidaceae [Chapter 6], Leguminosae, Meliaceae, Palmae [Chapter 10]) to understand their regeneration requirements within fragments for conservation and economic purposes.

Conservation Lessons

1. Forest fragmentation may prevent successful regeneration of canopy tree species.

2. The shape of a forest fragment dictates the extent to which edge effects will influence the growth and survivorship of seedlings near the periphery of the fragment. In the design of forest reserves it is recommended to avoid sharp edges.

3. Ecological processes in fragments probably do not reflect natural rates, so the results of experiments in fragments should be extrapolated to primary forest only with great caution.

4. Trees in fragments and the surrounding altered habitat may play an important role in regeneration by providing a source of seeds and attracting seed dispersers out of nearby primary forest.

Acknowledgments

This study would not have been possible without the financial and logistical support given by Consejo Nacional de Ciencia y Tecnología Mexico (CONACyT, grant no. 56519), the BDFFP, INPA-Smithsonian, and the MacArthur Foundation. The manuscript was written during a postdoctoral stay at the Instituto de Ecología, UNAM. I thank M. Martínez-Ramos for his continuous support during the development of this manuscript, R. Bierregaard for careful editing, and three anonymous reviewers for their helpful suggestions.

CHAPTER 12

Habitat Change, African Honeybees, and Fecundity in the Amazonian Tree *Dinizia excelsa* (Fabaceae)

CHRISTOPHER W. DICK

B ees are the predominant pollinators of tropical rainforest plants. In the Amazon basin, about 2,000 species of bees visit a majority of herbaceous and woody plant species. Although some tropical bee-pollinated herbs are pollinator specialists (e.g., Orchidaceae), most tropical trees are visited by generalist (oligolectic) bees and diverse small insects (Bawa 1990) and therefore may be resilient to changes in the species composition of pollinators caused by habitat change (Rathcke and Jules 1993; Bond 1994). Despite unprecedented rates of tropical deforestation and habitat fragmentation, however, only a handful of studies have examined the effect of habitat change on the pollination of tropical plants (see Renner 1997).

Deforestation alters the ecology of native bees by destroying nesting habitat and by changing the matrix between remnant food plants. Grazing by cattle dries ponds used for mud gathering and drinking, and the heat and desiccation of pastures can be physiologically stressful to forest-dwelling insects (Kearns and Inouye 1997; Kapos 1989; Chapter 18). Furthermore, deforestation may be accompanied by invasions of alien species. The most prominent alien pollinator in the Neotropics is the African honeybee (*Apis mellifera scutellata*) which thrives in anthropogenic habitat. Despite its ubiquity, however, little is known of its role in the pollination of native plants (Butz-Huryn 1997; Renner 1997).

Pollinator losses or replacements can alter the mating system and demography of host plants and are therefore important considerations in the conservation of remnant tropical plants (Schemske et al. 1994). The role of pollination in plant demographic growth depends on the extent to which seed set is pollen limited (Bierzychudek 1981), as well as the extent to which demographic growth is seed limited (Bond 1994). Pollinator-induced genetic changes can decrease the viability of remnant tropical trees because: (1) most tropical trees are self-incompatible and depend on animals for pollination (Bawa, Bullock, et al. 1985; Bawa, Perry, and Beach 1985); (2) tropical trees typically carry high mutation loads that lead to pronounced inbreeding depression (Alvarez-Bullya et al. 1996); and (3) tropical trees occur at very low population densities in undisturbed forest, so forest fragmentation results in extreme bottlenecks.

To illustrate the last point, the average population density of 127 leguminous tree species (with a DBH of at least 10 cm) in the

This publication number 279 in the BDFFP Technical Series.

Biological Dynamics of Forest Fragments Project inventory is 0.37 trees per hectare (unpublished BDFFP data; see also Chapter 6). Therefore, a typical 10 ha fragment will contain only two to four conspecific trees, and only a fraction of these will participate in reproduction. Such small populations are vulnerable to demographic and genetic stochasticity: chance events, such as fires or disease, can ravage local populations while genetic drift erodes genetic variation in the population over generations (Lande and Barrowclough 1987) and can fix mildly deleterious alleles (Lande 1994), thereby reducing population fitness. Because genetic drift is counteracted by gene flow, the effect of drift in remnant populations will be determined largely by the foraging behavior of pollinators and seed dispersers in anthropogenic habitats (Nason and Hamrick 1997). Several recent studies have shown that pollen-mediated gene flow to isolated tropical trees can be extensive (e.g., Chase et al. 1996). However, pollinators were not identified in these studies.

In this chapter I review several BDFFP studies of native bees and present an overview of the African honeybee invasion. I present data from ecological studies of the timber tree *Dinizia excelsa* (Fabaceae), which was visited primarily by African honeybees in disturbed habitats during the study period. I then discuss how the results apply to the ecology and conservation of other remnant tropical plants.

Bees at the BDFFP Study Sites

Native Bees

Three studies from the BDFFP document the species diversity of the bee community and its response to habitat fragmentation. In a survey of stingless bees (Meliponini), M. Oliveira, Morato, and Garcia (1995) and

Oliveira (Chapter 17) identified 54 species in 24 genera, of which *Trigona, Melipona, Partamona,* and *Tetragona* were, in descending order, the most species-rich genera. They collected 37 species in continuous forest, 22 species in 10 ha fragments, and 21 species in a 1 ha fragment. Although their study was not intended to evaluate the effect of fragment size on species diversity, the high diversity of bees in the 1 ha fragment is notable, indicating that even small forest fragments provide valuable habitat for native bees.

In the BDFFP study area, Powell and Powell (1987) scent-baited male euglossine bees in continuous forest, pasture, and reserves immediately following their creation. They report a positive correlation between fragment size and bee abundance for three species. They also concluded that the clearings of only 100 meters between fragments and continuous forest prevented euglossine bees from flying to the fragments, potentially disrupting the pollination of spatially isolated plants. These results are unusual, however, because euglossine bees can forage over more than twenty kilometers (Janzen 1971), and the same baits have attracted other euglossine species 1 km over water (Dressler 1982). The correlation between fragment size and euglossine bee abundance in this case may be better explained by edge effects (Becker, Moure, and Peralta 1991). If fire and desiccation alter the abundance of plants and nesting sites for a fixed distance into fragments, the smallest fragments will suffer the most until the vegetation grows back. Moreover, the response of male bees to scent traps may not reflect the foraging patterns of females that do most of the pollinating.

Becker, Moure, and Peralta (1991) bait-trapped euglossine bees five years after the Powell and Powell (1987) study, at which time the vegetation surrounding the isolated fragments had grown into dense stands of *Cecropia* spp. (Moraceae) and *Vismia* spp. (Clusiaceae) (plates 8 and 9). They captured sixteen species of euglossine bees in the genera *Eufriesea, Euglossa, Eulaema,* and *Exaerete* (Tribe Euglossini) and reported a decline in diversity only in the 1 ha fragments. A single species, *Euglossa chalybeata,* accounted for 85 percent of the collection. Interestingly, the abundance of bees in the 10 and 100 ha fragments exceeded continuous forest samples. However, this may have been a result of the small number of replicates. The species richness and abundance of euglossine bees in the BDFFP reserves was low in comparison with that in sites in Costa Rica, Panama, and Peru. This led Becker, Moure, and Peralta (1991) to conclude that host plants may be vulnerable to fluctuations in population size of these few species. However, long-term studies in Panama (Roubik and Ackerman 1987) have shown that euglossine populations are among the most stable insect populations known anywhere.

AFRICAN HONEYBEES

African honeybees (*Apis mellifera scutellata*) were imported to southern Brazil in 1956 to breed a tropical strain with improved honey production. The race *A.m. scutellata* has its origins in sub-Saharan Africa, where it inhabits woodland and savannas. It differs from European honeybees (*A.m. mellifera*) in its aggressive behavior, larger colony size, greater number of colonies per land area, and tendency to abscond (Michener 1975). Unlike European honeybees, which are often found dead in their hives during food shortages, colonies of African honeybees can travel more than 200 km to resource-rich areas (Kijatiira 1988, cited in Roubik 1989). In 1956, 26 queens and about 200 males escaped from

apiaries in Rio Claro, São Paulo, and spread across the Brazilian landscape. They initially covered about 320 km/yr and mated with all of the European honeybee queens in their path. The resulting hybrids retained morphological and behavioral traits of the African strain (Diniz and Malaspina 1996) and therefore are still called African honeybees. There were no colonies of *Apis* in the Amazon basin prior to the African honeybee invasion (Roubik 1989). African honeybees were reported in Belém in 1971 and in Manaus in 1974 (Prance 1976); they were observed in French Guiana by 1976 (Roubik 1978).

African honeybees seem to thrive in mosaic habitats because of their eclectic choice of nesting sites, their absconding behavior, and their use of primary forest, crop, and weedy plant species for nectar and pollen (Roubik 1989). African honeybees nest in large dead trees in pastures and in selectively logged forests, as well as in sites that European honeybees avoid, such as ground holes and termite mounds. There are an estimated 50 million to 100 million African honeybee colonies in Latin America (Winston 1992). Estimates of colony density range from 10/km² (Taylor 1977) to 108/km² (Kerr 1971, in Michener 1975). A census of hollow trees in lowland rainforest in the Yucatan yielded 15–36 colonies per square kilometer (Quezada-Euan and May-Itza 1996).

Studies of European honeybees provide insight into how African honeybees may affect the reproduction of native plants (Butz-Huryn 1997). However, there are important differences in the diet and foraging behavior of the races that must be taken into consideration. Villanueva (1994), for example, found that less than 50 percent of the pollen species in Yucatan, Mexico, were shared by European and African honeybees.

The foraging area of a single African honeybee colony of about 212 km² exceeds the foraging area of native stingless honeybees (12.5 km²) by more than an order of magnitude (Roubik 1989). Unlike native bees, African honeybees do forage over pastures. Although much of their foraging may be among flowers in the same tree, temporal variation in nectar availability may compel them to visit other plants (Frankie and Haber 1983).

COMPETITION WITH NATIVE BEES

A single colony of African honeybees may consume 40 kg of pollen and a few hundred kilograms of nectar per year (Roubik 1989). Given their high colony numbers and trophic overlap with native stingless bees (Wilms, Imperatriz-Fonseca, and Engels 1996; Roubik et al. 1986), it is plausible that African honeybees compete with native bees for limited floral resources. Roubik (1980) compared the foraging behavior of African honeybees and native stingless bees at artificial feeders and concluded that colony size, body size, and superior communication ability lead to a competitive advantage for African honeybees at rich, compact resources. Active aggression by honeybees against foraging native pollinators (interference competition) is known from a single case in which African honeybees were observed attacking native bees for a short period of time after their invasion (Posey and Camargo 1985).

Exploitative competition, in which the reduction of resources by one organism leads to displacement or exclusion of another, has been demonstrated for individual plants and populations. For example, Paton (1993) showed that honeybees removed more than 90 percent of the nectar of some bird-pollinated plants in Australia. However, on a community scale, Roubik and Wolda (in press) found no difference in the species composition or abundance of native bees in Barro Colorado Island, Panama, in the seven years before and ten years after the invasion of African honeybees.

Case Study of Dinizia excelsa

With a canopy that rises up to 60 meters above the forest floor (the main forest canopy is at 35 meters) and a girth that can exceed two meters, *Dinizia excelsa* Ducke is a giant among Amazonian trees (Ducke 1922). *D. excelsa* occurs in terra firme forest at natural population densities of about 0.3 individuals per hectare, which places it near the mean population density of 0.37 trees/ha of the 127 legume tree species of the BDFFP reserves (unpublished BDFFP data). The tiny yellow-green hermaphroditic and staminate flowers (calyx 1–1.5 mm) occur in terminal racemes (10–18 cm long) and produce nectar and fragrance. According to the classification of Bawa, Bullock, et al. (1985), these flowers reflect a generalist pollination mode that attracts small bees and diverse insects, and which occurs in about 30 percent of tropical rainforest tree species (Bawa 1990). Large indehiscent pods are produced at the end of the dry season and remain in the canopy for six months, until being dispersed by strong winds.

Parrots and macaws were observed foraging on the ripening pods of *D. excelsa* in the pastures and fragments of Colosso in September 1995. The most isolated trees in pastures served as roosting sites and food sources for macaws, which flew across open pastures at dawn and sunset. The single post-dispersal predator is an undescribed Bruchid beetle in the genus *Amblycerus* (Geoff Morse, pers. comm.).

D. excelsa seedlings thrive in high light (Varela and Santos 1992) and regenerate in abandoned pasture (unpublished data). Carbon dating has shown large individuals (about 1 m DBH) are usually younger than 300 years old (Chapter 7), making *D. excelsa* one of the fastest growing of the emergent timber trees that have been carbon dated (Chambers, Higuchi, and Schimel 1998). *D. excelsa* seedlings are common along the border of the 10 ha reserve in Colosso, along trails cutting through secondary growth, and along slightly shaded portions of the road near Colosso camp.

The heavy wood of *D. excelsa,* valued for furniture, accounts for about 50 percent of regional timber sales (Barbosa 1990). Because of its value for timber and shade, and because the trunks are difficult to cut, ranchers often leave large individuals standing in the middle of newly created pastures. Shade trees, along with individuals in continuous and fragmented forest, provided an experimental system with which to examine the ecology and breeding structure of remnant populations (Dick 1999).

Methods

MAPPING

A complete inventory was made of adult trees (at least 40 cm DBH) in the fragments and pasture to account for all potential pollen donors for the genetic studies (Dick 1999). The surveys were done, with the help of local woodsmen, in perpendicular transects about 50 meters apart. Surveys in continuous forest relied on existing trails. The objective of the surveys in continuous forest was only to find enough individuals to compare to the fragmented forest populations. All individuals were tagged and later mapped (Dick 1999).

PHENOLOGY

Phenological observations were made on trees in Colosso, Km 41, Porto Alegre, and

Dimona (see fig. 4.1; table 4.3) in 1995 to compare the relative intensity of flowering between individuals, and to document the floral overlap of individuals within populations. This information was combined with genetic data to elucidate the breeding patterns of these trees (Dick 1999). The phenological observations were made using binoculars at the base of each mapped tree. All individuals in a given site were visited. Stages of leaf and floral maturation were noted, including the timing of loss of old leaves, appearance of new leaves, appearance of floral buds, opening of flowers (anthesis), appearance of first pods, maturation of seeds, and dispersal of pods. Six visits were made to Colosso and Km 41; four to Porto Alegre and one to Dimona.

POLLINATION

To assess changes in pollination, I observed and collected insects on flowers of *D. excelsa* located in pastures, 10 and 100 ha isolated reserves, and continuous forest. Using rope-climbing techniques to gain access to the canopy (Tucker and Powell 1991; fig. 12.1), I climbed seven adult trees. Trees Km 41.11 and Km 41.01 were located in continuous forest at the Km 41 study plot (see fig. 4.1) and were separated by 700 meters. Col. 26 and Col. 19 were located in the 10 ha reserve in Colosso (Reserve 1202). Dim. 3.6 and P.A. 3.11 were located in the 100 ha reserves of Dimona and Porto Alegre. Col. 06 was located in the actively grazed pasture of Colosso and was separated from the nearest conspecific (Col. 13) by 600 meters of open pasture. All of the trees except Km 41.11 produced abundant flowers in 1995.

The studies were done at an average height of twenty-five meters. Most collections were made during periods of peak pollinator activity (7 to 11 a.m.), although some were made at other times of the day and

Fig. 12.1. Christopher Dick climbing an isolated *Dinizia excelsa* to sample insect pollinators.

night (table 12.1). I spent 62.5 hours in the canopies of the seven trees: 27 hours in continuous forest, 26.5 hours in fragments, and 9 hours in the isolated pasture tree Col. 06. Each visit lasted two to three hours. The insects were captured with nets, preserved with ethyl acetate, and identified by specialists.

FECUNDITY

Fecundity is an important component of fitness for many plant populations, and it can be influenced by changes in the physical environment and pollination. The fecundity of *D. excelsa* was measured as the total number of pods and seeds produced per tree. *D. excelsa* pods were counted from the ground using 10x binoculars in two continuous-forest sites (Cabo Frio and Km 41) and in the pasture and 10 ha populations at Colosso. The study trees were of comparable size. Small, nonreproductive trees were ex-

TABLE 12.1.   Insect Sampling in the Canopies of *D. excelsa*

Tree	Habitat	Date	Time	Hours
Km 41 11	forest	7/8/95	0930–1200	2.5
Km 41 11	forest	7/9/95	0630–1130	5.0
Km 41 11	forest	7/10/95	0830–1200	3.5
Km 41 11	forest	7/10/95	1530–1700	1.5
Km 41 11	forest	7/11/95	0800–1100	3.0
Col. 26	10 ha	7/13/95	0930–1200	2.5
Col. 26	10 ha	7/14/95	0730–0930	2.0
Col. 26	pasture	7/21/95	0900–1200	3.0
Km 41 01	forest	7/23/95	1000–1200	2.0
Km 41 01	forest	7/24/95	0730–1100	3.5
Km 41 01	forest	7/25/95	0800–1100	3.0
Km 41 01	forest	7/26/95	0800–1100	3.0
Col. 06	pasture	8/4/95	1340–1540	2.0
Col. 06	pasture	8/5/95	0800–1200	4.0
Col. 19	10 ha	8/6/95	0630–1200	5.5
Col. 19	10 ha	8/6/95	1530–2000	4.5
Col. 19	10 ha	8/7/95	0520–0720	2.0
Dim. 3.6	100 ha	8/19/95	1530–1730	2.0
Dim. 3.6	100 ha	8/20/95	0830–1230	4.0
P.A. 3.11	100 ha	9/7/95	0600–1000	4.0

TOTAL   62.5

cluded from the analysis. *D. excelsa* is an ideal species for studies of fecundity, because the pods are large, hang in the tree for many months, and are projected above the canopies of neighboring trees, making them easy to see.

At Colosso, *D. excelsa* occurs in pasture and in forest fragments, which include the 10 ha reserve, gallery forest, and forest edges. Pasture trees experience less root competition than do trees in forest fragments, and light may be more intense in pasture. However, these trees are exposed to fires, heat, and desiccation during the dry season when air temperature, air vapor pressure deficit, and drought exceed conditions in intact forest (Nepstad, Uhl, et al. 1996). In the analysis, I combined the Colosso subgroups (pasture and fragment trees), and I analyzed them separately to account for habitat differences. Seeds were counted from 2,177 fallen pods from trees in Colosso (n = 25), Km 41 (n = 45), and Cabo Frio (n = 23) to obtain estimates of seed set.

Results

Tree Distribution

A total of 145 adult trees were mapped (table 12.2). There were 43 individuals in fragments, 25 in pasture and gallery forest, and 90 in the continuous forest sites. The density of *D. excelsa* in the 232 hectares of forest fragments in the three ranches is 0.17 trees per hectare, or 1.0 adult tree per 5.9 hectares of terra firme forest. This estimate is about half the density of *D. excelsa* individuals larger than 10 cm DBH found in the 67 hectares of the BDFFP forest inventory. The estimate of 0.3 trees/ha places it close to the mean population density of 0.37 trees/ha calculated for all 127 leguminous tree species in the BDFFP reserves. Further, the spatial distribution of adults was

TABLE 12.2. Distribution of *Dinizia excelsa*

Reserve	Area (ha)	N	Density (tree/ha)
Dimona	100	11	0.11
Porto Alegre	100	16	0.16
Colosso	10	11	1.1
Porto Alegre	10	2	0.2
Dimona	10	1	0.1
Colosso	1	1	1.0
Dimona	1	1	1.0
Km 41	forest	45	n.e.
Cabo Frio	forest	30	n.e.

Notes: DBH > 40 cm in the BDFFP reserves. N.e. = not estimated.

clumped; the 10 ha fragment at Colosso had 11 adults compared with 2 adults in the Porto Alegre 10 ha fragment and a single adult in the Dimona 10 ha. This kind of spatial pattern is common for tropical tree species (Hubbell and Foster 1983).

Distance to nearest conspecific (DNC) was used as a measure of population density in Colosso (table 12.2). The Colosso 10 ha reserve has the highest density of adult trees (mean DNC = 50.5m [n = 12]), while pastures at Colosso have the lowest density (mean DNC = 235m [n = 18]). This is a highly significant difference (Welch ANOVA p < F = 0.002).

Phenology

Most individuals maintained leaves until the end of the wet season (April through June) at which time the leaves turn yellow and fall off. Flower stalks appear along with new leaves. Anthesis occurs during the dry season, from July through early September. Although individual flowers last only a few days, fresh flowers continually open on different inflorescences and branches, and individual trees remain in flower for up to one month. Large red pods are visible in the canopy in September, where they remain for up to eight months until the seeds mature. Pods with ripe seeds are dispersed by seasonal winds in May or June of the following year. Very few trees were observed with fruits in March 1995. This is consistent with the observations of Alencar, Almeida, and Fernandes (1979), which suggest that *D. excelsa* has biennial or superannual flowering.

The phenological observations suggest a slight difference in floral timing between Km 41 and Colosso. Sixty-three percent of the trees in Colosso had open flowers on July 22, 1995, with only 10 percent still in bud. One week later in Km 41 only 3 percent of the flowers were open while the rest were in bud. Within any one population there were only a few individuals whose flowering did not overlap with any other tree in the population. For example, Col. 03 finished flowering before any of its immediate neighbors began anthesis. Many insects, especially bees, were observed (through binoculars) on its flowers, but it did not produce any pods. This is consistent with genetic data (Dick 1999), which suggest that *D. excelsa* is partially self-incompatible. There was no marked difference in the mean flower production across sites.

POLLINATOR SHIFTS

D. excelsa of continuous forest were visited only by native insects. These included stingless bees (Meliponini), wasps, and tiny beetles (table 12.3). Stingless bees are probably the primary pollinators of trees in the continuous forest. They are common pollinators and were the only insects observed foraging extensively between branches. Native bees were represented by fourteen species in twelve genera; all of the specimens were smaller than *Apis mellifera*. The complete absence of larger bees (e.g., *Xylocopa, Euglossine*) is consistent with observations on other species with small, open flowers (Frankie et al. 1983). Along with bees, at least ten species of small beetles from eight families foraged in the floral cups

TABLE 12.3.   Bees and Beetles Collected from *D. excelsa* Flowers

Taxon	Size (mm)
Bees	
Anthophoridae (Tetrapediini)	9.0
Aparatrigona sp.	8.0
Apis mellifera	11.0
Cephalotrigona sp.	10.0
Frieseomelitta cf. *trichocerata*	7.0
Halictidae (sp.)	6.5
Melipona fulva	10.0
Melipona lateralis	10.5
Partamona cf. *nigrior*	6.0
Partamona cf. *pseudomusarum*	5.5
Partamona sp.	5.0
Plebeia sp.	3.5
Ptilotrigona lurida	8.0
Tetragona goettei	8.0
Trigonisca sp.	—
Trigona cf. *branneri*	6.0
Trigona sp.	6.0
Beetles	
Curculionidae sp. 1	5.0
Curculionidae sp. 2	2.0
Elateridae sp.	5.5
Mordellidae sp. 1	2.0
Mordellidae sp. 2	3.0
Melandrydiae sp. 1	4.0
Lycidae sp.	5.0
Cerambycidae sp.	10.0
Scarabaeidae sp.	6.0
Melolonthinae sp.	4.0

Notes: Several species of wasps (Vespidae) and moths (Lepidoptera) also were collected. Size indicates body length. Note that *Apis mellifera* is the largest of all the bee visitors. Bees were identified by M.V.B. Garcia. Beetles were identified by G. Morse.

of *D. excelsa*. These beetles are important pollinators for some plants (e.g., Armstrong and Irvine 1989). However, they were not observed foraging between inflorescences of *D. excelsa*; at most they may be vectors of self-pollination. The large predatory wasps were not considered potential pollinators.

Although most native insects left the canopy by sunset, the floral fragrance lingered for an hour after sunset, during which time moths (Noctuidae) were observed visiting the flowers of tree Col. 19 in the 10 ha fragment. The moths were observed in low numbers (two to three individuals per large flowering branch per thirty minutes). Even if the stigmas remain receptive at this time, moths did not visit frequently enough to contribute substantially to fecundity. Virtually nothing is known about the role of Noctuid moths as pollinators of tropical plants (Bawa 1990).

The same species composition and relative abundance of other native insects was maintained in the forest fragments. However, *D. excelsa* in all disturbed habitats were visited by large numbers of African honeybees, which were entirely absent in continuous forest. African honeybees were so copious in the 10 ha fragment trees that their buzz could be heard from the ground. No behavioral interactions were observed between African honeybees and native insects, despite their high densities. African honeybees were virtually the exclusive pollinating insects on the pasture tree Col. 06. That tree was not visited by any of the tiny beetles, and only a few individual bees from two native bee species (*Frieseomeitta trichocerata* and Trigonidae). African honeybees outnumbered a thousandfold the native bees on this tree.

The coexistence of native and alien bees in the 10 ha fragment suggests that African bees did not displace native insects from the pasture tree. Rather, the long distance of open pasture (and perhaps the canopy height of *D. excelsa*) may inhibit foraging by native insects, which nest in standing forest (Roubik 1989). The absence of African honeybees in the continuous forest in this and other studies (e.g., Aizen and Feinsinger 1994a; Roubik 1996) suggests that densities of African honeybees are lower in continuous forest.

The loss of small beetles in the pasture tree was probably of minor consequence for *D. excelsa*, because their foraging was limited to inflorescences on the same tree. Although "microbeetles" are important pollinators of dioecious trees in the nutmeg family (Myristicaceae) (Armstrong and Irvine 1989), an important difference is that nutmegs temporally limit the amount of nectar available to the microbeetles, which forces them to fly to other trees. *D. excelsa* provides floral resources continuously throughout the day. The effect of spatial isolation on the pollination of nutmegs and other beetle-pollinated trees merits further investigation.

Fecundity Differences

The difference in pod set between disturbed sites and continuous forest was striking. Colosso trees produced more pods than either continuous-forest population (fig. 12.2). Furthermore, there was no significant difference between the forest fragment and pasture subpopulations. This indicates that changes in the physical environment do not account for the increased pod set.

There were significant differences in the number of seeds per pod in Colosso (3.55; SE = 0.097; n = 592 pods), Km 41 (2.78; SE = 0.068; n = 815 pods), and Cabo Frio (3.65; SE = 0.077; n = 770 pods). But these samples may be biased because my technicians focused their collections on the pods that contained seeds, and different technicians worked at each site. Any such bias, however, would not change the overall result; the trees of Colosso produced more seeds than trees of continuous forest, regardless of whether they occurred in open pasture or in forest fragments. The main environmental change experienced collectively by these trees was their visitation by African honeybees, which is at least partially responsible for their increased fecundity.

It is plausible that increased fecundity will result in the regeneration of *D. excelsa* in disturbed habitat. *D. excelsa* pods can be dispersed widely across the open landscape

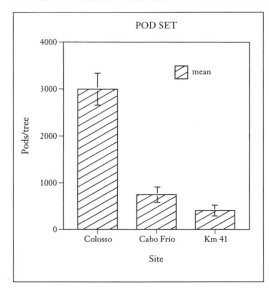

Fig. 12.2. Average number of pods per tree in three populations. Colosso (pasture and 10 ha fragment), n = 25; Cabo Frio (continuous forest), n = 23; and Km 41 (continuous forest), n = 45.

by strong winds. The seeds remain viable through long periods of desiccation, and seedlings thrive in abandoned pastures. By enhancing the fecundity of remnant populations and isolated trees, African honeybees increase the chance of recruitment in secondary habitats, and thereby promote forest regeneration.

GENE FLOW IN REMNANT D. EXCELSA

The high fecundity of remnant D. excelsa raises several intriguing genetic questions. Are the abundant seeds in the pasture trees the result of outcrossing or self-fertilization? Does the outcrossing rate vary across habitats? Finally, with respect to the outcrossed seeds from pasture trees, from where did the pollen come? Although honeybees forage over longer distances than do native bees, they typically forage among flowers in the same tree, and inter-tree foraging is considered rare (Butz-Huryn 1997). Ongoing work employing microsatellite DNA markers de-

veloped specifically for D. excelsa (Dick and Hamilton 1999) will provide answers to these questions.

IMPLICATIONS OF AFRICAN HONEYBEES FOR THE PLANT COMMUNITY

The residual, outlying D. excelsa in pasture and forest fragments were highly fecund, largely as a result of increased pollination by African honeybees. The replacement of native pollinators by African honeybees was a boon to remnant D. excelsa. But to what extent do these findings apply to other plant species?

Honeybees (including African honeybees) visit up to one-third of the plant species in local floras (Roubik 1996; Butz-Huryn 1997) but intensively visit and pollinate a much smaller proportion. Menezes, de Menezes, and de Camargo (1991) reported African honeybees on 47 out of 187 plant species in Brazilian wooded savanna (cerrado) south of the Amazon, although only nine species received 65 percent of the visits. To the plants they visit, honeybees can be major pollinators, or secondary pollinators that pollinate less effectively than native insects, or floral parasites that pollinate only incidentally (Butz-Huryn 1997). Honeybees are usually major pollinators of plants with simple flowers and rich nectar or pollen resources, such as many species in the Burseraceae, Sapotaceae, and Fabaceae, including D. excelsa. Honeybees are the major pollinators of a threatened endemic palm species in Madagascar (Neodypsis decaryi; Ratsirarson and Silander 1996) and of such commercial trees as cashew (Anacardium occidentale) (Freitas and Paxton 1996) and avocado (Persea americana) (Isham and Eisikowitch 1993).

Honeybees are usually secondary pollinators of plants that provide few rewards or offer non-nectar rewards of resins, oils, or fragrance. In a study of two open-pollinated shrubs in dry forest fragments of Argentina (*Propopsis nigra* [Mimosoideae] and *Cercidium australe* [Caesalpinioideae]), Aizen and Feinsinger (1994a) found that despite high visitation rates by African honeybees, the fragmented populations suffered slightly reduced seed set compared to continuous forest populations. In some cases, honeybees make up for lower efficiency by visiting flowers more frequently. In flowers of the dry forest annual *Kallstroemia grandiflora,* African honeybees transferred 2.5 times fewer pollen grains per visit than did the native bee *Trigona nigra* but compensated by making 2.65 times as many visits (Osorioberistain et al. 1997).

African honeybees may reduce the viability of plants with specialized flowers, especially if significant quantities of floral resources are extracted without pollination. One such class of plants has poricidal anthers that require sonic vibrations to release their pollen (buzz pollination). Bees curl around the anther and grasp the stamens tightly with their wings. By rapidly flexing their indirect flight muscles, they transmit sonic vibrations (buzzing) that expel pollen from the anthers (Buchman 1983). Although many native bees are capable of buzz pollination, honeybees are not. Buzz-pollinated *Mimosa pudica* (Mimosaceae) in French Guiana suffered a 26 percent decline in seed set when about 74 percent of the visitors were honeybees, compared to forest populations visited by high frequencies of native bees (Roubik 1996). In Roubik's study, the individuals with reduced fecundity also received fewer visits. Buzz-pollinated species are found in such prominent families of Amazonian trees as the Fabaceae, Lauraceae, Elaeocarpaceae, Solanaceae, and Melastomataceae (Buchman 1983).

The effect of pollinator disruptions on plant demography will vary widely among species and populations (Bond 1994). Self-incompatible species are expected to be more vulnerable to pollinator disruptions than are self-compatible species. Plants with highly specialized floral morphologies may be less able to endure changes in pollinator composition than species with simple flowers. For species with large fruits and seeds, nutrient or water resources may limit seed production more than pollination. Species with recalcitrant seeds (requiring moisture) may be limited by the availability of moist sites for regeneration in pasture-dominated landscapes. All of these expectations merit empirical examination.

CONSERVATION ROLE OF REMNANT TREES

The pattern of forest fragmentation in the ranches north of Manaus is not unique. Millions of hectares of Amazonian forest have been converted to pasture and abandoned (Neptstad et al. 1996). One observes the "BDFFP pattern" along major roads through Amazonas state—patches of disturbed terra firme forest, peninsular gallery forests, and solitary emergent trees surrounded by pastures or stands of cecropia and *Vismia* spp. Where *D. excelsa* does not occur, other emergent species take their place as shade trees in pasture. On the BDFFP ranch Porto Alegre "maçaranduba" (*Manilkara huberei;* Sapotaceae) plays this role. Pastures in Mexico are dotted by large fig trees (*Ficus;* Guevara and Laborde 1993). *Pithecellobium elegans* (Leguminosae) is a common pasture tree in Costa Rica (Chase, Boshier, and Bawa 1995). Pasture trees not only serve as genetic links between fragmented populations (Chase, Boshier, and Bawa 1995), but they also facilitate forest regeneration by providing roosts for frugivorous birds (Guevara and Laborde 1993) and epiphyte gardens (Hietzseifert, Hietz, and Guevara 1996).

Selective logging creates a different kind

of population fragmentation. Approximately 8,000 km² of Amazon forest are selectively logged each year in Brazil (Holdsworth and Uhl 1997). Loggers often cut out most of the reproductive trees in a forest. In the case of mahogany (*Swietenia macrophylla*) there may not be enough young trees left after logging to ensure the persistence of local populations (Gullison et al. 1996). Ecologically responsible logging operations like Mil Madeireira (see Chapter 7) leave many of the largest trees standing because they are hollow. After the first round of cutting, however, the hollow trees are systematically girdled to release light for regenerating seedlings (T. Meldik, pers. comm.).

Holdsworth and Uhl (1997) have shown that selectively logged Amazonian forests are susceptible to fires that result in a 44 percent mortality of all trees with a DBH of at least 10 cm. However, as this study has shown, large individuals of *D. excelsa* survive periodic fires and produce seeds that grow in desiccated, sun-drenched environments. These trees represent important gene banks that need to be conserved, especially when no other regeneration practices are being implemented. Moreover, the hollow trees provide nesting habitat for pollinators —both native and introduced.

Conservation Lessons

1. The diversity and abundance of insects that visit flowers of *D. excelsa* differ in pasture and fragments compared to continuous forest. Most native insects are absent from pasture trees. However, native insect abundance is maintained in 10 ha fragments. African bees were abundant in forest fragments and pasture but were not observed in continuous forest. African honeybees were nearly the exclusive pollinators of pasture trees and dominated the bee community in all disturbed habitats.

2. African honeybees are useful to only a small portion of Neotropical species. Only plants that offer rich nectar or pollen rewards are visited intensively by African honeybees. African honeybees can damage specialized flowers and may compete with specialized pollinators.

3. The loss of small beetles in pasture suggests that beetle-pollinated plants may be vulnerable to spatial isolation.

4. Because of its valuable timber, rapid growth, ease of seed storage, and resilience to desiccation and to pollinator shifts, *Dinizia excelsa* may be useful for silviculture and for the ecological restoration of Amazonian pastures.

5. The silvicultural practice of girdling large, hollow trees to release light may hinder forest regeneration. Such trees are often highly fecund, serve as gene and seed banks, and provide nesting sites for pollinators and roosting sites for seed-dispersing birds.

Acknowledgments

This work was supported by an NSF Predoctoral Fellowship, the Harvard Committee on Latin American Studies, the Department of Organismic and Evolutionary Biology of Harvard, and a Deland Award from the Arnold Arboretum. Antônio Ribeiro Mello, Romeu Moura Cardoso, Sebastão Salvino de Souza, and Alaércio Marajó dos Reis provided invaluable assistance in the field. I thank P. Ashton, S. Alberts, J. Chambers, R. Lewontin, D. Roubik, P. Stevens, and an anonymous reviewer for discussion and comments on the manuscript, and M. Nee for introducing me to the flora of BDFFP.

CHAPTER 13

Fragmentation and Plant Communities

Synthesis and Implications for Landscape Management

WILLIAM F. LAURANCE

A key component of the Biological Dynamics of Forest Fragmentation Project is a long-term study of forest dynamics, floristics, and biomass in fragmented and continuous forest. This work is crucial, for if fragmentation fundamentally alters forest vegetation, many aspects of the forest's ecology—its structure, microclimate, ecosystem functions, and the composition of animal communities—are also likely to be affected (Klein 1989; Didham, Lawton, et al. 1998; Laurance 1997).

Progress in the BDFFP plant ecology project was initially slowed by the overwhelming diversity of the Amazonian tree flora (see Rankin–de Mérona et al. 1992; Oliveira and Mori 1999), but there has been much progress in recent years, and some important patterns are emerging. Here I synthesize new findings and forward a general perspective on the effects of fragmentation on Amazonian plant communities. Parts of this synthesis are necessarily speculative, but they represent the best available knowledge.

There is no question that fragmentation is dramatically altering the ecology of Amazonian rainforests. These effects are most striking when one examines small (less than 10 ha) fragments, but even much larger (100–1,000 ha) fragments are clearly being affected. I briefly describe our study design, then highlight three general ecological phenomena—changes in forest dynamics, the erosion of floristic diversity, and the decline of biomass—that affect plant communities in forest fragments. I conclude by discussing some implications of these findings for landscape management and forest conservation.

Study Design

As detailed in Chapter 4, the BDFFP is a 1,000 km² experiment in which forest patches ranging from 1 to 100 ha in area were isolated in the early 1980s by clearing and often burning the surrounding rainforest (fig. 13.1). Today, these forest patches are encircled by cattle pastures or by 3 to 15 m tall regrowth forest, generally dominated by shrubby *Vismia* spp. and *Bellucia* spp. if the surrounding area was burned, or by taller *Cecropia* spp. if it was not burned (plates 8 and 9; see Chapters 24 and 25).

The core of the BDFFP Phytodemographic Project is 69 permanent, 1 ha study plots, within which more than 62,000 trees at least 10 cm in DBH have been studied since the early 1980s (Rankin–de Mérona, Hutchings, and Lovejoy 1990; Rankin–de Mérona et al. 1992). Thirty-nine of the plots are in frag-

This is publication number 280 in the BDFFP Technical Series.

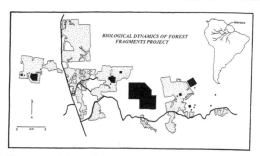

Fig. 13.1. Map of the BDFFP study area, showing locations of forest fragments and continuous-forest reserves (shaded) that contain permanent study plots. Shaded and unstippled areas are rainforest, while stippled areas are cattle pastures or regrowth forest. Thick, wavy lines are roads.

ments (four of 1 ha, three of 10 ha, and two of 100 ha), while the remainder are in nearby continuous forest (fig. 13.1). Plots in fragments are stratified so that edge and interior areas are both sampled.

Every few years, each plot is visited by field technicians to measure tree growth and to record any dead, damaged, or new trees. To date, taxonomic experts have identified about 1,300 species or morpho-species (species that are unidentified but appear morphologically different from anything already identified) of trees from our plots, including many previously undescribed species. The BDFFP reference collection has improved dramatically in recent years; almost 90 percent of the trees in the study have now been identified to at least genus and morpho-species level.

A recent addition to the phytodemographic study is lianas (climbing woody vines), which are also being sampled in each of the permanent plots (Laurance, Perez-Salicrup, et al., in press). Fragmentation studies on additional plant groups (e.g., palms: Wandelli 1991, Scariot 1998, Chapter 10; seedlings: Benitez-Malvido 1998, Chapter 11; Sizer and Tanner 1999), and basic studies of plant ecology or systematics (e.g., Ack-

erly, Rankin–de Mérona, and Rodrigues 1990; Mori and Becker 1991; Chapter 6; Vasconcelos 1991; Oliveira 1997) are being conducted by other scientists and students.

Changes in Forest Dynamics

Under typical conditions, rainforests have their own rhythm, with trees dying and regenerating at a moderate and roughly constant rate (Clark 1990). In any given year, for example, from 1 to 2 percent of the trees might die (Phillips and Gentry 1994). These natural dynamics are fundamentally altered by fragmentation. The most immediate effect of isolation is a dramatic pulse of tree death, with annual mortality rates commonly rising to 5 to 30 percent, especially near fragment margins (Lovejoy, Rankin, et al. 1984; Ferreira and Laurance 1997; Laurance, Ferreira, et al. 1998a). Large trees (more than 60 cm DBH) are especially vulnerable to fragmentation (Laurance, Delamônica, et al. 2000).

These tree deaths probably occur for two reasons. First, microclimatic changes near edges, such as reduced humidity and higher temperatures, create physiological stresses for many rainforest trees, especially within 15 to 60 m of edges (Kapos 1989; Williams-Linera 1990b). This is evidenced by trees near newly created edges often shedding many of their leaves—a common response to desiccation (Sizer and Tanner 1999) and increasing light levels (Bazzaz 1991). Microclimatic changes are probably responsible for the many trees that die standing, with no apparent damage, near fragment edges (Laurance, Ferreira, et al. 1998a). Such effects may be greatest in recently created fragments; edge-related changes in microclimate may be somewhat ameliorated over time as edges grow older and become "sealed" by secondary vegetation (Kapos et al. 1997).

Second, trees near forest edges are exposed to increased wind shear, turbulence, and vorticity, which often lead to elevated windthrow and forest-structure damage (Laurance 1991a, 1997; Chen, Franklin, and Spies 1992; Ennos 1997; Ferreira and Laurance 1997; see fig. 13.5b). As an edge effect, wind damage probably occurs over considerably larger spatial scales than changes in forest microclimate. For example, wind-tunnel models suggest that strong turbulence can persist downwind for at least 2 to 10 times the height of the forest edge (Savill 1983). In the BDFFP study area, rates of windthrow and tree damage are sharply elevated within 60 to 100 m of edges (fig. 13.2), with modest but detectable increases occurring up to about 300 m from edges (Laurance, Ferreira, et al. 1998a). Unlike microclimatic changes, wind damage is unlikely to lessen as fragment edges become older and less permeable, because downwind turbulence usually increases as edge permeability is reduced (Bull and Reynolds 1968; Savill 1983).

Available evidence suggests that increased tree mortality and damage are not merely ephemeral events that occur immediately after fragmentation; rather, the intensity of forest dynamics is elevated for long periods—perhaps indefinitely, the result of recurring windstorms or pervasive microclimatic changes in fragments. In the BDFFP study area, mortality and damage rates of trees have remained unusually high in fragments between 10 and 17 years old, while forest turnover (the mean of annual tree mortality and recruitment rates) has increased significantly with fragment age (Laurance, Ferreira, et al. 1998a).

Such fundamental changes in forest dynamics are likely to influence many aspects of the fragment's ecology. Microclimatic changes in fragments will be exacerbated if frequent treefalls rupture the forest canopy,

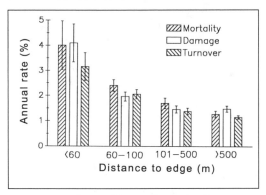

Fig. 13.2. Indices of forest dynamics for permanent study plots at varying distances from the forest edge.

allowing greater light and wind to reach the forest interior (Kapos et al. 1997). Because tree mortality rates are increased, the size distributions of trees will tend to shift toward smaller individuals. Disturbance-adapted vines, lianas, and weeds typically increase in abundance (Laurance 1991a, 1997; Oliveira-Filho, de Mello, and Scolforo 1997; Viana, Tabanez, and Batista 1997; Laurance, Perez-Salicrup, et al., in press), while growth rates of smaller trees will rise in response to increased light levels. In smaller fragments the complex architecture of the forest—from the ground to the treetops—can be markedly altered (Laurance 1991a; Malcolm 1994; Oliveira-Filho, de Mello, and Scolforo 1997).

Erosion of Floristic Diversity

These disturbances—in concert with the isolation of fragments from other areas of forest—may lead to substantial losses of plant diversity. There are several mechanisms by which such losses could occur.

First, increased disturbance rates in fragments, along with the close proximity of edge and matrix habitats, will clearly favor pioneer and successional species at the expense of old-growth species adapted for the

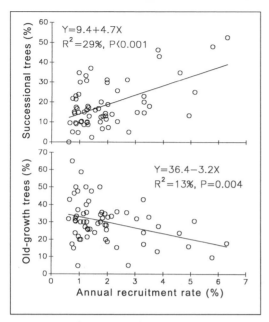

Fig. 13.3. Relationships between annual recruitment rate and proportions of successional and old-growth tree families in study plots.

forest interior (fig. 13.3). In the BDFFP, study plots with high rates of disturbance (high annual tree-recruitment rates) tend to have relatively high densities of trees in predominantly successional families (Annonaceae, Cecropiaceae, Clusiaceae, Euphorbiaceae, Malpighiaceae) and relatively low numbers of trees in mainly old-growth families (Burseraceae, Chrysobalanaceae, Lecythidaceae, Sapotaceae; Laurance, Ferreira, et al. 1998b). Increased litterfall and microclimatic changes near edges can reduce seed germination for some species (Bruna 1999). Epiphytes and orchids, which may be prone to edge-related drought stress, appear particularly vulnerable to fragmentation in Singapore (I. Turner, Tan, et al. 1994, I. Turner, Chua, et al. 1996).

Second, many rainforest tree species are rare, with average densities of less than one mature individual (at least 10 cm DBH) per hectare. Local populations of such species

may be too small to be viable without an ongoing influx of propagules from outside source areas, which are optimal habitats in which the species is most abundant and reproduction is occurring (Hubbell and Foster 1986; Condit, Hubbell, LaFrankie, et al. 1996). When fragments become physically isolated from other forest areas, the rate of arrival of propagules may decline sharply (Nason, Aldrich, and Hamrick 1997; Thebaud and Strasberg 1997), possibly leading to an eventual collapse of isolated populations. Because at least three-quarters of all rainforest tree species can be considered locally rare (Gentry 1990, Rankin–de Mérona et al. 1992), the losses of such species could seriously erode the floral diversity of forest fragments.

Third, tropical plants are involved in a rich array of mutualistic relationships with animals—for pollination, seed-dispersal, movement of mycorrhizal spores, and defense (as in ant-defended plants). In some cases these relationships are highly specialized, with strong interdependence between the plant and animal. Many orchid species, for example, are pollinated by a particular species of Euglossine bee (Powell and Powell 1987), and each fig has its own special Agaoinine wasp pollinator (Galil and Eisikowitch 1971). As the other chapters in this volume demonstrate, many animals decline or disappear in fragmented habitats (e.g., Chapters 16, 18, 20, and 21). If a crucial pollinator or seed-disperser should disappear, the loss could lead to reproductive failure and the eventual collapse of dependent plant species (Powell and Powell 1987; Aizen and Feinsinger 1994). Such phenomena have been termed ripple-effect extinctions, being likened to the waves emanating from a stone (the initial extinction) dropped into a pond (Gilbert 1980).

Fourth, invasions of exotic or generalist species from surrounding matrix habitats can alter the floristic composition of forest remnants and possibly lead to declines of native species (Janzen 1986a; Brown and Hutchings 1997; Chapters 10 and 14). A number of studies have demonstrated that disturbed forests are increasingly prone to plant invasions, probably because sunlight, which is normally a limiting resource, increases in the understory (e.g., Mooney and Drake 1986; Hobbs 1989; Hutchinson and Vankat 1997). In fragmented forests, the abundances of invading and disturbance-adapted species are often negatively correlated with distance from the forest edge (Laurance 1991a; Brothers and Spingarn 1992). In the tropics, forest edges are especially prone to invasions of exotic vines and lianas, which can overgrow and smother native vegetation (Stockard, Nicholson, and Williams 1985; Jenkins 1993; Laurance 1997; Viana, Tabanez, and Batista 1997).

Finally, various other factors could cause plant extinctions in fragmented landscapes. Deforestation is often a nonrandom process, so that the particular soils or microhabitats on which certain plant species depend may be absent or poorly represented in fragments (Laurance, Fearnside, et al. 1999). Fragments may be too small to sustain a complete mosaic of natural disturbances and seral vegetation, from pioneer to climax species (Pickett and Thompson 1978; Foster 1980). Distortions of food webs can occur in fragments, with predators and parasites declining and generalist herbivores increasing, and this may affect plant communities (Terborgh, Lopez, et al. 1997); on small islands in Lake Gatun, Panama, for example, the disappearance of rodent seed-predators has allowed large-seeded trees, which are highly competitive but normally vulnerable to seed predators, to dominate the islands (Leigh et al. 1993). Clearly, there is a wide range of mechanisms by which plant populations can decline or collapse in fragmented habitats.

Biomass Collapse

In addition to changes in forest dynamics, structure, and species composition, research in the BDFFP has revealed a new and unanticipated consequence of habitat fragmentation—a substantial loss of forest biomass (W. Laurance, Laurance, et al. 1997). This occurs as a direct result of increased tree mortality, especially of large trees (more than 60 cm DBH), which can contain a quarter or more of the forest's above-ground biomass (Laurance, Delamônica, et al. 2000). In Amazonian fragments, above-ground biomass declines substantially within 100 m of forest edges (figs. 13.4, 13.5a, and 13.5b), with some plots losing up to 36 percent of their total biomass. Although growth rates of lianas and new trees have increased in fragments (Laurance, Ferreira, et al. 1998b; W. Laurance, Laurance, and Delamônica 1998; Laurance, Perez-Salicrup, et al., in press), these increases have failed to offset biomass losses led by sharply elevated tree mortality (fig. 13.6).

The decline of biomass in forest fragments

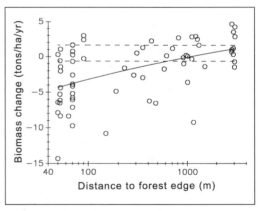

Fig. 13.4. Rate of change in above-ground tree biomass in BDFFP study plots as a function of distance of plots from forest edge. The solid line is an exponential curve fitted to the data, and the dashed lines are 95 percent confidence intervals for 26 forest-interior plots (more than 500 m from edge).

Fig. 13.5a. The corner of a recently isolated reserve shortly after the surround felled forest was burned. Note the defoliated trees along the forest edge. Photo by Richard O. Bierregaard, Jr.

Fig. 13.5b. A reserve edge (1202) more than ten years old shows that the canopy trees that stood just inside the fence have been replaced by low, dense second growth. Photo by Christopher Dick.

has important implications for the global climate. Upon decay, dead trees release such greenhouse gases as carbon dioxide and methane, which are known causes of global warming. Estimating the effects of forest fragmentation on greenhouse gas emissions is a complex matter, requiring assumptions about the sizes and shapes of forest remnants in fragmented landscapes, the biomass and carbon contents of undisturbed

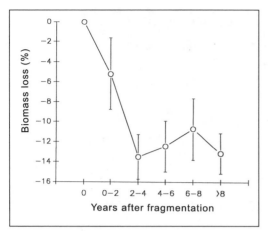

Fig. 13.6. Kinetics of biomass decline in sixteen forest plots that experienced substantial biomass losses (mean rate more than 3 tons/ha/year).

forests, the percentage of biomass lost in deforested areas, and other factors. An initial estimate, based on simulations of realistic deforestation patterns, suggests the worldwide fragmentation of tropical forests produces from 22 million to 149 million tons of carbon emissions annually, above and beyond that caused by forest clearing per se (W. Laurance, Laurance, and Delamonica 1998). These emissions are equivalent to clearing and burning from 0.15 to 1.1 million hectares of rainforest each year.

Implications for Landscape Management

A growing body of research indicates that the fragmentation of Amazonian forests will have deleterious effects at a wide range of spatial scales. On a local scale, fragmented forests exhibit a substantial loss of biodiversity (Lovejoy, Bierregaard, et al. 1986; Bierregaard et al. 1992). On a regional scale, landscape fragmentation hinders the movements of migratory species (Powell and Bjork 1995) and dispersal of plant propagules (Thebaud and Strasberg 1997), halts or

impedes gene flow (Nason, Aldrich, and Hamrick 1997), and causes ecosystem-level changes in nutrient flows, soils, hydrology, and climate (Salati and Vose 1984; Hobbs 1993). On a global scale, fragmentation contributes to the production of greenhouse gases, which are the principal cause of global warming (W. Laurance, Laurance, et al. 1997; W. Laurance, Laurance, and Delamonica 1998).

Aside from an obvious generic guideline —"Wherever possible, do not fragment tropical forests"—there are a number of specific recommendations for land management that arise from recent work in the BDFFP. Here I focus on two particular minima: minimum reserve sizes, and minimum widths for faunal corridors. In both cases my recommendations are based on the observation that edge effects often play a dominant role in the ecology of fragmented forests.

Minimum Reserve Sizes

Growing evidence suggests that different groups of organisms—plants, insects, vertebrates—exhibit very different spatial patterns of species diversity and endemism (Gentry 1992; Prendergast et al. 1993; Kerr 1997). As a consequence, a limited number of expansive nature reserves designed explicitly to accommodate the needs of carnivores or other large vertebrates—the so-called "umbrella strategy"—may perform poorly as a method for conserving the great diversity of plants and invertebrates, especially in tropical ecosystems, which often have high species diversity related to local habitat diversity (beta diversity) and regional species replacement (gamma diversity), and complex patterns of local endemism (Laurance and Bierregaard 1997a). For these organisms, substantial numbers of smaller reserves, stratified across major environmental gradients, are a far more effec-

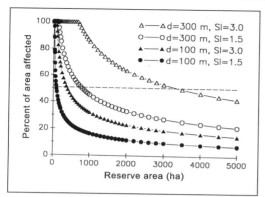

Fig. 13.7. A core-area model for rainforest fragments in central Amazonia, using empirical data on forest dynamics for strong (d = 100 m) and detectable (d = 300 m) edge effects. For each d value, curves were generated for fragments with moderately circular (SI = 1.5) and moderately irregular (SI = 3.0) shapes. The dotted line indicates the point at which 50 percent of the fragment's area is influenced by edge effects.

tive strategy (Pressey, Johnson, and Wilson 1994; Kerr 1997). Thus, networks of both large and small reserves are needed.

But how small is too small? At what point do the ecologies of small reserves become so seriously altered that their conservation values are likely to be minimal? One way of addressing this question is by applying the core-area model (Laurance 1991a; Laurance and Yensen 1991), which can be used to predict the unaffected core area within any forest remnant, provided that one knows its size, shape, and the distance (d) to which edge effects penetrate into fragment interiors. To develop a core-area model for Amazonian fragments, I used two empirically derived values for d: 100 m, which is the distance to which major changes in forest dynamics penetrate into fragments, and 300 m, which is the maximum distance to which accelerated forest dynamics have been detected (Laurance, Ferreira, et al. 1998a).

The core-area model suggests that detectable changes in forest dynamics (corresponding to d = 300 m) will become increasingly important once fragment area falls below 800–3,500 ha, depending on fragment shape (fig. 13.7). Major changes in dynamics (corresponding to d = 100 m) are likely to occur in fragments of less than 100–400 ha, again depending on fragment shape (fig. 13.7). Thus, reserves of only a few hundred hectares in area may be viable for organisms with very small area requirements (such as locally abundant invertebrates and plants), provided those species are not sensitive to moderately increased forest dynamics. For edge-sensitive species, or those that occur at lower densities and thus have larger core-area requirements, reserves of a few thousand hectares may be viable.

Obviously, a range of other factors, such as hunting pressure, the proximity of other reserves, and the nature of surrounding matrix habitats, will also affect the viability of small reserves, but the core-area model provides a simple, empirically based strategy for determining minimum reserve areas based on their susceptibility to edge effects. It must be emphasized that reserves of only a few thousand hectares in area will be far too small in the long term for many larger forest vertebrates, but for locally endemic plants, insects, and small vertebrates, small reserves can form an integral component of a regional reserve network. Small reserves are, nevertheless, especially prone to fires, logging, hunting, and other forms of exploitation from local populations, and for this reason they will often require intensive management and protection from external threats to remain ecologically viable (for further discussion, see Gascon, Williamson, and Fonseca 2000 and Chapter 29).

CORRIDORS AND CONNECTIVITY

The use of faunal corridors to link fragments and nature reserves has been widely advocated as a strategy to help counteract the deleterious effects of population fragmentation (Noss 1987; Hobbs 1992; S. Laurance and Laurance, 1999). In the Amazon, where deforestation is transforming landscapes throughout the region, several principles of corridor design can be inferred from ecological studies in the BDFFP and elsewhere.

First, corridors should be composed of primary forest. This is because many species that respond negatively to fragmentation—and thus could benefit from the use of corridors—are largely or wholly confined to primary forest (Lovejoy, Bierregaard, et al. 1986; Laurance 1991b; Stouffer and Bierregaard 1995a; Bierregaard and Stouffer 1997), although there are interesting exceptions (Malcolm 1991; Chapter 19). The need for primary-forest corridors highlights the dire urgency of preventing the complete disappearance of primary vegetation between both planned and existing nature reserves in the tropics (Laurance and Gascon 1997).

Second, wherever possible, even small gaps in corridors should be avoided. In the tropics, many forest-dependent birds (Bierregaard et al. 1992) and mammals (Laurance 1990; Goosem 1997) have been shown to avoid even narrow (50–100 m wide) clearings, especially if the clearing is maintained as open habitat (Stouffer and Bierregaard 1995a).

Third, studies of edge effects suggest that, wherever possible, corridors should be at least 300 m in width, to ensure that some relatively unaffected habitat persists within the corridor's interior. This recommendation follows from the observation that major changes in forest dynamics occur within the first 100 m of forest edges (Ferreira and Laurance 1997; W. Laurance, Laurance, et al.

1997; Laurance, Ferreira, et al. 1998a, 1998b). If the conservation goal is to provide a corridor with true forest-interior habitat, then a minimum width of 1 km is advisable, given that detectable changes in forest dynamics can occur up to about 300 m from the edge (Laurance, Bierregaard, et al. 1997; Laurance, Ferreira, et al. 1998a; see fig. 29.1).

Clearly, there is great scope for improvement of Brazilian legislation in relation to faunal corridors. Current legislation requires that buffer strips of forest be retained along streams, but the strips need be only 5–100 m wide (IBDF 1965), which is likely to be inadequate for faunal conservation (Laurance and Gascon 1997). Updating the legislation to incorporate these new guidelines—and ensuring their enforcement —could go a long way toward maintaining ecosystem connectivity in developed landscapes. It must be emphasized that the corridors described here are intended to link fragments separated by distances of only a few kilometers or less. Regional corridors spanning dozens to hundreds of kilometers will probably need to be far larger (perhaps 20 to 50 km wide) to remain viable.

Conservation Lessons

1. Fragmentation fundamentally alters the dynamics of Amazonian forests. Sharply elevated rates of tree mortality, damage, and turnover are apparently caused by microclimatic changes and increased wind turbulence near abrupt forest edges.

2. Altered dynamics affect many aspects of the forest's ecology, including its structure, composition, microclimate, and biomass. These diverse changes are likely to exacerbate the effects of fragmentation on disturbance-sensitive and forest-interior species.

3. The collapse of biomass in forest fragments could be a significant source of carbon emissions, potentially contributing to global warming. Model simulations suggest that fragmented tropical forests worldwide release from 22 million to 149 million tons of carbon annually, above and beyond that produced by forest clearing per se. These emissions are equivalent to clearing and burning an additional 0.15 million to 1.1 million ha of tropical forest each year.

4. In the Amazon, forest fragments of less than a few hundred hectares are likely to be substantially influenced by edge effects. Such fragments may have significant conservation values, but they are likely to differ ecologically from intact forest. Less striking but detectable changes in dynamics may occur in fragments of up to a few thousand hectares, especially if they are irregularly shaped.

5. Studies of edge effects suggest that faunal corridors linking nearby forest fragments in the Amazon should be at least 300–1,000 m in width, continuous, and composed of primary forest. There is great opportunity to improve ecosystem connectivity in fragmented Amazonian landscapes by modifying Brazilian legislation to require that wider swaths of vegetation be retained by landowners along streams and rivers. Improved enforcement of land-use legislation is also an urgent priority.

Acknowledgments

This chapter reflects the contributions of many individuals. Judy Rankin–de Mérona and Roger Hutchings initially directed the phytodemographic project, while BDFFP technicians invested thousands of hours collecting field data and many taxonomic experts kindly identified plant specimens. The efforts of Susan Laurance, Jakes Costas, Leandro Ferreira, Patricia Delamônica, Louis Coelho, and Sammya d'Angelo have dramatically improved the BDFFP herbarium collection and database. Claude Gascon and an anonymous referee reviewed a draft of this chapter. This work was supported by World Wildlife Fund-US, National Institute for Research in the Amazon, Smithsonian Institution, MacArthur Foundation, Andrew W. Mellon Foundation, and the NASA Long-Term Biosphere-Atmosphere Experiment in the Amazon.

Fragmentation Effects on Invertebrate Communities

We begin our section of chapters on invertebrates in fragmented landscapes with a study of a taxon for which the genetics are extremely well known while its ecology in nature has only recently been studied in detail—drosophilid fruit flies. In Chapter 14, Marlúcia Martins argues that the drosophilid fruit-fly guild, which relies on decomposing fruits, can serve as a valuable indicator of the strength of anthropogenic disturbance on a forest ecosystem. While overall the number of species using a similar resource can remain the same, the biomass and composition of the guild may change in response to changes in the environment. This echoes the familiar theme of a number of this book's chapters that it is not enough to simply measure species number in fragmented ecosystems.

Martins studied the fruit-fly guild in the experimental fragments of the BDFFP and the Mocambo Reserve in an even more fragmented landscape within the Belém city limits near the mouth of the Amazon.

At the BDFFP fragments, Martins found significant fragmentation and edge effects. Overall, the smaller reserves harbored fewer species, with rare, native, and species characteristic of forest being lost. Although the ranking of the most abundant species was the same in fragments as in continuous forest, their abundance increased dramatically. Strong edge effects were seen in a 100 ha reserve. The same group of species was dominant in the core and at the edge, albeit slightly less abundant, while other species were 100 times more abundant near the edge. Species number and diversity was also higher near the forest edge.

In the eastern study site, the changes in species richness and abundance were similar to those found at the BDFFP, only more accentuated. In particular, one exotic and invasive species, which has a more rapid life cycle than the native species, was much more prominent around Belém, perhaps because of the longer history of landscape modification and the greater impact of the nearby metropolitan area. The success of the invader may be an example of a second-order effect. Martins believes that fragmentation effects may have led to a paucity of small mammals, which would normally consume fallen fruit. This in turn means an ample supply of rotting fruit, which favors the rapidly breeding exotic species.

METAPOPULATIONS OF SOCIAL SPIDERS

For many, and not only the arachnophobes among us, one of the first images evoked by mention of a tropical rainforest is one of enormous tarantulas. The engraving of a bird-eating spider in Bates's *The Naturalist on the River Amazons* is an apt image of the topsy-turvy extremes (spiders are not supposed to eat birds!) to which evolution has wandered in tropical rainforests. Much less well known but equally fascinating are the social spiders. In a single colony, anywhere from 100 to 10,000 spiders weave a communal web that may be several cubic meters in volume. Social spiders have a division of labor within the colony, and they play host to a number of insect species that live in the webs with impunity, robbing prey caught in the spiders' web.

Social spiders (*Anelosimus eximius*) were the subject of Eduardo Venticinque and Harry Fowler's studies of metapopulation dynamics, reported in Chapter 15. "Metapopulation" is a term used to define a group of distinct subpopulations that are genetically interconnected by the movement of individuals between the subpopulations (Levins 1970). The concept adds an important refinement to island biogeography theory in that the source of potential colonists for an island or forest reserve is not only the nearby "mainland" but also the other islands themselves and follows from Brown and Kodric-Brown's (1977) notion of a "rescue effect" (see Chapter 2).

Venticinque and Fowler monitored the volume, numbers, and life span of spider colonies in a single 1 ha fragment, 10 ha isolates, continuous forest, and forest corridors. They showed that the spiders responded to both edge and fragmentation effects. The oscillation in colonies was associated with rainfall and was strongest at edges and weakest in the forest interior. Life spans varied between habitat type, with the forest interior having the longest life spans. The long life span of the colonies in the forest interior suggests an adaptation to periods of longer stability than are found in fragmented landscapes, as Martins suggested for fruit flies.

Larger colonies had stronger oscillations, and the fragment populations were synchronous. Finally, the population in the 1 ha reserve was the least stable and disappeared for a two-month period.

Venticinque and Fowler conclude that *A. eximius* in the landscape mosaic of the BDFFP does indeed behave as a metapopulation. They describe the spider as a "tramp" species, favoring clearings and forest edges in the wild, and said that it may be one of a subset of the Amazonian fauna that actually benefits when we fragment continuous forest.

ANTS IN MODIFIED LANDSCAPES

Ants abound across Amazonia. They are voracious predators in the form of army ants and highly evolved farmers in the form of herbivorous leaf-cutter ants, which feed on fungal gardens that they grow on masticated leaves in their enormous underground colonies. Ants represent a substantial portion of both the biomass and the biodiversity of tropical rainforests. Because they are relatively sedentary, taxonomically well known, ecologically diverse and important, and easy to catch, they make an excellent study organism with which to monitor the effects of our transformation of continuous forests into fragmented landscapes.

In Chapter 16, Heraldo Vasconcelos, Karine Carvalho, and Jacques Delabie review their work comparing the abundance and diversity of ants that live on the ground in undisturbed forest with those found in three types of disturbed habitats in the "matrix" around the BDFFP reserves. The matrix habitats they studied differed in the degree of original disturbance from areas where the forest had been felled but not burned at one extreme to areas where pasture had been planted and cattle raised for about ten years, at which time the soil is unable to support the *Brachiaria* spp. pasture grasses used in the region, at the other. This chapter is the first of several in this volume to report on studies of the matrix habitats surrounding the BDFFP's experimental plots—a research theme that has grown in importance since the ranchers essentially abandoned their cattle operations.

Not surprisingly, the conversion of mature forest to pasture has a profound effect on ant species. As expected, species richness declines (only about 17 percent of the primary-forest species are found in open areas), but at the same time the total abundance of ants increases. In abandoned areas, as a second-growth forest develops, the original ant fauna of primary forest begins to reassemble, with the rate of recovery strongly dependent on the severity of land use (number of years in pasture).

The authors conclude with a more traditional look at the effects of fragment size on the community of ground-dwelling ants in the BDFFP experimental plots. They surveyed the ants in 1 ha fragments and 1 ha subsamples of 10 and 100 ha isolates using pitfall traps and samples of the soil and leaf litter. There was no consistent relationship found between the number of ant species in the 1 ha subsamples and reserve size, but multivariate statistical analyses revealed similarities between the suite of species found in different-sized reserves and even on the different cattle ranches, suggesting noticeable differences in the species to be found only a few dozen kilometers apart, as well as an effect of either fragment size or the presence of a forest edge.

Given the heterogeneity of species composition observed at the scale of kilometers, the authors suggest that a series of "small" reserves sampling a variety of microhabitats would be more effective for the conservation of ants and some other invertebrates than a single very large reserve. But recognizing the area requirements of army ants (*Eciton* spp.) and some of the more sparsely distributed *Atta* spp. leaf-cutters, much larger areas, in excess of 1,000 ha, would certainly represent a minimum critical area for the preservation of ground-dwelling ants.

NATIVE BEES

As highlighted in Chapter 12, bees are the predominant pollinators for trees in the Amazonian rainforests. Márcio Luiz de Oliveira reports, in Chapter 17, on his surveys of two groups of native bees: the euglossine, or orchid, bees; and the meliponine, or stingless, bees, in primary forest and in BDFFP forest isolates.

The two groups, although ecologically dissimilar, are important pollinators in the primary forest. Euglossines are familiar to visitors to the rainforest, but their natural history and ecology is still poorly known. They do not have a hive with a queen and division of labor, nor do they produce honey. They are, however, important, if not essential, pollinators for many species of orchids. Male euglossines visit orchids to collect substances that they convert biochemically into the pheromones that they in turn use to attract mates. The group is fairly easy to sample because of the males' affinity for these chemicals, many of which are commercially available. Oliveira collected thirty-eight species, which combined with previous studies at the BDFFP bring the total number of species in the group to forty-seven—the highest single-site total for the South American tropical forests.

Stingless bees, in contrast, are honey producers and live in hives with the typical complex social structure found in other species in the same family (Apidae), including the Africanized honey bees (see Chapter 12). Oliveira collected fifty-four species of stingless bees, which represent a remarkable 21 percent of the number of species known in the Neotropics, and reports for the first time that these bees are the agents of seed dispersal as well as of pollination.

Censuses of stingless bees made in continuous forest, fragments, and deforested areas revealed, surprisingly, that species richness in the deforested areas (25 species) was higher than in any fragment and intermediate between species counts for fragments (9–22 species) and continuous forest (37 species). Once again, simple species numbers belie the threat of habitat fragmentation—11 species were found only in continuous forest, and several of the bees in deforested areas are probably recent invaders of the man-altered landscape, as Martins reported for fruit flies (Chapter 14). The persistence of the primary-forest species in a highly fragmented landscape is untested but unlikely.

Oliveira's comparison of the euglossine faunas of two areas of continuous forest not more than 10 km apart showed remarkable differences in the two species lists, suggesting that the seemingly homogeneous primary forest is, to the euglossine bees, a mosaic of very different microhabitats. This result echoes similar findings for other invertebrates (Chapter 15) and frogs (Zimmerman and Bierregaard 1986), although whether the scale of the microhabitat patches is large enough to argue for a scattered series of reserves as opposed to a single large one remains to be seen.

INVERTEBRATES AND THEIR VERTEBRATE PREDATORS IN FRAGMENTED FORESTS

Invertebrates dominate. In both numbers of species and sheer biomass, no other animal group comes close to matching the importance of invertebrates in the rainforest ecosystem. The herbivores among them are

the principal agents in the release of the solar energy captured by plants to the rest of the ecosystem. All major faunal groups, including invertebrates themselves, rely heavily on invertebrates for food. Changes in invertebrate densities or even species composition occasioned by habitat fragmentation are certain therefore to affect insectivorous vertebrates.

In the final chapter of this section on the effect of fragmentation on invertebrate communities, Chapter 18, Raphael Didham presents an exciting and integrative report that links studies of invertebrates, including his own dissertation field work carried out in the BDFFP reserves, with studies of vertebrates done in the very same fragments. The chapter provides an appropriate bridge to the next section on vertebrate communities and fragmentation and is an example of the value of integrated projects like the BDFFP. In particular, he speculates on how the changes he and others have seen in invertebrate populations in the BDFFP forest fragments may relate to the response of insectivorous birds to the same fragmentation events.

Didham sampled forest leaf litter for invertebrates at seven distances from the forest edge in different-sized fragments and continuous forest. Both total invertebrate density and the relative abundance of different species or groups of species showed strong edge effects. Density was as much as 150 percent higher near edges than deep in continuous forest. A fragmentation effect was also significant, with substantial (50 percent) reduction in insect abundance in 100 ha reserves when compared to continuous forest, although, somewhat surprisingly, within the fragments there was no discernible trend between area and invertebrate density, which suggests that area and edge effects may be confounding each other, or at least the researchers studying them.

As Stouffer and Bierregaard (1995a) have reported, the group of birds most susceptible to habitat fragmentation are the terrestrial insectivores, a group of birds that forage as they walk across the leaf litter. It is, at first glance, counterintuitive that this group should decline in the face of an overall increase in invertebrate density in the leaf litter. However, as Didham points out, while overall density of invertebrates increases, there is a pronounced change in the composition of invertebrates in the litter, which might mean a change in the quality of the prey available to insectivores.

In a fragmented landscape, Didham argues, there will be selection for generalist insectivores with no aversion to forest edges (which has been reported for understory birds in general at the BDFFP reserves by Quintela [1985]). His recommendation for a minimum reserve size is well in excess of 100 ha to preserve an unmodified litter invertebrate fauna, both for its own conservation value and as a resource base for insectivorous vertebrates.

CHAPTER 14

Drosophilid Fruit-Fly Guilds in Forest Fragments

MARLÚCIA B. MARTINS

Habitat fragmentation and subsequent isolation of forests has received much attention in conservation research, both worldwide and in tropical forests, including the Amazon. By definition, forest fragmentation reduces the original habitat and increases the isolation of remaining forest patches. Additionally, a series of microclimatic changes increase temperature and luminosity and decrease humidity in forest remnants (Kapos 1989).

In the case of the Amazon forest, high species diversity compounds the initial effects of fragmentation. Because many plant species occur at very low densities, deforestation produces a strong sampling effect; many plant species will not be present in the forest remnants because of their low density or patchy distribution. This may have cascading effects through faunal communities.

Habitat fragmentation is accompanied by the creation of a new matrix habitat that replaces the original forest. For example, in the case of a forest transformed into pasture, this new habitat has totally different temperature, humidity, vegetation cover. Many species not present in the original forest can opportunistically use the new matrix habitat and possibly invade forest remnants

(Martins 1989; Gascon et al. 1999 ; Chapters 19 and 20). The contact among these distinct faunas can initially increase species richness, but with time, the occurrence of competitive interactions and changes in forest conditions could lead many native species to local extinction (Chapter 2). This effect can be more or less intense in accordance with the level of susceptibility of the species to the alterations of the habitat (humidity loss, wind increase, temperature increase, qualitative and quantitative alterations in resources, invasion by new species [Bierregaard et al. 1992]).

The drosophilid fruit-fly community associated with decomposing fruits, as a group of organisms that use the same class of resources in a similar manner, constitutes a guild as defined by Root (1967). Although this guild concept groups species with similar ecological requirements without their necessarily being taxonomically related, in the case of drosophilids, taxonomic relationships do exist. Simberloff and Dayan (1991) argued that members of a guild are potential competitors for similar resources in a dynamic equilibrium. The composition of the guild, therefore, represents functional redundancy in an ecosystem, such that the structure of the guild can be more pre-

This is publication number 281 of the BDFFP Technical Series.

dictable than the abundance of the individual species or the guild's specific composition. The guild can be seen in this manner as a functional category shaped by adaptation for the exploitation of the same class of resources (Hawkin and MacMahon 1989).

Wolda (1992) suggested that populations are in a dynamic balance with the resource availability, that the communities are highly organized, presenting specific composition patterns with foreseeable changes, and that the number of individuals and their biomass can suffer great variations while species richness remains similar. This suggests that changes in guild structure caused by an anthropogenic influence, for example, can reflect changes in resource availability. Such a response would serve as an indicator of habitat disturbance.

Studies have shown drosophilids to be sensitive to fragmentation effects in a relatively short time period (van der Linde 1992; Davis and Jones 1994). Drosophilids can represent a significant portion of the small insect community, which, although nearly imperceptible to the human eye, represents a significant portion of Amazonian forest species diversity. The small insect community contributes effectively to the maintenance of the forest through pollination, seed dispersal, recycling of nutrients, and energy flux.

In this chapter I study drosophilid guild structure. I investigated the response of a drosophilid guild (drosophilids linked to decomposing fruits) to two different scenarios of habitat fragmentation in two fragmented forest landscapes to determine if changes in guild structure can serve as an indication of anthropogenic disturbance. The basic questions are as follows: Does fragmentation cause alterations in the characteristics of this guild? If so, what are these alterations? In the first case, results are reported from the Biological Dynamic of Forest Fragments Project north of Manaus (Martins 1985, 1987; see fig. 4.1). In the second case, results are reported from the extreme eastern Amazon, in the Mocambo Reserve within the Belém city limits (Martins 1996). Some other examples will be considered in less detail. In both studies, species abundance and guild composition were evaluated. A comparison of the response of the drosophilid guild in different fragmented landscapes permits us to detect common patterns that will aid in setting conservation guidelines for these invertebrates.

Central Amazon

METHODS

Studies in the central Amazon were performed at BDFFP reserves, in particular on the Esteio and Porto Alegre cattle ranches (see fig. 4.1, table 4.3). In 1980 and 1983, only two reserves were isolated (1104 and 1202, of 1 ha and 10 ha, respectively, on the Esteio ranch; see plate 1). On the Porto Alegre ranch, reserves 3114, 3209, and 3304 were not yet isolated. In 1986, a survey was conducted in the fragments of the Esteio ranch and in Reserve 3304 (by then isolated; see plate 2) on the Porto Alegre ranch. In April 1995 another survey of the area was performed with only one series of collections in 1 and 10 ha forest fragments of the Esteio ranch and in the 100 ha reserve on the Porto Alegre ranch (see Chapter 4, esp. table 4.3, for a more detailed history of the isolation of these reserves).

The climate of the region follows the "Afi" type of the Koeppen classification, having no monthly precipitation values below 40 mm and with mean temperatures between 26° and 27° C. Rains begin in January, reach their peak in March and April, and generally decrease between August and October (fig. 5.1).

In this study, adult drosophilids were collected at natural feeding and breeding sources, using local commercial fruit baits. The type of bait most commonly used was naturally fermented bananas, but surveys were also made using papaya and oranges.

Species abundance estimates were made with standardized banana baits, distributed uniformly in the habitat in 500 ml tin containers placed 10 cm above the ground. The same methods were used during the entire study period. Collections were made by netting fruit-flies over the baits twenty-four and forty-eight hours after they were set out.

The flies collected were stored in 70 percent alcohol, separated according to bait and day of collection, identified, and quantified in the laboratory. The interval between each sampling bout was one to two months.

For guild characterization, the Shannon and Fisher alpha indices and the Berger-Parker dominance (D) index were used. The similarity between habitats was estimated using the Jaccard index (see Krebs 1989).

Surveys in July 1986 were conducted during two consecutive days in each reserve (1104, 1202, and 3304) and in the secondary forest adjacent to the Esteio reserves, using naturally fermented bananas as bait placed in tin containers 50 cm from the ground.

In April 1995, only banana bait was used in the central areas of Reserves 1202, 1104, and 3304 to evaluate possible changes in the composition and relative abundance of the *Drosophila* species in these habitats. During this time, sampling was performed for three consecutive days in each reserve, using naturally fermented banana bait placed on the ground and in containers similar to those of previous surveys. Survey data for each bait station were totaled.

RESULTS

Occurrence of drosophilids on native fruits. The *willistoni* group (*Drosophila,* subgenus *sophophora*) represented 64 percent of the collections from native fruit on the forest floor. Species of this group occurred on all types of fruit and were the only taxa present on fruits of the families Guttiferae and Moraceae (*Ficus* sp.). The *tripunctata* group (*Drosophila,* subgenus *Drosophila*) represented 18 percent of collections and occurred on six of the nineteen types of sampled fruits. In the collections that were made on flowers dispersed on the forest floor, the *tripunctata* group was dominant, representing 89 percent of the individuals; it was present in all the flowers where drosophilids occurred. The abundance of the group was notable in the flowers of *Eschweilera odora* ("mata-mata"; Lecythidaceae), a species found quite frequently in the reserves. Drosophilids were found on only three types of fungi: *Auricularia delicata, Pleurotus* sp., and *Scuteger brasileiens.* These included flies of the subgenus *Hirtodrosophila* and, predominantly, of the genus *Zygothrica.*

In the 1983 surveys with different types of baits (oranges, bananas, and papaya), the *willistoni* subgroup once again dominated, representing 51 percent of the individuals, followed by *D. nebulosa* and *D. malerkotliana* with 21 percent and 12 percent of the individuals, respectively. The distribution of species occurrence did not significantly differ among bait types.

Comparison of abundance between fragments and continuous forest. Comparison of the relative species abundance patterns between these two habitats showed that among the most abundant species there were no changes in species ranking even though the total number of individuals collected in the fragments was about twice as

TABLE 14.1.   Abundance of *Drosophila* Taxa in Continuous Forest and Forest Fragments from a Seven-Day Survey in 1983, Three Years After Fragmentation

Taxa	Forest	Fragments Forest	Secondary	Total	Percentage
Subgroup *willistoni*	1,584	3,007	39	4,700	81.5
D. malerkotliana	45	275	30	332	5.7
D. sturtevanti	18	23	1	42	0.7
group *tripunctata*	10	25	4	39	0.7
D. nebulosa	5	63	195	263	4.5
D. cardini	0	39	80	119	2.1
D. latifasciaeformis	0	35	236	271	4.7
TOTAL	1,662	3,519	585	5,766	

TABLE 14.2.   Abundance of *Drosophila* Taxa in Forest Interior and Forest-Edge Habitats from a Five-Day Survey

Taxa	Forest interior	Forest edge	Total	Percentage
Subgroup *willistoni*	1,058	433	1,491	69.70
D. malerkotliana	234	149	383	17.90
D. latifasciaeformis	1	103	104	4.80
D. cardini	0	2	2	0.09
group *cardini*	0	11	11	0.50
D. sturtevanti	39	33	72	3.40
Group *saltans*	5	3	8	0.40
D. nebulosa	29	13	42	2.00
Group *tripunctata*	10	0	10	0.40
D. fumipennis	7	2	9	0.40
sp. 1	1	0	1	0.05
sp. 2	0	1	1	0.05
sp. 3	0	2	2	0.90
sp. 4	0	1	1	0.05
sp. 5	0	1	1	0.05
sp. 6	0	1	1	0.05
TOTAL TAXA	9	15	16	
TOTAL INDIVIDUALS	1,384	755	2,139	
Fisher's Alpha	1.2883	2.6525		
Shannon Index	1.126	1.854		
Dominance Index	0.76	0.57		

Note: Survey made in Reserve 3304 of Porto Alegre. Forest edge was one month old at time of survey.

large as in continuous forest (table 14.1). The predominant species in the second-growth forest (*D. latifasciaeformis*) was not present in continuous forest and appeared in low frequency in forest fragments. The subgroup *willistoni* was only the fourth most abundant species in second growth.

After complete isolation of one of the 100 ha reserves (3304), the comparison between forest edge and the reserve interior revealed intermediate changes in species abundance and diversity patterns in relation to the prior comparisons (table 14.2). The subgroup *willistoni* predominated in the interior as well as in the forest edge. Its abundance, however, was 2.5 times lower at the forest

TABLE 14.3. Abundance of *Drosophila* Taxa in the Different Habitats Surveyed on Esteio Ranch, 1980–83

Taxa	1202 center	1202 border	1202 edge	1202 total	1104	Secondary forest	Total	Percentage
Group *annulimana*								
D. araicas	1	0	1	2	1	0	3	0.03
Group *calloptera*	2	2	0	4	2	0	6	0.07
D. cannalinea (sp. 10)	0	1	0	1	0	0	1	0.01
Group *cardini* (sp. 7)	0	0	0	0	1	0	1	0.01
D. polymorpha (sp. 14)	0	0	0	0	1	0	1	0.01
sp. 5	0	0	0	0	5	0	5	0.06
D. cardini	1	14	60	75	11	320	406	21.09
D. cardinoides	0	0	0	0	1	0	1	0.01
Group *Dreyfusi* (sp. 24)	1	0	0	1	0	0	1	0.01
Group *guarani*	0	3	0	3	2	0	5	0.06
D. moju	0	1	0	1	0	0	1	0.01
D. mercatorum	0	2	0	2	0	4	6	0.07
D. ellisoni	0	1	1	2	1	0	3	0.03
sp. 31	0	1	0	1	0	0	1	0.01
sp. 32	0	0	4	4	0	4	8	0.09
sp. 16	0	1	0	1	0	0	1	0.01
sp. 3	0	0	0	0	1	0	1	0.01
Group *tripunctata*	3	13	1	17	54	5	76	0.90
Hirtodrosophila	0	0	0	0	2	0	2	0.02
D. latifasciaeformis	2	8	44	54	2	142	198	2.31
D. ananassae	0	0	1	1	0	2	3	0.03
D. malerkotliana	36	68	65	169	74	110	353	4.12
D. simulans	0	1	0	1	0	3	4	0.05
sp. 1 group *saltans*	0	0	0	0	1	0	1	0.01
Subgroup *sturtevanti*	109	217	55	381	240	58	679	7.92
D. prossaltans	0	1	0	1	0	1	2	0.02
Subgroup *willistoni*	988	2,419	299	3,706	2,676	121	6,503	75.94
D. nebulosa	6	38	40	84	127	59	270	3.15
D. fumipennis	0	0	0	0	4	1	5	0.06
sp. a	3	0	0	3	0	0	3	0.03
sp. b	0	0	0	0	0	1	1	0.01
sp. 2, n13	0	0	0	0	1	0	1	0.01
sp. 9	0	1	0	1	0	0	1	0.01
sp. 11	0	0	1	1	0	0	1	0.01
sp. 19	0	3	0	3	0	0	3	0.03
sp. 20	0	0	0	0	1	0	1	0.01
sp. 22	0	0	0	0	1	0	1	0.01
sp. 26	0	0	0	0	1	0	1	0.01
D. tuchaua sp. 27	0	0	0	0	1	1	2	0.02
sp. 29	0	0	0	0	0	1	1	0.01
TOTAL	1,152	2,795	572	4,519	3,211	833	8,563	
TOTAL SPECIES	11	19	12	25	24	16	40	
Fisher's Alpha	1.684	2.742	2.147	3.487	3.520	2.809	5.424	
Shannon Index	0.811	0.849	2.193	1.105	1.032	2.520	1.437	
Dominance Index	0.858	0.865	0.523	0.820	0.833	0.384	0.7594	

TABLE 14.4.   Jaccard Similarity Index
Values Between Habitats Surveyed on the
Esteio Ranch

Habitats	1202 center	1202 border	1202 edge	1104	Secondary forest
1202 center	1.00	0.36	0.30	0.34	0.35
1202 border		1.00	0.35	0.30	0.40
1202 edge			1.00	0.33	0.50
1104				1.00	0.26
Secondary forest					1.00

edge, whereas *D. latifasciaeformis* increased
in abundance more than 100 times from the
interior to the edge. The number of species
increased from nine to fifteen at the forest
edge, and the diversity, measured by the
Shannon diversity index, varied from H =
1.126 to H = 1.854. There was a slight increase in species diversity at the forest edge.
The similarity between forest edge and interior as measured by the Jaccard index was
J = 0.4375.

*The evaluation of diversity of the forest
fragments in the 1980s.* Overall, forty taxa of
Drosophilidae were encountered, including
species and morpho-species (individuals
apparently distinct from other species in the
sample but not yet reliably identified). The
number of taxa increased with increasing reserve size and with the degree of preservation of the habitat. Sixteen taxa were found
in second-growth forest, twenty-four in the
1 ha reserve, and twenty-five in the 10 ha reserve. The dominance index was higher for
fragments than for second-growth forest,
with the *willistoni* subgroup in the forest
areas reaching dominance levels above 80
percent (table 14.3). The Jaccard similarity
values indicated a greater similarity between the 10 ha forest edge and the second-
growth forest, and a lower similarity between the second-growth forest and Reserve
1104 (table 14.4). In general, similarity
among habitats was much lower, between

0.26 and 0.50, demonstrating a high level of
heterogeneity in species distribution.

Collections made in 1986. A total of 1,667
individuals, pertaining to five taxa and three
unidentified individuals, were collected in
continuous forest (table 14.5). Fragments revealed 3,526 drosophilids of eight taxa and
nine unidentified individuals. The same
eight taxa were found in the second-growth
area, although the number of individuals
was smaller—590 (table 14.5). Abundance
patterns were similar to those of the 1981
and 1983 collections.

*Abundance and diversity patterns in
1995.* A total of 7,371 individuals of thirteen
species were collected in 1995 (table 14.6).
Nine species were collected in continuous
forest, eleven in the 10 ha reserve and nine
in the 1 ha; eight species were common to
the three habitats. These results show only
slight differences in diversity and species
composition when compared to the 1983
surveys (table 14.7). Ten taxa were common
to the two periods in at least one reserve.
The Jaccard similarity index between
1981–83 and 1995 was 0.38 for the 1 ha reserve, 0.69 for the 10 ha reserve, and 0.60 for
Reserve 3304. These results suggest a much
larger change in species composition in the
1 ha reserve than in the larger reserves.

DISCUSSION

Although drosophilid fruit-flies use varied
resources, species composition by resource
type is fairly constant. In the forest, drosophilids associated with fruits are a distinct
guild from those associated with flowers or
fungi. It is expected that response patterns
obtained for one of these guilds will reflect
similar responses for other guilds.

Drosophilid species composition changed
dramatically between the forest and the
open areas. Forest fragmentation modified
specific composition and abundance pat-

TABLE 14.5. Abundance of *Drosophila* Taxa in the Habitats Surveyed in 1986 Within the BDFFP Landscape

Taxa	Continuous forest	1202 center	1202 border	1104	Secondary forest	Total	Percentage
Subgroup willistoni	1,584	656	1,599	823	39	4,701	81.10
D. malerkotliana	45	33	128	96	30	332	5.72
D. sturtevanti	23	6	11	6	1	47	0.81
Group tripunctata	10	4	8	13	4	39	0.67
D. nebulosa	5	4	39	20	196	264	4.55
D. cardini	0	28	8	3	80	119	2.05
D. latifasciaeformis	0	11	11	13	236	271	4.67
D. simulans	0	1	5	0	4	10	0.17
Others	3	3	0	3	4	13	0.22
TOTAL	1,670	746	1,809	977	594	579	

TABLE 14.6. Abundance of *Drosophila* Taxa in Continuous Forest and Forest Fragments in April 1995

	Location					
	Continuous Forest		1202		1104	
Taxa	Abundance	Percentage	Abundance	Percentage	Abundance	Percentage
Subgroup willistoni	3,460	89.10	1,865	86.20	1107	83.40
D. malerkotliana	163	4.20	39	1.80	99	7.40
D. nebuosa	11	0.30	14	0.60	6	0.40
D. latifasciaeformis	0		0		2	0.10
Group tripunctata	12	0.30	41	1.90	91	6.80
D. annulimana	4	0.10	7	0.30	1	0.07
D. guarani	0		1	0.04	0	
D. fulvimacula	13	0.30	20	0.90	1	0.07
D. repleta	0		21	1.00	0	
Group cardini	1	0.02	3	0.10	1	0.07
D. camargoi	0		1	0.04	0	
D. fumippenis	4	0.10	0		0	
D. sturtevanti	214	5.50	169	7.80	28	2.10
TOTAL	3,882		2,163		1,326	
TOTAL SPECIES	9		11		9	
Fisher's Alpha	1.099		1.513		1.298	
Shannon Index	0.671		0.914		0.956	
Dominance Index	0.89		0.86		0.83	

terns of the drosophilid guild. The main changes were a decrease in species richness with decreasing reserve size, producing guilds in fragments with intermediate composition between continuous forest and second-growth forest guilds.

The decrease in species richness was caused by the disappearance of rare, native, and characteristic forest species. Furthermore, patterns of dominance among the most common species were also altered. Species richness, however, showed a slight increase at the landscape level, probably resulting from a decrease in relative abun-

TABLE 14.7.   Relative Abundance of *Drosophila* spp. Present in Each of Three Sampling Periods in at Least One Forest Fragment on Esteio Ranch

Taxa	1202 Center			1104		
	81–83	July '86	April '95	81–83	July '86	April '95
Subgroup *willistoni*	85.7	87.9	86.2	83.3	84.2	83.4
D. malerkotliana	3.1	0.04	1.8	2.3	9.8	7.4
D. nebulosa	3.3	0.5	0.6	3.9	2.0	0.4
D. latifasciaeformis	0.2	1.4	0	0.06	1.9	0.1
Group *tripunctata*	0.2	0.5	1.9	1.7	1.3	6.8
Group *cardini*	0	3.7	0.1	0.6	0.3	0.07
D. sturtevanti	9.4	0.8	7.8	7.4	0.6	2.1
TOTAL INDIVIDUALS	1,152	746	2,163	3,211	977	1,326

dance of the originally dominant group in second-growth forest and the appearance of exotic opportunistic species, such as *D. malerkotliana.*

Fragment size also influenced species composition, as shown by the highest dissimilarity levels between the smaller fragment and continuous forest. Observed species losses, in this case, were of species most sensitive to changes in habitat quality, such as *D. camargoi, D. tuchaua,* and various species of the *repleta* and tripunctata groups.

The invasion of *D. malerkotliana* is well known, as well as its expansion rate in the Neotropics in the past twenty years (Sene and Val 1977; Martins 1989; Martins 1990; Sevenster 1992). Between the early 1980s and 1995 the abundance of this species grew modestly in continuous forest, but relatively more in the central area of the 10 ha reserve and in the smaller 1 ha reserve.

Results from Belém, Pará, and other eastern Amazonian sites can help one understand the invasion of *D. malerkotliana* and other impacts on species composition of the drosophilid guild in forest fragments.

Eastern Amazon

METHODS

Since 1986, drosophilid guilds have been surveyed in the Mocambo Reserve, a 5 ha terra firme forest fragment near the city of Belém on the delta of the Amazon River. The reserve is almost completely surrounded by streams, which expand and contract with the rising and falling of the Rio Guamá. A small section of the edge of the reserve abuts pasture.

Banana baits were used to estimate relative abundance of the drosophilid species in the area (for detailed methods, see Martins and Fonseca 1988; L. Oliveira 1993). I also identified the resources most frequently used by the drosophilids (Martins and Santos 1988) and evaluated the strategies of native fruit use by these species, focusing on the guild associated with the *Parahancornia amapa* fruit scattered over the ground as a model (Martins 1996).

RESULTS

As for the Manaus site, the *willistoni* subgroup was the dominant taxa among the emergent species from fruits, and many of the same species found in Manaus were also encountered among the emergent fruit species in the Mocambo Reserve.

TABLE 14.8. Relative Abundance (Percentage of Sample) of Drosophilids of the *willistoni* Subgroup and of *D. malerkotliana* in Different Studies in the Mocambo Reserve in Belém Brazil

Period	Type of sampling	*D. malerkotliana*	Subgroup *willistoni*	Number of individuals
1948–1952[1]	bait	0	84.88	28,452
May–June 1987[2]	bait	8.3	39.30	4,946
March 86–Dec. 87[3]	fruits	34.83	59.10	2,179
Jan.–Apr. 1990[4]	fruits	19.84	61.10	8,602
Jan.–Apr. 1991[5]	fruits	33.53	48.23	5,917
Jan.–Apr. 1992[5]	fruits	47.78	44.96	5,500
May 1992–Apr. 1993[5]	bait	22.92	71.84	6,017

Notes: [1]Pavan 1959. [2]Fonseca and Martins 1988. [3]Martins and Santos 1988. [4]Martins 1996. [5]L. Oliveira 1993.

Abundance and species richness patterns observed in Manaus were much more accentuated at this site. *D. malerkotliana* showed a huge increase in frequency in this area. In the 1950s, before the introduction of *D. malerkotliana,* twenty-five species of *Drosophila* were collected with banana bait (Pavan 1959) in this area. From 1987 to 1993, the number of taxa observed in the area was forty-two. However, for any given sample, a gradually smaller number of taxa was recorded. In 1986, abundance of *D. malerkotliana* and *willistoni* subgroup was 8.3 percent and 39.3 percent, respectively. In the 1992–93 survey, *D. malerkotliana* represented almost 23 percent of the samples, whereas the *willistoni* subgroup represented 70 percent. The remaining species accounted for a total of 7 percent, while in 1987 they represented about 50 percent in the surveys. Observations from the native fruits show the same pattern (table 14.8).

Between 1992 and 1993, the abundance of *Drosophila* species was estimated in a series of habitats representing a gradient of perturbation, including the Mocambo Reserve and its surroundings (L. Oliveira 1993). Perturbation levels were defined by the size of deforested area, the size of the remaining forested areas, and the proximity to human-occupied areas. Four areas in a 20 km radius were compared: one 23 ha *várzea* (flooded)

forest reserve; the Mocambo Reserve; the Ceasa forest—a 0.7 ha wooded area near a commercial farm; and a pasture adjacent to the Mocambo Reserve. Results showed a clear separation between the species composition of the pasture areas and the forested areas, similar to those observed for the western Amazon. These results also showed a reduction in species richness and a gradual decline in abundance of the *willistoni* subgroup as perturbation increased. *D. malerkotliana,* however, increased in abundance with increasing perturbation. Both taxa, however, were extremely rare or absent in the pasture area. The tendency of *D. malerkotliana* to become dominant in the more perturbed forest areas seems to represent a general trend, as it was also observed in the BDFFP area of central Amazon in the early 1980s, as well as in other forest fragments in the eastern Amazon, including Carajás, a continuous forest area; the margins of the Caruaca River, which are intensively explored for timber extraction; Paragominas, a series of forest fragments within an extensive area of pasture; Belém-Mocambo and Belém-Ceasa, fragments in close proximity with different levels of disturbance (table 14.9).

I can offer three explanations for these trends: fragmentation effects; the intensity of anthropogenic pressure on the fragments;

TABLE 14.9.   Relative Abundance (Percentage of Sample), Total Number of Individuals, and Number of Taxa of the *willistoni* Subgroup and of *D. malerkotliana* in Different Localities and Years in the State of Pará

Locality and years	*D. malerkotliana*	Subgroup *willistoni*	Total Drosophilidae	Number of taxa
Carajás, 1986	0.46	82.32	1,941	25
Margins of Caruaca River,				
Marajó Island, Aug. 1988	59.28	31.68	1,250	8
Paragominas, Jan. 1992	68.44	27.40	4,430	9
Belém-Mocambo, 1992–93	22.92	71.84	6,017	—
Belém-Ceasa, 1992–93	74.71	14.88	5,800	—

and the biological characteristics of the species. The first and last factors may best explain the results observed in the BDFFP reserves, because they are experimental fragments with minimal human impacts. All other areas have suffered from fragmentation impacts as well as from intense human pressure through hunting, timber extraction, and being in close proximity to rural or urban areas and to centers of exotic fruit commerce.

The biology of these drosophilid species was studied by Martins and Klaczko (unpublished data) and can be summarized in the following manner: *D. malerkotliana* showed demographic characteristics similar to those of the dominant native species, but it is more aggressive in the use of resources. This species possesses the shortest life cycle, consistently occupies resources at least 24 hours before all other species, concentrates its reproductive efforts in the initial phase of life, and prefers fruit in the early phase of decomposition. These characteristics confer a competitive advantage to this species, an advantage that could in part explain the success of its invasion (Shorrocks and Bingley 1994). This priority effect may result from the strong association of the species with the yeast *Kloerklera apiculata*, which is a pioneer among the yeast in decomposing fruit and is consumed in large quantities by this *Drosophila* species (Moraes,

Martins, and Hagler 1995; Martins and Klaczko, unpublished data). The emergence of adults also starts early from the newest fruits, whereas the development period, longevity, and period of oviposition do not substantially differ from species of the *willistoni* subgroup, and in fact fecundity is slightly inferior (Martins and Klaczko, unpublished data). Because of its relatively recent introduction, lower rates of predation and parasitism may occur on this species, but this remains to be documented.

As temporal heterogeneity in oviposition opportunities decreases and becomes more predictable, rapid life-cycle species will outcompete others and become dominant in more disturbed habitats (Sevenster and van Alphen 1993a, 1993b). For the species investigated in this study, this might occur because of an increase in available resources for some species (in this case fruit that falls and is not consumed because the usual frugivores are scarce in the disturbed area).

This three-factor model, suggested above, of ecological impacts of landscape change on the drosophilid guild seems to explain the observed responses both in the Mocambo Reserve and in the BDFFP areas. Both areas have been severely fragmented; human impact is almost absent at BDFFP, whereas it has been severe at Mocambo. The Mocambo Reserve has been exposed to intense hunting and harvesting of plant products for subsistence or commerce and changes in the intensity and period of stream inundation. This type of interference leads to the biological impoverishment of the forest and can alter its dynamics. One clear result has been the disappearance of a large fraction of the small mammals that consumed the fruit scattered on the ground. As a consequence, there is a greater availability of recently fallen fruit, which probably favored the increase in abundance of the rapid life cycle *Drosophila* species. This scenario may well explain the observed explosive population expansion of *D. malerkotliana,* with a consequent decrease in the total diversity and the loss of dominance of the subgroup *willistoni,* which used to be dominant in the Amazon forest (Dobzhansky and Pavan 1950; Pavan 1959). In the case of the BDFFP reserves, the process is similar, but the expansion of *D. malerkotliana* is much slower, and the reduction of diversity is manifested mainly through the disappearance of rare species and species more sensitive to microclimatic changes associated with fragmentation. Although geographic differences between eastern and central Amazonia may confound some of these results, there is little doubt that habitat fragmentation and anthropogenic impacts significantly alter the structure of drosophilid guilds. Many ecological processes in forest fragments, such as pollination, seed dispersal, and decomposition, will also be seriously affected as a result.

Conclusions

Evaluating the impact of forest fragmentation on faunal communities in tropical systems is difficult and requires the selection of indicator groups. One important recommendation is that the indicator groups be functional and representative within the ecosystem. The drosophilid guilds have been proposed as good indicators (Parsons 1991; van der Linde 1992; Davis and Jones 1994; Martins 1996), principally owing to the following attributes: they possess short life cycles in such a way that the effects, at the habitat level, can be observed independently of the historical effects; they are sufficiently mobile, so that the absence of one species may be attributed not to its inability to colonize, but rather to the adversity of the habitat for that species; and drosophilids are sufficiently numerous to detect possible differences among habitats (Davis and Jones 1994). Drosophilids are representative of a fauna of small insects associated with decomposing organic material; this guild can represent a reasonable fraction of the total diversity. If drosophilid guilds are good indicators of changes in natural landscapes, then these results point to important effects caused by fragmentation— for example, the importance of edge effects, the facility of exotic species introduction, the decrease in relative abundance of native species and in total species richness.

Conservation Lessons

1. Drosophilid species are sufficiently sensitive to alterations at the habitat level and, therefore, can serve as good indicators of habitat perturbation.

2. Forest fragmentation incurs the loss of drosophilid species. However, the magnitude of this loss is difficult to assess because the most sensitive species are naturally rare in the habitat and their loss can go unobserved.

3. Fragmentation and human impact change abundance patterns of species, favoring opportunistic species with rapid life cycles.

4. Landscape changes facilitate the introduction and expansion of exotic species in the Amazon forest, as is the case for *D. malerkotliana*.

5. The introduction of an exotic species can accelerate the process of species loss in a fragment.

6. The differing rates of expansion of *D. malerkotliana* in central Amazonian forest fragments and in the reserve in Belém suggest that the effects of human impacts are synergistic (or additive) to those of forest fragmentation. Human access to forest fragments can therefore hasten the negative effects of fragmentation.

7. Edge effects are very important in altering faunal composition in fragments, but the size of the reserve will determine the rate of species loss in these habitats.

CHAPTER 15

Local Extinction Risks and Asynchronies

The Evidence for a Metapopulation Dynamics of a Social Spider, Anelosimus eximius *(Araneae, Theridiidae)*

EDUARDO M. VENTICINQUE AND HAROLD G. FOWLER

The attention of the world has focused on the consequences of humankind's activities both on vast spatial scales, as we witness global climate changes (Shukla, Nobre, and Sellers 1990), and on local, small scales, where we see how our domination of the landscape affects populations of vertebrates and plants (Murphy 1989). At the same time, given the importance of invertebrates, especially in tropical ecosystems —they have more species, a greater biomass, and higher ecological diversity than any other group of animals (Wilson 1987b)—we have paid remarkably little attention to them in our studies of changing environments.

When an area of continuous habitat is fragmented, populations may continue to exist in these fragments as locally isolated populations or as metapopulations, groups of fragmented demographic populations connected by some level of genetic exchange (Levins 1970; Hanski and Gilpin 1991). The fundamental population-level processes in metapopulations are patch extinctions and colonizations, which are linked through migration. The rate of these processes will determine whether a given species will exhibit characteristics of a metapopulation after habitat fragmentation or, conversely, will be split into many small, completely isolated local populations with a high risk of local extinctions and consequent global extinction.

In practice, it is often extremely difficult to document local extinctions, as when a plant remains for some time without any above-ground vegetation, only to resprout when conditions are favorable, or when animals curtail their activities during periods of stress or unfavorable conditions. In such cases, without a better understanding of its biology, we might erroneously record the species as locally extinct.

Metapopulation structure and dynamics may be a natural response to habitat fragmentation and may be fundamental as a protection against global extinctions (Simberloff 1994). A number of metapopulation models exist (Hanski and Gilpin 1991), most of which produce similar predictions and dynamics. In conservation biology, metapopulation models (Hanski 1994) are frequently used as an alternative to classical island biogeographical models (MacArthur and Wilson 1967). A metapopulation model generally uses the equilibrium of colonization and extinction of island biogeographic theory, with an exception in that in the latter, the "continent" is considered immune to

This is publication number 282 of the BDFFP Technical Series.

extinctions, while in metapopulation models, the colonization source is the set of local populations (Levins 1969, 1970).

Marked seasonal fluctuations of insect and spider populations in tropical forests are well documented (M. Robinson and Robinson 1970; Wolda 1977, 1978; Lubin 1978, 1980; Fowler, Silva, and Venticinque 1993). Because of this, tropical environments select for short life spans (Vollrath 1986a). Among these taxa, colonies of the social spider *Anelosimus eximius* (fig. 15.1) have a potentially long life span and are present when most other spider species disappear or reach extremely low population levels (Vollrath 1986b). In our studies of this species in the experimentally fragmented habitat mosaic of the Biological Dynamics of Forest Fragments Project (see fig. 4.1; table 4.3), however, on various occasions we recorded only one colony in some localities, indicating strong risks of local extinction. Over longer time periods the risk of local extinction increases dramatically (Venticinque, Fowler, and Silva 1993). Nevertheless, because we have no clear definition of what constitutes a population of *A. eximius* —either one or contiguous webs—web-volume dynamics in colonies in different spatial locations may be a better indicator of population dynamics than are distinct web numbers. In this chapter we examine the dynamics of local populations of this spider at both single-web and summed-web volume levels in an environmental matrix in the central Amazon. Our goals were to examine synchrony in numbers and volume in spatially distinct aggregates and to measure the effect of an exogenous factor, rainfall, on the periodicity of population fluctuations.

The Social Spider *Anelosimus eximius*

Only 7 of the 105 described families of spiders have social species (D'Andrea 1987).

Fig. 15.1. A communal web of the social spider *Anelosimus eximius*. Photo by Eduardo M. Venticinque.

The highest levels of social behavior are found in the Dictynidae, Eresidae, Agelenidae, and Theridiidae, the latter of which includes *Anelosimus eximius* (Simon). *A. eximius* lives in colonies of 100 to 10,000 individuals (Nentwig 1985), marked by a highly skewed female-dominated sex ratio (Fowler and Levi 1979; Aviles 1986). Colonies have overlapping generations, with the communal sheet web, occupying several cubic meters (Brach 1975), located in trees or bushes a meter or more above the ground (Nentwig 1985). About 90 percent of prey captured by this knock-down web are larger than solitary spiders could subdue (Nentwig 1985).

Anelosimus eximius reproduces by the migration of solitary females, or by "sociotomy," the splitting of an established colony (Fowler and Levi 1979; Vollrath 1982; Christenson 1984). (Because of this reproduction by colony splitting, it is often difficult to classify colonies, or distinct webs, as separate entities.) The dynamics of reproduction have been linked to seasonality and habitat (Venticinque, Fowler, and Silva 1993).

In this chapter we examine the dynamics of population parameters with respect to habitat type and fragmentation in the experimentally isolated forest reserves of the BDFFP.

Population Dynamics

METHODS

We studied populations of *Anelosimus eximius* in three fragments (Colosso 1 ha [Reserve 1104], Colosso 10 ha [1202], and Cidade de Powell 10 ha [1207]; see fig. 4.1 and table 4.3), two edges between forest and cleared ranch land (trail to Cidade de Powell and trail to Florestal), and an area of continuous forest (Cabo Frio [3402]) from August 1988 through July 1990. Fixed transects were sampled at approximately twenty-day periods, with all colonies individually numbered and followed. For each sampling date, the size of the colony web was measured (width, height, and length), and its volume was estimated in cubic meters, calculated as an ellipsoid (Venticinque and Fowler 1998). The volume sum consists of the sum of all the colonies in a local population during each time interval. The approximate date of colony foundation and mortality was recorded when a new colony was observed on a sampling day or when a pre-existing colony was not found or was abandoned. Colony life span was thus the interval between foundation and disappearance.

Because we worked with the total number of colonies in each of the study areas, mortality and internal movement have different effects on census results. Fragments were surveyed completely, so webs were easily located. In our analyses, all webs in each isolated fragment were considered a local population. For webs at forest edges or in continuous forest, we considered each web representative of a larger, undefined local population, the demography of which was reflected by changes in the webs under study. All colonies alive at the first and last census were removed from our analyses. To compare life spans of colonies we used the Mann-Whitney test (Siegel 1975).

Time-series correlograms (Jenkins and Watts 1968) are commonly used to detect periodic behavior in a series of successive observations. This method consists of autocorrelating successive values at various time lags, or temporal differences, to detect if recurrent peaks occur at regular, albeit lagged, periods (Berryman and Turchin 1997). The correlograms were analyzed by month for the number of colonies and summed colony volume. Analyses used three-point running averages to smooth the series (Royama 1992) before calculating the correlograms and cross-correlograms. Tests for auto-correlation and second-order auto-correlation were run. In an auto-correlation, the value of each point in a series is related to the value of the point immediately preceding it ($y_2, y_1; y_3, y_2; y_n, y_{n-1}$). In a second-order auto-correlation, each point is related to a value two intervals behind it ($y_3, y_1; y_4, vy_2; y_n, y_{n-2}$) (Wilkinson 1990).

The correlograms were tested as a whole—that is, including all time lags. For this we applied the Bonferroni criterion to determine the significance level for each individual correlation (Sokal, Smouse, and Neel 1986). The whole correlogram was considered significant when the sum of the individual correlation probabilities was equal to or less than 0.05 (cf. R. Harris 1975).

To examine an apparent response to the rainy season, we conducted temporal cross-correlation analyses (Brockwell and Davis 1991) between colony numbers and volume and monthly rainfall. Cross-correlograms are an extension of spectral analyses (Jenkins and Watts 1968) and allow the identification of causal relationships between two time series. One time series is auto-correlated with another at varying lags to detect whether patterns of lagged correlations exist. Negative lag correlations indicate a relationship, perhaps of second order or delayed density effects (Berryman and Turchin 1997) between values of one series with anticipated values of the other. A zero correlation indicates no relationship, while positive lag correlations indicate that values of one time series are responding to the same periodicity as the other (Brockwell and Davis 1991).

Results

The number of colonies and their summed volumes demonstrated strong oscillations (fig. 15.2), especially for the volume occupied by the colonies. For colony volumes, rainfall was associated with strong collapses, and summed locality web volumes were more synchronous with rainfall than colony numbers. Edge populations had oscillations, one of which was caused by a high recruitment rate of solitary females in the seventh month of this study, followed by a high mortality rate of colonies in the eleventh and twelfth months (Venticinque, Fowler, and Silva 1993). The forest-interior population oscillated much less, especially after the tenth month (fig. 15.2). The shape of the distribution of life spans for all sampled populations indicated that most colonies reach advanced ages, while platykurtic (flat; negative kurtosis values in table 15.1) distributions were found in one forest fragment and one edge, indicating that age distributions were more uniform in these sites. The leptokurtic (skewed left; positive kurtosis values in table 15.1) distri-

bution found in the other sampled sites suggests high mortality in younger monthly age classes, pushing these toward significantly shorter life spans. A significant difference was found for the life span of colonies among all sites (Kurskal Wallis analysis of variance = 15.040, P less than 0.05). The forest interior population had significantly longer life spans than did edge populations (Mann-Whitney U-tests).

We found a greater colony life span in the continuous forest (Cabo Frio reserve; see figs. 4.1 and 4.3) and in one of our forest fragment populations (trail to the Cidade Powell reserve). The latter is a 10 ha reserve that was isolated in 1983. After isolation, the surrounding clearing was not maintained by the ranchers, leading to a dense secondary succession at its edges (see table 4.3). Our other forest fragment population was in the 10 ha Colosso reserve, which was continuously managed to remove secondary vegetation and had better-defined edges than did the Cidade Powell reserve. In the Colosso reserve fragment, colony densities were much higher than the 10 ha reserve at Cidade Powell (table 15.1), but all except one colony were found at fragment edges, while at Cidade Powell they were dispersed in equal numbers between forest edge and interior. The high densities of colonies on forest edges reflect the preference of the species for more open habitats. In a fragmented landscape there is an increase in the area of contact between forest and open habitats, such as pastures, and young second-growth forest providing this preferred habitat for the spiders.

In our study a relationship was detected between temporal oscillations and precipitation. Vollrath (1986b) suggested that *A. eximius* was less seasonal than sympatric solitary spiders, which greatly decrease in abundance at the end of the dry season. Using egg-sac production as an index of reproductive activity, we found that reproduction may cease during

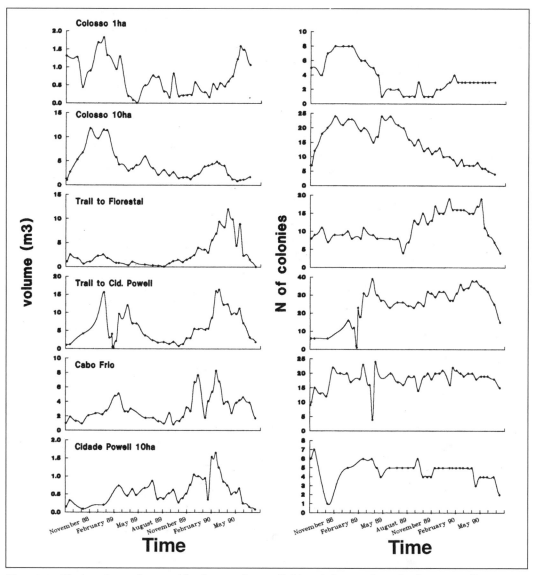

Fig. 15.2. Fluctuations of summed colony volumes (left) and the number of colonies in all localities. Each graph on the right corresponds to one on the left. Colosso and Cidade Powell are isolated reserves. Trails to the reserves of Cidade Powell and Florestal are forest edges. Cabo Frio is continuous forest.

the dry season in some sites but is continuous in others, suggesting the regulatory effects of habitat (Vollrath 1986b).

When we auto-correlated web numbers, a strong seasonality of population dynamics at a six- to eight-month periodicity emerged (fig. 15.3). With the exception of one edge population, all correlograms were significant. In the three largest populations (trail to Cidade de Powell, the 10 ha Colosso reserve forest fragment, and the continuous forest Cabo Frio Reserve), a stronger temporal dependence was detected—larger populations have a more pronounced periodicity. These

TABLE 15.1.   Basic Longevity Data for Individual Colonial Webs of the Social Spider *Anelosimus Eximius* in Six Different Spatial Locations in the Central Amazon

| Parameter | Forest edges | | | Forest fragments | | Forest |
	Trail to Cidade Powell	Trail to Florestal	Cidade Powell (10 ha)	Colosso (10 ha)	Colosso (1 ha) (1,000 ha)	Cabo Frio
Reserve number	none	none	1207	1202	1104	3401
Mean longevity	99	119	116	140	177	189
Skewness	1.60	0.755	0.011	1.324	1.392	0.875
Kurtosis	2.62	-0.475	-1.363	1.679	0.727	0.237
C. V.	99.4	81.5	71.9	94.1	105.2	86.0
Number of colonies	182	65	23	85	27	63

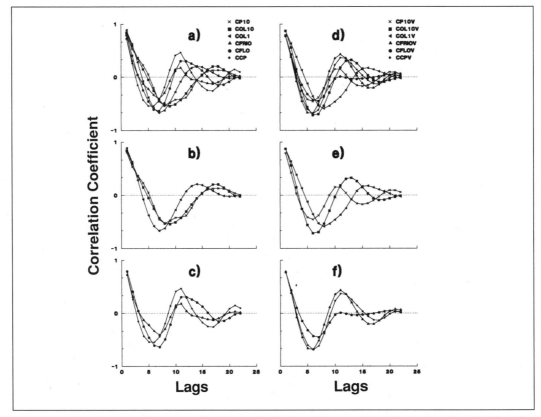

Fig. 15.3. Correlograms of the sum of volumes (a, b, c) and the number of colonies (d, e, f). (a) and (d) are localities, (c) and (f) are continuous forest and edges, and (b) and (e) are 1 and 10 ha forest fragments. CP10 = Cidade Powell, 10 ha; COL10 = Colosso, 10 ha; COL1 = Colosso, 1 ha; CFRIO = Cabo Frio, continuous forest; CCP = trail to Cidade Powell; and CFLO = trail to Florestal Reserve. "V" signifies volumes.

dynamics suggest that the number of colonies per locality depended on their level six months previously.

When colony volumes were analyzed, distinct patterns were found (fig. 15.3). For forest fragment populations (fig. 15.3d), a periodicity of 7 to 9 months was detected, with a secondary period of 17 to 18 months. A distinct periodicity (5–7 months, with a secondary periodicity at 10–12 months) was found for continuous forest and forest-edge populations. The forest-fragment populations showed strong synchrony, suggesting that the size of these fragments may regulate numerical population fluctuations (fig. 15.3b) and their total volumes (fig. 15.3e). However, this regulation may be exogenous because of higher fluctuations in weather variables, predation, kleptoparatism, etc.

Both monthly colony numbers and their volumes demonstrate that forest edges and continuous forests are synchronous, with alternative peaks and valleys at semiannual intervals. This suggests that seasonality may be more marked in these areas than in forest fragments, which demonstrated a higher variability in their correlograms (fig. 15.3).

When all possible cross-correlations were examined (fig. 15.4), most were decoupled, with little relationship apparent between colony numbers and previous or posterior rainfalls. The only exception was one forest-edge population (trail to Cabo Frio reserve). This pattern suggests that colony numbers are independent of rainfall. However, by considering each spatial aggregate of *Anelosimus eximius* as a unit and examining their summed web volumes, this pattern changes (fig. 15.5). For all cases, a pronounced synchrony was found with positive lags. This suggests that in months of elevated rainfall, colony volumes are higher (fig. 15.5). This pattern is apparent in an eleven-month cycle, with the exception of one edge (trail to the

Florestal reserve) and one forest-fragment population (10 ha Colosso reserve; fig. 15.5).

We found no coupling between the number of colonies and their volumes (fig. 15.5). The larger number of colonies resulted in larger colony volumes in only two localities (Cabo Frio and Cidade Powell; figs. 15.5e, 15.5f). This adjustment may be due to an increase in colony reproduction by splitting. This pattern also suggests that *Anelosimus eximius* may have evolved a strategy to permit its persistence at lower population levels under stressful conditions, such as the dry season. Instead of increasing the number of colonies with increased precipitation and associated prey availability, this species may be increasing colony volume with precipitation cycles. This adjustment may be through colony population decrease, or through fusion, much like that found in the population dynamics of social insects (Wilson 1971). Colony population dynamics may be viewed as individually stochastic because of a number of extrinsic factors, such as epidemics, availability of food, and catastrophic events, such as branch falls, which cannot be handled by deterministic models (Bhat 1984).

Anelosimus eximius as a Metapopulation

Local populations may or may not oscillate independently of each other. However, a metapopulation may be more vulnerable to a collapse when habitat patches are few or are more isolated. In our study, we found that *Anelosimus eximius* subpopulations fluctuated asynchronously. This asynchronicity may protect against a generalized population collapse and may be the result of different biotic and abiotic selective forces acting in the various subpopulations. Most metapopulation models suggest that increasing mortality is expected when

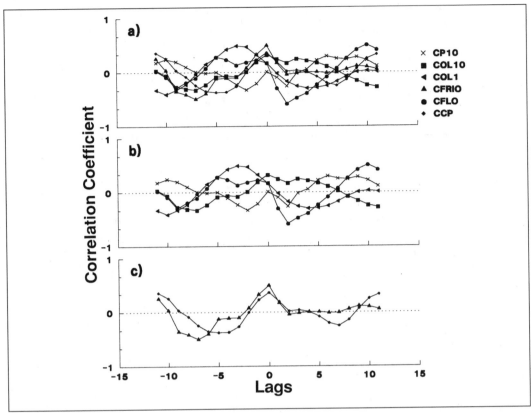

Fig. 15.4. Cross-correlograms of time series of rainfall and the number of colonies. (a) all localities, (b) Colosso 1 and 10 ha, Cidade de Powell 10 ha, and trail to Florestal, and (c) trail to Cidade de Powell e Cabo Frio. Legends are the same as in fig. 15.2.

metapopulation dynamics are closely tied to the spatial structure of the habitat patches. As habitat fragmentation increases, local populations may become completely disconnected from each other, and local extinctions may increase in frequency, leading to "non-equilibrium metapopulations" (Harrison 1991) and resulting, in the extreme, in regional extinction. For conservation biology, this implies that, as they become more and more isolated, local populations may need to be managed as separate entities and not as metapopulations.

Local population collapses and subsequent patch colonization are frequently cited as evidences for a metapopulation structure (Menges 1990; Stacey and Taper

1992). Many real metapopulations contain one or more local populations that never go extinct (Schoener and Spiller 1987; Gotelli 1991; Schoener 1991). This population structure, like the island-continent model (Harrison 1991), can result from natural differences in habitat quality and size, or from human fragmentation of a natural habitat. Because extinction and colonization affect only a subset of populations, these have little impact on regional population persistence but may greatly influence regional population distribution. Small or isolated habitat patches are less suitable than larger or intact habitat patches, which are functional analogs of continents (Fritz 1979; Smith 1980; Toft and Schoener 1983; Harri-

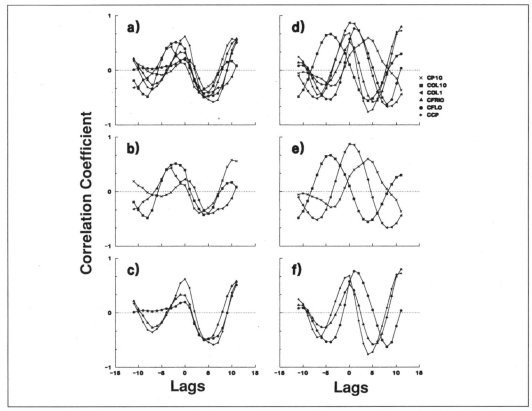

Fig. 15.5. Cross-correlograms of time series of the sum of colony volumes and rainfall, (a), (b), and (c); and of colony volumes and numbers, (d), (e), and (f). Legends are the same as in fig. 15.2. Graphs (b) and (e) represent fragments, (c) and (f) continuous forest and forest edges.

son, Murphy, and Ehrlich 1988; Peltonen and Hanski 1991), or as a constellation of small and large populations (Laan and Verboom 1990; Sjogren 1991a, 1991b; Thomas and Jones 1993). Because the environmental matrix of secondary growth and pastures in which the habitat fragments are embedded in our study is surrounded by an immense area of primary forest, a spatial inversion occurs, in which the population of *A. eximius* is protected from extinction because primary forest surrounds all of the fragments and the environmental matrix.

The disappearance of a species in a sample plot is often referred to as local extinction (den Boer 1970). At other spatial scales, this may denote the disappearance of a species at a scale smaller than its entire bio-

geographic range (Chew 1981). The strictest definition of local extinction is the disappearance of a genetic or demographic population (Harrison 1991). However, practical problems of using this restricted definition are obvious. For metapopulation dynamics, a more appropriate definition of a local population is that of a minimal unit that, once eliminated, should remain extinct for one or few generations. In our case, on one occasion in one forest fragment (1 ha Colosso reserve), *A. eximius* went extinct and remained so for about two months, which fits this definition. Importantly, the return of the species to the fragment demonstrated the colonization between subpopulations expected in a metapopulation situation.

Environmental stochasticity is the prime

force in local extinctions. This stochasticity, driven by temporal variation in food supply and other factors, directly influences colony growth rates. The magnitude of this variation depends on the life history of the organism in question. Environmental stochasticity and catastrophes are major selective forces for natural metapopulations. Many environmental factors that regulate population fluctuations and local extinctions act at regional scales. Rainfall is an obvious example that may affect simultaneously all subsets of a metapopulation, but with dampening effects of habitat. But if it is correlated with local extinctions, it reduces expected metapopulation persistence (Harrison and Quinn 1989; Gilpin 1990), although few cases have been tested (Ehrlich et al. 1972, 1980; Pollard, Hall, and Bibby 1986). This implies that dispersal is an efficient mechanism of metapopulation maintenance (den Boer 1970) and especially long-distance dispersal (Fahrig and Paloheimo 1988), possibly through habitat corridors, such as secondary forest. However, we have no evidence of this in our data, and little evidence exists in other studies (Fahrig and Paloheimo 1988). Among the many environmental variables that may enhance stochastic variation, we have identified, without reference to habitat patch, structural differences in vegetation, prey availability, predation, and kleptoparasitism. These will be discussed in turn.

Structural differences in vegetation between fragments, forest edges, and continuous forest as documented by Malcolm (1991, 1994) apparently influence colony mortality and dispersal. Vegetation substrate and plant density strongly influence web-building spider abundance and diversity (Greenstone 1984; D'Andrea 1987). We frequently observed colonies that were fragmented or destroyed by the fall of tree branches.

Prey availability also varies and probably has a strong selective effect on colony growth, reproduction, and mortality in species of *Anelosimus*. However, there was no direct relationship between sampled prey availability and mortality rates although an apparent relationship exists between colonizations and prey availability, as both prey availability and colonization rates are higher at distinct edges (Fowler, Silva, and Venticinque 1993).

A longer life span found for colonies of *A. eximius* in our continuous forest population suggests adaptations for survival in environments of longer periods of stability, when compared with fragments and edges. Environmental pressures in forest edges and fragments were apparent determinants of lower colonial life spans. Among these pressures, meteorological factors, at the micro and macro scale, such as wind speed and temperature and humidity variation, are more pronounced here than in continuous forest habitats (Kapos et al. 1997). Additionally, predation is a major source of colony mortality, especially in colonies founded by solitary females (Vollrath 1982; Christenson 1984). Among the major predators are army ants, monkeys, birds (Vollrath 1982), and other spiders and wasps (Lubin and Robinson 1982). We observed the army ant *Eciton burchelli* raiding a colony founded by a single female, while species of the wasps *Trypoxylon trypartgilum* and *T. punctinestex* are predators of *A. eximius* in the forest interior (Hubert Hoffer, pers. comm.). Additionally, kleptoparasites, other species of spiders or insects that invade or inhabit the communal sheet web and rob prey, are likely strong selective forces affecting colony growth rates (Christenson 1984; Venticinque 1995; Fowler and Venticinque 1996).

The persistence in a fragmented landscape of a species that does not depend on local patch dynamics but on metapopulation dynamics requires that four conditions are met: all habitat patches support locally reproducing populations; there is more than one population; habitat patches are not sufficiently isolated to inhibit recolonization; and local dynamics are sufficiently asynchronous to preclude simultaneous extinction of local populations at a regional level (Harrison 1994). Because *A. eximius* in our study area satisfied all criteria, this implies that it is a metapopulation and eliminates the alternatives of patch populations (Harrison 1994), island-continent populations (Hanski 1991; Schoener 1991), and non-equilibrium metapopulations (Harrison, 1994).

Implications of Metapopulation Dynamics

Anelosimus eximius can be considered a "tramp" species, preferentially colonizing forest clearings and edges. We were unable to document long-term local extinctions, principally because our isolated forest fragments were not distant from potential colonization sources. Even when we compared two 10 ha reserves, population dynamics were completely different because one reserve (Colosso) was surrounded by secondary vegetation (such that for *A. eximius* the reserve was really not isolated), while the other, where the second growth was repeatedly cleared, was more isolated. Two of our edge populations were along trails that linked reserves and were principally secondary vegetation, which functionally served as corridors linking forest populations. However, because these latter areas were not wide, population oscillations were more variable and extreme, probably because of increased edge effects.

Differing correlograms and cross-correlograms for our studied population subsets strongly suggest metapopulation dynamics, and different selection pressures in the lo-

calities studied. For *A. eximius* in the central Amazon, our 1 ha forest fragment had great difficulty maintaining local populations, indicating that fragments of this size can maintain populations only with a strong "rescue effect" through colonization from source populations (Brown and Kodric-Brown 1977). This reserve was continuously colonized because there was a more stable population in a 10 ha reserve 150 m distant, less than half of the maximal dispersal distance recorded for this species (Vollrath 1986b). The matrix habitat and the proximity to source populations are thus important factors affecting the metapopulation dynamics of *A. eximius*.

Evidence suggests that the mosaic of different successional stages within the Amazonian landscape is important for *A. eximius* metapopulation dynamics. The two-month, continuous extinction period that we recorded for the smallest forest fragment (1 ha Colosso reserve) also indicates that strong temporal fluctuations in populations should not be overinterpreted. This implies that the management of a successional mosaic within the Amazon may be more important for the maintenance of some animal species. Obviously, more long-term studies of the population dynamics of other animals are necessary to determine if *Anelosimus eximius* is the rule or an exception among the Amazonian invertebrate fauna.

Conservation Lessons

1. The abundance of colonies of the social spider *Anelosimus eximius* increases in fragments and along forest edges compared to continuous forest areas.

2. Population turnover also increases in fragments and along forest edges but is accompanied by a decrease in colony life expectancy.

3. The asynchronous population dynamics of this species in relation to precipitation may be a mechanism to decrease the probability of extinction of all colonies at the same time.

4. Local populations in small fragments suffer a high risk of going extinct, and they require a source of immigrants for survival. The level of immigration will depend on the surrounding matrix habitat and the distance to a source population.

5. Landscape configuration (distances to sources of colonists and the type of matrix surrounding the fragment) and fragment size are important for the conservation of this species.

Acknowledgments

Work was supported by the BDFFP, while Eduardo M. Venticinque was supported by an assistantship provided by the São Paulo State Foundation for Scientific Research (FAPESP—Grant 97/06692-7). We especially thank João de Deus Fragata, Carlos Alberto Silva, Ângela Imakawa, and Rosemary Vieira for assistance during our field work. Thanks to countless hours of conversation, Cláudio Von Zuben, Marina Fonseca, Heraldo Vasconcelos, and Sofia Campiolo greatly assisted in structuring our ideas. Claude Gascon, Richard O. Bierregaard, Jr., and two anonymous reviewers greatly assisted in making our report readable.

CHAPTER 16

Landscape Modifications and Ant Communities

HERALDO L. VASCONCELOS, KARINE S. CARVALHO, AND JACQUES H. C. DELABIE

According to Landsat satellite imagery of the forested portion of the Brazilian Amazon basin, deforestation in this region has increased threefold between 1978 and 1988 (Skole and Tucker 1993) and continued to increase in recent years (DOU 1996). The destruction of tropical rainforests by human activities threatens the conservation of global biodiversity, as these forests are home to at least half of the plant and animal species on Earth (Myers 1988b). Deforestation affects biological diversity not only through the destruction of the original forest habitat, but also through the isolation of forest fragments (which are almost inevitably created as a result of deforestation) and by exposing the remaining forested areas to edge effects. Edge, in this case, is defined as a boundary zone between forest and a cleared area (Lovejoy, Bierregaard, et al. 1986; Murcia 1995). Edge creation allows wind and light penetration into forest remnants, altering forest climate and vegetation structure (e.g., Kapos et al. 1997; Ferreira and Laurance 1997; Laurance, Bierregaard, et al. 1997). Such changes may be deleterious to many forest organisms, while favoring the invasion of non-forest species (Terayama and Murata 1990; Brown and Hutchings 1997; Chapter 14).

With a few exceptions (e.g., Greenslade and Greenslade 1977; Klein 1989; Souza and Brown 1994; Didham 1997a, 1997b), the effects of deforestation and tropical rainforest fragmentation on invertebrates are still poorly documented. Ants are a particularly suitable group on which to study these effects on invertebrates. They are relatively sedentary and responsive to changes occurring at small spatial and temporal scales, most species are easily sampled, and the taxonomy and ecology of the group is relatively well understood (Majer 1983; Agosti et al. 2000). Furthermore, given their preponderance in tropical forests, in terms of both biomass and species numbers (Fittkau and Klinge 1973; Wilson 1987b), ants play a major role in the structure and function of these complex and diverse ecosystems. In this chapter, we present the major findings of our research at the Biological Dynamics of Forest Fragments Project (BDFFP), where we have been studying the effects of forest disturbance on ants that forage or nest on the forest floor. We start by analyzing possible changes in ant community structure associated with relatively recent changes in land cover in central Amazonia (due to human activities), and then assess the possible impacts of forest fragmentation on these same communities.

This is publication number 283 in the BDFFP Technical Series.

Methods

A series of ant surveys was conducted in the study area of the Biological Dynamics of Forest Fragment Project, north of Manaus (see Bierregaard et al. 1992; Chapter 4; fig. 4.1; table 4.3), between 1993 and 1996.

Responses to Changes in Land Cover

To analyze the effects of changes in land use and land cover on the structure of ground ant communities, ants were collected in undisturbed, mature forest and in three areas where mature forest had been cleared eleven to thirteen years previously. Although these three areas were of about the same age, they had been exploited differently. The first area was abandoned immediately after the felling of the mature forest, because an abnormally early rainy season prevented the ranchers from burning to clear the land. At the time of the study, this area, hereafter referred to as old regrowth forest, was covered by a thirteen-year-old regrowth forest, dominated by pioneer *Cecropia* spp. trees, which formed a closed canopy 15–20 m in height (plate 8). The second area, hereafter young regrowth forest, was used as pasture for cattle grazing for about two years and then abandoned. A 10–15 m tall regrowth forest (about ten years old) dominated by pioneer *Vismia* spp. trees covered this area. The third study area was subjected to a high level of disturbance, as it had been exploited as pasture for about ten years and had been abandoned only one year prior to this study. Hereafter, this is referred to as abandoned pasture.

In each of these three areas, as well as in two control, mature forest areas, four plots of 20 × 20 m each were established. To collect ants, three methods, which tend to be complementary in terms of the species recorded, were employed. Collections were made using pitfall traps, sardine baits, and leaf litter samples (0.5 m² each). Nine samples of each collection method were taken from each plot.

Responses to Forest Fragmentation

As a first assessment of the effects of forest fragmentation on ant species richness and composition, an ant survey was made in one continuous forest area, and in eight fragments located in three fazendas of the BDFFP experimental landscape (table 16.1). In each fragment, as well as in the continuous forest studied (Reserve 1301), a 1 ha plot was established, generally in the center of the fragment. The plot in continuous forest was located 500 m from the edge between the forest and the cleared area, the same distance used to place our sampling plots in the 100 ha fragments. Within each plot we installed thirty-six pitfall traps, arranged within a square grid, and took thirty-six samples of leaf litter (0.5 m² each) and thirty-six soil samples (0.016 m³ each), from which ants were extracted manually.

Because of characteristics of the experimental design, there were limitations in the interpretation of the results derived from this study. In particular, we were not able to determine the relative influence of area effects and edge effects on ant communities, owing to the strong intercorrelation between these two predictors—the smaller the fragment, the closer the sampling plot was to the edge of this fragment. A new study was then initiated (Carvalho 1998; Carvalho and Vasconcelos 1999), in which it was possible to discriminate between effects resulting from distance to the forest edge and effects caused by forest isolation (fragment versus continuous forest). For this, ants were collected along transects located at exactly the same distances (5, 20, 40, 60, 100, 200, 300, 400, 500 m) from the edges of two continuous

TABLE 16.1. Characteristics of the Ground Ant Community in Forest Fragments Near Manaus

Site (*fazenda*)	Forest area (ha)	Species per ha (observed)	Species per ha (estimated)[1]	Species per Sample[2]	Species diversity (H')	Evenness (E)
Esteio	1	80	106	8.4	3.78	0.86
	10	91	124	8.9	3.98	0.88
	>10,000[3]	109	151	10.1	4.30	0.92
Dimona	1	94	129	9	4.09	0.90
	10	99	139	9	4 23	0.92
	100	111	149	11.1	4.28	0.91
Porto Alegre	1	98	140	8.5	4.10	0.89
	10	78	117	6.2	3.86	0.89
	100	79	112	7	3.87	0.89

Notes: [1]Based on sequential progressive sampling (Lauga and Joachin 1987). [2]Combined sample (one pitfall trap, plus a leaf-litter and a soil sample). [3]Continuous forest.

forests and two fragments of 100 ha. Because there was interest in determining the densities of ant nests on the forest ground, a different sampling method was used. Instead of using baits or traps, we examined decaying twigs and branches (less than 5 cm in diameter) for the presence of ant colonies. (A diverse assemblage of ant species nest in decaying twigs and branches [Byrne 1994; Carvalho 1998].) In all transects, twenty plots of 4 m² each were sampled; all existing dead twigs and branches (less than 5 cm in diameter) were examined.

To analyze the effects of forest fragmentation on ant species composition we used semi-strong hybrid multidimensional scaling, an ordination program that is more robust and effective than most other commonly used ordination techniques (Belbin 1995). As an association (or similarity) measure we used the "chord distance," an option in the analysis that is recommended when one wants to compare ratios rather than absolute values of variables (Belbin 1995). For the first of the two studies, the data (input matrix) contained the number of occurrences (for a maximum of 36 occurrences) for each ant species in each forest fragment or continuous forest studied. We

omitted all species that were recorded just once (65 species), resulting in a total of 162 species in nine forest areas. For the second study, the input matrix contained the number of occurrences (for a maximum of 20 occurrences) for each ant species, at each transect. In three of the 36 transects established along the four edges only two ant colonies were recorded. As these transects could not be well characterized in terms of species composition they were removed from the analysis.

We tested whether species composition (expressed as ordination scores) was affected by fragmentation using Analysis of Covariance (ANCOVA). For the first study, the covariate was forest area, and for this we considered the continuous forest to be a "fragment" of 10,000 ha. For the second study, the covariate was distance to forest edge, whereas "area" (in fact isolation; fragment or continuous forest) was treated as a fixed factor. The same models were used to test effects on species richness, and on the Shannon indices of diversity and evenness (see Krebs [1989] for a description of these indices).

Results

RESPONSES TO CHANGES IN LAND COVER

Conversion of mature forest into pasture areas caused a dramatic decline in ant species richness. Although the abundance of ants (as measured by the number of ants in the leaf litter) tended to increase with forest disturbance, the opposite trend was found for ant species richness (fig. 16.1). Thus, while 79 to 83 species were found in samples from mature forest and the old regrowth forest, only 36 were recorded in the abandoned pasture samples. Of these 36 species, only 13 were also found in mature forest. These results indicate not only a decline in ant species richness but also a significant change in species composition in the more disturbed areas (fig. 16.2).

Analysis of the ant communities in the two regrowth forests studied showed that once cleared areas are abandoned and forest regeneration proceeds, the original forest ant fauna gradually recovers. There is evidence, however, that the rate of recovery depends not only on the time since land abandonment, but also on the history of land use prior to its abandonment, in particular if the land had been used as pasture. Thus, although the forest established in a former pasture area had been fallow only three years longer than the one established in the nonpasture area, the number of ant species recorded in the latter (83 species) was about 35 percent greater than the one found in the former (62 species). Furthermore, the degree of similarity in ant species composition with mature forest was greater in the regrowth forest established in the nonpasture area than in the one established in the pasture area (fig. 16.2).

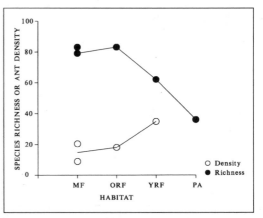

Fig. 16.1. Differences in ant species richness and density of ant workers between four habitats subject to different levels of anthropogenic disturbance. PA = 10 yr old cattle pasture, abandoned 1 yr prior to this study; YRF = "Young" (10 yr old) regrowth forest, established in a former pasture area; ORF = "Old" (13 yr old) regrowth forest, established in an area that was abandoned after clearing without burning. MF = undisturbed, mature forest.

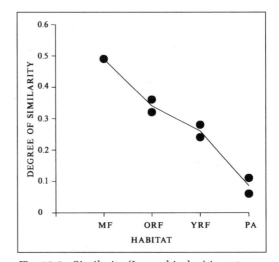

Fig. 16.2. Similarity (Jaccard index) in ant species composition between two mature forest areas (MF), and between these two forests and three other disturbed habitats. Abbreviations are as in fig. 16.1. Similarity with mature forest decreases as habitat disturbance level increases.

RESPONSES TO FOREST FRAGMENTATION

Between 78 and 111 ant species were recorded in each 1 ha plot established in the fragments of 1, 10, and 100 ha, and in the continuous forest. Based on sequential progressive sampling (Lauga and Joachin 1987) we calculated the estimated species richness, S, per hectare based on the following formula:

$$S = \frac{(S_1 * S_2 (P - 1))}{(S_1 * P - S_2)}$$

where S_1 is the mean number of species per sample, S_2 is the total number of species in all samples, and P is the total number of samples. We estimated that between 106 and 151 species could be found in these same areas (table 16.1).

There was no significant relationship between the size of the forest remnant and any of the three diversity indices analyzed, namely species richness and the Shannon indices of diversity and evenness (ANCOVA, $P > 0.05$; table 16.1). There was, however, a significant interaction between site effects and size effects. In two of the sites (ranches) studied the number and diversity of ant species increased as forest area increased, whereas in the third site (Porto Alegre) the opposite trend was found.

A significant relationship was found between the scores produced by the third axis of the ordination analysis (no effect detected for the other two axes) and fragment area ($F_{1,5} = 28.6$, $P = 0.003$), indicating a change, although not a dramatic one, in ant species composition with size of the forest remnant (fig. 16.3). In addition, there was an effect of site ($F_{2,5} = 19.7$, $P = 0.004$), which indicates some degree of heterogeneity in ant species composition between forest areas at Dimona, Esteio, and Porto Alegre, even though these localities were only 10–25 km apart (fig 4.1).

Ant species composition was also affected by distance to forest edge, as revealed by collections of ant nests in decaying twigs along transects in the edges of two continuous forests and two fragments of 100 ha. We found a significant relationship between distance to forest edge and the scores produced by the third ordination axis ($F_{1,31} = 18.5$, $P < 0.001$; fig. 16.4). Based on the positioning of the transects along axis 3 of the ordination analysis, we calculated that edge effects, as measured by changes in ant species composition, penetrate nearly 200 m into forest remnants in central Amazonia. Distance to forest edge, however, did not fully explain observed changes in ant species composition following forest fragmentation. Even though isolated and nonisolated forest areas were sampled at comparable distances to forest edge, similarity in ant species composition was greater between the nonisolated areas than between the isolated ones (table 16.2), suggesting that insularization effects also play a role in structuring ant communities from forest fragments.

Discussion

RESPONSES TO CHANGES IN LAND COVER

Clearing of mature forest for pasture establishment, and subsequent forest regrowth, strongly affected the richness, abundance, and community composition of ground-dwelling ants. This supports earlier findings from specific studies with leaf-cutting ants of the genera *Atta* and *Acromyrmex* (Vasconcelos and Cherrett 1995). Nest surveys in three sites of central Amazonia, including the BDFFP study areas, have shown that in mature forest the density of nests belonging to leaf-cutting ants is very low. The species found were *Atta cephalotes*, *Atta sexdens*, and *Acromyrmex hystrix*. In contrast, in re-

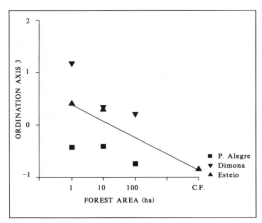

Fig. 16.3. Relationship between forest area (log transformed) and ordination axis 3. The ordination axis results from a three-dimensional solution ordination of ant species composition in nine forests. CF = continuous forest. Symbols represent different study sites (*fazendas*).

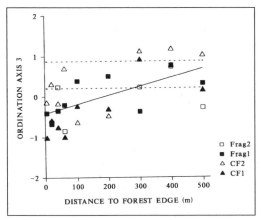

Fig. 16.4. Relationship between distance to forest edge and ordination axis 3. The ordination axis results from a three-dimensional solution ordination of ant species composition in 33 transects in two isolated 100 ha fragments (Frag1 [= 2303] and Frag2 [= 3304]) and in two continuous forests (CF1 and CF2) tracts. Dashed lines represent upper and lower 95 percent confidence intervals for mean in ordination scores for the three most interior transects (300 m), and the solid line represent the linear fit to the data set.

growth forests no nests of *A. cephalotes* and *A. hystrix* were found, even though these forests supported a density of *Atta* and *Acromyrmex* nests up to twenty times greater than that found in mature forest (Vasconcelos and Cherret 1995). The species typical of these more disturbed areas were *Atta laevigata* and *Acromyrmex laticeps,* which do not occur in mature forest.

Here, looking at the ground-dwelling fauna in general, we found that the number of species in the abandoned pasture was less than half the number found in undisturbed mature forest. This is a much stronger species loss than observed for other social insects, such as the meliponine bees (Chapter 17). Given the number of ant species found in abandoned pasture and the number common to nearby mature forest, it was estimated that a maximum of 17 percent of mature forest species are able to withstand the destruction of their original habitats (Vasconcelos 1999). As forest regenerates on the abandoned land, the original forest fauna gradually recovers. However, it was surprising to learn how strong was the effect of land use history on this recovery. Where forest was cleared and abandoned immediately afterward, recovery was relatively fast. Thirteen years later, the number of species found in this area was comparable to the one found in nearby mature forest, and many of the species were the same (figs. 16.1 and 16.2). By contrast, where forest was cut, burned, and planted to pasture, the number of species recorded ten years after pasture abandonment was about 30 percent lower than the one in mature forest, and the similarity in species composition to mature forest was relatively low. These differences are likely to reflect the differences in vegetation recovery in these same areas, which was slower where pasture was established than where it was not (Williamson, Mesquita, et al. 1998; Vasconcelos 1999).

TABLE 16.2.   Matrix of Jaccard Index of Similarity in Ant Species Composition Between Two Isolated and Two Nonisolated Forest Areas Near Manaus

	Continuous forest 1	Continuous forest 2	Fragment 1	Fragment 2
Continuous forest 1	(55)[1]			
Continuous forest 2	0.73	(39)		
Fragment 1	0.63	0.59	(38)	
Fragment 2	0.37	0.35	0.34	(16)

Note: [1]Number of species in the area.

Our study and others (Moutinho 1998) thus suggest that recovery of the original forest fauna after land abandonment depends on the type and intensity of land use (see also Chapters 19 and 20; for a botanical example, see Chapter 25). Land management practices that result in greater levels of habitat disturbance tend to slow recovery. Given these results it is important that we develop measures not only that reduce the actual rates of deforestation, but in addition, as a way to mitigate the deleterious effects of habitat destruction, that develop land management techniques that either minimize the initial loss of forest species or hasten their recovery. In this regard, some traditional systems of farming, such as coffee cultivation in Costa Rica (Perfecto and Snelling 1995), are better than the modern monoculture systems.

Despite a decrease in species richness with forest clearing, the total abundance of ants increased, at least in the younger successional areas. This may help to explain why forest edges, which are mostly composed of successional vegetation, support a greater density of ants and other leaf-litter invertebrates than the forest interior (Didham 1997b; Chapter 18). The observed changes in ant abundance, species richness, and species composition caused by deforestation are likely to have implications for forest regeneration in the abandoned areas. For instance, impaired regeneration of forest trees in some abandoned pastures in Ama-

zonia has been attributed in part to the deleterious effects of a leaf-cutter ant species not found in mature forest (Nepstad, Uhl, and Serrão 1990). *Solenospsis geminata,* a species that increased in abundance following forest disturbance (Vasconcelos 1999), is considered a keystone species in some areas, thanks to its major effects on the structure of both the arthropod and plant communities (Risch and Carroll 1982; Carroll and Risch 1984), and as such may have a crucial role in the process of forest regeneration.

RESPONSES TO FOREST FRAGMENTATION

No consistent changes in species richness were found in response to variations in fragment size (see table 16.1). However, this does not mean that large fragments do not contain more species than small ones (as they certainly do, given the high heterogeneity in species distribution), only that the number of species per unit area (also known as species density or "point" species richness) did not differ between small and large fragments. At two sites, point species richness increased, whereas in the third site (Porto Alegre) it decreased as forest area increased. Differences in the history of isolation (resulting in different matrix habitats) between fragments located at Porto Alegre and those at Dimona and Esteio may have accounted, at least in part, for these conflicting results. The area surrounding the fragments at Porto Alegre was not burned, and as a result, vegetation recovery, and recovery of the ant fauna, proceeded much faster in this area compared to those that were burned and planted to pasture (as discussed earlier), and this probably mitigated the effects of forest isolation as observed also for some vertebrates (Chapters 19 and 20).

Despite a lack of effect of fragment size on ant species richness, evidence was found that forest fragmentation does affect community composition. Both fragment size and distance to forest edge affected community composition, indicating that area and edge effects both play a role in structuring ant communities in Amazonian forest remnants. Part of the observed change in community composition was attributed to variations in litter depth, which increased markedly near forest edges (Carvalho and Vasconcelos 1999), although it is likely that changes in microclimate and vegetation structure were also involved.

Given the distance to which changes in ant community structure can penetrate into fragment interiors, we can conclude that even a 10 ha fragment will not be large enough to maintain an unaltered community of ground-dwelling ants, as this fragment will be largely affected by edge effects. Furthermore, for those species with large home-range requirements, such as the *Eciton* army ants, or those with patchy distribution and low densities, as is the case of the leaf-cutter *Atta cephalotes* in central Amazonia, much larger areas are required (Willis and Oniki 1978; Vasconcelos 1988; Harper 1989). This is not to say that even small fragments do not have a conservation value for ground-living ants, as several forest ant species, especially those with small home-range requirements and those that are not sensitive to fragmentation effects, are able to persist in these fragments. Whether populations of such species will be able to persist in the

long term, however, is unknown. Ideally, therefore, medium to large tracts of forest (hundreds to thousands of hectares) should be reserved and scattered throughout the landscape. Where deforestation has already occurred and only fragments remain, the conservation value of such areas can probably be enhanced through the establishment of buffer zones to minimize edge effects, and corridors between fragments to increase connectivity.

Our current knowledge about patterns in ant species richness and ant species distribution (including endemisms) in Brazilian Amazonia is limited. As these are important criteria in defining the location of conservation areas, selection of new areas needs to be based on information derived from other animal or plant taxa. However, the high degree of heterogeneity in ant species distribution over a scale of a few kilometers, as observed here and elsewhere (Wilson 1958; Majer 1983), gives support to the idea that for ants, and invertebrates in general (Brown and Hutchings 1997; Kerr 1997), a scattered system of smaller reserves (on the order of 1,000 ha) stratified across major environmental gradients is a better conservation strategy than is a single large reserve encompassing the same area.

Conservation Lessons

1. Ants are a species-rich group, filling a diverse and important suite of ecological functions in the rainforest ecosystem.
2. Deforestation significantly reduces the number of ants found in primary forest.
3. Fragmentation and edge effects have a pronounced influence on ground-dwelling ants. A forest patch of 10 ha would be entirely changed by edge effects, and certain species, such as army ants and some leaf-cutter ants, would

need areas of more than 100 ha to maintain stable populations even over the short term.

4. The taxonomy and distribution of ants are too poorly known regionally to be of much use in selecting target areas for conservation regionally, but the diversity witnessed across just a few kilometers of forest argues for establishing series of medium- to large-size preserves to adequately sample local diversity.

Acknowledgments

We are grateful to Antônio Casimiro, José M. Vilhena, and Aline Matsuo for their assistance during field and laboratory work. We also thank Claude Gascon, Bill Magnusson, and Bill Laurance for their comments and suggestions on various parts of this work. Financial support was provided by the BDFFP, INPA, Smithsonian Institution, and the Brazilian Council for Scientific and Technological Development (CNPq grant no. 300303/87–4 to HLV).

CHAPTER 17

Stingless Bees (Meliponini) and Orchid Bees (Euglossini) in Terra Firme Tropical Forests and Forest Fragments

MÁRCIO LUIZ DE OLIVEIRA

Stingless Bees

Stingless bees are well known and familiar to many in the tropics because of their defensive habit of biting and entangling themselves in people's hair and beards. They are social bees living in organized hives that have a queen, males, and workers (females) that perform various functions inside and outside the nest. The sole function of males is to mate with the queen. After mating, which takes place in full nuptial flight, the queen lodges in the hive and dedicates herself to laying eggs that can generate new queens, males, or workers. The workers labor the most, cleaning the interior of the beehive, collecting resin, clay, or other materials used in nest construction, or even collecting water, nectar (source of sugars), or pollen (source of proteins) for themselves or to feed to other members of the hive. The nectar, through an enzymatic process that takes place in the worker's stomach, is transformed into honey. The ecological importance of these bees, however, is not the production of honey but rather their role in pollinating various natural or cultivated plant species.

Orchid Bees

Unlike stingless bees, orchid bees (Euglossini) are still poorly known because they do not live in hives; do not have a queen; do not produce honey; and fly at a very high speed, almost always near tree canopies, where they mostly visit orchids. In many species the body is metallic green or blue, and the tongue is longer than the body. The geographical distribution of these bees is restricted to the Neotropics, from northern Mexico to northern Argentina (Dressler 1982), and only Jamaica and Trinidad in the West Indies.

Euglossine males and females visit flowers to obtain nectar in at least twenty-three plant families, whereas the females visit flowers of nine families for pollen and three others for resin (Roubik 1989). All orchid species of the sub-tribes Stanhopeinae and Catasetinae, as well as members of the other sub-tribes, are pollinated exclusively by euglossine males that visit flowers to collect floral fragrances (Williams and Whitten 1983), which they probably convert biochemically into the pheromones that they in turn use to attract mates. Pollen is also collected from plant families besides the Orchidaceae, including Araceae, Euphorbiaceae, and Gesneriaceae (Williams and Whitten 1983).

This is publication number 284 of the BDFFP Technical Series.

According to Dressler (1982), at least 625 species of 55 genera of orchids exist in the Neotropical region that do not produce nectar and whose pollen is not consumed by bees. Floral fragrances, in these species, would be the principal resource offered for attracting their pollinators, the euglossine males (Roubik 1989). These bees also take part in the pollination of the Brazil-nut tree (*Bertholletia excelsa* H. B. K.), along with bees of the genera *Bombus, Centris, Xylocopa,* and *Epicharis.* Brazil-nut trees have shown reduced fruit production in areas where their pollinators are not present in sufficient numbers to promote cross-pollination (Nelson, Absy, et al. 1985; Mori and Prance 1987).

Methods

Collections of stingless bees were made in isolated forest fragments of 1, 10, and 100 ha, as well as in continuous forest sites and deforested areas as part of the BDFFP, north of Manaus. Samples taken in the deforested areas were in the vicinity of the Colosso isolates (1202 and 1104) and camp. The second-growth vegetation surveyed was dominated by *Vismia* spp. (plate 9). (A detailed description of the study area can be found in Lovejoy and Bierregaard [1990] and in Chapter 4; see fig. 4.1 and table 4.3.)

The bees were collected in six ways: (1) by using honey as bait; (2) by collecting clay; (3) by collecting sweat; (4) in flight; (5) in their nests; and (6) in flowers. A 100 ha continuous forest area, subdivided into 20 × 20 m squares (Reserve 1501, the Lecythidaceae subproject study area described by Mori, Becker, and Kincaid [Chapter 6]), was used to survey stingless bee nests.

The euglossine fauna of two terra firme forest areas were compared. The two study areas were outside the limits of reserves 1401 and 1501 to avoid interference with ongoing experiments within the reserves. Collections were made fortnightly between September 1989 and August 1990, using fruit-fly traps modified according to Campos et al. (1989), containing synthetic fragrances and placed at two heights: near the canopy (12–15m) and on the ground.

Results

RELATIVE ABUNDANCE, RICHNESS, AND DIVERSITY

Fifty-four stingless bee species were found in the BDFFP areas, two of which are new species, *Plebeia* sp.n. 1 and *Plebeia* sp.n. 2 (table 17.1). A few stingless bees that, surprisingly, entered the traps for the Euglossini, were included here.

Thirty-eight species of Euglossini were found, three of which appear to be new species (table 17.2). In addition to the thirty-eight species encountered, previous collections made in the same region have recorded an additional nine species: *Euglossa platymera* and *Eufriesea surinamensis* (Kimsey and Dressler 1986); *Eufriesea laniventris, E. xantha,* and *E. purpurata* (Powell and Powell 1987); *Euglossa* cf. *amazonica* (Becker, Moure, and Peralta 1991); *Euglossa* cf. *securigera* and *E. liopoda* (Morato, Campos, and Moure 1992); and *Eufriesea theresiae* (Morato 1993). About 68 percent of the species encountered were represented by fewer than 20 individuals and can be classified as uncommon to rare. Examples include *Euglossa laevicincta, Euglossa* sp. 3, and *Eufriesea vidua,* of which only one individual was collected during the entire year. Also, *Euglossa bidentata, E. piliventris, E. prasina,* and *Euglossa* sp.3 were found in only one of the areas studied.

TABLE 17.1. Stingless Bees in the BDFFP Areas by habitat Type

Species	Isolates			Continuous Forest	Deforested Area	Total
	1 ha	10 ha	100 ha			
Aparatrigona impunctata (Ducke 1916)				1		1
Camargoia camargoi Moure, 1989	4	1		1	1	7
Celetrigona logicornis (Friese 1903)	2				1	3
Cephalotrigona capitata femorata (Smith 1854)				2		2
Duckeola ghilianii (Spinola 1853)	?[1]	?	?	?	?	?
D. pavani Moure, 1963				1		1
Frieseomelitta trichocerata Moure, 1988	1			3	1	5
Geotrigona subgrisea subnigra (Schwarz 1940)	?	?	?	?	?	?
Lestrimelitta limao (Smith 1853)					1	1
Leurotrigona pusilla Moure & Camargo, 1988	?	?		?	?	?
Melipona amazonica Schulz, 1905	2					2
M. captiosa Moure, 1962	2	1			1	4
M. fuliginosa Lepeletier, 1836				1		1
M. fulva Lepeletier, 1836	1	3		4	1	9
M. illustris Schwarz, 1932				1		1
M. lateralis Erichson, 1848		1	2	7	1	11
M. puncticollis puncticollis Friese, 1902		1		1		2
Nogueirapis minor Moure & Camargo, 1982			1	2		3
Oxytrigona obscura (Friese 1900)		1	1	3		5
Partamona sp.	?	?	?	?	?	?
P. mourei Camargo, 1980		1	4	1		6
P. cf. *nigrior* (Cockerell 1925)				1	1	2
P. cf. *pearsoni* (Schwarz 1938)	1	1			1	3
P. pseudomusarum Camargo, 1980	3	2	1	4		10
P. testacea (Klug 1807)	1			2	1	4
Plebeia sp. n. 1				1		1
Plebeia sp. n. 2				1		1
P. minima (Gribodo 1893)	1					1
P. margaritae Moure, 1962				1		1
Plebeia sp.				1		1
Ptilotrigona lurida mocsaryi (Friese 1900)	1	2		3	2	8
Scaptotrigona bipunctata (Lepeletier 1836)				2		2
S. fulvicutis Moure, 1964				2		2
S. aff. *Polysticta* (Moure 1950)				2	1	3
Scaura tenuis (Ducke 1916)					1	1
S. latitarsis (Friese 1900)	?	?	1	?	?	?
Tetragona clavipes (Fabricius 1804)					1	1
T. dorsalis (Smith 1854)	1			1	1	3
T. essequiboensis (Schwarz 1940)	3	1		1	1	6
T. goettei (Friese 1900)	2	2	1	2	1	8
T. handlirschii (Friese 1900)	4	1			1	6
T. kaieteurensis (Schwarz 1938)	1	2		1		4
Tetragonisca angustula (Latreille 1811)					1	1
Trigona branneri Cockerell, 1912		4		1	1	6
T. cilipes[2] Fabricius, 1804				1	2	3
T. crassipes (Fabricius 1793)	4	4	1	5	1	15
T. dallatorreana Friese, 1900	1			1		2
T. fulviventris Guérin, 1835	4	3	1	6		14
T. fuscipennis Friese, 1900	1	3		1	1	6
T. hypogea Silvestri, 1902		1	2	4	1	8
T. recursa Smith, 1863		3		3		6
T. williana Friese, 1900	1	2		5	1	9
Trigonisca dobzhanskyi (Moure 1950)	?	?	1	?	?	?
Trigonisca sp.	1					1
TOTAL	41	41	14	80	17	203

Notes: [1]? = Unknown collection site. [2]Dark wing tips, see Camargo 1988.

TABLE 17.2. Euglossine Bees Collected in Two Primary Terra Firme Forest Areas in the Central Amazon

Species	Reserve 1401	Reserve 1501	Percentage	Total
Euglossa analis Westwood, 1840	3	2	0.21	5
E. augaspis (Dressler 1982)	153	143	12.22	296
E. avicula (Dressler 1982)	136	143	11.52	279
E. bidentata (Dressler 1982)	—	3	0.12	3
E. chalybeata (Friese 1925)	121	231	14.53	352
E. cognata (Moure 1968)	2	14	0.66	16
E. crassipunctata (Moure 1970)	64	87	6.23	151
E. decorata (Smith 1874)	14	4	0.74	18
E. gainii (Dressler 1982)	8	8	0.66	16
E. ignita (Smith 1874)	14	10	0.99	24
E. imperialis (Cockerell 1922)	2	10	0.49	12
E. intersecta (Latreille 1938)	2	8	0.41	10
E. ioprosopa (Dressler 1982)	4	5	0.37	9
E. iopyrrha (Dressler 1982)	3	9	0.49	12
E. laevecincta (Dressler 1982)	—	1	0.04	1
E. mixta (Friese 1899)	27	40	2.77	67
E. modestior (Dressler 1982)	7	2	0.37	9
E. mourei (Dressler 1982)	23	19	1.73	42
E. parvula (Dressler 1982)	12	20	1.32	32
E. piliventris (Guérin 1845)	—	2	0.08	2
E. prasina (Dressler 1982)	—	2	0.88	2
E. retroviridis (Dressler 1982)	9	10	0.78	19
E. stilbonota (Dressler 1982)	311	474	32.41	785
E. viridifrons (Dressler 1982)	8	8	0.66	16
E. viridis (Perty 1833)	2	3	0.215	
E. sp. 1	—	3	0.12	3
E. sp. 2	2	—	0.08	2
E. sp. 3	—	1	0.04	1
Eulaema bombiformis (Packard 1869)	6	7	0.54	13
E. meriana (Olivier 1789)	42	38	3.30	80
E. cingulata (Fabricius 1804)	5	1	0.25	6
E. mocsaryi (Friese 1899)	17	11	1.16	28
Eufriesea pulchra (Smith 1854)	2	1	0.12	3
E. ornata (Mocsary 1896)	5	4	0.37	9
E. vidua (Moure 1976)	1	—	0.04	1
Exaerete frontalis (Guérin 1845)	50	30	3.30	80
E. smaragdina (Guérin 1845)	5	5	0.41	10
E. trochantherica (Friese 1900)	1	2	0.12	3
Total number of individuals	1,061	1,361		2,422
Total number of species	32	36		38

Bee Fauna Dissimilarity

Comparing the Euglossini bee faunas of the two continuous forest areas, Reserve 1501 had more individuals, as well as more species; nevertheless, Shannon diversity index values (H') were significantly greater in Reserve 1401 (1401, H' = 2.389, and 1501, H' = 2.257; t = 2.556, P = 0.005, d.f. = 1). Eq-uitability values J' (Ludwig and Reynolds 1988) were not different between the two areas (1401, J' = 0.689; 1501, J' = 0.630).

The two areas shared only thirty species (79 percent of the total species in the area) of euglossine bees (table 17.2), resulting in a dissimilarity coefficient of 0.22 between the two areas. This coefficient measures the dis-

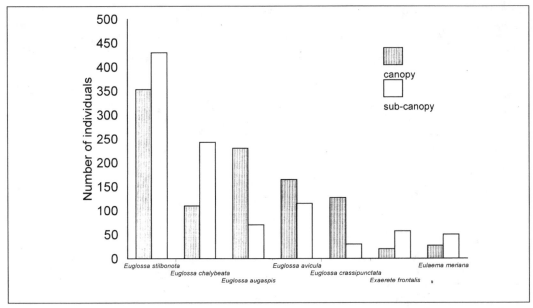

Fig. 17.1. Abundance of the most common Euglossini bee species in canopy and understory of terra firme continuous forest in central Amazon.

similarity between two samples and is based on the "chord distance" (Pielou 1984; Ludwig and Reynolds 1988). The maximum and minimum values for this distance are 1.41 and 0.0, respectively, in such a way that the greater the value obtained, the less similar the areas are. The value here obtained is relatively high, which is somewhat surprising, considering that the two areas are separated by a maximum of 10 km and are quite similar in their floristic composition; the value also shows considerable heterogeneity in the bee faunas over a fairly fine geographic scale (kilometers).

VERTICAL STRATIFICATION

In general, most of the euglossine species encountered in this study occurred in the forest canopy as well as in the understory. Species that were relatively uncommon or rare, such as *Euglossa laevecincta, E. prasina, E. viridis,* and *Euglossa* sp. 2, were captured exclusively in the canopy, whereas

E. piliventris, Euglossa sp.3, *Eufriesea vidua,* and *Exaerete smaragdina* were captured exclusively in the understory. Among the most abundant species, *Euglossa chalybeata, Exaerete frontalis,* and *Eulaema meriana* were much more common in the understory, whereas *E. augaspis* and *E. crassipunctata* were much more common in the canopy (fig. 17.1).

The number of individuals collected in the canopy was greater than in the understory, but the number of species was similar in the two strata. The diversity index value, however, was significantly greater in the canopy (H' = 2.346 and J' = 0.660) than in the understory (H' = 2.123 and J' = 0.607) (t = 4.419; d.f. 1; P less than 0.05).

EFFECTS OF FOREST FRAGMENTATION

Thirty-seven stingless bee species were found in continuous forest, 11 in 100 ha fragments (n = 1), 22 in 10 ha fragments (n = 4), 21 in 1 ha fragment (n = 5), and 25 in the deforested areas (table 17.1). Interestingly, total species richness in the four 10 ha fragments, or in the five of 1 ha, surpass the number of species found in the 100 ha fragment. Even more surprising is the higher species richness in the deforested areas than in any one fragment, surpassed only by the sum total of all the fragments (31 species).

Nest Density

Only 15 nests of nine stingless bee species were found in a 100 ha continuous forest area (table 17.3) at an average of about 0.15 nests/ha. This density is well below that encountered in other sites in the Neotropical region (table 17.4).

Seed Dispersal

In the terra firme forest near Manaus, at least five stingless bee species were observed visiting the ripe fruits of *Coussapoa asperifolia magnifolia* (Cecropiaceae) (Garcia, Oliveira, and Campos 1992). The workers of *Melipona seminigra merrillae, M. compressipes manaosensis,* and *Trigona williana* removed the mucilaginous epicarp of ripe fruits, along with the minuscule seeds, while *Aparatrigona impunctata* and *Trigona fuscipennis* removed only the epicarp of ripe fruits. Moreover, nests of four stingless bee species, *M. lateralis, M. fulva, M. seminigra seminigra,* and *M. rufiventris,* were constructed with a mixture of clay and mucilaginous epicarp containing seeds. On close analysis, these seeds were identified as those of *C. a. magnifolia,* although bees of these four species were never observed visiting fruits of *C. a. magnifolia.* Germination tests on these seeds were positive, suggesting that bees actually play a role in seed dispersal of these plant species.

Table 17.3. Stingless Bee Nest Density in a 100 ha Continuous Terra Firme Forest Area in the Central Amazon

Species	Nests/ha
Melipona captiosa	0.01
Partamona pseudomusarum	0.01
Ptilotrogona lurida mocsaryi	0.03
Scaptotrigona aff. *polysticta*	0.01
Tetragona dorsalis	0.01
T. goettei	0.01
T. kaieteurensis	0.02
Trigona crassipes	0.04
T. fulviventris	0.01

Discussion

Relative Abundance, Richness, and Diversity

The large number of stingless bee species found in the BDFFP areas (about 36,000 ha) is larger than the entire bee fauna of other areas, such as Madagascar, which has only four species; New Guinea with five; Australia with eight to ten; central Sumatra with 24; and Africa with 50 (Salmah, Inoue, and Sakagami 1990; Camargo and Pedro 1992). Only French Guiana has more known species, with 69 (Roubik 1989). It is also worth noting that the number of species found corresponds to about 21 percent of those previously registered for the Neotropical region as a whole (Camargo 1990).

Further, this abundance and richness of euglossine bee species is greater than anywhere else in Brazil (table 17.5). The 47 euglossine species known from the Manaus area is the highest total for the Neotropics, with the exception of Barro Colorado Island and Cerro Campana on Panama, where 53 species were recorded (table 17.6). It is likely that even richer faunas might be found in Amazonia as more and larger areas are surveyed. Recently, J. C. Brown (unpublished data) found 53 euglossine species in the state of Rondônia, but in a significantly larger area than that reported on in this chapter.

TABLE 17.4.   Stingless Bee Nest Densities in Neotropical Forests

Area (ha)	Nest density (per ha)	Species density (per ha)	Locality	Author
5.0	5.88	2.78	Panama	Roubik 1983
36.7	1.82	0.25	Costa Rica	Hubbell and Johnson 1977
1.0	1.00	1.00	Amazonas	M. V. B. Garcia (pers. comm.)
64.7	2.17	0.14	Panama	Michener 1946
100.0	0.15	0.09	Amazonas	M. Oliveira, Morato, and Garcia 1995

TABLE 17.5.   Published Studies Using Synthetic Fragrance Baits in Brazil on Euglossine Bees

State	Number of			Studies
	baits	specimens	species	
Amazonas[1]	4	76	10	Braga 1976
Amazonas[2]	3	992	15	Powell and Powell 1987
Amazonas[2]	3	290	16	Becker, Moure, and Peralta 1991
Amazonas[2]	4	1,242	27	Morato, Campos, and Moure 1992
Amazonas[2]	8	2,422	38	M. Oliveira and Campos 1995
Maranhão[2]	4	1,728	13	Gomes 1991
Paraíba	7	1,082	10	Bezerra 1995
Bahia	5	1,285	5	Raw 1989
Bahia	5	280	9	Aguilar 1990
Bahia	5	1,144	12	Neves and Viana 1997
Minas Gerais	4	896	11	Abrantes 1990
São Paulo	3	892	8	Rebelo and Garófalo 1991
Rio Grande do Sul	3	639	5	Wittmann, Hoffmann, and Scholz 1988

Notes: [1]Campina, campinarana, and forest area. [2]BDFFP areas.

BEE FAUNA DISSIMILARITY

The results highlight the spatial heterogeneity in bee species composition and the need to include different sites to estimate species diversity in a given area (see also Chapter 16). This spatial heterogeneity is surprising given that Euglossine bees are robust and capable of flying long distances (Janzen 1971; Raw 1989; Williams and Dodson 1972). However, Armbruster (1993) argued that although the tropical forest seems homogeneous to the human eye, from a bee's point of view the forest is a mosaic of diverse microhabitats characterized by the distribution and phenology of diverse plant species and by the sources of floral fragrances. Becker, Moure, and Peralta (1991) also emphasize that it is difficult to characterize the euglossine bee fauna of a forest by taking samples from only one location. It is possible that some euglossine species do not move between two proximate areas even though these bees possess great flying capacity and are able to cover large distances. Research in the Manaus region has shown that four euglossine species did not cross the 100 m of pasture that separated continuous forest from some forest fragments (Powell and Powell 1987). Moreover, Becker, Moure, and Peralta (1991) suggested that, on a small scale, the abundance of bees can vary completely in response to unknown factors, since they encountered substantial differences between sites separated by only 300 to 700 m sampled during the same day. Morato (1994) showed a reduction in the abundance

TABLE 17.6.   Published Studies on Euglossini Bees in the Neotropics That Used Synthetic Fragrance Baits

Country	Number of baits	Number of specimens	Number of species	Studies
Colombia (East Region)	3	160	42	Dodson et al. 1969
Costa Rica	5	961	27	Janzen et al. 1982
El Salvador	4	31	6	Dodson et al. 1969
Equador (East Region)	5	427	18	"
Guatemala	5	68	6	"
Guiana	23	713	45[1]	Williams and Dodson 1972
Honduras	14	147	13	F. Bennett 1972
Mexico (East Region)	5	203	9	Dodson et al. 1969
Nicaragua	4	73	4	"
Panama	16	27,874	53	Ackerman 1989
Peru	20	2,917	38	Pearson and Dressler 1985
Trinidad	5	244	13	Dodson et al. 1969
Venezuela (East Region)	4	89	18	"

Note: [1]Collections in flowers were included.

of euglossine males that were attracted by floral fragrances from the interior of the forest to the edge and to an adjacent deforested area. Morato (1994) concluded that, although the Euglossini are considered long-distance pollinators, open and deforested areas could constitute barriers to their dispersal, limiting pollination by bees in these areas. Lower gene flow in plant populations that require Euglossini as their principal or exclusive pollinator (e.g., orchids) could thus result.

The significant difference between the diversity of the euglossine fauna in the two areas is probably related to differences in floristic composition. However, microclimatic factors or the presence of parasites in nests might also play important roles (Folsom 1985). Finally, it should be mentioned that the dispersion of the synthetic fragrances used as bait is more effective in ventilated, well-lit, open vegetation areas (Folsom 1985). Thus, a greater number of sites should be sampled to correct for or to minimize these effects.

VERTICAL STRATIFICATION

Although the height difference between the understory and the canopy traps was only 12 to 15 m, it was sufficient to observe the preference of some species for one of these strata. It is possible that the use of traps at heights greater than 30 m, which is the average canopy height of these forests (Bierregaard and Lovejoy 1988) would show a more distinct stratification.

Various studies on tropical forests have found that the number of bee species in the forest canopy is much greater than in the understory. Erwin (1982) estimated that in tropical forests, the arthropod fauna (which includes bees) in the canopy should be two times greater than in the understory. In the tropical forests of Panama, Papua New Guinea, and Brunei, Sutton, Ash, and Grundy (1983) found a clear stratification and a marked preference of some insect groups, including Hymenoptera (which includes bees), for the canopy. In most cases, the insects were found between 20 and 30 m in the trees. Wolda and Roubik (1986) found that bee species of the genus *Rhinetula* and *Ptiloglossa*, which build nests in the ground, were mainly captured by 3 m high traps, whereas *Megalopta,* which build nests in wood, was more abundant at 27 m. These authors postulated that such factors as food and nest sites could influence the stratification of bees in a forest. Frankie, Vinson, and Barthell (1988) placed nest-traps for bees of the genus *Centris* at 0.5 and 2.5 m heights and found that a large percentage of these bees built nests at the latter height. Morato (1993) placed nest-traps for bees and solitary wasps at 1.5, 8, and 15 m, in the same forests as the present study, and found that most species built nests at heights of 8 and 15 m.

Bees and wasps seem to prefer the forest canopy, but the reasons for such a preference are unclear. Salmah, Inoue, and Sakagami (1990) have shown that the microclimatic conditions in primary forest vary between the interior of the understory and the edges of the canopy. The understory is characterized by cooler, darker, and more humid conditions, with very little fluctuation in these conditions. Forest canopy conditions show opposite trends and vary much more. These microclimatic gradients may produce a series of microhabitats in which various animals, including bees, have adapted themselves in different ways.

Of course, because orchids, from which Euglossini males remove floral fragrances, also show vertical distribution gradients (Braga 1987), it is possible that the Euglossini fauna simply respond to the distribution of orchids.

The understory faunas of the two forest areas showed a greater similarity to each other than did the canopy faunas. This similarity was greater than that between the understory and canopy of the same area (table 17.7), suggesting that most species show a clear preference for one stratum or the other.

TABLE 17.7.   Values of a Dissimilarity Index of Euglossini Bee Faunas Between the Forest Canopy and Understory in the Two Study Areas

	Canopy 1401	Understory 1501	Canopy 1501
Understory 1401	0.51	0.21	0.37
Canopy 1401		0.65	0.23
Understory 1501			0.48

Notes: Large values indicate less similar species composition. The index used varies between 0.0 and 1.41.

EFFECTS OF FOREST FRAGMENTATION

At first glance it would seem that the continued deforestation of terra firme forest area in the central Amazon would not cause any major impact on the stingless bee populations. This is misleading because 11 species were found exclusively in the continuous forest (*Aparatrigona impunctata, Cephalotrigona capitata femorata, Duckeola pavani, Melipona fuliginosa, M. illustris, Plebeia margaritae, Plebeia* sp., *Plebeia* sp. n. 1, *Plebeia* sp. n. 2, *Scaptotrigona bipunctata,* and *S. fulvicutis*). These species would be very sensitive to any deforestation or fragmentation of their habitats. Although *Melipona amazonica, Plebeia minima,* and *Trigonisca* sp. were found in the 1 ha fragments, which suggests some tolerance to fragmentation, we know nothing about the persistence of these species in such small areas through time. Long-term monitoring of these bee populations would be important.

Lestrimellita limao, Scaura tenuis, Tetragona clavipes, and *Tetragonisca angustula* were captured exclusively in deforested areas. These species probably represent recent invasions to the modified BDFFP landscape due to the appearance of farmland and pasture, and of young second growth (see also Brown and Hutchings 1997; Chapter 19). *Tetragonisca angustula,* for example, is a well-known species, perfectly adapted to large urban centers where it builds its nests in cavities of walls, posts, and the like.

The low species richness in the 100 ha fragment is more difficult to explain. Probably the niches occupied by some Meliponini species are somewhat discontinuous in forests (J. M. F. Camargo, pers. comm.), perhaps by chance alone being underrepresented in isolated forest patches.

NEST DENSITY

The nest density encountered is well below that reported for other regions (see table 17.4). Again this may reflect very patchy distribution of resources needed by bees in primary forests and is intriguing, given the high number of species in the area.

RISK OF EXTINCTION

The process of deforestation and the expansion of agricultural frontiers have the potential to reduce populations of Meliponini bee species, which may lead to extinction (Roubik 1983). In areas of natural low population density this risk is even greater. Kerr and Vencowsky (1982) estimated that a minimum of forty colonies are necessary in an area to maintain a stable population and decrease the risk of inbreeding. In areas of lower abundance, inbreeding favors the emergence of male diploids. In this case, workers kill the queen when the first diploid males emerge and kill other diploid males later (Carvalho et al. 1995). This behavior, of course, may mean the end of the colony.

SEED DISPERSAL

Vertebrate seed dispersal is a well-known phenomenon, and recently observations of seed dispersal by invertebrates has increased. LaSalle and Gauld (1993) showed that seed dispersal by ants occurs in at least eighty-seven genera of twenty-three plant

Fig. 17.2. *Melipona eburnea fuscopilosa* hive with seedlings of *Coussapoa* sp. growing out of a lateral wall.

families in Australia and that seeds of about 35 percent of all the herbaceous plant species and many tree species are dispersed by ants. A few years ago, Wallace and Trueman (1995) reported that workers of *Trigona carbonaria,* a stingless bee, are the effective agents of seed dispersal in the eucalypt tree, *Eucaliptus torelliana,* in the rainforests on the east coast of Australia.

Although not observed in the BDFFP area, it is common to find *Coussapoa* sp. seedlings sprouting in wooden boxes with *Melipona eburnea fuscopilosa* nests (fig. 17.2) or in natural *Partamona* sp. nests around Rio Branco, Acre, also in Amazonia. These findings confirm the role of stingless bees in dispersing seeds of *Coussapoa.* As this plant has epiphytic habits and strangling roots, it is probable that their seeds can germinate inside nests located in tree cavities and that *Coussapoa* seedlings subsequently spread out onto these trees, strangling them, as has been observed in nature.

Conservation Lessons

1. Euglossine bees have been recognized as keystone species, or, rather, species that play major roles in ecosystem structure and functioning. The removal or extinction of such species would lead to a

domino effect in nature (LaSalle and Gauld 1993). Gilbert (1980, cited by LaSalle and Gauld 1993) suggested that euglossine bees are among the most important "linking organisms" known in the Neotropical forests, because of their association with plants of all stages and strata in a forest.

2. The role of native bees in maintaining the tropical forest by means of pollination must be fully acknowledged. Tropical forests, unlike temperate forests, are characterized by a prevalence of dioecious plants (male and female in separate individuals) (Opler and Bawa 1978), and pollination tends to be performed by animals (Lovejoy and Rankin 1981). Bees as a group clearly play a major role in maintaining the forest, and the Euglossini represent 15–20 percent of these bee species (Ackerman 1985).

3. As deforestation in the Amazon continues at alarming rates (INPE 1998), the demarcation of forest reserves and parks be-

comes all the more necessary to protect natural resources. Because the euglossine bee fauna appears heterogeneous within a tropical forest, and the Meliponini bee fauna shows very low densities in many sites, the survival of many bee species, as well as the plants that they pollinate (or disperse), can be threatened without adequate protection, which includes large forest reserves that contain much of the spatial heterogeneity in natural resources needed by bees.

4. When planning and designing conservation areas, large forest areas are necessary to offer adequate conditions for the survival of those Meliponini species that have low nest densities. For example, *T. fulviventris*, which had only 0.01 nests/ha (see table 17.3), would require an estimated 4,400 ha as a minimum area capable of maintaining at least forty-four colonies to avoid the undesirable effects of inbreeding (Kerr and Vencowsky 1982; Carvalho et al. 1995).

CHAPTER 18

The Implications of Changing Invertebrate Abundance Patterns for Insectivorous Vertebrates in Fragmented Forest in Central Amazonia

RAPHAEL K. DIDHAM

ppreciation of the size of the soil and leaf litter invertebrate fauna has fueled speculation that the ground, not the canopy, may be the heart of diversity in tropical forests (Stork 1988; Rosenzweig 1995; Hammond, Stork, and Brendell 1997; Stork, Didham, and Adis 1997). Total invertebrate density and biomass is so great in the soils of tropical forests that terrestrial invertebrates mediate a significant portion of the total energy and nutrient fluxes through tropical ecosystems (Greenwood 1987; Wilson 1987a; Seastedt and Crossley 1984; Eggleton et al. 1996; Lawton et al. 1996). Invertebrates are also fundamental components of nearly all trophic levels. Consequently, any changes in the biomass or composition of the invertebrate community resulting from human disturbance will have far-reaching implications for tropical ecosystems. Here I consider the effects of forest fragmentation on the density and composition of the litter invertebrate fauna in central Amazonia.

Although the direction and magnitude of changes in invertebrate density in fragmented habitats are poorly understood, some general principles may apply (Didham 1997a). First, most fragmentation studies have shown a significant increase in invertebrate density at the forest edge (e.g., Helle and Muona 1985; Toda 1992; Fowler, Silva,

and Venticinque 1993; Didham 1997a, 1997b; Malcolm 1997b; but see Gunnarson 1988 and Ozanne et al. 1997). Second, invertebrate density increases with increasing area on real islands (Jaenike 1978) and a comparable response for habitat islands has been shown (e.g., Jennersten 1988; Aizen and Feinsinger 1994b; but see Martins 1989). The apparent contradiction of small habitat fragments having lower invertebrate densities despite an increasing proportion of edge habitat suggests that there may be significant area effects operating on invertebrate density, above and beyond simple edge effects. Third, compositional changes in the invertebrate fauna in fragmented habitats almost certainly have an important bearing on ecosystem processes and interactions with other organisms, although data are lacking (Klein 1989; Didham, Ghazoul 1996; Didham 1997a, 1997b; Didham, Hammond et al. 1998; Didham, Lawton et al. 1998).

The impact of invertebrate density on plant and vertebrate responses to disturbance is often a crucial missing component in habitat fragmentation studies (e.g., Forman, Galli, and Leck 1976; Askins, Philbrick, and Sugeno 1987; Bennett 1987; Bierregaard and Lovejoy 1989; Fonseca and Robinson 1990; Gibbs and Faaborg 1990; Blake 1991; Laurance 1991b, 1994; Newmark 1991; Soulé, Alberts, and Bolger 1992;

This is publication number 285 in the BDFFP Technical Series.

Leigh et al. 1993; Andrén 1994; Herkert 1994; Sarre, Smith, and Meyers 1995; Tellería and Santos 1995). Where invertebrate prey densities are not explicitly considered, it may be difficult to resolve the causes of population changes or local extinction in insectivorous vertebrate species (Wiens 1994; Stouffer and Bierregaard 1995a). In one example where invertebrate density was considered, Malcolm (1997b) found that the abundance and biomass of invertebrates increased at the forest edge and that there was a relative shift in invertebrate biomass from the canopy to the ground. These changes in invertebrate prey density in fragmented forests were highly correlated with the surprising increase in small mammal densities in small (1 ha and 10 ha) forest fragments and along edges in central Amazonia (Malcolm 1995, 1997a). The same conclusions apply to an increase in breeding bird densities in response to increased invertebrate density at forest edges in Finland (Helle and Muona 1985). In a complementary example, Harper (1989) explained the loss of ant-following birds from small forest fragments at the BDFFP study sites in central Amazonia by the absence of army ants and the consequent reduction in ant-flushed invertebrate prey density.

Of particular relevance to this study is the work of Stouffer and Bierregaard (1995a) on terrestrial insectivorous birds in fragmented forest in central Amazonia, again at the BDFFP study sites. Of thirty-five common insectivorous bird species in pre-isolation samples, the solitary terrestrial species group (seven species) and some members of the close functional group of low-level arboreal insectivores (five species) were most vulnerable to extinction. Stouffer and Bierregaard (1995a) clearly showed how these species declined in density from pre-isolation levels through nine years of post-isolation faunal collapse. They consider that

one of the most important unresolved factors contributing to the demise of the terrestrial insectivorous bird fauna, and the disintegration of mixed-species flocks in fragmented forest, may be changes in invertebrate prey density. This study, carried out in central Amazonia at the BDFFP, addresses these issues with an analysis of the terrestrial invertebrate fauna in continuous forest and subsequent changes in litter invertebrate density and proportional representation of different invertebrate groups in forest fragments, including four of those sampled by Stouffer and Bierregaard (1995a).

Methods

SAMPLING PROTOCOL

This study was part of the BDFFP, administered by Brazil's National Institute for Research in Amazonia and the Smithsonian Institution (for details of the study site see Chapter 4).

The sampling design was based on a comparison of two independent transects sampled at each of three locations: (1) deep within undisturbed continuous forest (more than 10 km from the nearest edge); (2) from the edge to the interior of continuous forest; and (3) from the edge to the interior of two 100 ha isolated forest fragments (fig. 18.1). The two 100 ha fragments (BDFFP Reserves 2303 and 3304; see fig. 4.1 and table 4.3) were located approximately 20 km apart (hereafter designated fragment edge-1 and fragment edge-2); the two continuous-forest edges, forest edge-1 and forest edge-2, were separated by a distance of 2 km (both on the western edge of continuous-forest Reserve 1401); and the deep-forest control plots, interior-1 and interior-2, were 2 km apart (both in Control Site 1501). All forest edges abutted well-maintained pasture (open, grazed pasture without second growth) and were west facing, with the exception of fragment edge-2, where the only edge abutting well-maintained pasture was north facing. Invertebrates were collected at seven distances along each of the six transects: 0, 13, 26, 52, 105, 210, and 420 m (fig. 18.1). This sampling protocol reflected the a priori expectation that changes in invertebrate community structure would be greatest near the forest edge.

In addition, to assess invertebrate populations in small forest fragments, two 10 ha fragments (Reserves 3209 and 1202) were sampled at 105 m from the edge (designated 10 ha-1 and 10 ha-2), and two 1 ha fragments (Reserves 2107 and 2108) were sampled at 52 m from the edge (designated 1 ha-1 and 1 ha-2).

INVERTEBRATE SAMPLING

At each of the forty-six transect sites, twenty random, 1 m² leaf litter samples were collected during the January to May 1994 rainy season. Daily sampling was randomly allocated between different transects and sites to prevent bias arising from daily and seasonal variation in activity patterns of invertebrates. All friable leaf litter was scraped rapidly from the quadrant and placed in a large bag-sieve to minimize invertebrate escape. The material was immediately sieved over a 9 mm mesh by vigorously shaking the bag-sieve for approximately five minutes. The fine, sieved litter containing invertebrates was then transported to the laboratory in individual cotton bags. Invertebrates were extracted using the Winkler method, whereby sieved leaf litter was carefully placed into coarse mesh bags, which were then suspended gently inside a large sealed cloth bag and hung for three days. As the leaf litter dried out, invertebrates sensitive to desiccation moved downward through the mesh bag and fell into a jar of alcohol below. The

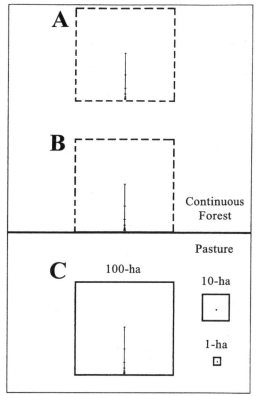

Fig. 18.1. Sampling protocol. Transects sampled at seven distances (0, 13, 26, 52, 105, 210, and 420 m) in: (A) continuous forest ("interior"). (B) edge of continuous forest ("forest edge"). (C) 100 ha fragments ("fragment edge"). Additional samples were taken at 105 m into 10 ha fragments and 52 m into 1 ha fragments. The entire design was replicated once. Distances between sites were 2 to 50 km, not the stylized arrangement illustrated here.

author and one field technician operated forty Winkler bags continuously for five months.

The Winkler method is sensitive to climate and collection methods and hence requires strictly standardized methods. I collected samples from plateau forest areas only (i.e., transects were not located in gullies or seasonally flooded areas) and only on dry mornings when there had been no rain the previous late afternoon or night. Leaf-litter sampling was discontinued if it rained. All samples were dried for three days, and no extra hand sorting of litter was performed. Despite these restrictions, the Winkler method is still inherently a "relative" trapping method, as with most invertebrate sampling methods, in that it does not sample all taxa with equal efficiency. It is a poor method for sampling micro-invertebrates such as mites and Collembola, so in this study I restrict analysis to macro-invertebrates. It is a particularly good method for the rapid and efficient extraction of beetles (Coleoptera) and ants (Hymenoptera: Formicidae) from large numbers of samples (Besuchet, Burckhardt, and Löbl 1987; Nadkarni and Longino 1990; Belshaw and Bolton 1994).

Results

INVERTEBRATE DENSITY IN UNDISTURBED CONTINUOUS FOREST

A total of 19,980 invertebrates (not including mites and Collembola) were collected in 280 m^2 of leaf-litter samples from undisturbed continuous forest, giving an overall invertebrate density of 71.36/m^2. The invertebrate fauna was dominated by ants, with 70 percent of total abundance, and adult beetles, with 11 percent of total abundance in continuous forest. Other important elements of the fauna included beetle larvae, flies (Diptera) and their larvae, wasps (Hymenoptera other than ants), moth larvae (Lepidoptera), and bugs (Hemiptera). Non-insect invertebrates were considerably less abundant, representing just 5 percent of total abundance.

Litter invertebrate density in central Amazonia was noticeably lower than published values for other tropical locations (e.g., Panama: 342/m^2 [Levings and Windsor 1983]; and Sulawesi, Indonesia: 227/m^2 [Hammond 1990]).

CHANGES IN INVERTEBRATE DENSITY AT THE FOREST EDGE

Total invertebrate density increased significantly with proximity to the forest edge (table 18.1 and fig. 18.2). The regression slopes for different forest edges did not differ significantly (i.e., nonsignificant treatment-by-distance, and replicate-by-distance interactions), although fragment edges gave a significantly better fit to linear regression models. The lack of fit of forest edge transects to linear regression was due largely to the aberrantly high invertebrate density at 420 m sites. Overall, however, total invertebrate density was conspicuously higher at all edges than in the forest interior.

Interestingly, the marked response of total invertebrate density to edge effects was not shown by all individual taxa (figs. 18.2 and 18.3). Ants and bugs showed significant overall edge responses ($F_{(6, 12)}$ = 4.28 and 4.29, respectively, both P > 0.016). However, the edge responses of other invertebrate groups varied considerably between forest edges, leading to significant interaction effects between treatment groups and between treatment replicates, but no robust statistical evidence for a ubiquitous edge effect on the densities of beetle adults, beetle larvae, or wasps (fig. 18.3) (all P < 0.05). Despite a great deal of noise in the data and variation between sites, note that no invertebrate group showed a decline in density with increasing proximity to forest edges.

CHANGES IN INVERTEBRATE DENSITY WITH DECREASING FRAGMENT AREA

There was a general trend toward lower invertebrate densities along fragment-edge transects than along forest-edge transects (figs. 18.2 and 18.3). For total invertebrate density (table 18.1 and fig. 18.2), and for the densities of all individual invertebrate groups except beetle adults (fig. 18.3), this

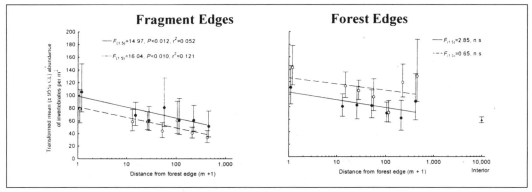

Fig. 18.2. Total leaf litter invertebrate densities at forest-edge and 100 ha fragment edge sites in central Amazonia. Solid circle = forest edge-1 and fragment edge-1 sites (solid fitted lines), open circle = forest edge-2 and fragment edge-2 sites (dashed fitted lines), solid triangles = interior sites. Forest-edge and fragment-edge points are means of 20×1 m² samples, and the Interior point is the mean of 280×1 m² samples.

trend was not significant ($F_{(1,2)}$, all $P < 0.05$). For beetle adults there was a significantly lower density of individuals at fragment-edge sites than at forest-edge sites, at all distances from the edge ($F_{(1,2)} = 23.55$, $P > 0.04$, fig. 18.3c). The mean density of beetles at forest-edge sites was 11.7/m², compared with 6.0/m² at fragment-edge sites—a reduction in density of approximately 50 percent. By comparison, the statistically nonsignificant decreases in density for other invertebrate groups were: beetle larvae, 38 percent; ants, 40 percent; wasps, 27 percent; bugs, 17 percent; and total invertebrate density, 36 percent. No group showed a mean increase in density in 100 ha fragments.

Of particular note is that invertebrate density (particularly of beetle adults and ants) was lower at the centers of 100 ha fragments than in deep continuous forest control sites, although the statistical significance of this finding is difficult to ascertain.

Although some evidence suggested a decrease in the density of certain invertebrate groups from continuous forest to 100 ha fragments, there was no evidence for a general positive relationship between invertebrate density and fragment area (see fig.

18.4). Rather than invertebrate density in 1 ha and 10 ha fragments being lower than in 100 ha fragments (at corresponding distances from the forest edge), the results were just the opposite; 1 ha and 10 ha fragments had invertebrate densities comparable to (or marginally higher than) 100 ha fragments.

Changes in the Proportional Representation of Invertebrate Groups in Fragmented Forest

The striking differences in edge responses of different invertebrate groups (figs. 18.2 and 18.3) give a clear indication that there were significant changes in the proportional representation of invertebrate taxa following forest fragmentation. Chi-square goodness of fit tests of the proportions of total abundance of ten dominant invertebrate groups at each forest-edge and fragment-edge site, relative to continuous forest, showed significant changes in proportional representation at all sites except 26 m from the edge of fragment edge-2 (all $P > 0.0015$, Bonferroni correction for multiple tests applied). The sign (+ or −) of significant deviations from continuous forest proportions indicated that, first, the

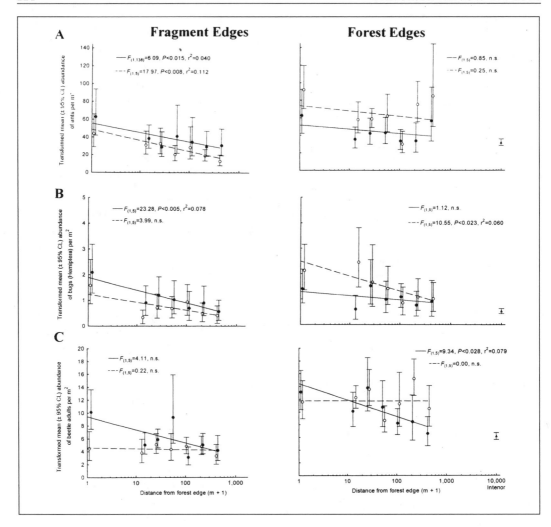

proportional representation of ants was consistently lower in all forest-edge and fragment-edge sites than in continuous forest. Second, beetle adults and beetle larvae were proportionally overrepresented at forest-edge sites but underrepresented at fragment-edge sites. Third, wasps and bugs had consistently higher proportional representation at all forest-edge and fragment-edge sites. These findings are in close agreement with the trends in fig. 18.3.

Conclusions

Invertebrate Density at the Forest Edge

Leaf-litter invertebrate densities at four forest edges in central Amazonia showed significant increases of 50 to 150 percent over average litter invertebrate density in deep continuous forest, with density often (but not always) being a decreasing log-linear function of distance from the forest edge. This pattern is typical of invertebrate edge responses (Helle and Muona 1985; Didham 1997a; Malcolm 1997b). Notably, however, the edge responses of individual invertebrate groups were highly variable, with

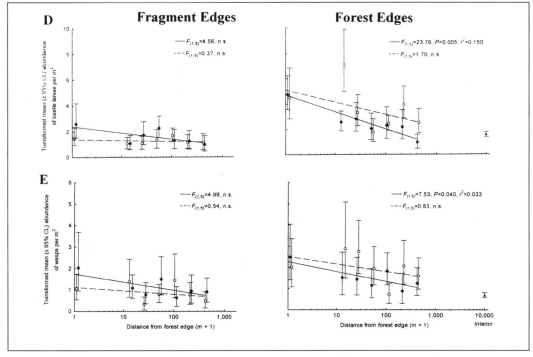

Fig. 18.3. Leaf litter invertebrate densities at forest-edge and 100 ha fragment-edge sites in central Amazonia: (A) ants; (B) bugs; (C) beetle adults; (D) beetle larvae; and (E) wasps. Symbols as in fig. 18.2.

TABLE 18.1. Analysis of Variance Comparing Total Leaf Litter Invertebrate Density in Two Fixed Fragmentation Treatments.

	df	MS effect	df effect	MS error	F error	P
Fragmentation treatment	1	28.773	2	5.466	5.264	0.149
Replicate (nested subgroup)	2	5.466	532	0.398	13.742	<0.001
Distance from edge	6	2.358	12	0.432	5.457	<0.007
Treatment × Distance	6	1.177	12	0.432	2.724	0.066
Replicate × Distance	12	0.432	532	0.398	1.086	0.369

Note: Fixed treatments are forest-edge vs 100 ha fragment. Each had two replicate sites (random, nested subgroups) sampled at seven fixed distances from the forest edge (0, 13, 26, 52, 105, 210, and 420 m) (n = 20 m² per site). P-values are statistically significant at P<0.05.

some groups showing no discernible trend at one or more forest edges. Statistical resolution is a particular problem in the analysis of such data, but importantly none of the edge responses showed that invertebrate density declined toward the forest edge, although such cases have been documented in temperate spruce forests (Gunnarson 1988; Ozanne et al. 1997).

Elevated invertebrate density at forest edges has important ramifications for the conservation of insectivorous vertebrates in fragmented ecosystems. Paradoxically, if total prey density is the principal limiting factor for insectivores, then one implication of higher invertebrate density at forest edges is that insectivores should be found at greater densities near tree-fall gaps or edges

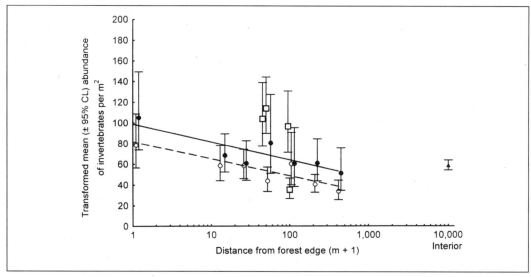

Fig. 18.4. Total leaf litter invertebrate densities in 1 ha (52 m from the edge) and 10 ha (105 m from the edge) forest fragments in central Amazonia (open squares), compared with densities at different distances from the edge in fragment edge-1 (solid circles), fragment edge-2 (open circles), and interior forest (solid triangles). Fitted regression lines as in fig. 18.2.

where prey is more abundant. Higher overall densities of both mammals and birds have been noted at the forest edge in some studies (e.g., Helle and Muona 1985; Malcolm 1995, 1997a), while others have shown overall bird density at the forest edge to be comparable to (Chapter 20) or lower than (Quintela 1985) that in the forest interior. However, while this may be true of total density, the densities of many terrestrial insectivorous bird species decline markedly in forest fragments (Stouffer and Bierregaard 1995a), despite increased total prey availability.

For total bird density to be maintained (or to increase) at the forest edge indicates some level of species replacement and an increase in the densities of species that are better able to persist at the forest edge. This points either to behavioral edge avoidance or to the importance of highly specialized prey item selection in determining the susceptibility of terrestrial insectivorous birds to forest fragmentation. There are probably important

tradeoffs between increased predation risk, ease of foraging, prey availability, and other factors that have led to a preconditioned avoidance response to edges.

INVERTEBRATE DENSITY AND FRAGMENT AREA

In addition to marked increases in litter invertebrate density at forest edges, there was limited evidence for area-dependent changes in invertebrate density, over and above edge effects. There was a significant downward shift in beetle density from continuous-forest edges to 100 ha fragment edges, suggesting that fragment size is an important determinant of invertebrate density. Overall, 100 ha fragments showed a 50 percent reduction in beetle density at all distances from the forest edge. A similar trend was evident for other invertebrate groups but was not statistically significant. In contrast to Malcolm (1997b), who attributed all area-dependent changes in invertebrate density to an increased proportion of edge habitat in small fragments, area-dependent changes in invertebrate density in this study were opposite those expected from the interaction of multiple edge effects.

The implications of area-dependent changes in invertebrate density are threefold. First, deep continuous forest insectivores may have reduced prey availability at the centers of relatively large forest fragments, where habitat conditions may otherwise be favorable. Second, and less intuitively, insectivorous edge specialists may also be adversely affected by fragment area. Edge species could face a similar 50 percent reduction in prey availability at the edges of 100 ha forest fragments, compared to continuous-forest edges. This leads to the prediction that insectivorous vertebrate density at edges should decline with decreasing fragment area. Third, the lower beetle density (and perhaps other invertebrate density) at the boundary (0 m) of 100 ha fragments than at the boundary of continuous forest suggests that invertebrate influx from outside the fragment may not be of prime importance in determining the increase in invertebrate density at forest edges. Naturally, it is difficult to determine whether the increased invertebrate density at forest edges is due largely to an influx of species from the surrounding second-growth matrix or to an increase in the density of forest species in response to increased productivity at edges.

One explanation for lower beetle density in 100 ha fragments might be that populations in small fragments cannot reach carrying capacity because mortality exceeds natality, leading to a positive relationship between density and area (Jaenike 1978; Andrén 1994). However, a positive density-area relationship does not hold for forest fragments smaller than 100 ha. Overall invertebrate densities in 1 ha and 10 ha fragments were higher than were densities at the same distances into 100 ha fragments. Inverte-brate density inside 1 ha fragments was more typical of 0 m invertebrate density in 100 ha fragments, which may highlight the increased edge influence in very small fragments. This is consistent with the additive nature of multiple edge effects (Malcolm 1994) but does not concord with a simple relationship between invertebrate density and fragment area.

PREY AVAILABILITY AS A SELECTIVE FORCE IN FRAGMENTED LANDSCAPES

If prey availability is an important determinant of terrestrial insectivorous vertebrate density in a highly fragmented landscape, then selection should act strongly for individuals to use more edge or edgelike habitat where prey abundance is highest. In evolutionary time there was probably relatively little edge habitat in the central Amazon region, but some did occur naturally in tree-fall gaps and larger blowdown areas (Nelson, Kapos, et al. 1994). Man-made edges are analogous in many ways to natural tree-fall gaps and edges. Many of the same processes that apply to the forest fragment edges studied here will also apply (less severely) to natural edges (but see Shelly 1988). Thus, edges are not an evolutionary novelty for insectivorous birds but do represent a significant new selection pressure in a fragmented landscape.

Species with behavioral preadaptations to edge or gap conditions are most likely to persist in fragmented habitats (see also Bierregaard and Lovejoy 1989; Stouffer and Bierregaard 1995a, 1995b; Malcolm 1997a). A good example of this is shown by the mixed-species bird flocks that are a common feature of Amazonian forests. These multispecies assemblages comprise ten to twenty understory and midstory insectivorous species that regularly travel and forage together through the forest (Powell 1985; Stotz 1993;

Stouffer and Bierregaard 1995a). Mixed-species flocks normally forage in an area of 8 to 12 ha, but they avoid exposed edges.

Following fragmentation, mixed-species flocks disintegrate within one year of isolation of 1 ha fragments, and within two years of isolation of 10 ha fragments (Bierregaard and Lovejoy 1988; Stouffer and Bierregaard 1995a). While the majority of flock species declined in abundance to near local extinction in these forest fragments, three species of "flock dropouts," *Glyphorynchus spirurus, Xiphorhynchus paradalotus,* and *Myrmotherula axillaris,* took on a solitary existence in which individuals used edges, and even foraged outside fragments, in areas that would normally be avoided when traveling in flocks (Stouffer and Bierregaard 1995a). Stouffer and Bierregaard (1995a) suggested that flock dropouts persist in fragmented forest because of greater behavioral plasticity. There is certainly evidence to suggest that this may be the case, given that invertebrate prey density is high at edges (so food limitation may not be a factor), and given that mixed-species flocks reassemble once sufficiently well-developed secondary vegetation buffers the fragments (Stouffer and Bierregaard 1995a). However, the possibility exists that some insectivorous species fail to persist in fragments because of specialized prey item selection (that is, limits on prey quality rather than prey quantity). Indeed, the proportional representation of major invertebrate taxa is altered in fragmented forest, and there is a major shift in invertebrate species composition (Didham 1997a, 1997b; Didham, Hammond et al. 1998; Didham, Lawton et al., 1998). A detailed study of insectivore diets would be required to determine the significance of these changes in invertebrate composition for insectivores.

Whether it is behavior or prey-quality limitation that determines edge avoidance in birds needs to be resolved, but whatever the

cause it is unlikely that most species' populations will be able to adapt to edge conditions within the time frame of population extinction, unless preconditioned for such a response. More positively, though, given close enough source populations, natural bird assemblages will re-form in forest fragments surrounded by even relatively young (10- to 15-year-old) second growth (Stouffer and Bierregaard 1995a). This testifies to the importance of an integrated network of forest fragments and large source areas and the need for a broad landscape approach to the conservation of fragmented populations in habitat mosaics (Wiens 1994; Pickett and Cadenasso 1995).

Conservation Lessons

1. Invertebrate density is higher near forest edges.
2. The densities of different invertebrate groups respond differentially to edge effects, giving rise to significant changes in proportional representation.
3. Invertebrate density declines from continuous forest to 100 ha fragments.
4. There is no evidence for a general positive relationship between density and fragment area because edge effects and area effects interact in small forest fragments that are composed entirely of edge habitat, causing an increase in invertebrate density.
5. Prey availability may be limiting at the centers of large (100 ha) forest fragments.
6. More typically, total prey quantity is generally not limiting near the forest edge or in small forest fragments. Thus, prey-quantity limitation is probably not the main cause of population decline in terrestrial insectivorous birds.
7. Insectivorous birds may be responding to changes in prey quality rather than prey

quantity, due to marked changes in the proportional representation of different invertebrate groups. This can be determined only by a detailed study of the diet of insectivorous vertebrates.

8. If prey-quality limitation is not a major component in the decline in abundance of terrestrial insectivorous birds in forest fragments, then undetermined factors, such as behavioral edge avoidance, predation, or competition, must also be important in explaining their susceptibility to fragmentation in central Amazonia.

9. The observed changes in invertebrate prey density favor generalist insectivores with behavioral preadaptations to edges, over more specialized, deep-forest species. Forest reserves will have to be considerably larger than 100 ha in order to preserve an unmodified litter invertebrate fauna, both for its own conservation value and as a resource base for insectivorous vertebrates.

Acknowledgments

I thank J. H. Lawton, R. O. Bierregaard, Jr., J. R. Malcolm, and an anonymous reviewer for useful comments on an earlier draft of the manuscript. Thanks to Claude Gascon for logistical support in Manaus. M. D. Tocher, S. J. Hine, and field technicians at the BDFFP provided assistance in the field. P. M. Hammond gave invaluable taxonomic advice and checked the species sorting. Funding was provided by the Commonwealth Scholarship Commission and the British Council, U. K., Natural History Museum, London, Smithsonian Institution, Washington, D.C., Instituto Nacional de Pesquisas da Amazônia, Manaus, NERC Centre for Population Biology, Silwood Park, U.K., and a University of Canterbury doctoral scholarship award, New Zealand.

Fragmentation Effects on Vertebrate Communities

Frogs in the Matrix and Continuous Forest

With the failure of the cattle ranches and abandonment of the pastures that isolated our experimental forest fragments, the cleared areas were fast reverting to second-growth forest. Changes in the vegetation surrounding our fragments, or the "matrix," have forced us to shift our attention from being narrowly focused on the fragments themselves to a broader perspective through which we see the biota of the fragments being profoundly affected by what is happening in the surrounding matrix habitat—the fragments cannot be understood without perceiving them as part of a larger landscape.

Frogs have been studied almost since the start of the BDFFP and are probably as well known in our study area as anywhere else in the Amazon. Tocher, Gascon, and Zimmerman (1997) presented the surprising result that small, isolated fragments consistently had more species than they had before isolation. As part of our efforts to understand this counterintuitive result in particular, and more generally how forest fragments interact with the landscape around them, Mandy Tocher, Claude Gascon, and Joel Meyer have studied frogs in continuous forest and in three different types of matrix habitat; they present their findings in Chapter 19.

Four lines of investigation were pursued: baseline censuses were conducted to learn what species would use different types of matrix habitat; pools in all habitat types were surveyed for signs of breeding; a focal species was followed; and larval survival in matrix habitat and continuous forest was compared. Sixty-one species were recorded in all the habitats, ranging from continuous forest to pasture. Although a surprising 71 percent of the frogs found in pasture were closed-forest species, a number of taxonomic families of frogs typical of closed-forest habitats were never recorded in this most disturbed habitat. Large-scale conversion of primary forest to pasture would certainly be devastating to the frogs in the region, but a landscape mosaic of pasture, second growth, forest fragments, and primary forest seems to be hospitable to a remarkable number of frog species. Frogs have proven to have more behavioral plasticity than expected and therefore to be the most resilient of any organism studied at the BDFFP to date.

A Bird in the Bush

Birds of the forest understory have been studied intensively since the project's inception. A large-scale mist-netting program was carried out for nearly fourteen years and demonstrated conclusively that 1 and 10 ha forest fragments lose a substantial portion of their original species complement (e.g., Bierregaard and Lovejoy 1988; Bierregaard and Stouffer 1997). As the mist-netting program was winding down in the early 1990s, Philip Stouffer and his students began to study birds in the matrix habitat around the fragments. In Chapter 20, Stouffer and Sérgio Borges review this work and present their conservation recommendations for birds in habitat mosaics.

Stouffer and Borges netted birds in the middle of fragments as well as at their edges, in continuous forest, and in different types of second-growth forest in the matrix around the fragments. Bird abundance and species richness were higher in 100 ha fragments and in continuous forest than in smaller fragments or secondary growth. This general pattern held for frugivores and four guilds of insectivores. Abundance of mixed-species flock insectivores and terrestrial insectivores was especially reduced in small fragments and secondary areas. In general, fragments of 1 and 10 ha had no more forest birds than did secondary areas, nor were they any more likely to have the species that appear to be most sensitive to disturbance. The authors suggest that fragments less than 100 ha have little conservation value for birds, because most of the species in these small fragments are also found in second growth.

As with frogs, second growth, especially cecropia-dominated areas (plate 8), permits movement across deforested areas for some species. This movement may be important for maintenance of populations in a fragmented landscape. Even so, many species of flock and terrestrial insectivores are rare in second growth and have low densities even in continuous forest. For population stability, this suggests that large areas of continuous forest are essential. Making more accurate predictions of area requirements will require more data on demography and movement patterns.

Primates

In Chapter 21, the first of two chapters on primates in the BDFFP study area, Kellen Gilbert and Eleonore Setz present a brief description of the ecology of the six diurnal primate species in the BDFFP study area and review research carried out by a number of researchers, Gilbert and Setz included, in fragments and continuous forest.

Three species—the black spider monkeys (*Ateles paniscus*), bearded sakis (*Chiropotes satanus*), and brown capuchins (*Cebus apella*)—require a large home range and cannot sustain even single groups, much less viable populations in reserves of 100 ha and smaller. Not surprisingly, the primates that are able to survive and reproduce in the small fragments have small range requirements and, in the case of howling monkeys (*Alouata seniculus*), can subsist on a large portion of leaves in their diet.

Because primates in central Amazonia are, with the exceptions of the brown capuchin and golden-handed tamarin (*Saguinus midas*), nearly completely arboreal, forest fragments separated by pasture even 100 m from continuous forest are true islands for these species. Forest corridors connecting otherwise isolated tracts of forest are sure to be important for these species in a fragmented landscape.

In Chapter 22 we conclude this section on vertebrates in fragmented landscapes with a summary by Wilson Spironello of his long-term studies of the brown capuchin (*Cebus apella*). For two years we watched in admiration as Spironello left camp in the predawn darkness to pick up his troop of capuchins where he had left them the night before and returned in the dark of night after putting them to bed in one of their favorite palm groves. His arduous and relentless field work provided us with not only an intimate account of the lives of this monkey in primary forest, but also the important data on its home range size that will be necessary to design preserves large enough for this species.

Even after fourteen months of intensive tracking with additional surveys in the following months, a precise figure for the home range of Spironello's troop could not be calculated. His best estimate was a remarkable 900 ha, well in excess of figures of 50–150 ha recorded elsewhere in Amazonia. Six other troops were encountered sporadically in the home range of the focal group; their use of the area was minimal, and they were tolerated on half of their encounters with Spironello's group. The monkeys were extremely catholic in their diets, including a lot of animal prey, but showed a preference for streamside areas, where they visited several species of palms to feed on fruit and to roost at night.

As we plan for conservation in tropical rainforests, the data that are most needed and rarely available are the home range requirements of species in the area to be protected. Spironello's data tell us two very important things: the minimum critical area required to maintain an effective population size of brown capuchins in the forests near Manaus would be about 23,000 ha; and more important, data from one part of the Amazon should be applied in other regions only with the utmost caution.

Community Composition and Breeding Success of Amazonian Frogs in Continuous Forest and Matrix Habitat Aquatic Sites

MANDY D. TOCHER, CLAUDE GASCON, AND JOEL MEYER

The main consequences of habitat fragmentation include the obvious loss of native habitat (Dodd 1990; Laurance and Bierregaard 1997a) and edge effects (Andrén and Angelstam 1988; Kapos 1989; Bierregaard et al. 1992; Gascon and Lovejoy 1998; but see Santos and Tellería 1992, Gascon 1993). The amphibian fauna of the BDFFP is well known, having been studied almost since the project's inception (Zimmerman and Rodrigues 1990; Gascon 1991; Tocher 1996), and constitutes a good base for determining how some of these species respond to changes in their habitat.

Much of the BDFFP study area, 80 km north of Manaus, was composed of continuous rainforest until the mid- to late 1970s, when government incentives for cattle ranching resulted in large tracts of forest being cleared for pasture. The net result has been the loss of forested habitat and the appearance in its place of pasture and second-growth forests, which form the "matrix" habitat surrounding the BDFFP forest fragments, in a region where there had been none, at least in recent history (Chapters 1 and 4). Over the years a series of farmlands has appeared along the main highway north of Manaus, linking the matrix habitat of the BDFFP to those much older, degraded tracts of land around the city to the south. These patches of matrix habitat can impede movement or dispersal of species from continuous forest (Lovejoy, Bierregaard, et al. 1986; Bierregaard et al. 1992), or can serve as corridors for open-area habitat species to move into areas not yet occupied.

Although the landscape has been drastically altered from continuous forest to pasture, many aquatic habitats remain in the man-altered landscape. In this sense it may also be possible for certain continuous-forest species that depend on aquatic habitats for reproduction to use this "new" habitat successfully if they are not physiologically or ecologically restricted to continuous-forest habitat. The degree to which populations of certain species can establish themselves in matrix habitat will influence the dynamics of populations in forest remnants within the same matrix habitat.

Existing studies in the Manaus area have mainly dealt with organisms that rely on the terrestrial component of the forest for all aspects of their life history (summarized in Bierregaard et al. 1992, and more recently in chapters in Laurance and Bierregaard 1997a and Gascon and Lovejoy 1998). Virtually no data exist on how amphibians respond to changes in their habitat (but see Gascon

This is publication number 286 of the BDFFP Technical Series.

1993; Tocher, Gascon, and Zimmerman 1997). Many amphibian species rely heavily on aquatic habitats for reproduction (Zimmerman and Bierregaard 1986), and some species will use available sites independent of their location in continuous-forest or matrix habitat (Gascon 1993). To understand how landscape changes in the Manaus area affect the distribution of amphibian species we surveyed different matrix habitat types and continuous-forest areas to determine amphibian species composition. We tested the ability of pool breeders to permeate the matrix habitat by comparing species composition at pools in three matrix types to continuous forest. Also, we used *Phyllomedusa tarsius* as a focal species to look at population size, movement patterns, and recapture rates in both matrix and continuous forest. Finally, we measured larval survival in matrix and continuous-forest pools to give a further indication of habitat quality.

Methods

COMMUNITY COMPOSITION IN DIFFERENT HABITAT TYPES

Study site and habitat types. The BDFFP complex is situated on three ranches: Porto Alegre, Dimona, and Esteio (see Chapter 4 for a description of the primary forest and the BDFFP). Four habitat types were surveyed—continuous forest and three different types of open-area habitat (nonprimary forest) that represent a gradient of human disturbance (areas that were cleared of the original continuous forest, used for pasture at varying intensities and durations, and are covered today by different vegetation types). The three open-area habitats were maintained pasture, second-growth forest that had been both cut and burned, and second-growth forest that had been cut but not burned (see Chapters 24 and 25).

Transects (trails surveyed by the techniques outlined below) within each open-area habitat covered on average 150 ha. Transects in continuous forest covered approximately 3,000 ha and were studied in 1983–89 (Gascon 1990; Zimmerman 1991). Additional transects in continuous forest (c. 500 ha) were performed by the senior author during 1992–94.

Maintained pasture occurred primarily on the Esteio ranch with one additional transect on Porto Alegre (see fig. 4.1 and table 4.3). Maintained pasture consisted of extensive clear-cut areas, usually grazed by cattle and Asian water buffalo, that had been subjected to repeated cutting and burning over a ten- to fifteen-year period. Both Esteio and Porto Alegre pasture were dominated by introduced grass species (Gramineae). The Esteio site had numerous pools (larger than 20 × 20 m) that were occasionally disturbed by buffalo. The Porto Alegre transect had no pools in its vicinity.

Second-growth forest with a history of repeated burn was dominated by *Vismia* spp., which formed a low, open canopy. *Vismia* spp. is a plant genus typically found in cleared and abandoned areas of the Amazon (plate 10; Lovejoy, Bierregaard et al. 1986; Chapter 25). The plants have narrow trunks with thin branches. Few continuous-forest plants grow under the *Vismia* (Williamson, Mesquita et al. 1998). Established 10- to 12-year-old second growth was surveyed for frogs and tadpoles, and additional surveys of breeding pools were made in younger (5- to 6-year-old) second growth on three occasions. All surveys were performed at Dimona and at two Esteio pools. Twelve-year-old cut second growth (without burn) was dominated by *Cecropia* spp., whose seedlings often germinate on decaying trunks (plate 8). Thus mature cecropia plants typically have tangled roots partly above ground leading to a tall trunk. The foliage is concentrated at the top of the trunk in an umbrella-like fashion. A single leaf and petiole from a cecropia plant can be 1.5 m long. Cecropia formed a deep litter layer with numerous regenerating primary forest plant species in the understory. This second-growth forest is floristically more diverse than *Vismia*-dominated forest (Williamson, Mesquita et al. 1998).

Continuous forest was composed of a high (often taller than 35 m) canopy with many subcanopy layers. The native forest was floristically very diverse (Rankin–de Mérona et al. 1992; Chapters 5 and 6), and the forest floor was covered with regenerating plant species.

The results of seven years of continuous-forest amphibian surveys at the BDFFP were used in this study (Zimmerman 1991; Gascon 1991). The senior author surveyed the same control plot during 1992–94.

FIELD SURVEYS

Because most frog species rely on a particular breeding habitat, no single survey method provided a complete inventory of the species present in each habitat type. Diurnal and nocturnal transect surveys as well as surveys at different types of aquatic habitats were used to inventory species composition in each habitat type.

Diurnal and nocturnal aural and visual surveys. Forest transects by aural and visual surveying have been shown to be the most efficient method of sampling frogs in the BDFFP sites (Zimmerman 1991). Transects were established throughout the areas to be surveyed and were walked repeatedly by the same observer or observers. Frogs were identified by call; species name, time, calling habitat, and number of callers were recorded. Simultaneously, a visual survey was conducted. For each frog encountered, species name, time, habitat, snout-vent length, and sex (where possible) were noted. Zimmerman (1991) used identi-

cal methods. Only presence-or-absence data are presented here.

Aquatic surveys. Within each habitat all available breeding sites were mapped. Tadpoles were surveyed by dip-netting the entire surface area of each pool. For large pools (greater than 20 m²), sufficient sampling effort (usually more than an hour) was exerted to ensure that no species were overlooked. Tadpoles that could not be identified in the field were preserved in 10 percent formalin for lab identification. During these surveys, the presence of juveniles was also noted. Tadpole species, larval stage, and abundance were noted. Habitat variables, including pool type, size, and presence of tadpole predators, were recorded. Pools were numbered for future comparison. Only species presence-or-absence data are presented here for tadpoles.

A common, semipermanent pool type found in continuous forest and all matrix habitats was used here for comparative purposes. This breeding habitat is referred to here as an upland forest pool. Upland forest pools are typically large (more than 20 m diameter), are prone to periodic drying in exposed sites even during the rainy season (i.e., low second-growth forest and pasture land), fill maximally during the rainy season, and often dry up when the rains cease. In sites where cattle have access, the pools appear to have a greater persistence (compared to, for example, continuous-forest upland pools), perhaps because cattle churn up and compact the soil underlying the pools making them more watertight. These pools were visited diurnally for tadpole identifications and nocturnally for breeding adults.

Plastic basins (50 cm diameter × 20 cm deep) are also useful for frog surveys (Gascon 1993) and were used here. Groups of three basins were placed randomly along transects in each habitat type. Each group was separated by at least 50 m. The basins were buried so their lips were level with the surface of the ground and mimicked peccary wallows, which are important breeding sites for frogs in continuous forest (Zimmerman and Bierregaard 1986) but are usually absent from open-area sites. The basins were used as a survey tool to attract frogs that preferentially breed in wallows and to measure breeding success of cohorts initiated within the basins. For artificial pools in each habitat, only presence-or-absence data of species of tadpoles as determined by dip-netting are presented here.

Survey protocol. For the purpose of this study, sites within the BDFFP landscape were surveyed during two consecutive rainy seasons (November–May 1992–93 and 1993–94), every two weeks during the first rainy season and monthly during the second. We surveyed extensive areas of the above habitats (matrix and continuous forest) for both adult and larval frogs using aforementioned techniques. Equivalent sampling was performed in all habitats (not including the seven-year continuous-forest survey of Zimmerman 1991 and Gascon 1991). Transects surveyed by Zimmerman were repeated as logistics permitted with many transects surveyed on more than twenty occasions (Zimmerman 1991).

In this study, species that were recorded in extensive surveys of continuous forest by Zimmerman (1991) are referred to as species of closed forest (table 19.1), whereas species that were added only after surveys in forest fragments, second-growth forest, and pasture are referred to as species of open areas. Within closed forest and open-area habitat categories there was considerable overlap, with some species (e.g., *Hyla minuta*) being more common in pasture than in continuous forest. Such species were still referred to as closed forest species, as they had been initially observed in continuous forest before

the existence of cattle ranches in the area (Zimmerman and Rodrigues 1990).

Focal-Species Population Comparisons—*Phyllomedusa tarsius*

Breeding population size and movement. Breeding populations of *P. tarsius* were sampled in twenty-one continuous-forest pools and nine matrix-habitat pools. The type of matrix habitat varied; it included maintained pasture as well as low second growth dominated by *Cecropia* spp. Continuous-forest pools were generally small (less than 20 m^2) and shallow and usually dried up in the dry season (Gascon 1991). Matrix habitat sites were much larger (at least 100 m^2) and are variable in persistence throughout the year. Vegetation structure surrounding the pools differed substantially between locations. Continuous-forest sites have a dense understory composed of seedlings of many different species with wide leaves (e.g., Melastomataceae). Matrix habitat sites were surrounded in some cases by grasses (i.e., Gramineae) as for those located in pasture, or had low second growth encroaching the water margin.

Sites were visited approximately every two weeks from November 1995 to June 1996, which corresponded to the bulk of the breeding season (Gascon 1991). During each visit, all individuals present (calling males and gravid females) were noted and marked with individually numbered bands. During subsequent visits, the presence of marked individuals was noted, as well as any new individuals present at each site. The bands were affixed to the upper arm region and proved more reliable than toe-clipping. Finally, we measured the perimeter length of each breeding site and used it as our estimate of pool size. Because pools were of different sizes, we tested for differences in chorus size between pool locations using

TABLE 19.1. Frog Species Encountered in Each Habitat Type Within the BDFFP Landscape

		Habitat type			
	Survey[1]	Pasture	Cut and burn	Cut only	Continuous forest
Leptodactylidae					
Adenomera andreae	U	x	x	x	x
Adenomera hylaedactyla	D	x		x*	
Ceratophrys cornuta	U			x	x
Eleutherodactyus fenestratus	U	x	x	x	x
Eleutherodactylus zimmermanae	U		x	x	x
Eleutherodactylus sp. A	D			x*	
Leptodactylus knudseni	U	x	x	x	x
Leptodactylus mystaceus	U	x	x	x	x
Leptodactylus pentadactylus	U		x	x	x
Leptodactylus rhodomystax	U	x	x	x	x
Leptodactylus riveroi	U		x		x
Leptodactylus stenodema	U		x	x	x
Leptodactylus petersii	U	x	x	x	x
Leptodactylus leptodactyloides	D		x		x
Lithodytes lineatus	U	x	x	x	x
Unknown tadpoles sp. A	D			x*	
Hylidae					
Hyla boans	U				x
Hyla brevifrons-like	U		x	x	x
Hyla calcarata	D		x		x
Hyla geographica	U		x	x*	x
Hyla granosa	U	x	x	x	x
Hyla lanciformis	D	x	x	x	
Hyla leucophyllata	U	x	x	x	x
Hyla marmorata	U	x	x	x	x
Hyla microcephala-like	U				x
Hyla minuta	U	x	x	x	x
Hyla nana	D	x			
Hyla parviceps	U		x		x
Osteocephalus buckleyi	U				x*
Osteocephalus leprieurii	D			x*	
Osteocephalus sp. A	U		x	x	x
Osteocephalus taurinus	U	x	x	x	x
Phyllomedusa bicolor	U			x	x
Phyllomedusa tarsius	U	x	x	x	x
Phyllomedusa tomopterna	U		x	x	x
Phyllomedusa vaillanti	U	x*			x
Phyrnohyas coriacea	U				x
Phyrnohyas resinifictrix	U				x
Scinax cruentoma	U	x	x	x	x
Scinax garbei	D	x			
Scinax rubra/bosemanni	U	x	x	x	x
Scinax sp. A	D	x		x*	
Microhylidae					
Chiasmocleis shudikarensis	U		x	x	x
Chiasmocleis sp.	U		x	x	x
Ctenophryne geayi	U				x
Microphylidae sp. A	D			x	x
Synapturanus miranderiboi	U				x
Synapturanus salseri	U		x	x	x

TABLE 19.1.   (continued) Frog Species Encountered in Each Habitat Type Within the BDFFP Landscape

		Habitat type			
	Survey[1]	Pasture	Cut and burn	Cut only	Continuous forest
Bufonidae					
Atelopus pulcher	U		x		x
Bufo granulosus	D	x	x	x	x
Bufo marinus	D	x	x	x	x
Bufo typhonius	U		x	x	x
Bufo typhonius-like	D		x		
Dendrophryniscus minutus	U		x	x	x
Dendrobatidae					
Colostethus marchesianus	U		x	x	x
Colostethus stepheni	U		x	x	x
Epipedobates femoralis	U	x	x	x	x
Pipidae					
Pipa arrabali	U				x
Centrolenidae					
Centrolenella oyampiensis	U				x
Unknown					
Unknown tadpole sp. A	D		x		
Unknown sp. A	U				x

Notes: [1]U = Species found in undisturbed forest in Zimmerman's (1991) surveys. D = species found in surveys of disturbed, matrix habitat. * = Species recorded less than five times in the respective habitat.

analysis of covariance with pool size as a covariant.

Tadpole survival in matrix and continuous-forest pools. We set up two concurrent experiments to test for differences in tadpole survival as a function of pool location (i.e., in continuous forest or in pasture matrix) and tadpole population (continuous forest versus matrix). Each experiment consisted of raising tadpoles in enclosures in pools in matrix habitat and continuous forest and monitoring survival. In one experiment we used tadpoles from continuous-forest populations, whereas in the other experiment we used tadpoles from matrix populations. To control for phenological effects we ran both experiments concurrently. Enclosures were constructed of fine nylon mesh sewn together in a box shape (Gascon 1995). Each enclosure measured 30 cm × 25 cm × 25 cm and was secured in its pool with wooden stakes. Three pools (replicates) in each habi-

tat type were chosen and two enclosures were placed in each (one enclosure for each population type). In each enclosure, ten stage 25 (Gosner 1960) tadpoles were placed. Experiments lasted for thirty days, at which point all surviving tadpoles were counted. We compared survival between habitat types for each tadpole population using t-tests. The design also allowed for comparison of survival between tadpole populations reared in the same habitat (i.e., matrix and continuous-forest sites separately).

Results

OVERALL COMMUNITY COMPOSITION AND HABITAT TYPE

Sixty-one species of frogs in seven families were observed in the BDFFP landscape (see table 19.1). Two species (Unknown tadpole sp. A and *Osteocephalus buckleyi*) were col-

TABLE 19.2. Number of Species Associated with Open Areas or Closed Forest in Each of the Four Habitat Types Sampled

| Habitat type | Number of species | | |
	Open areas	Closed forest	Total
Continuous forest	5	46	51
Cut	9	32	41
Cut/burn	6	34	40
Pasture	7	17	24

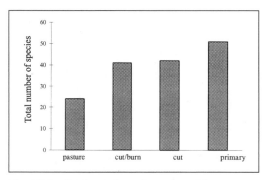

Fig. 19.1. Total number of species of frogs encountered in each habitat type.

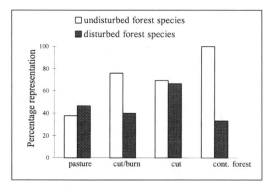

Fig. 19.2. Percentage of total undisturbed and disturbed forest species represented in the total number of species found in each habitat.

lected only as tadpoles. Fifty-one frog species were found in continuous forest (representing 85 percent of the entire assemblage), 41 in second growth that was cut (68 percent), 40 in cut-and-burned second growth (67 percent), and 24 in pasture (40 percent) (table 19.2; fig. 19.1).

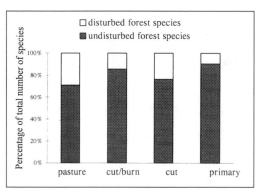

Fig. 19.3. Percentage of total number of species in each habitat that are characterized as undisturbed or disturbed forest species (see table 19.1).

Representation of closed-forest species declined with an increase in disturbance and as numbers of open-area species increased. In both second-growth habitats, closed-forest species occurred in similar numbers, whereas cut second growth had more open-area species than any other habitat (table 19.2; fig. 19.2). Despite the decrease in representation of forest species as disturbance intensity increased, 71 percent of the species present in pasture were closed-forest species (fig. 19.3). Similarly, a high percentage of species encountered in both second-growth types were closed-forest species (76.2 percent and 85.4 percent, cut and cut-and-burned second growth, respectively).

All families had similar representations in both second-growth types (table 19.3). No species of microhylidae, pipidae, or centrolenidae were ever recorded in pasture. Only one species, *Hyla nana* (Hylidae), was found exclusively in pasture (see table 19.1).

POOL-BREEDING SPECIES AT UPLAND FOREST POOLS

Surprisingly, the number of core upland forest-pool breeders that used pools in continuous forest was not significantly different

TABLE 19.3.   Representations of Frog
Families in Both Second-Growth Forest
Types of the BDFFP Landscape

Family	Secondary growth type	
	Cut only	Cut and burn
Leptodactylidae	13	12
Hylidae	16	15
Michrohylidae	4	3
Dendrobatidae	3	3
Centrolenidae	0	1
Bufonidae	4	6
Pipidae	0	0

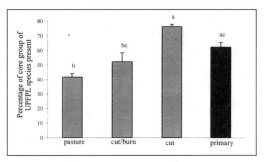

Fig. 19.4. Percentage of the core group of
upland forest-pool species encountered in
each habitat type. Only data from the aquatic
sites surveys were considered here.

from the number that used these in the two
matrix types (fig. 19.1). Fewer species were
present at upland forest pools with increasing disturbance, with pasture pools supporting on average only 42 percent of the
pool breeders (fig. 19.4). Distance of the upland forest pools from continuous forest was
not related to number of species present (fig.
19.5).

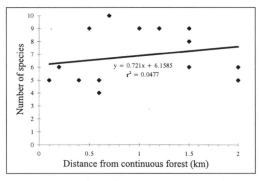

Fig. 19.5. Number of species encountered in
upland forest pools in all types of matrix
habitat as a function of distance of pool to
continuous forest.

POPULATION COMPARISONS—*PHYLLOMEDUSA
TARSIUS*

The total number of *P. tarsius* individuals
encountered in matrix pools was 150 (of
which 141 were males), whereas 48 individuals (of which 45 were males) were
recorded at pools in continuous forest. Overall mean population size per pool for males
and females was greater in matrix habitat
sites (mean 17.0 individuals) compared to
continuous forest (mean 3.2 individuals).
However, when pool size was taken into account, no differences existed in the number
of individuals observed per site in each
habitat (analysis of covariance [ANCOVA],
$F = 0.360$, $P = 0.557$, df = 1,15).

Twenty-two percent of all captures were
previously marked individuals. Only 16
percent of all individuals marked in matrix
habitat were recaptured at least once, compared to 18.8 percent for individuals

marked at continuous-forest sites (table
19.4). Considering only individuals that
were recaptured at least once, the mean
number of recaptures per individual was
very similar for the two habitats (1.38 for
matrix habitat versus 1.33 for continuous
forest). Finally, only 24 percent and 17 percent of all recaptures, for matrix habitat and
continuous forest sites, respectively, were at
different pools than those where the individual had been marked. No across-habitat
movement was observed.

TADPOLE SURVIVAL AND HABITAT TYPE

Survival was high in both habitat types for
both tadpole populations (table 19.5). Al-

TABLE 19.4.   Mark-Recapture and Movement Characteristics of Populations of *Phyllomedusa tarsius*

	Matrix habitat	Conintuous forest
Total number of individuals marked	150	48
Total number of recaptures	33	12
Total number of recaptured individuals	24	9
Proportion of individuals marked that were recaptured	0.16	0.188
Total number of movements	8	2
Mean number of recaptures (per recaptured individuals only)	1.38	1.33
Proportion of total recaptures at different pools (number of movements/total number of recaptures)	0.24	0.17
Mean distance of movements (m)	216	364

TABLE 19.5.   Survival of *Phyllomedusa tarsius* Tadpoles in Enclosures in Open-Area Habitat and Continuous Forest Habitats.

	Mean survival in	
Source of tadpoles	Open-area habitat	Continuous forest
Open-area habitat	9.05 (0.83)	7.00 (3.40)
Continuous forest	7.35 (1.35)	8.50 (0.71)

Notes: Mean survival values are numbers of tadpoles out of the original ten that were alive at the end of the experiment. Numbers in parentheses are one standard deviation of the mean.

though there was a tendency for tadpoles of matrix habitat and continuous forest to have higher survival in their "native" habitats, an analysis of variance (ANOVA) showed the difference was not significant ($F = 0.007$, $P = 0.937$). Likewise, no difference in mean survival was found between tadpole populations reared in the same habitats ($F = 0.166$, $P = 0.694$).

Discussion

Frog species composition varied considerably between open-area habitats and continuous forest sites. This has also been shown for other faunal groups, such as ants (Chapter 16). In this study, which covered the entire BDFFP landscape, we recorded seventeen frog species that were not found in previous surveys of continuous forest (Zimmerman 1991; Gascon 1991). We noted a decrease in total number of species as the level of disturbance increased (the human-induced processes of deforestation that result in alteration of pristine forest), in particular a drop in representation of closed-forest species, especially those associated with streams or specialized breeding habitat. Despite this, closed-forest species still made up half the total number of species present in the most disturbed habitat. Furthermore, surveys of calling males and determination of breeding patterns of wallow and upland forest-pool breeders showed that matrix is exploited as a quality breeding habitat, and in some cases frogs appear to be more successful in matrix habitat.

No one habitat was found to have the full complement of frog species present in the BDFFP landscape. However, continuous forest had more than twice the number of species found in pasture, and it is clear that disturbance results in a decrease in the overall number of frog species. This agrees with a general pattern of faunal response to disturbance in the BDFFP landscape for ants (Chapter 16), small mammals (Malcolm 1991), and birds (Borges 1995; Chapter 20).

Many of the open-area frog species (e.g., *Scinax rubra*) are now known to be present in continuous forest (see table 19.1). Some of these species (e.g., *Adenomera hylaedactyla*) may have migrated into the BDFFP landscape from areas closer to Manaus, where disturbance has a longer history.

Although strong evidence is presented for a decrease in species richness with increasing disturbance, very few species were found exclusively in continuous forest. In fact, we found 82 percent of all species recorded in continuous forest in one or more of the open-area habitats. Other taxonomic groups also show similar high proportions of forest species using matrix habitats (Gascon et al. 1999). For frogs, notable exceptions have specific breeding requirements or are rare in the BDFFP landscape. For example, although *Phrynohyas resinifictrix* is a common species in forest reserves in the area, it has specialized breeding requirements. *P. resinifictrix* breeds only in large water-filled tree holes (B. Zimmerman, pers. comm.), which are rare in matrix forest and pasture. Consequently, *P. resinifictrix* was never recorded in either of the two matrix types or pasture. Similarly, *Osteocephalus* sp. A is rare in matrix forest and absent from pasture where its preferred breeding habitat, water-filled bromeliads and palm axles (Jungfer, pers. comm.), is also rare. Thus species with very specialized breeding habitats are unlikely to be present in the matrix as true residents.

The high representation of closed-forest species in the most disturbed habitat-pasture needs further explanation. Many frogs are known to return to their natal breeding pools (see review in Sinsch 1990) regardless of how the surrounding area has been disturbed. But due to the likely longevity of many of the frog species in the BDFFP, relative to the time pools have been surrounded in pasture, a more likely explanation is that some frog species show remarkable plastic-

ity in reproductive behavior, allowing them to adapt to different ecological conditions for breeding.

Cases in point are the two species of *Phyllomedusa* (*P. tarsius* and *P. tomopterna*) that were commonly found in open-area habitats. As with most species of this genus, females will lay eggs in broad leaves of plants at the pool margin that are either folded over (*P. tomopterna*) or joined together (*P. tarsius*) to surround the eggs. In many open-area sites that were surveyed, no such plants existed at the pool margins. Eggs were laid in clumps of grass blades at the water margin, leaving much of the egg mass exposed to desiccation or predation. Regardless, juveniles of both *P. tarsius* and *P. tomopterna* were discovered in the pasture. Further, neither measure of reproductive success for *P. tarsius* (chorus size or larval survival) differed between open-area habitats and continuous forest, indicating that not only is this species capable of using open areas but that it does so successfully.

Certainly, species-specific differences in behavioral plasticity can explain some of the observed distributions, but not all forest species that were found in pasture were necessarily breeding successfully. For some closed-forest species, adults may be attracted to pasture pools, which are not suitable for their breeding. As a result, pasture pools may act as a breeding sink, robbing the local population of breeding individuals.

Both second-growth forests were equally "permeable" to wallow and upland forest-pool breeding species. However, repeated burning, such as that inflicted on pasture habitat, has a greater impact on the frog community than does the lessened burning regime of the cut-and-burned matrix surveyed here. In general, pasture sites were found to be depauperate in closed forest species in comparison to cut-and-burned areas, which offer more physical cover (vegetation) to resident frogs than the more open pasture habitat.

Physical cover offered by cut-and-burned matrix may also explain the difference in species composition found between pasture and second-growth forests. These data are similar to Tocher's (1998) findings for the entire frog assemblage of the BDFFP (i.e., not just frogs that utilize upland forest pools and wallows). In contrast Phelps and Lancia (1995) showed less impact of forest loss on an amphibian assemblage in South Carolina. Although many species decreased in abundance in cleared areas, relatively few frog species were lost completely. Our results are in sharp contrast, however, with those studies showing more serious impacts of landscape changes on amphibians (Corn and Bury 1989; Petranka, Eldridge, and Haley 1993; Grant et al. 1994; Dupuis, Smith, and Bunnell 1995). Thus predictions as to the effects of forest loss on the resident frog community appear site specific.

The core group of upland forest-pool breeders identified by Zimmerman (1991) all had one thing in common—they most often bred at upland forest pools. Interestingly, although fewer of these species were present at pasture pools, well over half of them were. It is difficult to know whether they were resident in the pasture, or whether they returned to breeding pools from surrounding forest (Sinsch 1990). No relationship was found between the number of species present at pools and the distance to the nearest continuous forest, so these data provided no clues to resolving this issue. However, although no individuals were found to move between habitat types, some individuals were recaptured at different breeding sites within the same habitat types, suggesting that movement is possible.

There was a high presence of upland forest-pool species in second-growth forest, and on average more species were present at cut second-growth forest pools than were present in continuous-forest pools (although breeding success of all of these species in open areas was not evaluated).

These findings, in particular the demonstrated ability of upland forest-pool species to move through open cleared areas, including pasture, have obvious management implications. Some frog species are able to alleviate the negative effects of fragmentation, such as reduced and isolated populations, because of their ability to permeate, or disperse into, and exploit the cleared areas. In fact, many species of different taxonomic groups are capable of using the matrix habitat, and this ability is strongly correlated to their persistence probability in forest fragments (Laurance 1991b; Gascon et al. 1999). Furthermore, many species of frogs can use and move along linear forest remnants along streams in open areas, further increasing their chances of surviving in a modified landscape (Lima and Gascon 1999).

It is not surprising that cut (but not burned) second-growth forest has high species richness and a high permeability to upland forest-pool breeding species, considering its relatively moderate disturbance history and structural similarity to continuous forest (Williamson, Mesquita, et al. 1998; Chapter 24). However, with respect to community composition, and also when all breeding groups were investigated independently, no real differences in composition, abundance, or breeding success were found between the two types of second growth (in stark contrast with birds [Chapter 20] and ants [Chapter 16]). It therefore seems that regardless of burning history, the eventual frog community in established second-growth forests will be similar.

In maintained pasture the scenario is very different; fewer frogs are present due to the extreme lack of cover and perhaps the occurrence of repeated burnings. The implications of these joint findings for conservation are huge. When continuous forest is cut and burned, as was the case for the BDFFP landscape, all frogs, with the exception of fossorial species, must surely be killed. Individuals encountered in each habitat type (excluding continuous forest) in this study have certainly moved into these areas from surrounding continuous forest. This suggests that many frog species have the capacity to respond to loss of primary habitat, at least in the short term; using a modified, disturbed landscape. It is important, however, to emphasize that more severe disturbances, such as repeated cutting and burning over decades, may result in an overall impoverished frog community. Results presented here portray but one possible scenario of tropical deforestation with moderate to low levels of disturbance and should not be extrapolated to areas where the geographical extent and intensity of disturbance are greater.

Conservation Lessons

1. Although more detailed surveys are needed in open-area habitat of all types, it is clear from these data that continuous-forest (including primary forest and second growth) loss will result in an impoverished amphibian fauna in cleared areas. Many of the continuous-forest species were not observed outside continuous forest.

2. Species found exclusively in open-area habitats are capable of reaching these disturbed areas from source populations in relatively short periods of time. In our study areas the forest had been cleared for less than thirteen years. Species now associated with these open areas were never encountered in continuous forest

(Zimmerman and Rodrigues 1990; Gascon 1991) and so are hypothesized to be immigrating from other disturbed areas, probably close to Manaus. Disturbed areas along the main highway probably serve as corridors for these species to invade new matrix habitat.

3. Many species previously characterized as primary-forest species (Zimmerman and Rodrigues 1990; Gascon 1991) are capable of using other types of habitats for reproduction (see also Gascon 1993). This capacity does not seem to be restricted to certain taxonomic or ecological groups because most genera in table 19.1 have representatives in all types of habitats. However, forest destruction will have overall negative effects on the full complement of primary forest species as many are not capable of using open-area habitats and are thus at risk in fragments (Gascon et al. 1999).

4. These results have clear implications for population dynamics and conservation biology. The differential response of amphibian species to landscape changes precludes any generalizations as to the effects of habitat loss or fragmentation. Rather, species-specific requirements seem to dictate how individual species respond to habitat changes (also see Gascon 1993; Tocher, Gascon, and Zimmerman 1997). The presence of calling individuals of *P. tarsius*, among others, and the observed movement of individuals between pools in matrix habitat is indicative of some potential for dispersal through open-area habitat to other patches of primary forest. For such species there may be a lesser need for corridors linking patches of forest as the populations in open-area habitats can act as stepping stones or sources of colonists for dispersal.

Acknowledgments

This study was supported by a grant from the Smithsonian Institution, Washington, D.C. Logistic support came from Instituto Nacional de Pesquisas da Amazônia; the Zoology Department, University of Canterbury, New Zealand; and the Department of Conservation, New Zealand. Ocírio Pereira, Antonio Cardosa, and R. Didham were a great help in the field. Mike Winterbourn provided valuable comments on this manuscript.

CHAPTER 20

Conservation Recommendations for Understory Birds in Amazonian Forest Fragments and Second-Growth Areas

PHILIP C. STOUFFER AND SÉRGIO H. BORGES

Recent anthropogenic deforestation in central Amazonia has left a mosaic of native forest fragments, pasture, and second growth surrounded by extensive primary forest. At the BDFFP site this has provided an opportunity to study how birds use a modified landscape. This is an important examination, because it is the only way to predict what will happen as Amazonia is further perturbed. It is also important to address this question now, while there is relatively little disturbance, because the full complement of forest birds is present in the ample areas of undisturbed forest that surround disturbed areas.

Most of the previous work at the BDFFP site has documented the changes that occur in fragments during and after fragmentation (e.g., Bierregaard and Lovejoy 1988, 1989; Bierregaard 1990; Stouffer and Bierregaard 1995a, 1995b). That work emphasized studies of the changes over time in bird abundance after a series of forest plots were isolated by deforestation. Our approach here is less focused on fragments per se. Instead, we examined bird abundance in fragments, on the edges of fragments, in second-growth ("secondary") areas, and in continuous forest. This was meant to give a snapshot of where understory birds are in this landscape.

With the exception of one fragment, all fragments had been isolated for at least seven years at the time this study began (see table 4.3). We suspect that the birds we recorded are those that breed in fragments and second-growth areas or move through this landscape. This is a bird assemblage distinct from that found in the first few years after isolation, when fragments were colonized by birds fleeing the local deforestation (Bierregaard and Lovejoy 1989). Birds that were "trapped" in fragments by isolation probably perished or emigrated within the first few years after isolation (e.g., Bierregaard and Lovejoy 1989; Stouffer and Bierregaard 1995a). The secondary areas had been originally cleared at least seven years before this study began. Almost no forest birds move across very young second growth, but birds begin to move through these areas as vegetation becomes developed, and by seven years of growth many species use this matrix (Stouffer and Bierregaard 1995a; Borges and Stouffer 1999).

Using this landscape, we asked two very general questions. First, how do abundance and species richness of understory birds

This is publication number 287 of the BDFFP Technical Series.

vary among fragments, continuous forest, and second growth? Second, how do social and foraging guilds differ in their abundance in this landscape? Beyond these general questions, we also describe how bird abundance is affected by edges and variation in second-growth vegetation. Based on these results, we make management recommendations and suggest research priorities that would be most useful for determining how to maintain stable populations of forest understory birds in human-modified landscapes in the Amazon.

Methods

We sampled all of the eleven BDFFP experimental forest fragments except 2107 (see fig. 4.1, table 4.3); the histories of these fragments are described in Chapter 4 and in Lovejoy, Bierregaard, et al. 1986. Because of differences in land use, the vegetation surrounding the fragments differed substantially. The Colosso fragments (1104 and 1202), isolated in 1980, were surrounded mostly by active cattle pasture and low (less than 3 m) second growth. Tall (more than 10 m) cecropia surrounded the fragments at Cidade Powell (1112 and 1207), isolated in 1983, and three-fourths of the border of the 100 ha fragment at Porto Alegre (3304) (plate 8). One side of 3304 was bordered by an inactive cattle pasture. The second growth around the other fragments at Porto Alegre (3114 and 3209) and the fragments at Dimona (2108, 2206, and 2303) had been cut just before our sampling began. Our control plot was in the BDFFP's control area (1501) at Km 41 of the ZF-3. The second-growth sampling was done at the following reserve clusters: Cidade Powell, Florestal, Colosso, Porto Alegre, and two sites at Dimona. This sample includes areas dominated by both *Vismia* spp. and cecropia, but in all cases

the vegetation was at least 2 m tall (plate 9; see Borges 1995; Chapters 24 and 25).

The data included here are based entirely on mist-net sampling. All netting was done with the same type of net (NEBBA-type ATX, 36 mm mesh, 12 × 2 m), always set with the bottom of the net at ground level. Lines of eight nets were used to sample the interior of 1 ha fragments. Lines of sixteen nets were used to sample the interior of 10 ha fragments (one line per fragment) and 100 ha fragments (two lines per fragment). We sampled borders of all fragments with sixteen nets distributed as a line of four nets on each of four sides of the fragment. In continuous forest, we used three lines of sixteen nets and four lines of four nets in a 100 ha sampling area. In secondary-forest sites we used two lines of thirty nets oriented parallel to a forest border, with one line 50 m from the edge and the other line 250 m from the edge (see Borges 1995 and Borges and Stouffer 1999 for more details). At all sites we opened nets from 0600 to 1400 for one day of sampling at a time. We sampled fragments and continuous forest sites eight days each, at roughly six-week intervals, between October 25, 1991, and September 29, 1992. The secondary forest sites were sampled six times each, at one-month intervals, from March to October 1993. This effort amounted to 1,536 net-hours ("nh") in each 1 ha fragment, 2,048 in each 10 ha fragment, 3,072 in each 100 ha fragment, 4,096 in continuous forest, and 2,880 at each secondary-growth site. Combined, the effort was 41,856 net-hours.

We use capture rate (captures/1000 net-hours) as a standard index of abundance. (Hereafter, capture rates will be presented without units.) Same-day recaptures are excluded from all analyses. In most cases we compared sites using an analysis of variance (ANOVA), after log-transforming capture rate to reduce the correlation between variance and mean within samples (Sokal and Rohlf 1981). For significant ANOVA results we compared site means based on the least significant difference (LSD) with a P of 0.05 or less (Sokal and Rohlf 1981). We used each net line as a separate sample for continuous forest and 100 ha fragments. It is certainly not correct to assume that these are independent samples, but even so, the variation among continuous forest samples, despite being separated by only 500–800 m, was comparable to that in the spatially distinct smaller fragments and secondary areas (see Results). Variation at a scale of 500 m is expected, because many of the species we netted have patchy distributions and have territories of less than 20 ha (Stouffer, unpublished data; see also Terborgh, Robinson, et al. 1990; Cohn-Haft, Whittaker, and Stouffer 1997).

For calculations of species richness we standardized the sampling effort to be analyzed by randomly selecting nets to include in the sample. From fragments, we used a randomly selected sample of sixteen nets (eight edge nets and eight interior nets) from each 1 and 10 ha fragment. Two samples of sixteen nets (eight edge nets and eight interior nets) were taken from each 100 ha fragment. Four sixteen-net samples (the entire sample) were taken from continuous forest. From secondary growth sites, we used a sample of twenty-one nets at each site. This amounted to four samples of 1,024 net-hours from each fragment size and continuous forest, and six samples of 1,024 net-hours from second growth.

Movement of birds between fragments was rare, so to identify the species that moved we used a larger sample of mist-net data from the fragments, beginning March 2, 1988. At this point the fragments were all isolated except 2303. This larger sample is used only for our analysis of bird movements.

Results

VARIATION IN CAPTURE RATES

At all sites combined we recorded 3,643 captures of 147 species. The most common species in the combined sample were captured in continuous forest, second growth, and some or all fragments. The most common species were *Pithys albifrons* (322 captures), *Pipra pipra* (210 captures), *Phaethornis superciliosus* (186 captures), and *Mionectes macconnelli* (175 captures; table 20.1, fig. 20.1). Abundance quickly tailed off after these common species; twenty-eight species were netted only once. Some of the species netted only once are probably quite rare at our site (e.g., *Phaeothylpis rivularis, Nonnula rubecula, Philydor pyrrhodes,* and *Frederickena viridis*), although others were fairly common but usually occurred above the level of the nets (e.g., *Trogon viridis, Tyranneutes virescens,* and *Piprites chloris*). Some of the other apparently rare species are common in open pastures with small patches of scrubby growth but seldom use older second growth or forest edges (e.g., *Crotophaga ani, Columbina passerina, C. talpacoti,* and *Leptotila verreauxi*). In general, the analyses that follow focus on the species that are better sampled and more representative of the sites where we worked. Cohn-Haft, Whittaker, and Stouffer (1997) provided a more comprehensive but less quantitative discussion of the avifauna at the sites.

The most general comparison to make among sites is the overall capture rate. This ranged from about 60 captures/1000 net-hours in second growth and 1 ha fragments to over 160 captures/1000 net-hours in continuous forest (fig. 20.2). Mean capture rates along edges were similar to the capture rates in the interior of the same fragments. Based on a Wilcoxon matched-pairs test, interior capture rate did not differ from edge capture

TABLE 20.1.   Number of Captures of the Ten Most Commonly Netted Species in
Continuous Forest, Fragments, and Second Growth, Including Edge and Interior Nets

Species	Continuous forest	100 ha	10 ha	1 ha	Second growth
Pithys albinfrons	91	46 (3)	+	+	165 (1)
Glyphorynchus spirurus	53	27 (8)	38 (2)	16 (6)	+
Hylophylax poecilinota	43	41 (5)	29 (5)	+	+
Gymnopithys rufigula	35	32 (6)	+	+	60 (4)
Pipra pipra	34	60 (1)	25 (7)	12 (8)	79 (3)
Pipra serena	29	+	+	+	+
Thamnomanes caesius	21	+	+	+	+
Thamnomanes ardesiacus	21	+	+	−	+
Myrmotherula gutturalis	20	+	+	−	+
Phaethornis superciliosus	18	45 (4)	59 (1)	34 (1)	30 (10)
Mionectes macconnelli	+	56 (2)	37 (3)	26 (3)	42 (6)
Percnostola rufifrons	+	23 (9)	22 (8)	31 (2)	88 (2)
Myrmotherula longipennis	+	23 (9)	+	−	+
Hypocnemis cantator	+	+	18 (9)	14 (7)	40 (7)
Phaethornis bourcieri	+	+	16 (10)	11 (10)	+
Thamnophilus murinus	+	+	+	12 (8)	34 (9)
Automolus ochrolaemus	+	+	+	+	59 (5)
Myrmotherula axillaris	+	+	+	+	35 (8)
Ramphocelus carbo	−	29 (7)	34 (4)	18 (5)	+
Ammodramus aurifrons	−	+	28 (6)	19 (4)	+

Notes: The ranking is by abundance in continuous forest. For other sites, the number in parentheses is the rank. Species captured but not among the ten most common species in that site type are indicated by a "+." Species not captured in that site type are indicated by "−."

Fig. 20.1. The White-plumed Antbird (*Pithys albifrons*), an obligate army-ant follower, was the most abundant bird in the mist-net samples. Photo by Richard O. Bierregaard, Jr.

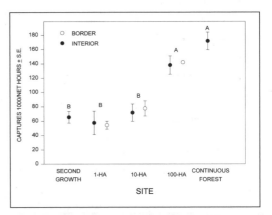

Fig. 20.2. Capture rate on borders and in interior of fragments, as well as in second growth and continuous forest. Replicates are sites or net lines. Sites with the same letter are not significantly different by ANOVA.

rate. We combined interior and edge samples to examine differences among the sites. Capture rate was significantly higher in continuous forest and 100 ha fragments than in smaller fragments and secondary growth (ANOVA, P > 0.001).

The number of species captured in a standardized sample effort also varied significantly among sites (fig. 20.3). Once again, species richness was significantly higher in continuous forest and 100 ha fragments than in second growth, 1 ha fragments, and 10 ha fragments. Without standardizing the sampling effort, more species were captured in second growth (98) than in continuous forest (72), but the sampling effort was 17,280 net-hours in six second growth sites compared to 4,096 net-hours in a single area of 100 ha in continuous forest. The variation among replicates was higher in 10 ha fragments than in other sizes (fig. 20.3). Fragment 1202, which was surrounded by cattle pasture and low second growth, had only 27 species netted. In contrast, 52 species were netted in 1207, which was surrounded by tall, cecropia-dominated second growth.

SPATIAL VARIATION AMONG
FORAGING GUILDS

We examined the sample in more detail by considering four foraging and social guilds of insectivorous birds that dominate the understory avifauna in continuous forest (see table 20.1; also Stouffer and Bierregaard 1995a). For each of these four guilds there were significant differences among sites (fig. 20.4). In general, abundance was slightly reduced in 100 ha fragments compared to continuous forest, but small fragments and secondary growth had less than 25 percent of the abundance recorded in continuous forest.

Ant followers were more abundant in continuous forest, 100 ha fragments, and

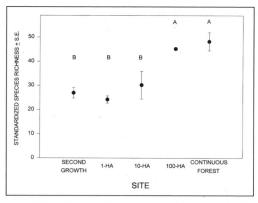

Fig. 20.3. Standardized species richness at all sites. Sites with the same letter are not significantly different by ANOVA.

second growth than in 1 and 10 ha fragments (fig. 20.4, table 20.2; ANOVA, P > 0.001). These birds were very rarely netted in 1 and 10 ha fragments, although they were fairly common in second growth (see table 20.1). For obligate mixed-species flock species (see Powell 1985) and terrestrial insectivores, continuous forest and 100 ha fragments had much higher capture rates than the other sites (ANOVA, both P > 0.001). These birds were nearly absent except in 100 ha fragments and continuous forest. Flock dropouts—three species that are present in most mixed-species flocks but also regularly occur outside of flocks, especially outside of continuous forest—had a slightly different pattern. They varied significantly among sites (ANOVA, P > 0.001) but declined less strongly in 1 and 10 ha fragments and second growth compared to obligate flock species. The difference between flock dropouts and flock obligates in 1 and 10 ha fragments supports our distinction between these groups; flock dropouts appear to have smaller area requirements than do flock obligates (see also Stouffer and Bierregaard 1995a). *Thamnomanes ardesiacus, Myrmotherula gutturalis,* and *M. longipennis,* three flock obligates common in continuous forest and 100 ha fragments,

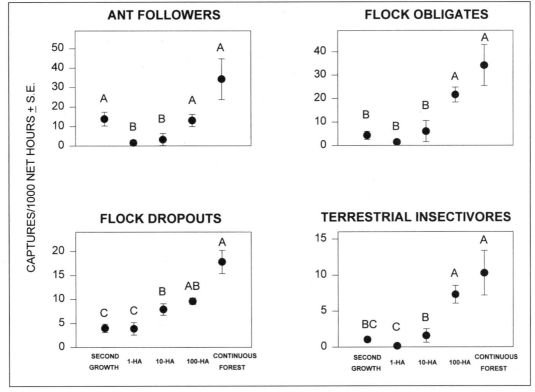

Fig. 20.4. Capture rates for four guilds of obligate insectivores. See table 20.2 for the species included in each guild. Sites with the same letter are not significantly different by ANOVA.

were never netted in 1 ha fragments (see table 20.1).

In addition to the obligate insectivores, we also performed analyses for guilds that we suspected would be less affected by fragmentation (fig. 20.5; see also Stouffer and Bierregaard 1995b, 1996; Bierregaard and Stouffer 1997). Hummingbirds were most abundant in 10 ha and 100 ha fragments (table 20.3) (ANOVA, P = 0.007). They were significantly less common in second growth, even though second growth contained all the forest species and an additional species, *Phaethornis ruber*, that was found only in secondary areas. The group we classified as understory frugivores (table 20.5) are actually among the most omnivorous birds at our site. Some of these species were common everywhere (e.g., *Pipra pipra* and *Mionectes macconnelli;* see table 20.1),

although abundance was twice as high in 100 ha fragments and continuous forest as in smaller fragments and second growth (ANOVA, P > 0.001; fig. 20.5, table 20.4).

Finally, we analyzed a broad group we called simply nonforest species (table 20.5). This group includes frugivores, such as *Manacus manacus;* granivores, such as ground doves and sparrows; omnivores, such as most of the flycatchers; and strict insectivores, such as wrens and the antbird *Myrmeciza atrothorax*. These are species that are almost never found in continuous terra firme forest around Manaus, although they are sometimes found along edges or in secondary forests (see also Cohn-Haft, Whittaker, and Stouffer 1997). None of these species was captured in continuous forest, and few were captured in second growth; they were most common in fragments. Be-

TABLE 20.2.   Social and Foraging Guilds of Forest Insectivores

Family	Species	Common name
Flock dropout insectivores		
Dendrocolaptidae	*Glyphorynchus spirurus*	Wedge-billed Woodcreeper
	Xiphorhynchus pardalotus	Chestnut-rumped Woodcreeper
Formicariidae	*Myrmotherula axillaris*	White-flanked Antwren
Obligate flock insectivores		
Dendrocolaptidae	*Denconychura longicauda*	Long-tailed Woodcreeper
	Denconychura stictolaema	Spot-throated Woodcreeper
	Campylorhamphus procurvoides	Curve-billed Scythebill
Furnariidae	*Philydor erythrocercus*	Rufous-rumped Foliage-gleaner
	Automolus infuscatus	Olive-backed Foliage-gleaner
	Xenops minutus	Plain Xenops
Formicariidae	*Thamnomanes ardesiacus*	Dusky-throated Antshrike
	Thamnomanes caesius	Cinereous Antshrike
	Myrmotherula gutturalis	Brown-bellied Antwren
	Myrmotherula longipennis	Long-winged Antwren
	Myrmotherula menetriesii	Gray Antwren
Tyrannidae	*Myiobius barbatus*	Sulphur-rumped Flycatcher
	Rhynchocyclus olivaceus	Olivaceous Flatbill
Vireonidae	*Hylophilus ochraceiceps*	Tawny-crowned Greenlet
Ant-following insectivores		
Dendrocolaptidae	*Dendrocincla merula*	White-chinned Woodcreeper
Formicariidae	*Pithys albifrons*	White-plumed Antbird
	Gymnopithys rufigula	Rufous-throated Antbird
Terrestrial insectivores		
Furnariidae	*Sclerurus mexicanus*	Tawny-throated Leafscraper
	Sclerurus rufigularis	Short-billed Leafscraper
	Sclerurus caudacutus	Black-tailed Leafscraper
Formicariidae	*Myrmeciza ferruginea*	Ferruginous-backed Antbird
	Formicarius colma	Rufous-capped Antthrush
	Formicarius analis	Black-faced Antthrush
	Grallaria varia	Variegated Antpitta
	Myrmothera campanisona	Thrush-like Antpitta
	Hylopezus macularius	Spotted Antpitta
	Conopophaga aurita	Chestnut-belted Gnateater

cause of the enormous differences in variance among the sites, we were unable to analyze these data with ANOVA. Even without further analysis it is clear that these species were both more common and more variable from site to site in fragments than in second growth. This is probably because they were generally captured on fragment edges. Of 187 captures of the nonforest species in fragments, 171 (92 percent) were on edges. Nonforest species were most common on edges adjacent to young second growth (some-

times less than 1 m tall), while samples from edges by older second growth were more like those from second-growth net lanes. The most common nonforest species, *Ramphocelus carbo* and *Ammodramus aurifrons*, became especially common along young edges (see table 20.1).

MOVEMENT AMONG FRAGMENTS

We analyzed recorded movements of banded birds between fragments based on our sam-

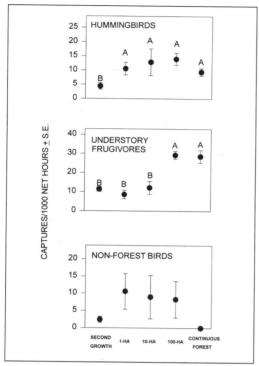

Fig. 20.5. Capture rates for hummingbirds, understory frugivores, and nonforest birds. See tables 20.2, 20.3, and 20.4 for lists of the species included. For hummingbirds and understory frugivores, sites with the same letter are not significantly different. No statistics were performed on the nonforest birds.

ple and other netting done beginning in 1988. Species that were not banded, including all hummingbirds, are not included. Of 1,879 captures of banded birds, 44 (2.3 percent) were of individuals that had previously been netted in another fragment. Of these birds, 25 had moved between fragments connected by cecropia (including 15 between the 1 and 10 ha fragments at Cidade Powell; see fig. 4.1, table 4.3), while 19 had moved between fragments connected by Vismia. Based on the total number of captures of previously banded birds in fragments surrounded by *Vismia* (1,144) and cecropia (735), bird movements were more likely between fragments connected by cecropia (Log-likelihood test, G = 13.81, P > 0.001). Although 19 species changed fragments, 5 species (*Pithys albifrons* [8], *Glyphorynchus spirurus* [6], *Gymnopithys rufigula* [4], *Percnostola rufifrons* [4], and *Cyanocompsa cyanoides* [3]) accounted for 25 of 44 movements. No terrestrial insectivores and only 3 of the 14 species of flock obligates moved between fragments.

TABLE 20.3. Hummingbirds (Trochilidae)

Species	Common Name
Phaethornis superciliosus	Long-tailed Hermit
Phaethornis bourcieri	Straight-billed Hermit
Phaethornis ruber	Reddish Hermit
Campylopterus largipennis	Gray-breasted Sabrewing
Florisuga mellivora	White-necked Jacobin
Thalurania furcata	Fork-tailed Woodnymph
Hylocharis sapphirina	Rufous-throated Emerald
Amazilia versicolor	Versicolored Emerald
Amazilia fimbriata	Glittering-throated Emerald
Topaza pella	Crimson Topaz
Heliothryx aurita	Black-eared Fairy

TABLE 20.4.   Forest Species with Mixed (Frugivorous and Omnivorous) Diets

Family	Species	Common name
Pipridae	*Corapipo guttaralis*	White-throated Manakin
	Pipra erythrocephala	Golden-headed Manakin
	Pipra pipra	White-crowned Manakin
	Pipra serena	White-fronted Manakin
	Schiffornis turdinus	Thrush-like Manakin
Tyrannidae	*Mionectes macconnelli*	Macconnell's Flycatcher
Muscicapidae	*Turdus albicollis*	White-necked Thrush
Thraupinae	*Tachyphonus surinamus*	Fulvous-crested Tanager
	Tachyphonus cristatus	Flame-crested Tanager
Cardinalinae	*Cyanocompsa cyanoides*	Blue-black Grosbeak

TABLE 20.5.   Nonforest Species

Family	Species	Common name	Foraging mode
Cuculidae	*Crotophaga ani*	Smooth-billed Ani	I
Formicariidae	*Myrmeciza atrothorax*	Black-throated Antbird	I
	Thamnophilus punctatus	Slaty Antshrike	I
Pipridae	*Manacus manacus*	White-bearded Manakin	F/O
Tyrannidae	*Myiarchus ferox*	Short-crested Flycatcher	I/O
	Tyrannus melancholicus	Tropical Kingbird	I/O
Troglodytidae	*Thryothorus coraya*	Coraya Wren	I
	Troglodytes aedon	House Wren	I
Vireonidae	*Cychlaris gujanensis*	Rufous-browed Peppershrike	I
Thraupinae	*Ramphocelus carbo*	Silver-beaked Tanager	O
Emberizinae	*Ammodramus aurifrons*	Yellow-browed Sparrow	I/G
	Volatinia jacarina	Blue-black Grassquit	G

Notes: I = insectivore; F = frugivore; O = omnivore; G = granivore.

Discussion

PATTERNS OF BIRD ABUNDANCE
AND RICHNESS

Continuous forest and 100 ha fragments have much greater richness and abundance of understory birds than do 1 ha and 10 ha fragments and second growth (figs. 19.1 and 19.2). These results are not surprising; they would be predicted from both the patterns of species loss and general decline in abundance described earlier from the fragments (e.g., Stouffer and Bierregaard 1995a; Stratford and Stouffer 1999) and from theoretical predictions about species richness as a function of island size (Darlington 1957; Chapter 2). In general, about half the number of species should be lost as area is reduced by

90 percent. This prediction is only partially supported, however. Although the advantage of additional area for species richness is apparent between 10 ha and 100 ha fragments, there is no difference between 1 ha and 10 ha fragments. The smallest fragments actually have more species than would be expected. The species in these fragments, in turn, do not differ from the species in the vast areas of second growth that surround the fragments. This suggests that the same group of bird species (including some adaptable forest species and some edge and second-growth species) uses the small fragments and secondary areas about equally, a conclusion supported by the data from individual guilds (see below). Island biogeographic theory breaks down for those

species that do not strongly distinguish between "island" and "sea."

Despite the unexpectedly high abundance and richness in 1 ha fragments compared with that in 10 ha fragments, it should not be overlooked that there is simply much greater bird abundance in the 100 ha fragments and continuous forest. If the net sample accurately reflects overall bird activity, it would appear that the ecological role of understory birds is much different in small fragments and second growth than in continuous forest. Based on the data from guilds (fig. 20.4), patterns of insectivory should be especially altered, because insectivores of all types are reduced. Omnivores may perform some of the same functions as the lost frugivores and insectivores, but their increase is much less than the decrease in the forest birds, and it largely occurs on edges (see below). It remains to be determined how removal of insectivores will affect insect communities, patterns of herbivory and perhaps plant communities, or how seed dispersal is affected by loss of frugivores (e.g., Marquis and Whelan 1994; Dial and Roughgarden 1995; Chapter 18). Some of the ecological roles of understory birds may be filled by small rodents, marsupials, and bats, which show much less variation in abundance between continuous forest and fragments than do birds (Malcolm 1997a; E. Sampaio, pers. comm.)

Differences Among Foraging Guilds

The guilds that were most reduced in small fragments were generally those that had been shown to be vulnerable to fragmentation. Ant-following birds were nearly absent in 1 ha and 10 ha fragments, although they were somewhat more common in second growth (fig. 20.4). This corresponds to our earlier data showing that they move through tall cecropia-dominated second growth

(Stouffer and Bierregaard 1995a). Obligate flock species and terrestrial insectivores were very strongly reduced in 1 ha and 10 ha fragments as well as in second growth. These appear to be the birds that are most restricted to large patches of forest. This was previously suggested for terrestrial insectivores (Stouffer and Bierregaard 1995a; Canaday 1996), and their local extinctions in the BDFFP fragments have been empirically shown (Stratford and Stouffer 1999). Data from these same fragments showed that most groups of litter arthropods are more abundant along edges and in small fragments than in the interior of large fragments, so the lack of litter-foraging birds is not related to reduced prey (Chapter 18). Although flocks persist in 10 ha fragments surrounded by tall cecropia (Stouffer and Bierregaard 1995a), it appears that they do not regularly use second growth (see also Borges 1995; Borges and Stouffer 1999). The three species of flock dropouts were present in all landscape elements, although even they were significantly reduced outside of large fragments.

We did not analyze separately a large subset of the obligate insectivores found in forest that do not forage in flocks, at ant swarms, or on the ground. These species are difficult to combine into social or foraging guilds, and most are too uncommon to analyze species by species. Further, many of these birds often forage above the level of nets, so they are poorly sampled by this technique (see also Stouffer and Bierregaard 1995a; Remsen and Good 1996). This diverse assemblage of thirty-five species includes those that are relatively common in small fragments and some secondary areas, such as the gap-specialists *Percnostola rufifrons* and *Hypocnemis cantator* (see table 20.1; see also Stouffer and Bierregaard 1995a). Other species appear to be restricted to continuous forest sites (e.g., *Myrmothe-*

rula guttata) or are much more common in continuous forest and 100 ha fragments (e.g., *Hylophylax poecilinota*).

Several forest-dwelling species with both frugivorous and omnivorous diets, such as *Mionectes macconnelli* and *Pipra pipra,* were among the most common birds in small fragments and second growth, but even this guild showed a significant decline outside of continuous forest and 100 ha fragments (fig. 20.5). This decline was driven by such species as *Turdus albicollis, Corapipo guttaralis,* and *Schiffornis turdinus,* which were occasionally netted almost everywhere but were much less common outside of continuous forest and the 100 ha fragments (Bierregaard and Stouffer 1997). Recapture data suggest that *T. albicollis* and *S. turdinus* are more sedentary than the other frugivores, most of which have a lek-based breeding system. Although classified here as frugivores, *M. macconnelli* and *P. pipra* are relatively omnivorous and opportunistic. Birds such as these are probably good examples of those that do not distinguish between small islands of forest and the sea of second growth. Even for these species, however, increasing landscape change will likely have a negative impact. As far as we know, *Pipra pipra* leks only in large areas of forest; nearly all of the birds captured elsewhere are females or males that have not yet molted into their adult, breeding plumage. This may also be the case for *M. macconnelli* (Willis, Wechsler, and Oniki 1978), but we cannot age or sex them.

Although we did not directly address the role of frugivorous birds in facilitating regeneration, the forest frugivores that we captured outside of forest or on edges are probably important for dispersal of seeds of some forest plants (e.g., Gorchov et al. 1993; Silva, Uhl, and Murray 1996). Most of the plants that these small frugivores disperse, however, are likely to be understory shrubs

rather than forest trees (Loiselle and Blake 1990). For example, small frugivorous birds probably never disperse seeds of palms or Sapotaceae and are therefore not involved in fragmentation effects on palm communities (Chapter 10) or Sapotaceae seedlings (Chapter 11). Also, most forest frugivores are not likely to cross large open areas, so their effect is likely to be strongest along edges or in second growth that has already become established (see Silva, Uhl, and Murray 1996).

Understory hummingbirds showed relatively little difference among landscape elements, but even these highly mobile species were slightly less common in 1 ha fragments (fig. 20.5). Even so, their ubiquity in the landscape confirms earlier analyses of long-term changes and seasonal movement patterns at the BDFFP site (Stouffer and Bierregaard 1995b, 1996). At net level, the common understory hummingbirds may be most abundant on the borders of fragments.

COMPARISON OF EDGE, INTERIOR, AND SECOND GROWTH

The general pattern of bird abundance was similar on edges and inside fragments for all fragment sizes (fig. 20.2). This is somewhat surprising, given that many forest species are restricted to the interior of fragments (Stouffer and Borges, unpublished data) and avoid edges of continuous forest (Quintela 1985). Fragment edges suffer massive loss of live tree biomass (W. Laurance, Laurance, et al. 1997; Chapters 9 and 13); this structural change certainly affects the suitability of fragments for some forest species. These structural effects seem swamped by the effects of the matrix for 1 ha fragments; the highest species richness and capture rates were in fragment 1112, which is among the most altered fragments. In contrast, the more intact 1 ha fragment 1104 had low richness and capture rate.

Birds from several sources make up for the loss of forest species on edges. First, fragments surrounded by very young second growth have a variety of nonforest species, especially *Ramphocelus carbo* and *Ammodramus aurifrons,* which colonize edges but rarely penetrate into the interior (see table 20.1, fig. 20.5). Second, some forest birds, such as flock obligates and ant followers, will move across borders of 10 ha and 100 ha fragments adjacent to tall cecropia. Third, two species of forest-gap specialists, *Hypocnemis cantator* and *Percnostola rufifrons,* are more abundant on edges than in the interior of fragments or in continuous forest (Stouffer and Borges, unpublished; see also Stouffer and Bierregaard 1995a). We note again that the edges sampled were structurally diverse, ranging from areas of active pasture to *Vismia* and cecropia greater than 10 m tall. Net samples on the older edges were most like those from our second-growth sites. This heterogeneity among edges explains the great variation among sites for nonforest birds. The low abundance of nonforest species in second growth was due to the absence of species that colonize edges adjacent to cleared areas but do not persist after the second growth reaches about 1.5 m, especially *Ramphocelus carbo, Ammodramus aurifrons,* and *Troglodytes aedon* (see table 20.1; see also Johns 1991 and Loiselle and Blake 1994).

For most guilds, secondary areas had about the same capture rates as the small fragments. Obviously, bird communities in second growth are very dynamic, so the results here apply only to second growth between eight to twelve years old. Younger growth, especially younger *Vismia,* and active cattle pastures contain only a handful of open-area birds, such as *Columbina passerina, Leptotila verreauxi, Sporophila castaneiventris, Ammodramus aurifrons,* and *Leistes militaris,* species that were very rare

in the second growth we sampled. More detailed analyses, however, have shown differences between the two general types of second growth (Borges 1995; Borges and Stouffer 1999). Cecropia-dominated growth is richer in forest bird species (and is floristically more diverse than *Vismia*-dominated forest [Williamson, Mesquita et al. 1998]), while *Vismia*-dominated growth is colonized by more nonforest species. Recaptures of banded birds that moved between fragments also show that cecropia-dominated growth is used preferentially.

LESSONS FOR LAND USE

All of these analyses demonstrate the importance of large (more than 100 ha) fragments for forest understory birds. Smaller fragments have almost no conservation value for birds (although this is not to say that they cannot contain populations of other forest taxa—e.g., Turner and Corlett 1996). The small fragments we studied were very close to continuous forest; they would be increasingly devoid of forest birds with increasing distance. Tall second growth, especially cecropia, provides an important corridor for understory bird movement (see also Stouffer and Bierregaard 1995a), although our sampling showed relatively few forest birds there. Also, some of the most area-sensitive birds, such as terrestrial insectivores, are those that rarely move through second growth, at least based on known movements between fragments by banded birds. There is no reason to believe that small fragments (less than 100 ha) are important as stepping-stones for bird movements, analogous to small oceanic islands, because most birds are no more common in them than in second growth. Active cattle pasture and very young growth, however, are used by very few forest birds (Johns 1991; Stouffer and Bierregaard 1995a; Silva,

Uhl, and Murray 1996). In this respect, some birds seem as isolated by open areas as arboreal primates (Chapter 21) or other notoriously forest-restricted species.

Our fundamental recommendation for bird conservation is trite: fragments much greater than 100 ha are critical for persistence of understory birds. It is important to point out that presence of species in our samples by no means indicates presence of a breeding pair, much less presence of a stable population. Available data suggest that most Amazonian forest birds, even small songbirds, have low densities and large territories relative to temperate zone birds — often only a few pairs/100 ha (see Terborgh, Robinson, et al. 1990; Thiollay 1994; Stouffer 1997), so stable populations will require areas of forest in the tens of thousands of hectares (based on population viability analyses; Stouffer, unpublished data). Data from a few other large Neotropical fragments (87–1,400 ha) also suggest that some small understory birds were lost after isolation (Leck 1979; Willis 1979). Thus most species in fragments, even 100 ha fragments, are almost certainly demographically doomed without regular colonization by birds that emigrate from larger populations. Because movement through second growth is limited for some species, fragments must be connected by tall second growth for smaller fragments to be recolonized after local extinction. Even so, it must be recognized that species differ greatly in their ability to recolonize.

Research Priorities

Future research should examine the behavior and population processes of forest birds in second growth and large fragments. Predicting persistence in the landscape requires knowledge of local demography as well as patterns of movements between patches

(Brown and Kodric-Brown 1977; Harrison 1991). It may be that fecundity never exceeds mortality in fragments, indicating that they function only as a sink for surplus individuals produced in continuous forest (Harrison, Murphy, and Ehrlich 1988; S. Robinson, Thompson, et al. 1995). Should this be the case, conservation of very large areas of forest would become especially important (see Rolstad 1991 for a review of area requirements for nontropical bird populations).

A second area that merits attention is the suitability of various types of vegetation for movement corridors or foraging areas. Some birds have little behavioral plasticity to tolerate physical structures that are new to them (e.g., Greenberg 1983, 1989). This appears to be the case for other highly mobile taxa that have much different abundances between forest and adjacent secondary areas (e.g., Powell and Powell 1987; Klein 1989; Laurance 1991b; Chapter 21). Understanding these preferences could lead to better management of secondary areas, which are an increasingly important component of the Amazonian landscape (e.g., Moran et al. 1994). Previous attempts to model animal movements were probably too simplistic because they considered only one category of deforested habitat (e.g., Dale et al. 1994). For birds, and probably other taxa, the use of these habitats varies rapidly as vegetation structure changes (Loiselle and Blake 1994; Stouffer and Bierregaard 1995a). Models could become more realistic with better data on spatial and temporal variation among second growth and habitat-specific movement patterns.

Conservation Lessons

1. Abundance and species richness of small understory birds are much lower in 1 ha and 10 ha fragments and in eight- to

twelve-year-old second growth than in 100 ha fragments and continuous forest.

2. Abundance of forest insectivores, especially those that forage on the ground or in mixed flocks, is especially reduced outside of continuous forest and 100 ha fragments.

3. Small fragments (1 ha and 10 ha) and second growth almost certainly cannot support populations of many forest birds.

4. Movements of birds between fragments are rare; second growth is a significant barrier to many species.

5. Long-term persistence even in 100 ha fragments is unlikely for species with low densities and poor dispersal through second growth.

6. Population stability for these taxa probably requires very large areas (more than 10,000 ha) of continuous forest.

7. Future research should examine movement patterns and demographic processes of forest birds to determine the value, if any, of large fragments on the order of hundreds of hectares and secondary areas to overall population stability.

CHAPTER 21

Primates in a Fragmented Landscape

Six Species in Central Amazonia

KELLEN A. GILBERT AND ELEONORE Z. F. SETZ

Increased anthropogenic activity in the tropics has resulted in areas of deforestation, habitat fragmentation, and the isolation of forest fragments. What happens to the flora and fauna in this new mosaic of isolated forest remnants and secondary growth is only beginning to be understood. Examining the effects of fragmentation on the medium- and large-sized mammals is often the first and easiest place to begin, given their relative ease of observation and measurement (Wilson and Wilson 1975; Bernstein et al. 1976; Malcolm 1988, 1990).

Early studies of primates and habitat fragmentation focused on the effects that selective logging had on primate populations (Wilson and Wilson 1975; W. Wilson and Johns 1982) and the effects on primate species diversity in areas disturbed by agricultural activities (Bernstein et al. 1976). More recently, habitat loss caused by widespread deforestation has been taken into account to explain decreases in primate diversity and declining populations in forests throughout the Paleo- and Neotropics (White 1994a, 1994b; Ferrari and Diego 1995; Hsu and Agoramoorthy 1996; O'Brien and Kinnaird 1996; Garcia and Mba 1997; Thoisy and Richard-Hansen 1997). A consistent finding in these primate population studies is that if human degradation of forest environments either directly or indirectly continues unchecked, the survival of most of the endemic primates is threatened.

Primates in the isolated and continuous forest reserves of the BDFFP have been studied off and on since the very early years of the project. Some of the same groups have been observed in reserves both before and after isolation (Rylands and Keuroghlian 1988). The results from BDFFP studies of primate ecology and behavior in isolated reserves and a continuous forest reserve serving as a control provide unique and valuable data about what happens to primate populations in a fragmented landscape over time.

In this chapter we provide an overview of the primate research that has been conducted in the BDFFP reserves. Specifically, we discuss the primates found in the reserves, the results of research in forest fragment reserves and the continuous forest, and reports of anecdotal information. We compare the Manaus primate populations, using estimates of biomass, to those of another well-studied Neotropical site, Manu National Park, in Amazonian Peru. Finally, we discuss the current status of the primate population in the reserve area and directions for future research.

This is publication number 288 of the BDFFP Technical Series.

Primates of the Reserves

Six primate species inhabit the reserves of the Biological Dynamics of Forest Fragments Project: red howling, or howler, monkey (*Alouatta seniculus*); black spider monkey (*Ateles paniscus*); tufted, or brown, capuchin (*Cebus apella*); bearded saki (*Chiropotes satanas*); white-faced saki (*Pithecia pithecia*); and golden-handed tamarin (*Saguinus midas*). The night monkey, *Aotus* sp., may occur in the region but has been observed only once in a reserve (M. van Roosmalen, pers. comm.). With the exception of the golden-handed tamarin, all six species are found in moist forest habitats from French Guiana to eastern Peru, where they have been recorded and studied at Cocha Cashu in Manu National Park (e.g., Terborgh 1983). Among these six species, the white-faced saki is the rarest (Rylands and Keuroghlian 1988). Other primates in the Manaus region, the bare-faced tamarin (*Saguinus bicolor*) and the squirrel monkey (*Saimiri sciureus*), are not found as far north as the reserves. The density of primates in the reserves is markedly lower than in other Neotropical moist forests (Mittermeier and van Roosmalen 1981; Terborgh 1983; Chapter).

Black Spider Monkey

Although the black spider monkey, known locally as *macaco aranha,* is the largest primate in the region, the least is known about its behavior and ecology in the BDFFP reserves. Because groups are widely dispersed, have large ranging areas, and are very fast moving, spider monkeys are difficult to study, and there has been no research conducted on this species in the reserves. Most of our information is based on anecdotal evidence.

The black spider monkey is characterized by its pink face and black pelage. As in other forests, spider monkeys in the reserves appear to be mainly frugivorous (van Roosmalen and Klein 1988). They are completely arboreal and are observed mostly in the middle to upper canopy. In our study areas, spider monkeys travel in small groups, typically of two to three individuals. Loud contact calls, which may help individuals track dispersed group members, are heard in the continuous forest reserves of Km 41, Km 34, and also around Porto Alegre (see fig. 4.1; table 4.3).

In the BDFFP study area, spider monkeys are found only in the continuous forest. Immediately after isolation of the 10 ha and 100 ha reserves, spider monkeys left the reserves (Rylands and Keuroghlian 1988). Recently (1992 to present) workers have observed spider monkeys in older second growth (at least 12 meters tall) dominated by *Cecropia* spp. at the *fazenda* (ranch) Porto Alegre.

Their frugivorous diet combined with large body size and large ranging area requirements, particularly in a fruit-poor area (see Chapter 5), probably result in a 100 ha isolated fragment being too small to sustain a spider monkey group (Rylands and Keuroghlian 1988). Frazão (1992) studied a troop in the continuous forest at Km 41 and followed the monkeys through an area of 1,000 ha. This is consistent with the finding that of all the *Ateles* species, the black spider monkey appears to be the most restricted to habitats of undisturbed high forest (van Roosmalen and Klein 1988).

Red Howling Monkeys

The red howling monkey, known in the Brazilian Amazon as *guariba,* is the most common primate in the reserves. Howling monkey socioecology has been studied in both forest fragment and continuous forest reserves (Neves 1985; Neves and Rylands 1991; Gilbert 1994).

Red howling monkeys are large; adults range from 4.2 to 9.0 kg (Rowe 1996). Their brightly colored coats range from orange to dark purple. Adult males have deep red to black beards. Their awesome (and presumably territorial) howling bouts, which can be heard over kilometers, are a characteristic sound of Neotropical forests. Adult males begin each predawn howling bout accompanied by the adult females howling in the background (Gilbert, pers. obs.).

Howler groups are found in all of the isolated 10 ha and 100 ha reserves, as well as the continuous forest (Rylands and Keuroghlian 1988; Gilbert 1994). There is no significant difference in the group size (mean = 6) or density of groups between reserve sizes (Gilbert 1994). Howlers live in groups of two to ten individuals. In the reserves, all groups were multi-male/multi-female in organization, with one alpha male per group.

The daily activity pattern of howlers consists of feeding, resting, and traveling. They display little overt social behavior. Grooming is rare. Older infants and juveniles are the most active members of the groups. Adult males spend on average 1 to 4 percent of the daily activity scanning the area, usually while sitting in the upper canopy (Gilbert 1994). This may be scanning for predators or other adult male howlers; or it could be prompted by the presence of an observer. Members of a group sleep together in the upper canopy. In the isolated reserves, howlers forage and rest on the forest edges. These behaviors displayed in exposed areas suggest that adult howlers are not under great pressure from avian predators (Gilbert 1994).

Howlers are the most folivorous of the primates in the reserves, with about 60 percent of the diet consisting of immature and mature leaves (Neves and Rylands 1991; Gilbert 1994). The fruit portion of the diet is dominated by Sapotaceae fruits. Nearly all of the seeds ingested, including those as large as 3 cm in length, are passed whole in the feces (Gilbert 1994).

Howlers may be one of the main large seed dispersers in the reserves. In an ongoing study at Km 41 and the isolated 100 ha and 10 ha fragments, Ellen Andresen is examining the role of secondary seed dispersers at howling monkey defecation sites. She has found that howling monkeys produce the large amounts of high-quality dung needed to maintain a healthy dung beetle community (Klein 1989). The dung beetles disperse seeds found in the dung and may effectively hide seeds from secondary seed predators, such as rodents (E. Andresen, pers. comm.).

As howling monkeys are the primate species most likely to persist in forest fragments, and in inflated densities, relationships among size of fragment, population density, and disease were investigated. Gilbert (1994) found a higher prevalence of gastrointestinal parasites in groups in the 10 ha reserves than in 100 ha reserves or continuous forest groups. There was also a positive correlation between parasite load and howler density. Overall, the monkeys were infected with eight species of parasites. This is the highest number of parasite species reported in free-ranging howling monkeys (Stuart et al. 1990; Stuart, Strier, and Pierberg 1993). A cost of living in a fragmented habitat, though not measured in terms of fitness, may be an increase in the number of individuals with chronic parasitic infections (Gilbert 1994).

CEBUS APELLA

The tufted, or brown, capuchin, locally known as *macaco prego,* is a common primate in the continuous forest and has been intensively studied at the Km 41 continuous forest reserve by Wilson Spironello (Spironello 1991, 1999; Chapter 22). This medium-sized monkey weighs from 1.3 to 4.8 kg (Rowe 1996). Capuchins live in large multi-male/multi-female groups of eight to twenty individuals. Capuchins spend a large part of their daily activity foraging and traveling. They use all levels of the forest and occasionally come to the ground. The group is widely dispersed while foraging and very vocal. The home range of a study group at Km 41 (1501) was 860 ha (Chapter 22).

Capuchins are the most omnivorous of the primate species in the reserves. They forage on a variety of fruits (especially palms), insects, palm pith, honey, and occasionally small mammals (Terborgh 1983; Spironelo 1991; Rowe 1996). They manipulate certain hard fruits, such as Lecythidaceae, and pound out the lipid-rich seeds (Gilbert, unpublished data). Spironello (1999) has investigated the role capuchins play as dispersers of Sapotaceae fruits in the continuous forest.

Capuchins appear to be restricted to the continuous forest, although one group is consistently observed in the 100 ha reserve at Porto Alegre (3304; see plate 2). This reserve has not been completely isolated. A narrow corridor connecting this reserve to tall secondary growth and the continuous forest may allow the capuchins to include the reserve in their ranging area.

BEARDED SAKI

The bearded saki, known locally as the *cuxiú*, is distinguished in the reserves by its bushy, nonprehensile tail and knoblike swellings on the temporal region of the head. Adults range in weight from 1.9 to 4.0 kg (Rowe 1996). Sakis travel in large groups of ten to thirty individuals. They have large home and daily ranges.

Bearded sakis are mainly frugivorous, with a diet consisting of immature fruit and seeds (Frazão 1992). In addition to fruit and seeds, bearded sakis eat a variety of insects and will break apart wasp nests to eat the larvae (Frazão 1991).

Saki groups are restricted to the continuous forest. Immediately following isolation of the Porto Alegre and Dimona reserves, bearded saki groups left the area (Rylands and Keuroghlian 1988). No groups have recolonized any of the isolated reserves. An adult male associated with a howling monkey group in the 10 ha reserve at Dimona (2206) for five months and then left the reserve (Gilbert 1994), and a group of three (adult male, adult female, and an infant) was observed in the 10 ha reserve at Colosso (1202) for two weeks in 1995 (Gilbert, unpublished data). Their relatively large adult body size, large home range requirement, and reliance on medium to large fruit all contribute to a sparse bearded saki population.

WHITE-FACED SAKI

The white-faced saki, known locally as the *parauacú,* is the smaller of the two saki species and is relatively well studied in the reserves (Setz 1993, 1994). Adults weigh from 0.78 to 2.5 kg (Rowe 1996). White-faced sakis are one of the few sexually dichromatic New World primate species. Adult males are black with a white face, although some adult males in the reserves have chestnut-colored faces. Females are gray-brown.

Sakis live in groups of two to five individuals, with one adult male. They are most active in the lower to middle levels of the forest. They move quickly, incorporating long jumps from branch to branch. Like bearded sakis, white-faced sakis are frugivores and seed predators. In more than 1,000 hours of observation, Setz (1994) found that in the dry season fruit made up

61.5 percent of the diet, while in the wet season 91 percent of the diet consisted of fruit. One group in a 10 ha forest fragment ate parts of 190 plant species (Setz 1993).

White-faced sakis are found in the 10 ha and 100 ha reserves and the continuous forest. Groups have recolonized the 10 ha isolated fragments since the initial isolation (Lovejoy et al.1986; Gilbert 1994). Because of a relatively small home range requirement, a saki group can survive in an isolated fragment.

GOLDEN-HANDED TAMARIN

The golden-handed tamarin, known locally as the *sauim,* is the smallest primate (adult females weigh 432 g and adult males 586 g [Rowe 1996]) in the reserves. They are characterized by a black pelage with golden brown hands and feet.

Tamarins live in groups of four to ten individuals in home ranges of 10 ha to 30 ha (Lovejoy, Bierregaard, et al. 1986). Females typically give birth to twins, which are then carried by the adult male and juvenile females of the group. Their diet consists of fruit and insects, and they appear to benefit from the increased densities of highly clumped trees with small fruits and arthropods that are found in secondary growth (Lovejoy, Bierregaard, et al. 1986). They appear to be most active in the middle to upper canopy, displaying quick running and leaping locomotion.

Swallow-tailed kites (*Elanoides forficatus*) were observed in association with a group of golden-handed tamarins in a 100 ha reserve in search of insects that the tamarins were disturbing (A. Whittaker, pers. comm.). Larger raptors, such as Harpy Eagles (*Harpia harpyja*), Crested Eagles (*Morphnus guianensis*), or Ornate Hawk-Eagles (*Spizaetus ornatus*), may pose a serious threat as predators to the small tamarins.

Unlike the tamarin species at Cocha Cashu (Terborgh 1983), tamarins in the reserves have not been observed in interspecific associations with other primates.

While tamarins are commonly seen in the reserves and along the road to the *fazendas,* there have been no studies of these primates in the reserves. Tamarin groups are found in all of the isolated reserves and the continuous forest, although they are uncommon in the continuous forest (Gilbert 1994). They appear to prefer edges and secondary growth and can be seen foraging in tall capoeira around the isolated reserves. Tamarins also move across short distances on the ground.

Primate Research at the BDFFP Reserves and the Effects of Forest Fragmentation

Despite their relatively large size, diurnal activity, and charisma, primates in the BDFFP reserves have not been studied to the degree that other vertebrates, such as birds and frogs, have been. Most of the information on the primate population in the reserves comes from a series of censuses conducted in the 1980s. Baseline information on the diversity and density of primates in the reserves was collected both pre- and postisolation, from 1983 through 1985 (Rylands and Keuroghlian 1988). The most recent surveys of the isolated forest fragments were conducted in 1993 and 1995 (Gilbert, unpublished data).

To survey the reserves, the researchers used the repeated line transect method (Rylands and Keuroghlian 1988; Schwarzkopf and Rylands 1989; Gilbert 1994; Gilbert, unpublished data). Observers walked the existing trail system in each reserve, beginning at 0600 h and again at 1500 h. All primates observed were identified by sex, age class, and group size and composition, if possible. Distance to the primates from the trail was recorded. To survey a continuous reserve (1501), the repeated line transect method was used in a 100 ha area (Gilbert, unpublished data). Some of the reserves surveyed in the 1980s have now been isolated from the continuous forest.

The immediate effect of isolation was the loss of the large frugivores from 10 ha and 100 ha forest fragments (tables 21.1 and 21.2). These species—the black spider monkeys, bearded sakis, and brown capuchins—require a larger ranging area. The primates that are able to survive and reproduce in the small fragments have small range requirements and, in the case of howlers, can subsist on a large portion of leaves in the diet and are not constrained by the availability of fruit (Neves 1985; Rylands and Keuroghlian 1988; Neves and Rylands 1991; Gilbert 1994).

Recolonization of the isolated forest fragments has not occurred rapidly. The results of a survey of all of the isolated reserves and one of the control reserves (1501) conducted in 1993 showed primate recolonization in four of the eleven isolated reserves over the past twelve years. Three 10 ha isolates (1202, 2206, and 3209) had one group each of white-faced sakis and golden-handed tamarins. One 100 ha isolated forest fragment also had a group of tufted capuchins, although the group appears to move in and out of the reserve (Gilbert 1994, unpublished data). No other groups of the large frugivorous species have recolonized 100 ha forest fragments. In 1995, only one 10 ha reserve (1202) retained its three resident species—howlers, tamarins, and white-faced sakis. During the survey this reserve also had an adult male bearded saki (Gilbert, unpublished data).

Schwarzkopf and Ryland's (1989) intensive examination of the relationship between forest characteristics and primate distribution in four isolated 10 ha fragments is

one of the few studies of primate ecology in the BDFFP forest fragments. They found more primate species in the more structurally complex isolated reserves, defined as those with a high mean number of trees and lianas, a low mean percentage of trees greater than 10 cm DBH, and with streams within the reserve boundaries. As with some bird species (Chapter 20), the presence of tall secondary growth around a reserve was associated with the presence of tamarins in the isolated reserve. There was no relationship between the distance from an isolated reserve to primary forest and the primate species richness in it (Schwarzkopf and Rylands 1989).

Estimates on the density of primates in the reserves remain incomplete. Results from the 1995 survey appear to show a small increase in density from 1983–84 in the 100 ha reserves and the continuous forest but a decline in the 10 ha reserves (table 21.3). With the small number of reserve replicates and unequal sampling efforts it is all but impossible to make statistical comparisons.

Discussion

What have we learned about the primates in central Amazonia and about the effects of forest fragmentation on this population? It is clear that a 100 ha isolated forest fragment is too small to sustain groups of three of the six local primate species. In other fragmented landscapes with resident primate species, such as Kibale and Uganda, the more terrestrial catarrhines and chimpanzees may be able to include isolated fragments in their ranging areas (Onderdonk and Chapman 1996). The primates in central Amazonia, with the exceptions of the tufted capuchin and golden-handed tamarin, which do come to the ground occasionally, are nearly completely arboreal, restricting their movement

TABLE 21.1.　Primate Species in the BDFFP Isolated Forest Reserves and the Continuous Forest (Reserve 1501) by Year of Survey

Reserve	Area (ha)	Year of survey			
		1983–84	1985–86	1993	1995
1202	10	Alouatta	Alouatta Saguinus	Alouatta Pithecia Saguinus	Alouatta Pithecia Saguinus
1204	10	—	—	—	Alouatta
1207	10	Alouatta Saguinus	Alouatta Saguinus	Alouatta Saguinus	Saguinus
2206	10	Alouatta Saguinus	Alouatta Pithecia	Alouatta Pithecia Saguinus	Alouatta Saguinus
3209	10	Alouatta Saguinus	Alouatta Saguinus	Alouatta Pithecia Saguinus	Alouatta
1302	100	—	—	—	Alouatta Pithecia Saguinus
2303	100	—	—	Alouatta Pithecia Saguinus	Alouatta Pithecia Saquinus
3304	100	Alouatta Pithecia Saguinus	—	Alouatta Pithecia Saguinus Cebus	Alouatta Pithecia Saguinus Cebus
1501	Cont. forest	Alouatta Pithecia Saguinus Cebus Chiropotes Ateles	—	Alouatta Pithecia Saguinus Cebus Chiropotes Ateles	Alouatta Pithecia Saguinus Cebus Chiropotes Ateles

Notes: A dash indicates that no data were available from that reserve for that year.

TABLE 21.2.　Primate Groups, by Species, Observed in the 10 ha, 100 ha, and Continuous Forest Reserves in the 1995 Survey

Species	Fragment size		
	10 ha	100 ha	>1,000 ha
Black spider monkey (*Ateles paniscus*)	A	A	P
Red howling monkey (*Alouatta seniculus*)	P	P	P
Tufted (Brown) capuchin (*Cebus apella*)	A	P	P
Bearded saki (*Chiropotes chiropotes*)	A	A	P
White-faced saki (*Pithecia pithecia*)	P	P	P
Golden-handed tamarin (*Saguinas midas*)	P	P	P

Notes: P = Present, A = Absent.
Source: Gilbert, unpublished data.

TABLE 21.3. Comparison of the Average Estimated Primate Population Sizes in Isolated Forest Reserves of Different Sizes and Continuous Forest (1501) for 1984 and 1995

Species	10 ha		100 ha		>1,000 ha	
	1984	1995	1984	1995	1984	1995
Black spider monkey	0	0	0	0	1.0	3.0
Red howling monkey	7	4.7	5.8	12.5	10.5	18.0
Tufted capuchin	0	0	0	4.5	2.2	6.0
Bearded saki	0	0.7	0	0	5.5	8.0
White-faced saki	5.0	1.0	1.7	4.0	0.7	1.0
Golden-handed tamarin	5.5	2.7	25.6	4.5	3.9	7.0

Notes: All 1984 data from Rylands and Keuroghlian (1988). All 1995 data from Gilbert (unpublished data).

TABLE 21.4. Primate Biomass (kg/km²) at the BDFFP Continuous Forest Reserve 1501 and Cocha Cashu, Peru

Species	BDFFP	Cocha Cashu
Ateles paniscus	21	175
Alouatta seniculus	126	180
Cebus apella	39	104
Cebus albifrons	—	84
Chiropotes satanas	35	—
Pithecia monachus	—	2
Pithecia pithecia	6	—
Saimiri sciureus	—	48
Aotus trivirgatus	—	28
Callicebus moloch	—	17
Saguinus fuscicollis	—	17
Saguinus imperator	—	5
Saguinus midas	2	—
Cebuella pygmaea	—	1

Sources: Gilbert, unpublished data, Terborgh 1983.
Average adult weights were used (Terborgh, 1983, Rowe 1996).

to forested areas. As a result, forest fragments separated by as little as a pasture 100 m from continuous forest are true islands for these species.

We have also learned that even with relatively few data, the density of primates in the reserves is markedly lower than in other Neotropical moist forests (Mittermeier and van Roosmalen 1981; Terborgh 1983; Rylands and Keuroghlian 1988; Gilbert 1994; Chapter 22). The reasons density differs are unclear. Resource availability in the reserves, in particular fruit availability and

abundance, certainly needs to be considered in order to understand differences in primate density across sites.

It may be more useful to compare primate communities between sites by examining biomass differences. Primate biomass of the BDFFP reserves is quite lower than that of Cocha Cashu, Peru (table 21.4). For the three species that are common to both sites and are also the largest in body size, the biomass differences are striking, particularly for the black spider monkey, the largest frugivorous species at both sites. With all species combined, the total primate biomass at Cocha Cashu is 649 kg/km², and in the BDFFP reserves, 229 kg/km². Again, quantitative data on fruit availability from the reserves may help explain the paucity of primate species.

The opportunity for primate studies in the BDFFP reserves is great. Most studies have focused on groups in the continuous forest and in one of the 10 ha isolated forest reserves. There are many other reserves with primates to be studied, and there are hundreds of questions about primate ecology and behavior in central Amazonia to be asked. We urge researchers and, in particular, students to take advantage of the unique research opportunity and examine the dynamics of forest fragmentation on central Amazonian primates.

Conservation Lessons

1. Forest areas need to be larger than 100 ha, and ideallỹ at least 1,000 ha, to sustain viable populations of the six primate species found in central Amazonian terra firme forest. Long-term stability of populations in such small reserves would require genetic exchange with neighboring populations.

2. Forest complexity, measured by plant species diversity and the variety of plant structures, should be considered in forest management decisions regarding the conservation of primates in central Amazonia.

3. The role of forested corridors connecting isolated forest fragments to the continuous forest in protecting the primate population in areas of habitat fragmentation needs to be examined.

CHAPTER 22

The Brown Capuchin Monkey (*Cebus apella*)

Ecology and Home Range Requirements in Central Amazonia

WILSON R. SPIRONELLO

Reported ranging patterns of primate groups vary according to the area where they have been studied. Ranging patterns have been correlated with metabolic necessities, sleeping sites, social factors, and seasonal variation in availability and distribution of food resources (McNab 1963; Clutton-Brock 1977; Rasmussen 1979; Terborgh 1983; van Roosmalen 1985; Newton 1992; Nunes 1995; Zhang 1994, 1995a; Izar 1999). Primates are flexible, however, in diet and area used; they are more likely to respond to the production cycles of some food plants than to season (Chapman 1988). In addition, even group home ranges vary over time (Izar 1999). Thus, home range size and ranging patterns among primates may depend on social aspects and feeding-behavior strategies, which may change according to availability of food resources in different areas of forest.

The destruction of natural habitat in central Amazonia terra firme forest near Manaus has been increasing rapidly, and, as a result, the forest around the city has been broken up into fragments in a pattern seen wherever and whenever industrialized humans encroach upon extensive forest habitat (see Chapter 3). It has been demonstrated in the BDFFP reserves that habitat fragmentation can affect the microclimate and various aspects of the ecology of plant, insect, avian, and mammalian communities (Kapos 1989; Lovejoy and Bierregaard 1990; Malcolm 1990; Rankin–de Mérona, Hutchings, and Lovejoy 1990; Benitez-Malvido 1995; Souza and Brown 1994; Camargo and Kapos 1995; Scariot 1996; and many of the chapters in this volume). While studies at the BDFFP site (and elsewhere in the tropics) are beginning to elucidate patterns of disturbance and susceptibility (e.g., Laurance and Bierregaard 1997a; Laurance, Bierregaard, et al. 1997), the ecological implications of fragmentation remain largely unknown for the majority of plant and animal species in these fragments, and even basic biological data are still lacking for most of the plants and animals of the primary forest at this site. This is especially true for medium-sized and large animals, such as primates.

One factor governing fragmentation effects on primates is their home range size. Previous studies suggested that primates in the forests north of Manaus have larger ranges than elsewhere in Amazonia (Emmons 1984; Rylands and Keuroghlian 1988; Malcolm 1990; Peres 1993; E. Frazo, pers. comm). This study was designed to gather

This is publication number 289 of the BDFFP Technical Series.

basic biological information about the brown capuchin monkeys (*Cebus apella apella*), locally known as *macaco prego,* and to investigate whether the home-range requirement for this species is larger than at other Amazonian sites, as has been suggested. The aims of the study were fourfold: to estimate the home range for a troop of brown capuchins; to estimate how many groups overlap the resident group's home range; to test whether difference in ranging patterns of brown capuchins could be related to the different food sources used in their diet, location of the plant sources, and shelters; and to point out what ecological parameters differ from other Amazonian sites where brown capuchins have been studied.

Study Site

The study was carried out in the BDFFP continuous-forest reserve (located at Km 41 of the penetration road ZF3; see fig. 4.1). Mean total annual rainfall over an eight-year period, including the years of this study, was 2,673 mm (SD 314 mm), with a monthly minimum of 109.5 mm (SD 47.2) in September and a maximum of 380 mm (SD 61) in March (fig. 22.1). The wet season is from December to May and the dry season from June to November. The mean minimum and maximum annual temperature ranged from 21° C to 22.5° C and from 31° C to 33.5° C, respectively. Elevation in the area varies between 80 and 110 m, and the soils are clay, deep, acidic, and well drained, and generally poor in mineral nutrients (Sombroek 1966; Irion 1978; Chapter 23; see also Lovejoy and Bierregaard [1990] and Chapter 4 for details on the BDFFP forests).

The area is covered mainly by dense terra firme forests intersected by numerous small streams (Pires and Prance 1985; Prance 1990). High forest occurs in well-drained

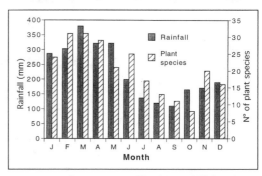

Fig. 22.1. The average of both rainfall (mm) and number of plant species (ripe fruit) recorded monthly at Km 41 reserve.

soils, creek-side forest occurs along perennial streams on poorly drained soils, and swamps are dominated by arborescent palms on waterlogged soils. In poorly drained soils and swamps the palm community is dominated by *Oenocarpus bataua* var. *bataua* Henderson, *Euterpe precatoria* Martius, *Socratea exorrhiza* (Mart.) H. A. Wend., and *Mauritia flexuosa* L. (Kahn and Castro 1985).

The BDFFP study site contains in excess of 1,000 tree species from 55 families, of which the most numerically abundant are also the most species rich, with 87, 49, and 45 species, respectively, in the Sapotaceae, Burseraceae, and Lecythidaceae (Rankin–de Mérona, Hutchings, and Lovejoy 1990; Rankin–de Merona et al. 1992; Chapter 5; T. Pennington, pers. comm.). This forest is poor in epiphytes and shrubs (Gentry and Emmons, 1987; Prance 1990), and the understory has an abundance of stemless palms, such as *Attalea* spp. and *Astrocaryum* spp. (Kahn and Castro 1985). Canopy trees are 30–35 m in height, and emergents are 35–55 m; the mean tree density (with a minimum DBH of 10 cm) is 637 trees/ha, and the mean basal area is 31 m²/ha (Rankin–de Mérona et al. 1992).

Regional studies have shown that the main flowering period in the region is during the dry season, with a peak between August and October, and the fruiting period occurs during the wet season between December to May (Alencar, Almeida, and Fernandes 1979; Alencar 1998). In a two-year study period at the Km 41 forest (where this study was carried out), fruiting peaked between January and June (Oliveira 1997; fig. 22.1).

Methods

Data analyzed in this study are based on fourteen months of observations. An additional thirteen months were used only to verify variations in home range size. The study group was followed for 183 days, including 45 incomplete days. Observation time varied between 61 and 156 hours per month, accounting for a total of 1,402 hours. Group composition varied from nine to fourteen individuals. This variation was due to three births, four immigrations, and three emigrations. Overlap between groups was analyzed based on observation during the late period, but additional data also were recorded between 1995 and 1997.

Based on preliminary observations of a troop of brown capuchins, a 100 × 100 m grid of trails was cut in the forest, demarcating 1 ha squares. As I followed the monkey troop over the course of the study, the trails grid was expanded whenever the troop led me into new portions of their range. At fifteen-minute intervals the location of the group, defined as roughly the center of the individuals in the troop, was mapped. Daily range length was calculated by measuring the straight-line distance moved by the group on the map, and daily range was estimated counting the total number of 1 ha grid squares used by the group each day. The cumulative range was based on new areas (1 ha squares) entered during the study. In addition, to refine my home range estimate, the peripheral squares where capuchins were seen were divided into smaller squares of

0.25 ha, and only the smaller squares visited by the capuchin group were included in the home-range estimate.

Over the course of the study the troop was recorded in 5,467 scan samples, ranging from 249 to 629 scans in a month. If the whole troop was in a quadrat, that quadrat was scored 1.0 for that scan. When the group occupied more than one 1 ha quadrat during a given scan (a frequent occurrence), the occupation of each area was calculated as a factor of the number of quadrats the troop was in at that scan—if the group was in four quadrats, each quadrat was scored 0.25. In addition, to refine my home range estimate, the peripheral squares where capuchins were seen were divided into smaller squares of 0.25 ha, and only the smaller squares visited by the capuchin group were included in the home range estimate.

Data on feeding behavior were recorded by the scan-sampling method at ten-minute intervals (Altmann 1974). Observations on feeding habits were subdivided into three different food categories: ripe fruits, seeds, and flowers (especially nectar); animal matter (insects, arthropods, small vertebrates, and invertebrate and vertebrate eggs); and others (sap, branches, roots, stalks, and leaves). The feeding group score for each food source was based on the number of times the capuchin group was recorded using a given food during scan periods (one record per scan). The most important plant resources were defined as those representing 4.5 percent or more of food-group scan records monthly. In addition, the group's marginal area was represented by squares with low scan occupancy (less than 0.2 percent, or 10 scans, n = 5,467).

The availability of the food sources used by capuchins was based on the number of plants visited monthly. General fruit availability in the area, however, was also based on long-term phenology studies in two terra firme forests north of Manaus (Alencar, Almeida, and Fernandes 1979), as well as on work carried out at the study site (Oliveira 1997).

Data Analysis

The Spearman rank-order coefficient (r_s) was used to analyze the effect of observation effort on home range size, to correlate frequency of the feeding group records in each 1 ha square with frequency of the occupancy of the group in such square, and to correlate the proportion of food items and abundance of food plants with day range, day range length, and home range patterns of the brown capuchins. A chi-square test was used to test preferences for palm trees in swamp forest versus canopy trees on highland forest. The Margalef diversity index (Mg) was used to estimate the diversity of the plant food sources used by capuchins through the year; the choice was based on the index reliance (see Magurran 1988). The Kruskal-Wallis distribution was used to compare seasonal differences in diversity of food use, and Student's t-test was used to compare daily range lengths across seasons (Zar 1984; Siegel and Castellan 1988).

Results

Home Range Size and Length and Daily Ranges

Monthly home ranges varied between 212 ha and 300 ha. A total of 915 ha was covered by the group after twelve months (fig. 22.2). The estimate of the group's home range decreased to 852 ha (60.8 ha/ind or 1.6 ind/km²) after excluding the 0.25 ha squares on the range periphery that were not used by the capuchins. The home range observed in different months was not significantly related to the observation effort (r_s = 0.42, n =

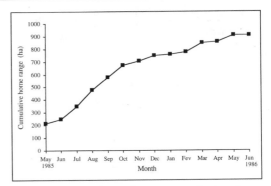

Fig. 22.2. Cumulative home range during the fourteen months of the study.

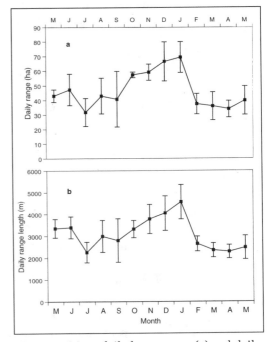

Fig. 22.3. Mean daily home range (a) and daily range length (b) of the study group from May 1985 to May 1986, vertical lines indicating standard deviations.

14, P > 0.05) during the period of this study. In the second year, however, the observation effort was 58 percent lower, and, as a consequence, the area estimated for the resident group was also 42 percent smaller (530 ha).

Daily ranging varied between 34 ha (SD = 9) in July and 70 ha (SD = 12) in January (fig. 22.3a). The daily range length was between

2,236 m (SD = 470) also in July and 4,560 m (SD = 460) in January, and the highest average values occurred between October and January (fig. 22.3b). No clear difference existed between the dry (June to November) and wet seasons (December to May) (t-test: F = 5, P > 0.05).

FOOD SOURCES

The capuchin's diet consisted of plants (82 percent) and animal matter (18 percent). Plant sources consisted mainly of ripe fruits, followed by seeds, and flowers (mainly nectar). Among these resources, fruits of Palmae (19.5 percent), Sapotaceae (13.5 percent), and Euphorbiaceae (8.5 percent) were the most important families. The diversity of plant sources used in capuchins' diet varied seasonally, being higher in the wet season (December to May; Kruskall-Wallis test: H = 5.769, P < 0.02; fig. 22.4). The availability (plants visited/ha) of food sources used by the study group showed peaks between July and September and between February and May (fig. 22.4). Ripe fruits were used throughout the year, especially in the late wet season, whereas immature seeds were consumed more from May to December, and nectar from October to February. Animal matter was consumed more in the early dry and late wet seasons.

HABITAT SELECTIVITY

Capuchins preferentially spent 39 percent of their time in forest areas along perennial streams ($X^2 = 16.3$, P < 0.001). This was in part because they relied heavily on *Oenocarpus bataua* fruits, but also because the monkeys used *O. bataua* and *Mauritia flexuosa* as sleeping sites. These palm trees are found in poorly drained soils along streams. *O. bataua* was used on 137 (92 percent) nights, compared to 11 (7 percent) nights for *Mauritia flexuosa* and 2 (1 percent) nights

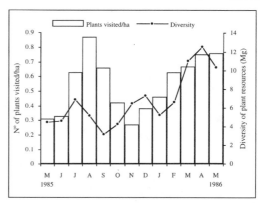

Fig. 22.4. Diversity (Mg) and abundance of food sources visited by capuchins monthly.

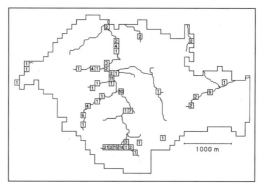

Fig. 22.5. Distribution of shelters and frequency of use by the group during the study.

for other canopy trees. *O. bataua* occurred at a density of 44 trees (DBH of at least 15 cm) recorded in a 100 ha plot. In contrast to this density of 0.44 palms/ha, Rankin–de Mérona et al. (1992) reported a density of about 180 other trees (DBH of at least 20 cm) per hectare. Despite this relative scarcity of palms, capuchins showed a very strong preference for them as sleeping sites. The resident group used shelters located in the central, southeast, and southern areas of their range quite often (fig. 22.5).

RANGING PATTERNS

The seasonal ranging patterns were significantly related to the most important food

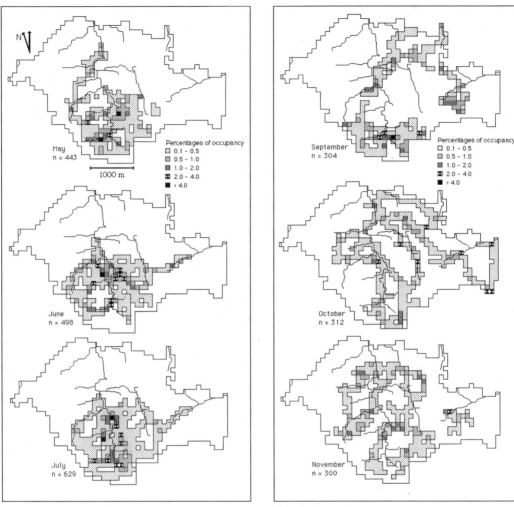

Fig. 22.6 (A) Seasonal variations in percentages of occupancy of the area used by the capuchins between May and July 1985 (late-wet and early dry season). Number of scan records and creeks (lines) are also shown. (B) Seasonal variations in percentages of occupancy of the area used by the capuchins between September and November (late-dry season). Number of scan records and creeks (lines) are also shown.

sources, such as *Maximiliana maripa* (ripe fruits), *Cupania* sp. (Sapindaceae, ripe fruits), *Goupia glabra* (Celastraceae; ripe fruits), *Oenocarpus bataua* (immature seeds and ripe fruits), *Micrandropsis scleroxylon* (Euphorbiaceae, immature seeds), *Ptycho-petalum olacoides* (Olacaceae, ripe fruits), *Duguetia* sp. (Annonaceae, immature seeds), *Norantea guianensis* (Marcgraviaceae, nectar), *Pouteria oblanceolata* and *P. fulva*

(Sapotaceae, ripe fruits), and *Dicranostylis holostyla* (Convolvulaceae, ripe fruits; table 22.1). Furthermore, the home range used by the study group varied monthly according to the kind of food sources consumed. In the late wet and early dry seasons (between May and July) the capuchins relied intensively on palm fruits, and the group spent more time in restricted areas using fruits of *Max-imiliana* sp. and *Oenocarpus* sp. in riverine

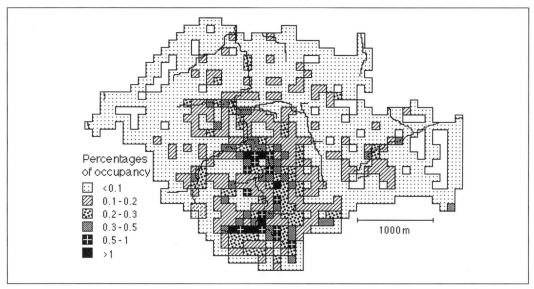

Fig. 22.7. The total occupancy of the study area used by the capuchin group during 13 months. Different intensities correspond to percentages within the total observation, and white areas represented squares not visited by the group.

areas. The quadrats with heavy use (more than 2 percent of recorded scans [fig. 22.6a]) outside of the riverine forests frequently contained these two palm species. The late dry and early-wet seasons (between September and November) were periods of scarcity of ripe and palm fruits, during which capuchins spread out and used marginal areas quite often (fig. 22.6b).

Home range size and distance traveled daily were positively correlated with consumption of seeds and flowers and negatively correlated to both foraging on animal matter and availability of food sources (table 22.2). Foraging on animal matter, however, was not correlated with specific habitats in the forest but was related to grids with twelve major plant sources visited by capuchins during the year ($r_s = 0.71$, $P < 0.001$, $n = 56$). The proportion of total occupancy of the area visited by the group in squares with roosting trees was also correlated with distribution of preferred shelters in riverine forests ($r_s = 0.39$, $P < 0.01$, $n = 42$). In addition, the proportion of the occupancy during

the year showed that the capuchins used more intensively (over 0.2 percent of scans) squares located in central and southern parts of the area (fig. 22.7), which seemed to be the group's core area.

OVERLAP BETWEEN GROUPS

Five other groups shared the area with the resident group and were observed on twenty-eight occasions. The minimum group sizes estimated for these additional groups were: 20 (group a), 15 (group b), 7 (group c), 9 (group d), and 6 (group e) individuals (fig. 22.8). The mean group size overall of the six groups in the study area was estimated as 11 (SD = 5.5) individuals. Although the focal capuchin group used 915 ha, the total grid system in continuous forest accounted for up to 1,082 ha. As five other groups overlapped their home range with the resident group, it seems that at least 66 monkeys were recorded in the entire gridded study area. Nevertheless, these groups surely used areas outside the grid system, and much more data are

TABLE 22.1. Spearman Rank-Order Correlation Analysis Between the Time Capuchins Spent Feeding and Percentages of Occupancy in Squares During Different Months

Month	Species	Feeding group (percentage)	Percentage occupancy	No. plants	r_s	n (squares)	P
May '85	*Maximiliana maripa* (Palmae)	18.3	22.4	1	—	1	—
June	*Maximiliana maripa*	9.2	14.7	4	—	3	—
	Cupania sp. (Sapindaceae)	8.4	15.6	5	—	5	—
	Oenocarpus bataua (Palmae)	19.4	29.6	34	0.79	27	0.001
July	*Maximiliana maripa*	15.7	19.3	3	—	—	—
	Oenocarpus bataua	10.6	58.5	77	0.5	29	0.010
	Goupia glabra (Celastraceae)	4.6	6.8	7	—	4	—
Aug.	*Maximiliana maripa*	12.0	13.1	3	—	3	—
	Oenocarpus bataua	13.4	34.4	92	0.89	27	0.001
	Micrandropsis scleroxylon (Euphorbiaceae)	23.7	54.7	94	0.63	56	0.001
Sept.	*Micrandropsis scleroxylon*	59.5	78.5	110	0.66	65	0.001
	Ptychopetalum olacoides (Olacaceae)	10.9	25.4	14	0.71	13	0.010
Oct.	*Micrandropsis scleroxylon*	10.2	23.3	30	0.39	26	0.050
Dec.	*Duguetia* sp. (Annonaceae)	9.6	20.5	25	0.64	22	0.002
	Pouteria oblanceolata (Sapotaceae)	6.6	11.8	12	0.91	12	0.001
Jan.	*Norantea guianensis* (Margraviaceae)	8.8	17.7	17	0.78	15	0.001
Feb.	*Norantea guianensis* var. *gracilis*	12.9	35.3	33	0.72	31	0.001
April	*Pouteria* cf. *collina* (Sapotaceae)	10.9	30.6	43	0.41	41	0.010
May '86	*Pouteria* cf. *collina*	10.8	29.3	18	0.77	14	0.005
	Dicranostylis holostyla (Convolvulaceae)	8.7	24.5	18	0.57	18	0.020

Note: For small samples the analysis was excluded, and only feeding time and percentages of occupancy records are shown.

TABLE 22.2. Spearman Correlation Coefficient Between Food Items and Availability of Resources (plants visited/ha) in Relation to the Home Range and Daily Range Length Used by Capuchins Among the Thirteen Months

Food items	Home range		Daily range length	
	r_s	P	r_s	P
Seeds	0.61	0.05	0.60	0.05
Nectar	0.59	0.05	0.59	0.05
Animal matter	−0.66	0.02	−0.71	0.01
Availability of resources (Food plants visited / ha)	−0.66	0.02	−0.69	0.01

Note: n = 13 for all tests.

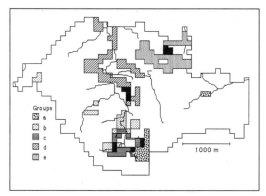

Fig. 22.8. Area of range overlap among the other capuchin groups present at the resident group home range; dark square represents overlapping areas. The minimum group size estimated for these groups were twenty (a), fifteen (b), seven (c), nine (d), and six (e).

needed to establish their home range size and areas of overlap with the core group.

In addition, only six encounters between groups were observed during the fourteen months of the main study. In three encounters the capuchin groups tolerated each other, giving only some alarm calls. The adult males of the resident group chased the others away in the three other encounters (once in the roosting area).

Discussion

Home Range Size

Depending on the method used to analyze home range, the area for animal groups may be overestimated, especially for species that use small areas (Rudran 1978; Olson 1986). Using smaller grid sizes (0.25 ha) at the periphery of the group's home range, my estimate of the home range decreased from 915 to 852 ha. On the other hand, the group visited these peripheral areas in a period of low ripe-fruit availability between October and November, when my observation effort was the lowest, so the 852 ha may actually underestimate the area used in that period. Support for this figure being an underestimate is that the home range estimated for the resident group in the following year reached 530 ha after 584 hours of observation, compared to the previous estimate of 915 ha calculated after 1,400 hours following the study group.

At other Amazonian sites (Manu National Park in eastern Peru, for example) the home ranges used by capuchin groups were estimated differently, with the areas ranging from 50 to 150 ha (Terborgh 1983; Terborgh and Janson 1983; Janson 1984; J. Robinson and Janson 1987). Low observation effort, season of the study, and annual changes in fruit availability, as well as the home range analysis, may explain some of the difference between the home ranges estimated for these primate groups, but it is likely that there are real ecological differences between the study sites that result in large differences between area requirements by the same species in different parts of their range.

Overall density of primates in BDFFP reserves is one of the lowest documented in Amazonia to date (Rylands and Keuroghlian 1988). At the Manaus site a group of bearded saki monkeys (*Chiropotes satanas chiropotes*) also may use large areas of forest (about 1,000 ha; E. R. Frazão, pers. comm.), whereas the areas estimated for bearded sakis in other localities were 200 and 350 ha, respectively (Ayres 1981; van Roosmalen, Mittemeier, and Milton 1981). The large home range used by the resident group of capuchins in this study (c. 800–900 ha) also indicated that capuchins are not in high density (1.6 individuals/km²) in this area. But because other capuchin troops seemed to include the core study group in substantial parts of their range, group requirements seem to be smaller at the community level. This is supported by densities of 2.2 ind/km² for capuchins recorded in the

BDFFP forest areas by Rylands and Keu-roghlian (1988).

Ranging Patterns

The distribution of preferred roosting shelters, especially *O. bataua* trees, was correlated with the ranging pattern of the capuchins, and the strong selectivity for specific shelters seems to be a predator avoidance, or refuge, strategy. However, the presence of food sources associated with the roosting sites suggests that a feeding strategy cannot be excluded (Spironello, unpublished data). Capuchin groups also regularly used these palms as roosting trees in French Guiana, where Zhang (1995b) suggested the palms were chosen for predator avoidance as well as for comfort and social contact.

The ranging patterns of capuchins also were related to location of the preferred food plants, but some of these resources were rare species with irregular fruiting periods, such as *M. maripa* (Palmae) and *P. fulva* (Sapotaceae). Moreover, even seasonal plants changed their fruiting period in different years. *Micrandropsis scleroxylon,* for example, fruited between August and October 1985 but fruited three months later in 1986. Fruiting of *P. oblanceolata* peaked in November and December 1985, whereas this species was recorded fruiting mainly in March 1996–97 (Spironello 1999). Also, the number of fruiting trees within the same species population varied among years. *Micropholis guyanensis* and *M. g. duckeana* (Sapotaceae), for example, registered a lower density of fruiting trees in 1995–96, respectively 0.1 to 1.2 trees/ha, than in 1996–97, respectively 0.3 to 1.5 trees/ha (Spironello 1999). As a result, ranging patterns appeared to vary according to the availability of resources in a specific area among years.

The study group showed different strategies using the area when fruits were scarcer.

During late wet and early dry seasons, for example, the group used few individuals of the rare plants and plants with a clumped distribution. As a result they traveled less and used smaller areas than when they consumed mainly immature seeds and nectar, which include abundant and evenly distributed species, respectively. Similar ranging behavior was documented for capuchins during periods of low fruit availability in French Guiana (Zhang 1995a). In two Amazonian sites, seeds and flowers were secondary food sources for some primate species that fed mainly on ripe fruits (Terborgh 1983; Peres 1994b). In fact, *Cebus* species adopt different strategies, decreasing or increasing their daily home range, to exploit the food sources during periods of fruit shortage (J. Robinson 1986; Izar 1999).

The availability of insects in some forest types has been reported to be high during the wet season (Charles-Dominique et al. 1981; Pearson and Derr 1986). Although capuchins foraged for animal matter more in the late wet season, they also consumed similar amounts during the early dry season —between June and July. Foraging was related to daily range length, but the availability of insects did not seem to affect the ranging patterns of capuchins in the study area, even though they spent more time looking for invertebrates in areas around plant resources. In both Guyana and Brazil's Atlantic forests, invertebrate distributions or availability also appeared not to affect the ranging of this species (Zhang 1995a; Izar 1999). Probably terra firme forests are poorer in availability of large insects than are other forest types, such as flooded forests. Among flooded forests the mixed-water inundation (white and black) forest is the richest in species of canopy insects, whereas the white-water (*varzea*) seasonally flooded forest is richest in numbers of individuals (Erwin 1983). In fact, although terra firme forest has a high diversity of insects, they are on average less diverse than are flooded forests (Erwin 1983).

Comparison with Other Amazonian Sites

Foodplain forests in Manu, Peru, yield more edible fruits than do terra firme forests, and palms composed 60 percent of fruit productivity in this forest type (Phillips 1993). At Manu the high diversity and abundance of mammals in an alluvial forest on richer soils, for example, may be related to the high annual production of fruits and seeds and high habitat diversity (Janson and Emmons 1990). The low density of small animals and primates at Manaus, therefore, could be related to the low fruit availability on poorer soils (Emmons 1984; Gentry and Emmons 1987). Thus, terra firme forests appear to be poorer in fruits than are habitats with natural disturbance, such as lowland alluvial and flooded forests on rich soils (Ayres 1989; Peres 1991, 1994a; Phillips 1993), and likely also poorer than mosaic terra firme forests in Guyana, where capuchins used smaller home ranges and traveled between 42 percent and 63 percent less than at Manaus in periods of low availability of fruits (see Zhang 1994, 1995a).

The larger home range used by capuchins at Manaus suggests that this forest type is poorer in food sources than are other Amazonian sites; the composition of the forest tree species at the family level, as well as some environmental parameters, also seem to differ between the sites where capuchins have been studied (table 22.3). The study site home range was even larger than that found for other subspecies and species studied elsewhere (tables 22.3, 22.4). In the Atlantic forest, capuchins (*Cebus apella nigritus*) also used larger home ranges than did the brown capuchins (*Cebus apella apella*)

in the Manu and Nouragues sites (Izar 1999). Izar pointed out that that variation was explained by lower fruit productivity in the Atlantic Forest.

An additional factor governing home range for primates is the availability of key food plants (Terborgh 1986). The relatively small home range used by capuchins at Manu and Urucu forests is likely explained by a high density of palms (see Terborgh 1983; Peres 1994a). In the BDFFP reserve, palm fruits most consumed by capuchins had lower density than did palms used by them in other localities (Spironello 1991). The low abundance of palms in the area, particularly of *O. bataua,* may also contribute to the low density of capuchins north of Manaus (Spironello 1991). In fact, palm productivity at the BDFFP study site was roughly sixty times less than in Urucu terra firme forests (Spironello 1999). Therefore, long-term studies focusing on fruit production and including different microhabitats in the forest are needed to explain such variations in primate densities among forest areas in Amazonia. For example, Izar (1999) suggested that fruit productivity during the periods of fruit shortage and not total annual fruit production is related to home range requirement for *Cebus apella.* In Atlantic forest areas capuchins used larger forest areas than recorded anywhere other than this study (Rímoli and Ferrari 1997; Izar 1999.)

Land-Use Patterns and Their Implications for Other Primate Species

North of Manaus the forest has been cut down to allow for the expansion of agricultural projects, cattle ranching, logging operations, and colonization, creating many small forest fragments in what will likely be a model for the future of much of the currently extensive rainforest of the region. Most, but not all, of the fragments left in the

TABLE 22.3. Home Range Used by the Brown Capuchin Monkeys (*Cebus apella*) in Different Amazonian and Brazilian Southeastern Sites in Comparison to Ecological Parameters

Study site	Group size	Home range (ha)	Area/ monkey (ha)	Period (month)	Forest type	Top families	Soils fertility	Rainfall (mm)	Sources
Colombia									
La Macarena	16	>260	>16	2	Lowland (mosaic, with marked dry season)	Moraceae, Leguminosae, and Palmae	Moderate to rich	—	Izawa 1980
Peru									
Cosha Cashu	10	80–15	15	9	Lowland alluvial (mosaic)	Leguminosae, Moraceae, Sapotacae, Annonaceae, Lauraceae, and Bombacaceae	Moderate to rich	2000	Terborgh 1983; Janson 1984; Foster 1990
French Guiana									
Nouragues	15	350	23	8	Terra firme (mosaic)	Caesalpinaceae, Lecythidaceae, Sapotaceae, Chrysobalanaceae, and Burseraceae	Moderate	3000	Sabatier 1985; Sabatier and Prévost 1990; Zhang 1995a
Brazil									
Amazonian sites									
BDFFP	14	850	60	27	Terra firme	Lechythidaceae, Leguminosea, Spaotaceae, Burseraceae, Violaceae, and Annonaceae	Poor	2670	This study
Urucu	16	>250	>16	20	Terra firme	Lecythidaceae, Sapotaceae, Meristicaceae, Moraceae, Chrysobalanaceae, and Palmae	Poor	3200	Peres 1993
Southeastern sites									
Caratinga[1]	27	400	15	12	Atlantic Forest (with marked dry season)	—	Rich	1250	Mendes 1989; Rímoli and Ferrari 1997
Saibatela[2]	15	465	31	32	Coastal Atlantic Forest (very humidity)	Myrtaceae, Leguminosae, Rubiaceae, Lauraceae, Moraceae, and Euphorbiaceae	Moderate	4200	Galetti 1996; Izar 1999

Notes: [1]*Cebus apella nigritus* (fragmented forest, Minas Gerais). [2]*Cebus apella nigritus* (continuous forest São Paulo).

TABLE 22.4.   Home Range Use by *Cebus* Species in Different Localities

Species	Group size (ha)	Home range (ha)	Area/ monkey (ha)	Period (months)	Country	Sources
Cebus albifrons	35	115	3.3	6	Columbia	Deflero 1979
	9	>150	16	7	Peru	Terborgh 1983
Cebus capucinus	15	80	5	18	Panama	Freese and Oppenheimer 1981
	20	50	2.5	15	Costa Rica	Freese 1976
Cebus olivaceus	20	260	13	14	Venezuela	J. Robinson 1986

modified landscape are small, and some do not include creek-side forest. Even some larger fragments do not include streams. Although some primate species, such as white-faced saki (*Pithecia pithecia*) and red howling monkeys (*Alouatta seniculus*), have been surviving in fragments of 10 ha or smaller (Neves 1985; Neves and Rylands 1991; Lovejoy, Bierregaard, et al. 1986; Setz 1992, 1993; Chapter 21), this study indicated that brown capuchin monkeys need much larger home ranges in central Amazonian terra firme forests. Additionally, creek-side forests were shown to be important feeding and roosting areas for this species.

Franklin (1980) suggested that to maintain the evolutionary potential of populations a reasonably effective population size should be 500 individuals. To reach this value, at least 46 groups should be necessary to maintain viable populations of brown capuchins north of Manaus, and this in turn would require a forest area of 23,000 ha, based on Ryland and Keuroghlian's (1988) estimate of 2.2 ind./km².

Conservation Lessons

1. The minimum required areas for primate species vary dramatically from one region of the Amazon to another, with the largest areas needed on the poor soil areas of the central Amazon.

2. Future plans to maintain viable brown capuchin populations in central Amazonian terra firme forest therefore should be no smaller than 23,000 ha and include swampy palm-forest areas.

Acknowledgments

I thank my field assistants for their essential help during the development of the study, especially José Eremildes G. da Silva and Alfredo F. Cotta. I also thank Drs. T. E. Lovejoy, A. B. Rylands, R. A. Mittermeier, R. O. Bierregaard Jr., and L. C. de Miranda Joels. My thanks also to Dr. M. G. M. van Roosmalen for identification of fruits, and D. J. Chivers and V. Kapos for useful comments on the manuscript. I am indebted to the Biological Dynamics of Fragments Forest Project, the Instituto Nacional de Pesquisas da Amazônia, the World Wildlife Fund–US, and the Smithsonian Institution for logistic support and financial assistance.

PART IV
Management Guidelines

Introduction

In this concluding section of our book, we present six chapters that offer important data, guidelines, and philosophical perspectives for the planning and management of developed areas in Amazonia. The perspectives of the chapters runs the gamut from analyses of variation in soil quality at a micro-habitat scale to the interpretation of remotely sensed satellite data and include studies of the abandoned pastures that surround the BDFFP experimental forest isolates.

SOIL AND DEVELOPMENT IN AMAZONIA

Perhaps the greatest paradox of tropical rainforests is that over much of its expanse, the world's richest terrestrial ecosystem has its roots in some of the world's poorest soils. Looking at the verdant expanse of forest that carpets equatorial South America from the Andes to the Atlantic, it is hard to imagine that the soils underlying all that green could be anything but extremely fertile. In fact Brazil's resettlement program of the 1970s, based around the Transamazonia highway system, was predicated on such an erroneous assumption. The true nature of the soils of the region doomed the project from the outset, and the forest has now reclaimed

great, abandoned stretches of this highway system.

In Chapter 23, Philip M. Fearnside and Niwton Leal Filho present a rather encyclopedic primer on tropical soils while describing the soils of the roughly 1,000 km² BDFFP study area in detail. Sections of this chapter will be a bit intimidating for those whose last chemistry class is a fading memory, but the broad perspectives offered into a topic so fundamental to the ecology and conservation of the region make this a most important chapter. The soils underlying the BDFFP study area are typical of vast areas of Brazilian Amazonia that are going to be developed for forestry and agriculture. Fearnside and Leal Filho's results, therefore, have clear regional importance.

Soils underlying much of the region have low fertility, are acid, and have high levels of toxic aluminum ions. Steep topography, generally very high clay content, and low values of available water capacity mean that the agricultural potential over much of Amazonia is low and the prospects for resilience are not bright given the reality of changing land-use patterns in the region.

Understanding the soils under the forest will permit planners to predict the environmental impact of converting these areas to pasture and other uses, including the potential for release of greenhouse gases and the probable response of forest to fragmentation and climatic change. Fearnside and Leal Filho conclude that most of the Amazon would produce little (as has most of the area that has already been cleared) if converted to agriculture or ranching, and they argue for land uses that maintain forest cover as intact as possible. They point out that the value of the environmental services provided by the standing forest, hard as this is to quantify, surely far exceeds the returns that can be expected from clearing it for agriculture or ranching, a conclusion echoed by Niro Higuchi in Chapter 26.

Managing the Matrix

Chapter 24 focuses strictly on the "matrix" habitat around the BDFFP. Because plants are sessile organisms, immigration into a forest fragment depends either on seed dispersal across the entire matrix habitat surrounding the fragment, or a reproductively successful plant establishing itself in the matrix close enough to the fragment that the next seed-dispersal event can reach the fragment.

Gislene Ganade reports here on her experiments with seedling establishment in abandoned pastures where regrowth has already begun. She first studied the "seed rain" that falls in regrowth areas (where pioneer species have already established a canopy), showing that fewer seeds of fewer species make it into abandoned pastures than under the mature forest. However, for seeds that do arrive in the regrowth areas, her experiments showed that germination and seedling growth rates are substantially higher than in primary forest. Ganade's work strongly suggests that mostly because of increased light availability and decreased herbivore pressure, the conditions in areas of second growth are favorable for the establishment of native tree species. It appears

that it is seed dispersal that limits the rate of regeneration of primary forest species on abandoned pasture land.

In the central Amazon, as opposed to the southeastern region in the states of Pará and Maranhão, much of the land that has been cleared in the past two or three decades was abandoned after only light use and has reverted quickly to second-growth forests. Ganade shows how these large areas can be managed to provide an income stream for the landowners while serving as a genetic bridge between plant populations in a fragmented landscape. Simply scattering seeds of species that provide valuable woods, fruits, and oils, for example, under the canopy of the pioneer trees in abandoned pastures can turn these areas into economically productive property. Most importantly, it serves as an alternative to re-clearing these areas in an attempt to reestablish pasture. Such a scenario can be managed at the landscape level to promote conservation while providing income for landowners. Areas of second growth targeted for this enrichment process would be chosen around existing fragments to serve as buffers against edge effects in primary forest remnants and could provide corridors for faunal dispersal across the landscape mosaic.

Fire and Rainforest Regeneration

Natural fires exert an influence on the primary rainforests of central Amazonia only under extreme and very rare climatic conditions. In contrast, in the hand of Man, fire is an essential tool in converting primary forest to agriculture or pasture land. It both clears the felled canopy trees from the ground as well as releasesing essential nutrients into the soil. G. Bruce Williamson and Rita C. G. Mesquita, in Chapter 25, describe the ways fire affects second-growth vegetation and suggest that, in certain circumstances, it can be a tool for managers attempting to foster succession and species richness in abandoned farm and ranch land.

Trees of the primary rainforest have none of the adaptations commonly found in forests that are regularly subjected to fire. Williamson and Mesquita show that species of *Vismia,* one of the two common genera of pioneer trees that invade abandoned pasture, can survive moderate levels of burning, whereas species in the other pioneer genus, *Cecropia,* are killed in areas where fire is repeatedly used to maintain pastures. *Vismia* species, in contrast, can resprout from the base of the stem and in some species at least from lateral roots (see Chapter 24). Despite the appearance that *Vismia* spp. "thrive" in burn-managed pastures, the authors show that it is, at best, surviving and will disappear if the frequency of fire and weeding is sufficiently high.

The authors cautiously report that a moderate to low intensity burn in an area where *Vismia* spp. are becoming established can increase the diversity of the successional flora after the fire. The larger shoots of *Vismia* spp. will be burned and replaced by smaller shoots from lateral roots or the base of the stem. If the fire was not so hot that it killed the seeds in the soil, the removal of the *Vismia* spp. canopy provides an opportunity for a number of other pioneer and early second-growth species to become established.

Williamson and Mesquita provide two prescriptions for accelerating the rate of succession and steering it to higher diversity. If a commitment is made and action taken with one or two years of pasture abandonment, seeds of several important successional species can be sown and trees planted to provide focal areas for seed dispersal by bats and birds, as also described by Ganade (Chapter 24). In the case of areas that have been abandoned for three or more years, a prescribed burn can, under the appropriate conditions, foster succession and increase the diversity of the second-growth flora. This recommendation is proffered guardedly, because there are a number of scenarios under which a prescribed burn could have a negative effect. If the burn is too hot or the area burned is too large, the fire can impede the process of succession in the abandoned areas, and may escape the prescribed area and damage more established second-growth areas and the edges of primary forest adjoining the abandoned pasture.

Logging in Amazonia

As Jay R. Malcolm discusses in Chapter 27, forest fragmentation can occur at vastly different scales, from enormous areas cleared by industrial-scale agribusinesses to selective logging, where the forest might be better described as "perforated" rather than fragmented. In Chapter 26, Niro Higuchi presents a broad perspective on the past, present, and future of logging in the Amazon and argues that selective logging can be economically productive without causing irreparable ecological damage.

Throughout the 1970s and 1980s, when the Brazilian government was subsidizing cattle ranching, mining, and hydroelectric projects, timber operations were a by-product of these activities. In the 1990s, the scenario has switched, with few or no subsidies going to large-scale agricultural projects. With an increase in the market value of Amazonian hardwoods, timber operations are now subsidizing clearing for ranching, especially in southeastern reaches of the region (the Brazilian states of Pará and Maranhão). Additionally, international timber companies, seeing an end of harvestable forests (as well as their welcome in some areas) in sight in southeast Asia, are looking to the vast expanses of Amazonia for their next harvest of hardwoods. The vastly untouched state of Amazonas is a particularly attractive target. Amazonia has until now provided a mere 10 percent of the worldwide market, but this is likely to change soon, such that Amazonia will be the biggest supplier of hardwoods on the international market in this century.

Recent assessments of current logging practices suggest that across the Amazon basin a combination of poorly trained professionals, lack of infrastructure, and regional political infighting have led to a completely chaotic situation. As Higuchi points out, there are adequate laws and regulations in place as well as the technological knowledge to extract timber through selective logging from the region in a sustainable manner. If all the timbering projects that have been developed claiming to use sustainable harvesting techniques had indeed done so, we would now be getting most of our tropical hardwoods from second-growth forests.

Higuchi argues for a regional ecological and economic zoning of the region, which would have to include an accurate assessment of the true costs and benefits of any planned extractive activities. His "bottom line" assessment of the 50 million ha that have already been deforested in Amazonia is that the region has been impoverished and needlessly lost vast natural resources.

MODELING EDGE EFFECTS IN
LANDSCAPE MOSAICS

Chapter 27, by Jay R. Malcolm, combines spatial models of edge effects at a landscape scale across vastly different scales of disturbance with a test case from Amazonia based on bird studies in the BDFFP study areas and a selectively logged area in French Guiana. In this chapter, Malcolm discusses

how edge and island biogeography effects can be confused, showing that there are situations in which one or the other will be irrelevant; when the extent of deforestation is vast, edge effects are probably insignificant, whereas in a selectively logged area, where connectivity between remnant patches is high, the now traditional island biogeography model is unlikely to be informative because colonization of patches would be an everyday event.

Previously, Malcolm (1994) had developed a model that accounted for edge effects coming from not just a point at the edge, but from the whole reach of the edge within a certain distance of the point in question. The model therefore was able to predict that a point near a corner of a forest fragment would be more severely affected than one near a long, linear edge, because of the additivity of edge effects. In the model presented here, Malcolm extends this concept into the clearing itself, arguing that strength of edge effects will be related to the extent of the clearing around the forest in question. He then models the physical process of heat flow in a fragmented landscape, arguing that many of the edge effects we have measured in forest fragments (Kapos et al. 1997) are associated with an increase in temperature and that heat flow may be an apt analogy for many biological processes that permeate a fragmented landscape. One of the intriguing results of the model is that the most profound effect on the environment comes at the very first stages of deforestation, as seen for example in a selectively logged forest, but at the same time, these habitats are still more similar to continuous forest than those where the "grain" of deforestation is coarser.

Although this chapter may be challenging for readers without a strong theoretical background, the insights it provides into habitat fragmentation are worth the extra effort it may take to digest. Malcolm takes us beyond simplistic models of edge effects and metapopulations (sets of relatively isolated populations that exchange genetic material as defined by Levins [1970]), combines the two, and shows us how these concepts are both important and tractable in the field.

REMOTE SENSING OF THE LANDSCAPE

The enormous extent of the Amazon has until now been its greatest protection. Across the basin, deforestation has been going on with few restraints and often with governmental incentives or encouragement. Despite this voracious and decades-long assault, roughly 85 percent of the forest remains at least structurally intact. Those who predicted in the 1970s that the Amazon rainforests would be gone by the year 2000 did not understand just how hard it is to even get to most of the basin, much less cut it all down.

Just as its very vastness and remoteness have protected it in the past, these characteristics will pose substantial problems as the nations of the Amazonian basin attempt to monitor logging and mineral concessions as well as colonization projects, and prevent intrusions into their national parks and other protected areas, many of which have only one or two guards responsible for hundreds of thousands of hectares of forest. Recent advances in our ability to interpret remotely sensed data (mostly from satellites) have provided tools with which even the most distant corners of Amazonia can be monitored and offer a solution to the most substantial logistical problem confronting those who would develop and protect the region in a sustainable fashion.

Because we have documented the changes in land cover and land use in a 1,000 km² landscape over nearly two decades, the BDFFP site provides an invaluable opportunity to fine-tune remote sensing analyses. Miles Logsdon, Valerie Kapos, and John B. Adams have worked with data covering the BDFFP study area for almost a decade. In Chapter 28, our last data-based chapter, they describe these new analytical techniques and discuss how they can be applied across the region.

Using our land-use data to bench-mark their analyses, the authors develop a series of "metrics" that reflect the degree to which a landscape has been fragmented. They used images of the study area taken in four different years and were able to quantify changes in the number of patches, amount of edge and core-habitat area, and the degree to which different habitat types have become juxtaposed across the landscape. The number of patches decreased and their average size increased, while overall habitat diversity decreased. All of these changes are associated with the abandonment of pastures and subsequent second-growth forests that have developed in their place. Logsdon and his colleagues were also able to monitor changes in different land-cover types, such as primary forest, pasture, or recent slash. Their results reflect the dynamic nature of a fragmented landscape that has become an unexpected, but key, element in research in the BDFFP forest fragments.

CHAPTER 23

Soil and Development in Amazonia

Lessons from the Biological Dynamics of Forest Fragments Project

PHILIP M. FEARNSIDE AND NIWTON LEAL FILHO

Soil quality is obviously fundamental to the potential of any area to produce and sustain agricultural yields. If planning decisions are made to encourage agriculture in areas with soils that are not suited to this purpose, then crops can be expected to fail. Planning decisions opening vast areas to settlement are frequently made in a near vacuum of information on soils. For example, when the POLONOROESTE Project opened the state of Rondônia to settlement through a World Bank–financed regional development program centered around paving the BR-364 (Cuiabá–Porto Velho) Highway, the soils information available for Rondônia (RADAMBRASIL 1978, vol. 16) was based on only eighty-five soil cores. Similar, ill-founded projects are recurring today quite near the BDFFP reserves; on June 24, 1997, president Fernando Henrique Cardoso announced in his weekly radio program "Palavra do Presidente" that 6 million hectares along the BR-174 (Manaus-Caracaraí) Highway would be opened to settlement, and he suggested that the area farmed there would be "so colossal that it would double the nation's agricultural production" (de Cássia 1997). Despite almost certain hyperbole in both expected production and area likely to be settled, the intention of initiating a major

program on the BR-174 Highway appears to be real. The announcement of the BR-174 settlement program came as a surprise, as the paving of the highway (in 1996 and 1997) had been presented as a surgical cut through the forest that would allow the city of Manaus to trade with Venezuela and have access to that country's ports.

Neither studies of soils and agronomic potential nor assessment of environmental impact were done prior to the announcement of the BR-174 agricultural project. As is true for most of Brazilian Amazonia, information on soils used in planning is essentially restricted to the results of the RADAMBRASIL Project, which mapped soil, vegetation, and other features based on side-looking airborne radar (SLAR) imagery in the early 1970s (RADAMBRASIL 1976, vol. 10; 1978, vol. 18). Original imagery was at a scale of 1:250,000 and was published at a scale of 1:1,000,000. Areas with the same appearance on the images were mapped as the same, with field checking to identify vegetation and soil in many of the units. The BR-174 area has been identified by the Amazonas state government as a priority for economic-ecological zoning (Pinheiro 1997).

The portion of the BR-174 to be opened

This is publication number 290 in the BDFFP Technical Series.

for agriculture lies over the Guianan Shield and can therefore be expected to be more fertile than those in the BDFFP reserves. However, the Landless Rural Workers' Movement (MST) has invaded some ranches in the SUFRAMA (Manaus Free Trade Zone Superintendency) Agriculture and Ranching District (not those under study by the BDFFP), raising the possibility that some of these ranches may be distributed to small farmers for agriculture (Pacífico 1997).

Soil Classification

The BDFFP reserves are located 80 km north of Manaus (see fig. 4.1). The site is approximately 50 to 100 m above mean sea level; mean annual temperature at Manaus is 26.70 C and mean annual rainfall (30-year average) is 2,186 mm with a three-month dry season lasting from July to September (Lovejoy and Bierregaard 1990; see fig. 5.1). The site of the BDFFP reserves is typical of much of central Amazonia.

The RADAMBRASIL maps classify the BDFFP reserves as allic yellow latosols, which are Allic Haplorthoxes (Oxisols) in the U.S. soil taxonomy and haplic or xanthic Ferralsols in the FAO/UNESCO system (see Beinroth 1975). The classification as a latosol relates to the type of clay minerals present.[1] The relative amounts of clay minerals composed of silicate (kaolinite), iron (goethite), and aluminum (gibbsite) determine the structural stability, the natural fertility, and the effect of applying fertilizers (Sombroek 1966).

Oxisols are the most common soil in the Amazon basin, covering 220 million ha, or 45.5 percent of the total (including areas outside of Brazil); most of the remainder is covered by Ultisols (such as the red-yellow podzolic soils of the Brazilian system), covering 142 million ha, or 29.4 percent (Cochrane and Sanchez 1982). Oxisols are distin-

guished from Ultisols by lack of higher clay content with depth; Ultisols have at least a 20 percent increase in clay content in the lower (B) horizon (U.S. Department of Agriculture, Soil Survey Staff 1975). At a high level of generality, Ultisols are considered less appropriate than Oxisols for mechanized agriculture due to susceptibility to soil compaction and their frequent occurrence on more steeply sloping terrain (Sanchez 1977). However, the broad range of characters within either of these great groups (as illustrated, for example, by the results of the present study) indicate the need for caution in applying such generalizations to specific management decisions (see Fearnside 1984).

Oxisols and Ultisols (i.e., yellow latosols and red-yellow podzolics) often occur in close proximity in Amazonia and frequently intergrade with each other (Sombroek 1966). Profiles in nearby sites confirm this (Ranzani 1980; Chauvel 1983). However, since both Oxisols and Ultisols are similar in being acid, infertile soils that are indistinguishable from each other without information on differences in the B horizon (which is outside the rooting zone of most crops), taxonomic differences between them generally have little relevance for agriculture as practiced in Amazonia.

The BDFFP reserves are located about 50 km south of the edge of the Guianan Shield (the Balbina Hydroelectric Dam is located at the edge of the Shield). The reserves are within the basin bounded by the Brazilian and Guianan Shields. This basin covers approximately 1.2×10^6 km^2 in Brazil, or about 25 percent of Brazil's 5×10^6 km^2 "Legal Amazon" region. It is known as the Alter do Chão Formation (formerly called the Barreiras Formation). The soils are derived from sedimentary deposits from the bottom of a shallow sea that occupied the center of the Amazon Basin during the Tertiary (Falesi

1974; Jordan 1985). The soils derived from these sediments have been exposed to heavy rainfall and high temperature over most of the approximately 60 million years since the region was uplifted at the time that the Andes mountains were formed; most nutrients have therefore been leached out of the soil (Sombroek 1984). Younger soils, such as those derived from igneous rocks in the Guianan and Brazilian Shields, have higher fertility than those of the BDFFP area, although they too are far from being classed as fertile for agriculture.

The general characteristics of the origin of soils are closely associated with their overall fertility. In the case of the BDFFP area soils, this history rules out occurrence of high-fertility soils, with the exception of patches of anthropogenic black soil (*terra preta do índio*), which have not been found in the project area. However, great variation in some characters occurs within an area, such as the BDFFP reserves, that appears to be uniform at the scale considered by the RADAMBRASIL survey (the maximum level of detail that can presently enter land-use planning decisions); in reality, not even this level of detail is considered when many actual decisions are taken. Perhaps the clearest example of this was completely ignoring existing soil maps when government decisions were taken on the location of settlement areas in agriculturally unpromising areas of Rondônia (Fearnside 1986a).

Soil Variability

The BDFFP soil survey offers a unique opportunity to assess fine-scale variability in Amazonian soils and the potential significance of this variability for development planning. Fine-scale variability can be expected to affect agricultural success; it also affects the natural vegetation, both in influencing what trees (and thereby other life

forms) occur and the stress placed on those trees when they find themselves isolated in forest fragments (see Chapters 9 and 13).

Variability in initial soil quality, as well as in other factors, is a key determinant of human carrying capacity (Fearnside 1986b). In the Transamazon Highway colonization area, for example, soils vary from fertility levels similar to those found in the BDFFP area to considerably higher levels in Alfisol (*terra roxa*) areas, as well as in much smaller patches of anthropogenic black soil (*terra preta do índio*) (Fearnside 1984). The occurrence of fertile patches, even if small in area, is important in the success of colonist agriculture, as is the ability of farmers to identify these patches by knowledge of the tree species that typically grow on them (Moran 1981).

Economic-Ecological Zoning

Brazil's 1988 constitution calls for zoning of the entire country, after which land-use decisions for each area would have to be made in accordance with the zone assigned to it. This provision was supported by the "environmental caucus" of constitutional delegates and was viewed as a great victory for increasing the amount of protection against excessive environmental impacts of development modes, such as those that result in tropical deforestation (see Chapter 3). However, the question of zoning soon became a controversial one, with those who implement zoning at the level of state governments often viewing it more as a means of "opening up" areas to development than as a means of containing development excesses (see Fearnside and Barbosa [1996] for examples from Roraima).

Following the October 1988 constitution, an inter-institutional struggle ensued for control of the zoning at the federal level; the principal contenders were the Brazilian Institute for Geography and Statistics (IBGE), the Brazilian Enterprise for Agriculture and Ranching Development (EMBRAPA), and the Ministry of Science and Technology (MCT) (Régis 1989). The matter was settled by a 1991 presidential decree giving responsibility to the Secretariat for Strategic Affairs (SAE), which had been known as the National Information Service (SNI)—a much-feared internal espionage agency. This institutional setting can potentially affect both the priority assigned to environmental concerns and the degree of popular participation in the zoning process.

The methods to be used in the zoning have been the subject of long controversies. A series of very general maps was prepared at a scale of 1:2.5 million using data from the RADAMBRASIL surveys; additional data collection was not undertaken for these "diagnostic" maps, but some additional data collection is foreseen in state-level zoning projects. The zoning projects and the strengthening of the state-level agencies that have begun to carry them out is being done with assistance from the G-7 Pilot Program to Conserve the Brazilian Rainforest (with funds administered by the World Bank).

Increasing the level of detail of data on characteristics such as vegetation and soils is obviously important to increase the degree of confidence in zoning conclusions. It is also important to achieve a better understanding of how to scale down from general zoning maps to the fine-scale variation that exists on the ground, where development actually takes place. One approach proposed to interpreting soils information in conjunction with information on vegetation, topography, and other features is the "land units" approach (Sombroek 1966; see Sombroek et al. 1999). Land units use naturally co-occurring sets of these characters as the basis for defining the categories into which the landscape is divided for purposes of planning.

Importance of Soil Properties

SOIL TEXTURE

One of the most important characteristics of the soil is its texture—for example, the balance between sand and clay fractions. Very sandy soils are poor for plant growth because they lack sites (provided by clay particles) for holding cations (positively charged molecules such as Ca^{2+}, Mg^{2+}, K^+, and Na^+, which plants need for nutrition). Consequently, cations are leached away, leaving the soil infertile, and because sandy soils do not hold water well, they expose plants to drought stress during dry periods. Clays have more sites for cations, largely because clay content is positively correlated with organic matter, when comparing surface samples at different locations (as distinct from comparisons made between different depths in the soil profile). Clay also holds cations directly, independent of its effect on organic matter. Organic matter is especially important in contributing to cation exchange capacity (CEC) in soils with clay minerals of low activity, such as the Oxisols and Ultisols of Amazonia (Lenthe 1991). Cation exchange capacity is a measure of the negatively charged sites on the surfaces of clay and humus molecules; these sites may be occupied by either nutrients (Ca^{2+}, Mg^{2+}, K^+, and Na^+) or competing ions in acid soils (H^+, Al^{3+}, and Fe^{2+}, but only the first two of these are included in the CEC).

Clays also hold water better in the smaller pores in the soil matrix. However, very heavy clays can have disadvantages because they have a higher proportion of very small micropores from which removing water requires an undue exertion of force by the plant roots (i.e., water requires a tension of more than fifteen atmospheres to remove). The large percentage of water at tensions above fifteen atmospheres (bars) in very clayey soils, which can be as high as 30 percent by weight, has a positive role as well,

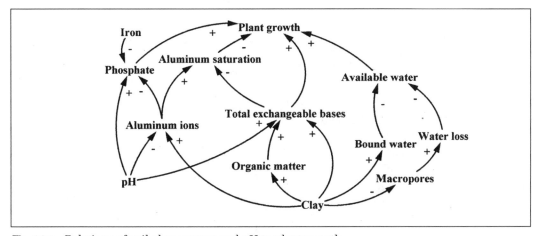

Fig. 23.1. Relations of soil clay content and pH to plant growth.

increasing the stability of organic matter, contributing to the positive relationship between clay content and organic matter (Bennema 1977). The relationship between clay and plant growth can be complicated by the effect of aluminum ions, which are toxic to most plants in high concentration. The allic (high aluminum) clays that are common in Amazonia lead to a strong positive correlation between clay content and aluminum. Because most of the effects of clay are beneficial to plants, one can expect a positive association between aluminum and plant growth even though the effect of aluminum per se is negative.

Excessively heavy clays can also pose a physical impediment to the growth of tree roots. Soils for citrus trees, for example, are considered best if they have a mix of sand and clay rather than being pure or almost pure clay. For cacao (*Theobroma cacao*), the best soils have 30–40 percent clay, 50 percent sand, and 10–20 percent silt (Smyth 1966, cited by Alvim 1977). The various effects of clay content on plant growth can be summarized in causal loop diagrams (fig. 23.1), where the sign at the head of each arrow indicates the direction of change in the quantity at the head of the arrow given an increase in the quantity at the tail of the arrow.

Clay content of the surface soil is relevant to the susceptibility of soils to erosion, particularly laminar (sheet) erosion (see Fearnside 1980a). Clay, being composed of particles of the smallest dimensions, is more easily carried away by runoff water than are coarser particles, such as sand. Clay can make the soil less permeable, leading to greater amounts of runoff for a given amount of rainfall, but in well-aggregated clay soils, such as those in the BDFFP area, drainage can be good despite high clay contents. In typical yellow latosols (Oxisols) the soil aggregates are not covered with silicate clay skins or linings, and the soils are little subject to gully erosion (Sombroek 1966). Oxisols with low iron content (a feature indicated by the yellow color of the BDFFP reserve soils) are the ones most susceptible to topsoil deterioration when exposed to sun and rain under agriculture, making them more susceptible to erosion when the surface becomes impermeable and runoff consequently increases (Bennema 1977).

AVAILABLE WATER CAPACITY

"Available water capacity," also known as "available water," is a measure of the amount of water that the soil can hold in a form that can be extracted by plant roots. It is calcu-

lated as the difference between field capacity (total moisture content of soil after gravity water has been allowed to drain away)
and wilting point (the moisture content at
which plants wilt and do not recover on
rewetting) (Young 1976). Water held in fine
pores cannot be extracted from the soil at the
pressures that plants are able to exert
through their roots. Field capacity is determined at 0.33 atmospheres (bars) of tension,
and wilting point at 15 bars.

For several reasons, the available water capacity calculated by this standard procedure
is really only an index of availability. One
reason is that the ability of plants to extract
water from the soil varies among species,
which may differ from the sunflower—the
plant taken as a standard by soil scientists.
A second is that the 0.33 bar pressure used
as a standard operational definition of field
capacity underestimates water holding capacity. A pressure of 0.1 bar is believed more
appropriate, meaning that the standard
method leads to an underestimation of
available water capacity by about 35 percent
(Sanchez 1976). A third is that intact soil
cores taken from the side of a soil pit are the
ideal material for the measurements, and results using dried material from traditional
soil samples (as in the present study) provide only an indication of these parameters
in the field.

Some indications exist that available
water capacity may pose a restriction on use
of water by the forest in soils similar to those
in the BDFFP reserves. A study of un-
deformed cores in INPA's "Model Basin"
(about 25 km southwest of the center of the
BDFFP study area) indicated that, while
pores make up 60 percent of the soil volume
at 10 cm depth, only 17 percent of the water
contained in the pores is available to plants,
while 50 percent is bound water and 33 percent is lost at a pressure less than 0.33 bar
and drains away by gravity (S. Ferreira
1997).

PHOSPHORUS

Phosphorus (P) is a limiting element for agricultural crops and cattle pasture in Brazilian
Amazonia. For example, P fertilization is the
key to increasing pasture grass growth in
technical packages formulated by EM-
BRAPA (e.g., Koster et al. 1977; Serrão et al.
1979). Phosphorus is low in virtually all
soils in Brazilian Amazonia, even including
relatively fertile ones, such as the terra roxa
(Alfisol) occurrences in settlement areas
along parts of the Transamazon Highway in
Pará and the BR-364 Highway in Rondônia
(see Fearnside 1984, 1986b). Furthermore,
the prospects are poor for maintaining vast
areas of agriculture dependent on phosphate
fertilizer in Amazonia because of limited deposits of rock phosphate in Brazil and
worldwide; virtually all of Brazil's modest
supplies of this mineral are located outside
of Amazonia (see de Lima 1976; Fenster and
León 1979; Beisiegel and de Souza 1986;
Fearnside 1997a, b, 1998a).

It is available phosphorus (PO_4, actually
the anion H_2PO^{4-}), rather than total P, that is
most directly related to plant growth. Phosphorus availability in latosols is generally
very low because most of it is in highly insoluble Fe and Al compounds (Kamprath
1973a). Phosphorus availability in agriculture is usually stated in terms of phosphoric
anhydride (P_2O_5); considering analyses
made with the North Carolina extractant
(0.05 N HCl and 0.025 H_2SO_4) that is standard in Brazil, 1 milliequivalent (m.e.) of
PO_4^{3-}</100 g of dry soil is equal to 103 parts
per million (ppm) of P or 23.7 mg of
P_2O_5/100 g of dry soil (e.g., Vieira 1975).
While the total soil P obviously sets a limit
on the amount of available phosphorus that
can be present, a variety of soil characteristics and processes determine the available
phosphorus present within this limit. Values
of pH below 5.5 are generally associated
with marked decrease in phosphorus avail-

ability (Young 1976). Organic carbon and Fe_2O_3 both are positively related to available phosphorus (P_2O_5) in Brazilian Oxisols (Bennema 1977). Since both carbon and iron are associated with high clay content, the granulometric structure of the soil relates to plant growth through phosphorus, as well as through other nutrients and through water. Micorrhizae are important in mobilizing phosphorus into available forms (St. John 1985). Micorrhizal associations have been found in many, but by no means all, of the few Amazonian trees that have been examined (St. John 1980).

pH, ALUMINUM, AND CATIONS

Little information exists on the response of Amazonian trees to different soil characters. One important exception is the case of cacao, which is a native Amazonian tree that has been the subject of much more research than other species due to its economic importance. It is noteworthy that, while most Amazonian trees appear to be highly tolerant of very acid soils, soil pH (soil reaction) is the best predictor of cacao yields (Hardy 1961; see Fearnside 1986b). Under the acidic conditions that prevail in most Amazonian soils, soil pH is the single most important influence on yields for many crops, such as maize (see Fearnside 1986b; fig. 23.1). In agricultural settings, soil pH at the time of planting is dependent not so much on the initial pH of the soil as on the quality of the burn that the farmer obtained when preparing the land for planting (Fearnside 1986b).

Cation exchange capacity of soils rich in iron and aluminum oxides (such as those of the BDFFP reserves) is pH dependent (Young 1976). In addition, low pH leads to increased concentrations of ions that are toxic or useless to plants (Fe^{2+}, Al^{3+}, and H^+), which occupy some of the reduced number of binding sites that do exist. "Base satura-

tion" is the percentage of binding sites that are occupied by useful cations, or exchangeable bases (Ca^{2+}, Mg^{2+}, K^+, and Na^+, the sum of which is "total exchangeable bases"). Cation exchange capacity is normally calculated as the sum of exchangeable bases, Al^{3+}, and H^+, all in units of m.e./100 g of dry soil (Guimares, Bastos, and Lopes 1970). Cation exchange capacity, total exchangeable bases and base saturation are, in practice, often calculated without data on Na^+ (e.g., Young 1976); the results are little altered as sodium ions are normally present in low and fairly constant quantities (around 0.01 m.e./100 g of dry soil).

The importance of low pH to agriculture is often due to its close relationship to toxic aluminum ions (Al^{3+}). Aluminum ion concentration normally has a negative logarithmic relationship with pH (see Sanchez 1976; Fearnside 1984). It is generally believed that aluminum saturation, rather than the absolute concentration of aluminum ions, is most closely related to plant growth (Primavesi 1981). Aluminum saturation can be computed in two ways, with or without including exchangeable acidity (H^+). Including H^+, the sum of Al^{3+} and H^+ is divided by the total exchangeable bases plus Al^{3+} and H^+, and the result is expressed as a percentage (Sanchez 1976). This is the same as 100 percent minus the base saturation, so only one or the other need be used in analyzing relations with plant growth when defined in this way. In Brazil, however, aluminum saturation is usually calculated without H^+, it being Al^{3+} divided by the sum of exchangeable bases and Al^{3+}, expressed as a percentage (Brazil, SNLCS-EMBRAPA 1979). Aluminum saturation calculated in this way of over 25 percent indicates inhospitable soils for cacao (Alvim 1977).

Crops vary widely in their sensitivity to aluminum toxicity. In terms of aluminum saturation with H^+, maize can tolerate up to

60 percent aluminum saturation, while
sorghum is restricted even at very low levels
of aluminum saturation (Sanchez 1976).
Aluminum ions also have a detrimental effect as one factor affecting fixation of phosphorus into unavailable forms (Kamprath
1973b). Aluminum toxicity itself acts partly
through phosphorus, as aluminum tends to
accumulate in the roots and impede uptake
and translocation of both P and Ca to the
aerial portion of the plant (Sanchez 1976).
Aluminum in the soil solution depends not
only on pH but also on organic matter content of the soil; aluminum ions decrease as
organic matter increases because organic
matter forms strong complexes with aluminum (Sanchez 1976).

The difference between pH in potassium
chloride (KCl) and water, or delta pH, indicates the charge status of an oxide system (a
soil in which entire clay particles consist of
iron and aluminum oxides, or allophanes, or
a soil in which stable coats of these oxides
cover silicate particles) (Sanchez 1976). If
pH in KCl is less than that in water (delta pH
is negative—the usual case), then there is a
net negative charge (cation exchange capacity), whereas if the reverse is the case there
is a net positive charge (anion exchange capacity). The magnitude of the charge affects
the soil pH at the isoelectric point, or zero
point of charge. This, in turn, determines the
cation exchange capacity (CEC) at any given
soil pH level.[2] The organic matter content of
the soil has a strong influence on these relationships and resulting CEC (Sanchez
1976; Bennema 1977). Delta pH is often
used as an indicator of organic matter. If pH
in KCl is higher than pH in H_2O (delta pH is
positive), then organic matter is low. Delta
pH is normally positively associated with
C:N ratio in Amazonian soils, especially
when disparate soils are compared with
wide differences in delta pH values (Tanaka
et al. 1984).

Nitrogen

Nitrogen is traditionally considered to be
the major nutrient deficiency in tropical
agriculture (National Academy of Sciences
1972; Webster and Wilson 1980). Leguminous trees are able to fix nitrogen with the
aid of symbiotic bacteria, which probably
gives members of this superfamily a competitive advantage over species in families
that lack this capability. This helps explain
why legumes are a common group in the
BDFFP reserves, but hardly a dominant
one—Burseraceae, Sapotaceae, and Lecythidaceae are all more common (Rankin–de
Mérona, Hutchings, and Lovejoy 1990). In
a model developed by the BIONTE (Biomass
and Nutrients) project for INPA's "Model
Basin" (located on the same soil type about
25 km south of the BDFFP reserves), N was
assumed to be limiting for the forest as a
whole (Biot et al. 1997). In the BDFFP reserves, total N was found to be positively
correlated with forest biomass (Laurance,
Fearnside, et al. 1999).

Sollins (1998) reviewed the literature on
the relation of soils to lowland rainforest
composition and suggested that soil factors
in decreasing order of importance would be
P availability; Al toxicity; depth to water
table; amount and arrangement of pores of
different sizes; and availability of base-metal
cations; micronutrients (e.g., B, Zn); and N.
Nitrogen is listed last because most lowland
tropical soils are relatively N rich. On the
other hand, studies in tropical montane
forests in Venezuela, Jamaica, and Hawaii
have found tree occurrence and growth to be
related to N (Tanner et al. 1998). Work on
tropical montane forests is more advanced
than that in lowland forests, as some montane forest work includes manipulative experiments rather than exclusively correlation. Tanner et al. (1998) speculate that
many lowland forests are limited by P and
many montane forests by N.

ZINC

Zinc deficiency can be caused by fixation by crystalline sesquioxides (Bennema 1977). Since sesquioxides (Fe_2O_3 and Al_2O_3) are the hallmark of latosols (Oxisols), zinc deficiency may pose a problem for plant growth. Zinc solubility bears a negative exponential relation to pH (pH values represent an exponent), and increases rapidly at soil pH values below 5 (Coelho and Verlengia 1972). Critical values for zinc in agriculture are between 1 and 2 ppm when extracted with 1 N HCl (Cox 1973a).

MANGANESE

Manganese is a micronutrient required in small quantities by plants (Young 1976). However, at high levels it is toxic and, like aluminum toxicity, can inhibit plant growth. The availability of Mg to plants is closely tied to pH, with highest solubility at low pH. Manganese toxicity is a common problem for legume production in Brazilian agriculture on acid soils (Cox 1973a).

OTHER ELEMENTS

The above discussion has been restricted to elements believed to be potentially limiting for forest growth and for agriculture and ranching implanted on soils like those in the BDFFP reserves. The data set for the BDFFP reserve soils includes information on sulfur and copper. Boron, molybdenum, and chloride were not analyzed. Total potassium was not analyzed; this measure is ordinarily not correlated with plant uptake and thus is primarily just of academic interest (Cox 1973b).

The BDFFP Soils Survey

OVERVIEW OF THE DATA SET

The soils data set contains surface samples from 1,693 locations under forest in the BDFFP reserves (272 of which have additional samples taken at later dates), plus 41 soil profiles and 1,693 soil density cores. This allows the effects of "fine scale" variation to be investigated in a way that would not be possible at the scale of soil maps used for zoning and other planning exercises, such as those based on the RADAMBRASIL surveys. The BDFFP soils data indicate substantial variability within an area that is all mapped as having identical soil on the RADAMBRASIL maps. The differences among sampling locations in the BDFFP reserves are sufficient to have an impact on agriculture were these areas (or areas like them) cleared and planted. The differences are also likely to affect the distribution of tree species within the forest, the biomass of the forest, and the susceptibility of the forest to drought stress if exposed to climatic variability (such as that provoked by the El Niño phenomenon), to the effects of edge formation through fragmentation, and to the effects of probable climatic changes such as global warming and reduced rainfall due to loss of evapotranspiration.

Field Methods

Surface samples (0–20 cm) were taken using a screw-type soil auger, each sample being a composite of 15 cores taken at haphazardly chosen locations within a 20 m × 20 m quadrat. Quadrats were delimited by permanent markers at the corners (PVC pipes with numbered aluminum tags), thereby allowing the same locations to be resampled in additional studies.

A soil density sample was taken at the center of each quadrat where surface samples were taken. Volumetric cylinders 20 cm in length and 6.9 cm in diameter were used. The cylinder fits into the end of a soil corer on the shaft of which a movable weight slides up and down. The weight was raised and dropped repeatedly to pound the cylinder into the ground. Once in place, the cylinder was removed by digging around it with a digging tool, taking care that the cylinder is removed completely full of soil (which requires that the cylinder not be pulled out from above). The soil was removed by hitting the outside of the cylinder with a blunt instrument. Samples were placed in double plastic bags for transport to the laboratory and weighing.

A profile to 1.5 m depth was taken at the center of a subset of the quadrats using a "Dutch-type" soil auger. Each profile was divided into eight layers, the first seven samples corresponding to layers 20 cm thick (to a total of 1.4 m) and the last sample representing a 10 cm layer. The samples were laid out on a plastic tarp beside the hole to allow identification of horizons (by color, texture, and plasticity). These methods follow those of Vieira and Vieira (1983). The samples were then placed in separate plastic bags for transport to the laboratory. The sites of the profiles were identified by PVC pipes with numbered aluminum tags.

Laboratory Methods

SAMPLE PREPARATION, SOIL COLOR, TEXTURAL APPEARANCE, AND CONSISTENCY

When surface and profile samples arrived from the field they were classified in the wet state for assessment of color (using a Munsell color chart), wet consistency (stickiness and plasticity), and textural appearance. These descriptions followed the methods of Vieira and Vieira (1983).

The samples were then dried in a solar dryer, followed by 24 hours in an electric oven at 105° C. After this drying, charcoal fragments were removed by hand, weighed, and stored. The dry sample was then ground by hand (using a board and rolling pin), passed through a 20 mm and then a 2 mm mesh sieve. Stones and lateritic concretions removed in the sieves were weighed and recorded. A 300 g subsample of the dried soil was stored in a glass container in a voucher collection. Additional 300 g subsamples were prepared for the INPA (National Institute for Research in the Amazon, Manaus, Amazonas) and CENA (Center for Nuclear Energy in Agriculture, Piracicaba, São Paulo) soils laboratories. Analyses of granulometric characters, bulk density, organic carbon, available water capacity, and pH in H_2O and KCl were done at INPA; total N, total C, Al^{3+}, H^+, Ca^{2+}, Mg^{2+}, K^+, PO_4^{3-}, Cu, Fe, Zn^+, and Mn^{2+} were done at CENA. The INPA soil texture results were checked by comparison of 10 blind samples that were also analyzed by the EMBRAPA soils laboratory in Manaus; no significant differences were encountered. (The CENA laboratory sends ten samples every two months to one of a group of cooperating Brazilian laboratories to obtain an independent check on its chemical analyses.)

Soil Texture

Textural (granulometric) analysis separated material into four fractions: coarse sand (grains 0.2–2.0 mm diameter); fine sand (grains 0.05–0.2 mm diameter); silt (particles 0.002–0.05 mm diameter); and clay (particles less than 0.002 mm diameter) (Vieira and Vieira 1983). The particle-size limits in these classes observe the U.S. Department of Agriculture standards rather than those of the International Society of Soil Science. The pipette method was used (Brazil, SNLCS-EMBRAPA 1979).

The clay fraction was determined by mixing soil in NaOH solution in an electric agitator. The mixture was then placed in a 1 liter graduated sedimentation cylinder, mixed by successive inversions, and allowed to settle. After a period of time determined from a table depending on the temperature, a 25 ml sample is drawn at a depth of 5 cm, using a pipette. The material was weighed after drying in an electric oven at 105° C.

Total sand (coarse plus fine) was determined by sieving with a 0.053 mm mesh, followed by washing, oven drying, and weighing the material. This material was then sieved in a 0.2 mm mesh sieve. The fine sand was weighed and the coarse sand determined by difference from the total sand. Silt was determined by difference of the total sample weight from the sum of the other three fractions.

Bulk Density

The samples were dried to constant weight in an electric oven at 105° C. Fine roots and charcoal were removed from these samples immediately after dry weight determination. Roots were then washed, oven dried, and weighed. Charcoal was stored after weighing.

ORGANIC CARBON AND ORGANIC MATTER

The modified Walkley-Black method was used at INPA, in which volumetric measurements are made by potassium bichromate and titration with ferrous sulfate. The method (Walkley and Black 1934) included the modifications employed by EMBRAPA (Brazil, SNLCS-EMBRAPA 1979). Organic matter was derived by multiplying the result by the constant 1.72 (Brazil, SNLCS-EMBRAPA 1979), a common practice (Young 1976). The C:N ratio is organic carbon (Walkley-Black) divided by total nitrogen.

AVAILABLE WATER CAPACITY

Field capacity and wilting point were determined indirectly on a pressure membrane apparatus (Soil Moisture Equipment Co., Santa Barbara, California), field capacity corresponding to moisture content under a suction of 0.33 atmospheres (bars) and wilting point to the corresponding value at 15 bars. The difference in the amount of water held by the soil between these two points is the available water capacity. Moisture content and available water capacity are expressed as percent water by weight on a dry basis (Klar 1984). Note that, following the standard procedure, the soil was dried and sieved before being placed in the pressure plate rings, thereby altering its structure.

pH IN WATER AND KCL

Soil reaction (pH) in distilled H_2O and in 1 N KCl solution was measured with a pH meter. The ratio of oven-dried soil to water or KCl solution is 1:1 on a volumetric basis (20 ml soil in 20 ml of water or solution). Work in tropical soils elsewhere has found soil reaction to increase by about 0.3 pH units in dried samples as compared to field conditions; pH is unaffected by storage once dried (Gillman and Murtha 1983).

MACRO AND MICRONUTRIENTS

The CENA method for micronutrient determination has been described by Zagatto et al. (1981) and Jorgensen (1977). Samples were digested using the EMBRAPA method: 0.0025N HCl and 0.005 H_2SO_4 (Brazil, SNLCS-EMBRAPA 1979). K was measured by atomic emission spectrophotometry in an air-acetylene flame, using a Perkin-Elmer AAS 306; Al^{3+}, Ca^{2+}, Cu, Fe, Mg^{2+}, Mn^{2+}, Na^+, and Zn^+ by atomic absorption spectrometry with plasma induced in argon, using a computerized spectrometer system

TABLE 23.1. Variation in Soil Characters (0–20 cm) in the BDFFP Reserves.

Soil character	Units	Mean	SD	Minimum	Maximum	N
Granulometric Characters						
Clay	%	54.7	13.5	18.0	68.8	54
Silt	%	21.2	4.3	8.3	32.5	54
Fine sand	%	5.6	3.9	1.3	18.1	54
Coarse sand	%	18.5	12.5	4.4	56.6	54
Available Water						
Soil moisture content at 1/3 bar	% H_2O by weight	31.8	6.8	13.0	41.2	45
Soil moisture content at 15 bars	% H_2O by weight	24.3	5.8	9.1	31.9	45
Available water capacity	% H_2O by weight	7.6	2.0	3.3	12.0	45
Topography						
Slope	%	10.8	8.9	1.4	38.7	36
Carbon						
Organic matter	%	2.1	0.7	0.8	3.3	50
Organic C (Walkley Black)	%	1.6	0.3	0.8	2.2	40
Total C	%	1.96	0.45	1.27	3.07	51
C/N ratio	dimensionless	9.9	1.6	8.4	17.0	38
Soil reaction						
pH in H_2O	pH units	4.0	0.3	3.4	4.4	53
pH in KCl	pH units	3.8	0.2	3.2	4.3	53
Delta pH	pH units	−0.3	0.1	−0.5	0.0	53
Primary nutrients						
N (total)	%	0.2	0.0	0.1	0.2	38
PO_4^{3-}	m.e./100 g dry soil	0.030	0.005	0.022	0.041	38
K^+	m.e./100 g dry soil	0.060	0.011	0.032	0.077	38
Secondary nutrients						
Ca_2^+	m.e./100 g dry soil	0.058	0.026	0.015	0.131	38
Mg_2^+	m.e./100 g dry soil	0.076	0.031	0.013	0.125	38
Na^+	m.e./100 g dry soil	0.052	0.018	0.026	0.106	18
S	ppm	13.0	1.4	10.6	15.0	13
Micronutrients						
Cu	ppm	0.33	0.12	0.10	0.54	24
Fe	ppm	137	31	77	185	24
Zn	ppm	1.48	0.78	0.61	2.99	24
Mn	ppm	1.81	0.49	0.87	2.49	24
Other ions						
Al^{3+}	m.e./100 g dry soil	1.63	0.29	1.03	2.22	38
"Al sat. w/ H^+, w/ Na^+"	% of CEC	89.4	1.0	86.3	90.5	18
"Al sat. w/ H^+, w/o Na^+"	% of CEC-Na^+	92.4	1.7	87.8	96.2	38
"Al sat. w/o H^+, w/ Na^+"	% of Base sat. + Al^{3+}	85.2	1.2	82.4	86.5	18
"Al sat. w/o H^+, w/o Na^+"	% of Base sat. + Al^{3+}	89.5	2.2	84.3	94.4	38
H^+	m.e./100 g dry soil	0.70	0.12	0.39	0.85	38

TABLE 23.1.   (continued) Variation in Soil Characters (0–20 cm) in the BDFFP Reserves.

Soil character	Units	Mean	SD	Minimum	Maximum	N
Cation measures						
Cation exchange						
capacity (w/ Na⁺)	m.e./100 g dry soil	2.7	0.2	2.5	3.0	18
Cation exchange						
capacity (w/o Na⁺)	m.e./100 g dry soil	2.5	0.4	1.7	3.3	38
CEC of clay (w/o Na⁺)	m.e./100 g dry soil	5.1	1.6	3.7	10.5	38
CEC of clay (w/o Na⁺)						
w/ C correction	m.e./100 g dry soil	14.4	1.7	10.9	18.2	38
Total exchangeable bases						
(w/ Na⁺)	m.e./100 g dry soil	0.3	0.0	0.2	0.4	18
Total exchangeable bases						
(w/o Na⁺)	m.e./100 g dry soil	0.2	0.1	0.1	0.3	38
Base saturation (w/ Na⁺)	% of CEC	10.6	1.0	9.5	13.7	18
Base saturation (w/o Na⁺)	% of CEC	7.6	1.7	3.8	12.2	38

(Jarell-Ash Plasma Atomcomp Direct Reading Spectrometer), and S (as barium sulphate) by turbidimetry, using a flow injection system. PO_4^{3-} was determined in an autoanalyzer using the molybdenum blue method (Jorgensen 1977).

TOTAL NITROGEN

Total nitrogen was determined with Kjeldahl digestion, using a mixture of H_2O, Li_2SO_4, and concentrated H_2SO_4 to break down the organic matter and transform the nitrogen in the sample into NH^{4+} (Parkinson and Allen 1975). The distillate was titrated with H_2SO_4.

TOTAL CARBON

Total carbon (%C) was determined at CENA by the "dry" method, which converts the forms of soil carbon into CO_2 by combustion at 1100° C. The gas was then sent to a standard sodium chloride cell, where the difference in electrical conductivity between this solution and one carbonated with CO_2 was detected, and the result is expressed in milligrams of carbon (Cerri, Eduardo, and Piccolo 1990).

Results

Results presented here are restricted to the 54 noncontiguous ha in which tree surveys were carried out in the BDFFP project (Rankin–de Mérona et al. 1992; Chapter 5). The reserve area lacks patches of the best Amazonian soil types (anthropogenic black soil *terra preta do índio* and Alfisol *terra roxa estruturada*), and of the worst types such as white sand campina soils (podzols or Spodosols). Nevertheless, the data indicate a substantial variation in hectare-level means of soil quality indicators, despite the area's uniformity at the level of general soil maps (table 23.1).

Clay content (mean value = 54.7 percent) varies greatly from 18.0–68.8 percent; because clay content is closely tied to several indicators of soil fertility, the general fertility level also varies substantially. The mean soil would be classed in the middle of the "clay" category (Vieira and Vieira 1983). The soil has a significant percentage of silt-sized particles (mean = 21.2 percent), making the soil differ from some Amazonian soils where particles are concentrated in sand and clay fractions, with very little in between. It is possible that incomplete dispersion of clay aggregation could result in an apparent abundance of silt-sized particles when, in fact, this size class is present in much smaller amounts (Thierry Desjardins, pers. comm.). The statistical analyses were therefore performed lumping the clay and silt categories.

The terrain is undulating, with a mean slope of 10.8 percent. A portion of the area with steep slopes would be prone to erosion problems, in addition to not being appropriate for mechanized agriculture. Land-use capability classifications consider slopes of 0–2 percent to be without limitations, 2–8 percent to be slightly limited, 8–30 percent to be moderately limited, and more than 30 percent to be very limited (Benites 1994). By these criteria, of the 373 points with slope measurements in the data set used in the current study, 57.1 percent had no limitation, 23.6 percent had slight limitation, 23.1 percent had moderate limitation, and 2.7 percent had severe limitation from slope. In addition, the clay:sand content ratio of 2.6 (or 3.1, considering clay and silt) in the BDFFP reserves is considered an impediment to mechanized agriculture, the maximum value of this ratio considered appropriate being 2.0 (Vieira and Vieira 1983). However, this classification of mechanization limitations based on clay content represents an average for all of Brazil; W. Sombroek (pers. comm.) believes that, in the case of yellow latosols in Amazonia, such as those in the BDFFP reserves, soils with clay contents higher than the limits indicated by these standards could be used for mechanized agriculture.

The soil is quite acid, with a mean pH in water of 4.0 (range 3.4–4.4). However, it should be remembered that what is of interest to agriculture is not the level of nutrients in the soil under forest, but rather the level that will be present when the forest is cleared and burned. Especially in the case of pH, the values are increased by an amount

that varies with burn quality (Fearnside 1986b).

The delta pH (mean = −0.3) indicates a net negative charge (a cation exchange capacity). This also confirms a reasonable level for soil organic matter by the standards of tropical agricultural soils.

Aluminum saturation (excluding H^+), with a mean value of 92.4 percent, is clearly high, while base saturation (without Na^+) is low, with a mean value of only 7.6 percent. Al saturation (without H^+) less than 50 percent is classified as low and more than 50 percent is classified as high (Vieira and Vieira 1983).

Total carbon had a mean value of 1.96 percent. Organic carbon, as determined by the modified Walkley-Black method, had a mean value of 1.58 percent. Organic matter had a mean value of 2.72 percent. While these values may appear low for agricultural soils, they are typical of Amazonia.

Total N levels are moderate, with a mean value of 0.16 percent; values less than 0.1 percent are considered low with a probable response to fertilization in crop plants, while values in the 0.1–0.2 percent are in the moderate range where responses are possible (Young 1976). The C:N ratio (mean value = 9.9) indicates a reasonable but not ideal quantity of nitrogen available to plants (values over 15 indicate little N in available forms).

Phosphate levels are very low (mean value of PO_4^{3-} = 0.030 m.e./100 g of dry soil). PO_4^{3-} is considered "insufficient" for crop plants with levels less than 0.097 m.e./100 g of dry soil (equivalent to less than 2.30 m.e. of P2O5/100 g of dry soil); "fair" levels of PO_4^{3-} are 0.097-0.253 m.e./100 g of dry soil (equivalent to 2.30–6.00 m.e. of P_2O_5/100 g of dry soil), while "good" soil has more than 0.253 m.e. of PO_4^{3-}/100 g of dry soil (more than 6.00 m.e. of P_2O_5/100 g of dry soil) (Vieira and Vieira 1983).

Plate 1. The 1 and 10 ha isolated forest fragments (1104 and 1202) after the October 1980 burn. Photo by Richard O. Bierregaard, Jr.

Plate 2. The first isolated 100 ha study site (3304) on the *fazenda* Porto Alegre in 1983. The corridor connecting the reserve borders a stream that leaves the reserve and was severed in 1984. Photo by Richard O. Bierregaard, Jr.

Plate 3. Primary forest canopy north of Manaus, Brazil, stretches unbroken from the BDFFP study area to the Amazon River. Photo by Richard O. Bierregaard, Jr.

Plate 4. A 1 ha isolate (2107) at the *fazenda* Dimona several years after its isolation. Note the second growth established in the near right corner following substantial loss of canopy trees due to windthrows. Photo by Richard O. Bierregaard, Jr.

Plate 5. Typical primary forest understory in BDFFP study area. Photo by Carol Gracie.

Plate 6. Recently felled forest drying before the 1984 burn surrounds a 1 ha reserve (2108) at the *fazenda* Dimona. Photo by Richard O. Bierregaard, Jr.

Plate 7. The ranchers burn the felled forest near the Colosso camp in October 1980. Photo by Richard O. Bierregaard, Jr.

Plate 8. A nearly single-species ocean of *Cecropia sciadophylla* surrounds a 10 ha forest fragment (3209) on the *fazenda* Porto Alegre about five years after the 1983 clearcut. Early rains prevented the ranchers from burning this area. Photo by Richard O. Bierregaard, Jr.

Plate 9. Well-established second growth dominated by *Vismia* spp. near the Colosso camp and isolates. Photo by Rita C. G. Mesquita.

Plate 10. Most of the researchers, graduate students, and field interns participating in the BDFFP assembled at the Gavião camp in about 1981. Photo by Richard O. Bierregaard, Jr.

Plate 11. The major base camp at the control forest study site, Km 41. Photo by Philip C. Stouffer.

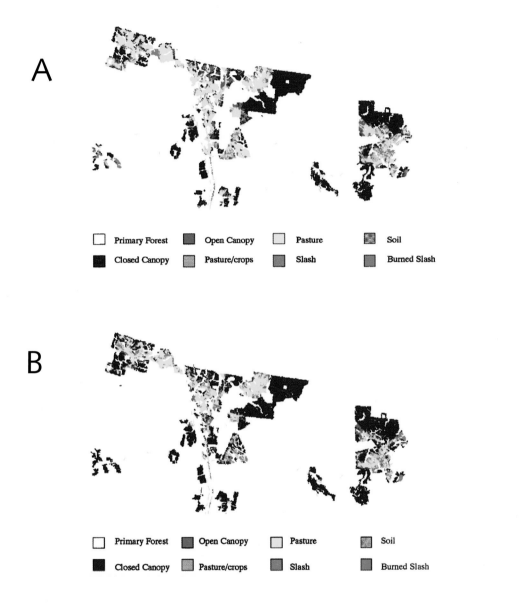

A

Primary Forest
Closed Canopy
Open Canopy
Pasture/crops
Pasture
Slash
Soil
Burned Slash

B

Primary Forest
Closed Canopy
Open Canopy
Pasture/crops
Pasture
Slash
Soil
Burned Slash

Plate 12. (A) 1988 BDFFP land cover patches; (B) 1989 BDFFP land cover patches.

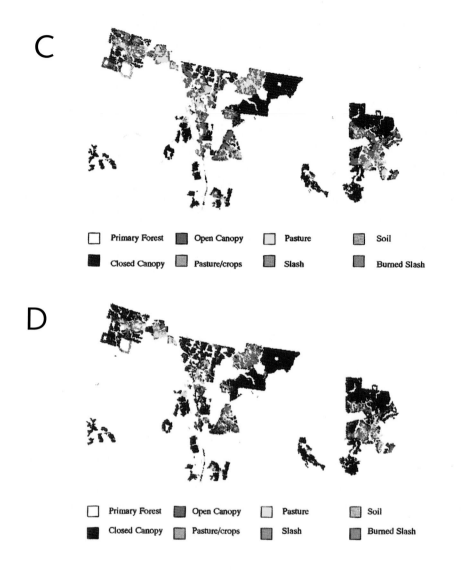

C

	Primary Forest		Open Canopy		Pasture		Soil
	Closed Canopy		Pasture/crops		Slash		Burned Slash

D

	Primary Forest		Open Canopy		Pasture		Soil
	Closed Canopy		Pasture/crops		Slash		Burned Slash

Plate 13. (C) 1990 BDFFP land cover patches; (D) 1991 BDFFP land cover patches.

Exchangeable potassium (K+) is low, with a mean value of 0.06 m.e./100 g of soil; values below 0.2 m.e./100 g of dry soil are considered to be low, with fertilizer response probable in crop plants (Young 1976). A land-use capability rating system used in Brazil (Vieira and Vieira 1983) classifies K+ levels less than 0.11 m.e./100 g of dry soil as "insufficient," 0.11 to 0.37 as "fair," and more than 0.37 as "good." The absolute minimum level required by crop plants is considered to be 0.10 m.e./100 g of dry soil (Boyer 1972). Of the nutrient cations, potassium is the one which is in greatest demand by crops (Webster and Wilson 1980). Lack of exchangeable potassium may therefore affect plant growth on these soils.

Calcium ion concentrations are very low (mean value = 0.058 m.e./100 g of dry soil). For crops, less than 1.50 m.e./100 g of dry soil is "insufficient," 1.50–3.50 m.e./100 g of dry soil is "fair," and more than 3.50 m.e./100 g of dry soil is "good" (Vieira and Vieira 1983).

Magnesium ion concentrations are fair (mean value = 0.076 m.e./100 g of dry soil). For crops, less than 0.50 m.e./100 g of dry soil is "insufficient," 0.50–1.00 m.e./100 g of dry soil is "fair," and more than 1.00 m.e./100 g of dry soil is "good" (Vieira and Vieira 1983).

The cation exchange capacity (expressed as the CEC of clay) is very low, with a mean value of 14.4 m.e./100 g of clay after correction for carbon. With a correction for carbon (each 1 percent of carbon is considered to correspond to 4.5 m.e. of CEC), a CEC of 24 m.e./100 g of clay is the dividing line between low and high activity clay (Vieira and Vieira 1983). The value of 24 m.e./100 g of clay is considered low for agriculture (Benites 1994).

Base saturation higher than 50 percent is considered as indicating high fertility, provided that CEC of the clay is more than 24 m.e./100 g of clay; base saturation less than 50 percent indicates moderate fertility like these (Benites 1994; see also Vieira and Vieira 1983). The mean value of 7.6 percent (without Na+) in the study area is obviously low, as is the maximum value of 12.2 percent. The Na+ levels found appear to be unusually high (mean = 0.052 m.e./100 g dry soil); combined with very low values for Ca2+ (mean = 0.058 m.e./100 g dry soil), Na+ makes an unusually large contribution to exchangeable bases. However, due to the much less complete data set for Na+, cation measures excluding sodium were used in the statistical analyses.

The relationships among soil characters in the BDFFP data set were calculated for total exchangeable bases, organic matter, aluminum ions, and available water capacity (table 23.2). Some of the other relationships shown in figure 23.1, although known to hold generally, did not emerge as significant regressions in the BDFFP data set. Total exchangeable bases was not predicted significantly better by including pH (or $\log_{10}$ pH) in the regression, although higher pH values are known to be associated with greater amounts of cations such as Ca2+ and Mg2+ (see Fearnside 1986b). Phosphate was not significantly related to pH, Al3+, and Fe in the data set.

Lessons for Development

Agricultural Potential

Our knowledge of the BDFFP reserve area greatly exceeds the level of knowledge that can be expected to be available for zoning decisions in the rest of Amazonia. If we cannot come up with well-founded recommendations for this area, then we cannot expect such suggestions to be obtainable for the vast majority of the rest of the region.

Wide differences of opinion exist as to whether a given fertility level means that an

TABLE 23.2.   Relationships Among Soil Parameters

Total exchangeable bases (without Na^+)

$$TEB = 1.86 \times 10^{-3} CS + 3.88 \times 10^{-2} OM - 5.20 \times 10^{-2}$$

($P < 0.00001$, $r^2 = 0.84$, $n = 38$)

where: TEB = Total exchangeable bases (without Na^+); (m.e./100g dry soil)

CS = Clay + silt (%)

OM = Organic matter (%)

Organic matter

$$OM = 2.13 \times 10^{-2} CS + 1.12$$

($P < 0.00001$, $r^2 = 0.73$, $n = 38$)[1]

where: OM = Organic matter (%)

CS = Clay + silt (%)

Aluminum ions

$$Al^{3+} = 7.66 \times 10^{-3} CS - 7.70 \times 10^{-1} \log10\ pH + 5.64$$

($P < 000001$, $r^2 = 0.78$, $n = 38$)

where: Al^{3+} = Al^{3+} (m.e./100g dry soil)

CS = Clay + silt (%)

pH = pH in water (pH units)

Available water capacity

$$AWC = 5.24 \times 10^{-2} CS + 3.66$$

($P < 0.01$, $r^2 = 0.43$, $n = 45$)

where: AWC = Available water capacity (% H_2O by weight)

CW = Clay + silt (%)

Note: [1]With elimination of one outlier.

area should be opened up for agriculture, or that it should be maintained under forest cover. The data make clear that soil fertility is low and impediments to agriculture, such as aluminum toxicity, are substantial. This suite of problems leads some authors to conclude that soils such as these should not be used for agriculture. Irion (1978), for example, concludes for soils on the Barreiras (or Alter do Chão) formation, such as those of the BDFFP reserves, "extensive cultivation . . . is impossible, as the quality of the soil is insufficient. Any such cultivation, aiming at export of agricultural products, would result in an impoverishment of the soil within a few years, thus making the soil agriculturally unusable for many decades." Van Wambeke (1978) warns of the potential for "irreversible destruction of soils and the creation of fertility deserts." On the other hand, Serrão and Homma (1993) look at the same situation and conclude that 70 percent of the land in Brazilian Amazonia is "appropriate

for crop production," and assert that "regions with low fertility and acidic soils have not been transformed into deserts, as some have foreseen. . . . On the contrary, such regions have been very dynamic in terms of agricultural development" (Serrão and Homma 1993). Whether these regions can be described as "very dynamic" is open to question; certainly the cattle ranches on which the BDFFP reserves are located are less than dynamic, having been unable to turn a profit or even keep open the land originally cleared for cattle pasture.

Amazonia in general, and particularly the state of Amazonas, has been protected from deforestation by impediments that include poor soils, human diseases (particularly malaria), and difficulty of access from densely populated portions of Brazil. But disease is no longer the barrier that it once was; for example, if opening areas today required the sacrifice of lives that opening the Madeira-Mamoré railway in Rondônia did at the beginning of the twentieth century, then the threat of deforestation would be much less. Poor access is also rapidly changing; many settlers have gone directly from Rondônia to Roraima, bypassing the state of Amazonas through which they passed, a phenomenon explained both by somewhat better soils on the Guianan Shield in Roraima and by inducements offered by the government of Roraima. The paving of the BR-319 Highway linking Rondônia with Manaus, expected between 2001 and 2003, will open the floodgates to an influx of prospective settlers to central Amazonia, including the general area of the BDFFP reserves.

What happens when an area like the BDFFP reserves is cleared depends on what is done with the land. Some options are better than others from the standpoints of maximizing sustainability and minimizing environmental impact. For example, agroforestry, or cultivation of crops in mixed plantings that include trees, is better in many respects than either annual crops or cattle pasture (see Chapter 26). However, severe limitations exist for widespread use of agroforestry in the vast areas of degraded lands that have already been created in the Amazon region, let alone the even larger areas that would exist if more forest were to be cut in the belief that agroforestry can solve the problems of sustaining production to support the human population (Fearnside 1998b; see also Chapter 24).

The political discourse surrounding the announcement of new settlements in Amazonia is invariably permeated with images of permanent prosperity emanating from the agricultural systems to be implanted in the cleared areas. However, governmental decisions to cut the forest and implant agricultural settlements in areas with soils like those of the BDFFP reserves, which are typical of vast expanses of Amazonia, imply one of two things: either the government must be willing to provide or subsidize regular inputs of fertilizers for the indefinite future (a highly improbable scenario), or the decision makers must accept responsibility for trading the forest for a landscape of degraded cattle pastures and second-growth stands. Unfortunately, this responsibility is frequently avoided by recourse to histrionic devices such as the platitude that agriculture or ranching will be sustainable with "adequate management," the implication being that, if the project fails at some future date, then it will be the farmers who are to blame for not having applied "adequate management."

Two of the principal crops currently in fashion for promotion in Amazonia are not likely to be successful in areas like the BDFFP reserves. One is oil palm (e.g., Smith et al. 1995). In areas with a significant dry season (such as the Manaus area) the yield of oil drops quickly in comparison with optimal yields. In Brazilian Amazonia the two locations with optimal climatic conditions for oil palm are the areas near Belém (Pará) and Tefê (Amazonas). Plantations in the Belém area suffer the effects of shoot rot disease (Fearnside 1990a). Because most of the costs of establishing and maintaining plantations are similar on a per-area basis regardless of the yield, plantations located in suboptimal areas are unlikely to be competitive.

The other crop that figures heavily in current discourse on Amazonian development is soybeans. Unfortunately, the social benefits from this crop are small for the local population, and the mechanized agriculture that it requires is capital and chemical intensive. Although mechanization in areas like the BDFFP reserves is not impossible, the area would be less than ideal for these methods because of undulating terrain and high clay content of the soil.

Low soil fertility can be alleviated by application of fertilizers. However, a series of severe constraints limits the extent to which agriculture based on a supply of nutrients from fertilizers can be extended to wide areas in Amazonia. The best example of this is the history of the "Yurimaguas technology" developed for continuous cultivation in Amazonian Peru (Sanchez et al. 1982; see Fearnside 1987, 1988; Walker, Lavelle, and Weischet 1987). Despite a long list of subsidies ranging from chemical inputs to free soil analyses and technical advice on a field-by-field basis, this high-input management package did not gain popular acceptance in the area. The limits of physical resources, such as phosphate deposits, as well as financial and institutional restraints, make widespread use of such systems unlikely (see Fearnside 1997a).

Application of fertilizers is only one method for addressing soil fertility limitations. One must consider the extent to which the agricultural prospects of areas like the BDFFP reserves would change if other kinds of technical advances were to occur. For example, recent progress has been made on removing aluminum saturation limitations through development of transgenic crop plants (Barinaga 1997; de la Fuente et al. 1997). It is not inconceivable that phosphorus limitations could be relaxed by development of crop plants with appropriate micorrhizal associations. Nitrogen limitations of various non-leguminous crops may be relaxed through pseudosymbiotic relationships with a variety of types of nitrogen-fixing bacteria, an area in which significant advances have been achieved in Brazil through the work of Johanna Döbereiner (e.g., Döbereiner 1992).

Today, however, it is not viable to sustain agriculture and ranching in vast areas of Amazonia due to limits of markets, phosphate deposits, and funds. Therefore one should not count on a "technological fix" to solve the problems of sustainability until such time as the technological advances in question are actually achieved. Not "counting one's chickens before they hatch" is a universal precautionary principle that is a basic rule for avoiding unwise adventures in myriad situations in addition to this one. One should also remember that the environmental costs of forest loss must be considered, and that these costs do not change much if the agriculture implanted is productive or not (see Fearnside 1997a).

The soils in the BDFFP reserves are clearly infertile; indicators of soil fertility such as pH, cation exchange capacity, total exchangeable bases, and PO_4^{3-} are low, while aluminum saturation is high. Under such circumstances, it is logical to maintain these areas under forest rather than convert them to short-lived low-productivity land uses. But to what extent would the situation be different if the soils were more productive? What level of soil quality would make it worthwhile to sacrifice the forest? There are no simple answers to these questions. Rational decision-making will require assessment of the value of both the agricultural production that can realistically be expected from the area and the environmental cost of sacrificing the forest.

So what is the lesson? We believe that the main lesson is that the returns from converting areas like the BDFFP reserves to agriculture or ranching are minimal when compared with the true value of the environmental services of the intact forest. Even though the amount that countries like Brazil may one day be able to collect on the basis of supplying these services is much less than the true value of the services, the returns from agriculture and ranching are also meager when compared with the amounts that might, in fact, be collected (Fearnside 1997c; Chapter 3).

PROSPECTS FOR RESILIENCE

The relationships found among the various soil characters follow patterns that are consistent with what is generally known about tropical soils. These relationships (fig. 23.1 and table 23.2) give an indication of some of the changes that are likely to occur in the face of such human interventions as selective logging, formation of edges through forest fragmentation, and removal of forest for implanting agriculture and ranching.

In the case of logging, changes normally found include soil compaction along logging tractor paths (Veríssimo et al. 1992). Nutrients are exported in the biomass removed (Ferraz et al. 1997), and soil cations are lost through leaching and runoff (Jonkers and Schmidt 1984).

Outright removal of the forest has major impact on soil nutrients and structure (see review in Fearnside 1986b). Soil characters important for plant growth can be expected to change together; as they adjust to a new equilibrium they will tend to maintain the same relationships among each other that were found in the forest soil. Burning raises pH and provides inputs of nutrients to the soil, but cultivation usually leads to losses of organic matter and clay that gradually reduce the ability of the soil to hold cations. Some of the changes, such as loss of organic matter content and increase in soil compaction, can be reversed through periods under fallow. How much this capability for recovery of the soil, as well as capability for recovery of the forest, compensate for the impacts of deforestation is an important area of debate. Unfortunately, the theoretical possibility of recovery if areas are abandoned for many decades or even centuries has little relevance to the real impacts that are caused by the continued advance of deforestation in the region.

"Resilience" is a term that is fashionable in discussions of Amazonian development; it refers to the ability of a system to recover its original characteristics if perturbed (e.g., Smith et al. 1995). Are the soils of the BDFFP reserves likely to be "resilient" if they are converted to pasture and degraded? Reasons to doubt this include the already low levels of key elements that would only be further depleted. Among the changes are loss of clay and carbon, together with the cations that are tied to these components. Carbon can be regenerated under extended fallows, but, in practice, it is doubtful that fallows would be left for sufficient periods for this to happen (Fearnside 1996a).

Different people can look at the same set of facts and arrive at radically different conclusions. Serrão, Nepstad, and Walker (1996), for example, take the regrowth of secondary forest in abandoned pastures to indicate that "the Amazonian upland forest ecosystem is fairly resilient to current uses," whereas others disagree (e.g., Fearnside and Guimarães 1996).

DISTRIBUTION OF NATURAL VEGETATION

On a broad scale, a relationship between soil and tree species and biomass is evident, such as the difference between *campina* and *campinarana* vegetation with low biomass that grows on white sand soils in the Manaus area, versus the upland forests with high biomass on more clayey soils like those in and around the BDFFP reserves. Within more similar soils, such as those in the reserve system, differences are more subtle but are likely to be present.

Studies in various parts of the tropics indicate relationships between tree species occurrence and soils. In a review by Sollins (1998) of eighteen studies in tropical lowland forests, relations between species occurrence and soil drainage regimes were common. Notable is the study by Lescure and Boulet (1985) of 16.8 ha of forest in French Guiana, which found that 69 percent of the tree species showed response to soil drainage conditions.

Only three of the eighteen studies reviewed by Sollins (1998) were able to detect relationships of tree species occurrence to chemical properties: Ashton and Hall (1992) found relationships with P and cations in Sarawak and Brunei; Clark et al. (1995) found relationships with P, Al, and pH in Costa Rica; and Van Shaik and Mirmanto (1985) found a relationship with pH in Sumatra. The three studies that reported relationships with chemical properties were those with the largest ranges of values for the chemical indicators of fertility (the independent variables). Sollins (1998) believes that the limited range of soil fertility within the areas in which studies have been attempted to relate species occurrence to chemical properties is a primary reason for the elusiveness of demonstrating significant relationships. Other confounding factors include seasonal variation in some key fertility indicators, especially available phosphorus, cations, and pH. Both of these restrictions apply to the BDFFP data set as well.

Soil fertility and water relations can be expected to be related to seedling survival and to the growth rates of trees. Species associations with soils may be related to forest biomass if species characterized by larger individuals are found on certain soils. Any soils' effect on the occurrence of species that become large canopy emergents would have a great effect on biomass because a large part of forest biomass is often held in only a few very large individual trees (Brown, Mehlman, and Stevens 1995; Clark and Clark 1995). Independent of species effects, general favorability of the soil for plant growth can be expected to be positively associated with forest biomass. An ordination analysis relating biomass to soils in the BDFFP reserves indicates that 53 percent of the variation in biomass is explained by a gradient in clay content (and positively associated levels of total N, organic C, exchangeable bases, K^+, Mg^{2+}, Ca^{2+}, Al^{3+}, H^+, and CEC (Laurance, Fearnside, et al. 1999).

POTENTIAL FOR RELEASE OF GREENHOUSE GASES

Carbon stocks in soil and in forest biomass in Amazonia are very large because of high per-hectare loading and the vast areas of Amazonian forest still standing. Release of these stocks is a significant contributor to global warming at present rates of deforestation (Fearnside 1996b, 1997d), while the large remaining stocks mean that the potential future importance of avoiding emission is even greater (Fearnside 1997c). Converting each hectare of forest to the equilibrium landscape that replaces it releases approximately 8.5 t of soil carbon to the atmosphere considering soil to 8 m depth over fifteen years, or 7.4 t C/ha to a depth of 1 m (Fearnside and Barbosa 1998).

The fine-scale spatial variability in soil carbon distribution indicated by the BDFFP soil survey suggests that a substantial amount of uncertainty is inherent in studies of soil carbon change based on "chronosequences," where the effects of land uses (such as pasture) are inferred from comparison of roughly simultaneous samples taken at a series of sites with different land-use histories (see Fearnside and Barbosa 1998). Substantial differences among these sites could be due to natural spatial variation rather than to the effect of land use.

PROBABLE RESPONSE OF FOREST TO FRAGMENTATION

The BDFFP has provided undeniable evidence that creation of edges results in greatly increased mortality of trees located near the edges (Lovejoy, Rankin, et al. 1984; Laurance, Ferreira, et al. 1998a; Chapter 9). These edges have drier air than the forest interior (Kapos 1989; Kapos et al. 1993). Tree mortality near the edges leads to a "biomass collapse," releasing carbon to the atmosphere (W. Laurance, Laurance, et al. 1997).

Soils are likely to play a role in mortality of trees under these conditions. Soils with greater amounts of sand can be expected to retain less water, leading to greater drought stress when conditions are dry due to edge proximity. However, sandy soils are associated with valley bottoms, which would be expected to have some additional water as compared to higher locations.

PROBABLE RESPONSE OF FOREST TO CLIMATIC CHANGE

Climatic variations and changes of several types exist or are predicted. The El Niño–Southern Oscillation is a periodically recurring phenomenon that has occurred for many millennia, albeit at lower frequencies in the period before 1976 (Nicholls et al. 1996). El Niños result in droughts that can have significant impact on Amazonian forest (Tian et al. 1998). Tree species that are specialized on wetter sites are particularly vulnerable to drought stress. For example, such species in the forest on Barro Colorado Island, Panama, suffered extraordinarily high mortality during the 1982–83 El Niño event (Hubbell and Foster 1990).

Archeological evidence suggests that catastrophic fires have occurred in Amazonia at the times of major El Niño events four times over the past 2000 years: 1500, 1000, 700, and 400 B.P. (Meggers 1994b). Human action could now turn less intense El Niño events, such as the 1982–83 and 1997–98 events, into major catastrophes. Such less

intense events are much more frequent than major ones, but these would be added to the effects expected from climatic changes such as reduction in rainfall from lowered evapotranspiration caused by continued deforestation (Salati and Vose 1984) and the effect of temperature and rainfall changes caused by global warming (see Fearnside 1995). While neither of these changes is expected to result in radical reductions in rainfall, the effect is added to that of natural variability such as that caused by El Niño events and such disturbances as logging and edge formation. Logging is rapidly increasing in Amazonian forests, leading to more flammable conditions from accumulation of slash and accidentally killed trees in the logged-over forests (Uhl and Buschbacher 1985; Uhl and Kaufman 1990; Chapter 26). The continued advance of settlement and deforestation in the region means that many more opportunities exist than in the past for fires to enter adjacent forests. These effects are added to those of climate variability and climatic change, increasing the risk of fire entering standing forest (Nepstad, Moreira, and Alencar 1999; Barbosa and Fearnside 2000).

Soils play a role in the forest's response to these events. As in the case of edge-related drought stress, trees on soils with little available water are more likely to succumb to extreme events.

Conservation Lessons

1. The soils of the BDFFP reserves are typical of vast areas of Brazilian Amazonia that are likely to come under increasing pressure for deforestation.

2. These soils have low fertility, are acid, and have high levels of toxic aluminum ions. They also have limitations from rolling topography, high clay content,

and low available water capacity. The soils data indicate that they would produce little if converted to agriculture or ranching, and point to land uses that maintain forest cover intact as preferable.

3. Although the value of the environmental services provided by standing forest currently provides no source of cash income, the potential value of these services far exceeds the returns that can be expected from cutting the forest for agriculture or ranching.

Notes

1. The defining criterion for latosols is that the B horizon must be "latosolic," as opposed to "textural" (i.e., clayey). The criterion for identifying latosolic character in the Brazilian soil classification system is that the ratio of silicon and aluminum oxides (SiO_2:Al_2O_3) must have a value less than two (see Sombroek 1966). Values below two generally indicate that the silicate clay minerals have a 1:1 lattice structure (Sombroek 1966). The ratio refers to the number of sheets of silica tetrahedra to sheets of alumina octahedra, and is one of the primary determinants of the properties of the clay (Young 1976). The lower amount of silicate in these soils, as compared to those with a 2:1 lattice structure, is a consequence of removal of silicon from the soil column (along with most weatherable minerals, including important nutrients for plants) over millions of years of leaching. In the absence of analytical data on SiO_2 and Al_2O_3, clay mineralogy of subsoil horizons (which are low in organic matter) can be deduced from the cation exchange capacity (CEC) of the clay fraction, which is calculated as the CEC of the soil divided by the percent clay, multiplied by 100; values below a cutoff point of 16 to 20 m.e./100 g clay indicate absence of 2:1 lattice minerals (Young 1976).

2. The pH of the extracts used in laboratory analyses affects the values obtained for the cations that make up the CEC and, since the determinations are not normally made at the natural pH of the original soil, the values only represent an index of the true CEC. In Brazil, the extracts used for the determinations are buffered to pH 7.0 (Brazil, SNLCS-EMBRAPA 1979). Values for CEC determined at the pH of the soil (effective CEC) are much lower than those determined either at pH 7.0 or at the pH 8.2 standard sometimes used in the United States (Sanchez 1976).

Acknowledgments

Field collections were done by Irene Tosi Ahmad, Celso Paulo de Azevedo, Ronaldo Gomes Chaves, Fernando Moreira Fernandes, Jorge Gouveia, Marcelo Vilela Galo, Michael Keller, Evaldo Moreira Filho, Manuel de Jesus Barros Nogueira, Fernando José Alves Rodrigues, Joel Costa Souza, and Wilson Roberto Spironello. Sample preparation was performed by Félix de Almeida and Raimundo Matos. Soil analyses performed at INPA were done by Rosinéia Gomes da Silva, Newton Falcão, and Eduardo White Martins. Takashi Muroaka coordinated analyses at the Centro de Energia Nuclear na Agricultura (CENA), in Piracicaba, São Paulo. The EMBRAPA soils laboratory in Manaus analyzed samples for comparison with the INPA soil texture results. Susan G. Laurance and William F. Laurance helped with data coding and checking. Thierry Desjardins, Wim Sombroek, Ed Tanner, Summer Wilson, and two anonymous reviewers commented on the manuscript. Financial support was provided by World Wildlife Fund–US and the Smithsonian Institution, the National Institute for Research in the Amazon (INPA PPIs 5-3150 and 1-3160), and the National Council of Scientific and Technological Development (CNPq AI 350230/97-98 and 523980/96-5).

CHAPTER 24

Forest Restoration in Abandoned Pastures of Central Amazonia

GISLENE GANADE

Forest clearing and habitat fragmentation in Brazilian Amazonia have escalated since the 1970s due to commercial activities such as logging, mining, and cattle ranching, but also as a side effect of the expansion of the road system (Fearnside 1989b; see also Chapters 3 and 26). Of these activities, the establishment of pastures for cattle ranching is, nevertheless, the main factor responsible for the contraction of Amazonian forests (Buschbacher 1986; Nepstad, Uhl, and Serrão 1991). In central Amazonia, many pastures that were created during the 1980s have been abandoned after only a few years of use. Abandoned pastures were then invaded by pioneer woody species, and the landscape was transformed into a combination of continuous forest and forest fragments embedded within a mosaic of pasture land and regrowth areas, or secondary forests. In this chapter I focus on the natural processes involved in the conversion of abandoned pastures to regrowth areas and the role that regrowth vegetation may play in directing forest restoration. I will show results of experimental studies on seed and seedling establishment undertaken in regrowth areas of the BDFFP in central Amazonian Brazil (fig. 4.1) and discuss the possibilities of managing this large and complex landscape in a way that both stimulates forest preservation and improves the land-use prospects for ranchers.

Surrounded by an Ocean of Pastures

Forest fragments in Amazonia are isolated in an ocean of green pastures. Plants fall into the class of organisms that can be most jeopardized by this isolation because they are highly dependent on the efficiency of their dispersal agents to migrate to new sites. For example, some wind-dispersed tropical seeds may not be able to arrive in patches of forest much beyond 100 m away (Augspurger 1984). Additionally, many animal dispersers may be unable to promote seed migration between spatially isolated areas (Janzen and Vásquez-Yanes 1991). Therefore, species with low mobility can become confined to fragments and dependent on rare events of migration to disperse and exchange genetic material with neighboring populations.

In this scenario, in which some plant species are fated to drown in the green ocean that surrounds them, a simple question comes to mind. Do pasture areas function as a harsh ocean prohibiting the establishment of forest species? Some studies

This is publication number 291 in the BDFFP Technical Series.

have reached the conclusion that they do. Pasture grasses can be strong competitors for space, water, and nutrients, considerably reducing the chance of native tree establishment in abandoned pasture lands (Uhl, Buschbacher, and Serrão 1988; Nepstad, Uhl, and Serrão 1991; Miriti 1998). Furthermore, intense management of tropical pastures can impoverish and erode the soil, decreasing the likelihood of subsequent natural forest invasion. Evidence supporting this trend has been noted in Colombia, Puerto Rico, and eastern Amazonia, where forest restoration of abandoned pastures could hardly be achieved without human intervention (Uhl, Buschbacher, and Serrão 1988; Nepstad, Uhl, and Serrão 1990; Aide and Cavelier 1994; Aide et al. 1995).

There are, nevertheless, some situations in which pastures can be readily invaded by forest species. This arises because the likelihood of natural forest invasion in pastures is inversely related to the intensity of land use prior to abandonment (Uhl, Buschbacher, and Serrão 1988). Lightly managed pastures, therefore, have a higher chance of being invaded by forest species than do more intensively managed ones. That is the case for many pastures surrounding forest fragments in central Amazonia. In this region, pastures were created following governmental incentives, but their low productivity together with a lack of specialized personnel and technology led to large pastoral areas being abandoned within a few years of establishment. As no serious slash and burn practices continuously occurred in many of these areas, forest recovery was promptly initiated, beginning with the invasion by plant species that typically occur in forest gaps. These "pioneer" species possess life-history characteristics that have predisposed them to invade disturbed sites. Isolated patches of forest thus became surrounded by a new and dynamic matrix of pioneer vegetation that may offer opportunities for the recruitment and migration of plant species previously confined to forest fragments.

Restoration from Abandoned Pastures to Regrowth Areas

The initial recovery from abandoned pastures to forest occurs when pioneer tree species establish themselves within pasture lands. Their seeds are brought to abandoned pastures by flying frugivores, i.e., bats and birds, which can travel long distances transporting large quantities of seeds in their guts (Nepstad, Uhl, and Serrão 1991). The most common pioneer species to become established in pasture areas of central Amazonia are in the genus *Vismia* (Clusiaceae). A survey of abandoned pastures located around forest fragments near Manaus, Brazil, have shown that three species of *Vismia*—V. *japurenses,* V. *guyanensis,* and V. *cayenensis*—represent more than 50 percent of the trees dominating these regrowth areas (Ganade 1996). It seems fair to ask why *Vismia* species are so successful in invading these pastures.

Vismia species are good colonizers of open sites and exhibit fast growth rates once established as seedlings. They produce seeds in large numbers that are usually dispersed by animals with wide territorial ranges. *Vismia* species can germinate promptly when light and heat are available, and their seeds are able to wait in the soil until good conditions for germination are reached. These features are, however, common to most pioneer species, and there are a considerable number of species in this region that could fulfill these requirements. Some unique traits, however, may partly explain the great success of *Vismia* over other species in invading pasture areas. As pointed out in Chapter 25, some *Vismia* species can survive moderate levels of fire and are able to spread vegetatively. So a single adult tree that survives a fire, or a small seedling germinating in a tiny opening of grass, might be able to initiate the formation of an island of trees after few years. Pasture grasses are generally shade intolerant, so that, as *Vismia* trees grow and produce litter, the grasses become overshadowed and die, releasing resources for young vegetatively spread shoots. Second, the success of *Vismia* in colonizing abandoned pastures can also be explained by its ability to compete with pasture grasses for soil resources.

Working in the matrix around the BDFFP fragments, Miriti (1998) experimentally demonstrated that the presence of grasses reduces the establishment of a wide range of forest species in pasture areas of central Amazonia. This pattern could be partially explained by grasses rather than forest seedlings being superior competitors for soil resources in fertilized pastures. Indeed, to improve pasture persistence, phosphorus, the most scarce soil nutrient in the Amazonian region (Vitousek and Sanford 1986; see also Chapter 23), must be added either by direct fertilization or by burning forest biomass prior to pasture establishment (Fearnside 1990a). The ability of *Vismia* to invade pastoral areas suggests that *Vismia* species may be better able than other pioneer tree species to compete with grasses in phosphorus-fertilized sites.

To measure the importance of phosphorus I conducted an experiment in which all standing vegetation of a regrowth area was cleared and seedling recruitment was monitored in unfertilized and phosphorus-fertilized plots (Ganade 1996). Phosphorus addition enhanced the abundance of seedlings of the pasture grass *Brachiaria* sp. (Poacea) and at the same time decreased the relative abundance of pioneer species of the genus *Bellucia* (Melastomataceae). The abundance of *Vismia* species, on the other hand, increased simultaneously with the abundance of the pasture grasses in phosphorus-enriched plots. These results show that pasture grasses may not be able to suppress the establishment of *Vismia* species in phosphorus-fertilized fields as they do with other pioneers, such as *Bellucia* spp. Thus, the ability of *Vismia* species to spread vegetatively, and their potential of invading pastures with an increased phosphorus content in the soil, may partially explain why *Vismia* is more successful in colonizing young pasture areas than other pioneers.

Six years after pasture abandonment, pioneer trees completely dominate the landscape, forming what I have been calling here "regrowth areas." These areas rapidly spread through abandoned pastures, radically changing the structural features of the matrix in which forest fragments are embedded. This phase of forest regeneration, in which pioneer species invade abandoned pastures, occurs relatively fast, when one bears in mind that up to two hundred years may be required for abandoned fields of Amazonia to reestablish a vegetation with attributes that resemble the species richness and total biomass levels of pristine forests (Saldarriaga and Uhl 1991). Therefore, the invasion of regrowth areas by species characteristic of pristine forests may be one of the crucial factors controlling the speed in which abandoned pastures will revert to primary forest. Understanding the processes controlling their invasion is, then, a key issue for the implementation of programs of forest restoration and landscape management (Luken 1990).

Colonizing Regrowth Areas

The establishment of primary forest species in regrowth sites is a function of seed arrival. Seeds can be transported by abiotic agents such as wind and water, although most tropical forest species depend on animals for dispersal (Howe and Smallwood 1982). Seed arrival, thus, depends on the ability of dispersal agents to transport seeds to regrowth sites (Nepstad, Uhl, and Serrão 1990; Janzen and Vásquez-Yanes 1991). Because animals that disperse the large seeds of primary forest species rarely venture into disturbed areas, seed arrival is frequently seen as a barrier to forest restoration in degraded habitats (Nepstad, Uhl, and Serrão 1990; Janzen and Vásquez-Yanes 1991; Robinson and Handel 1993).

To test whether seed shadow is an important limiting factor in the establishment of primary forest species in regrowth areas, I studied the patterns of seed arrival in a primary forest and an adjacent regrowth site by collecting falling seeds in 40 seed traps of 1 m² for 14 months. Fifteen traps were placed at each site in a 100 × 100 m (1 ha) area. Traps were 1 m² and 0.5 m above ground and made with a nylon cloth that held seeds as small as 1 mm in diameter. Tanglefoot® was spread on each of the four legs that supported the traps to deter climbing insects. Surveys were conducted at biweekly intervals for the first six months. In a second phase, arriving seeds were left to accumulate in traps for eight months, at which time all seeds, fruits, and germinated seedlings in the traps were collected and "morphotyped" (separated into unidentified but recognizable species).

A total of 96 seed morphotypes from 17 plant families were collected in the forest site, while only 9 seed morphotypes were collected in the regrowth site throughout the study period. Most seeds arriving in the regrowth site were pioneer trees of the genera *Vismia* and *Bellucia,* and these were most probably produced by trees already established in the site. Moreover, the local richness of seed morphotypes falling per square

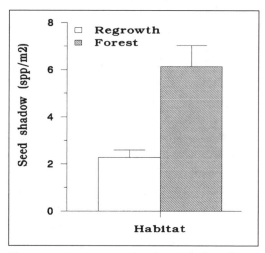

Fig. 24.1. Average number (+ 1 SE) of seed morphotypes falling in regrowth areas (unshaded bar) and forest (shaded bar) from January 1994 to March 1995.

meter within the regrowth site was 2.5 times lower than the local richness of seed types arriving in the adjacent forest (fig. 24.1). These results show that although there are many seeds being produced in the primary forest, their chance of arrival in regrowth sites is low and seed dispersal may be a key factor slowing the rates of conversion of regrowth areas into tropical forest.

Natural Seedling Recruitment

If seed arrival is a key factor constraining plant establishment in regrowth areas, natural seedling recruitment should be lower in regrowth sites when compared to forest sites. To test this hypothesis I studied natural seedling recruitment for fifteen months in eight pairs of 2.0 × 1.0 m plots randomly assigned within a 2 ha area of each forest and regrowth site. All seedlings up to 1 m in height that were present in the plots before the beginning of the study were removed to promote free space for arriving species. Seedling diversity is represented by the relationship between log-transformed

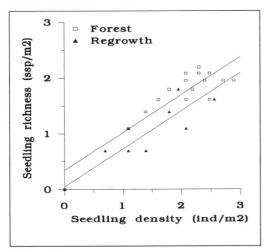

Fig. 24.2 Relationship between number of species and seedling density in forest (open rectangles) and regrowth areas (solid triangles).

(ln[n+1]) species richness and plant density, also log-transformed. There was a significant relationship between species richness and density in both forest and regrowth sites ($y = 0.03 + 0.678x$, $r^2 = 0.80$). An analysis of covariance (ANCOVA) showed that, for a given plant density, plots in the forest site have on average 1.7 times more species than do plots in the old-field site (fig. 24.2; $F = 11.48$, $P < 0.01$, $df = 1,29$).

Natural seedling recruitment in regrowth areas was very low and dominated by vegetative sprouts of *Vismia*. On the other hand, seedlings emerging in the forest site constituted a diverse range of shade-tolerant woody species. These species represented members of the families Leguminosae, Arecaceae, Lecythidaceae, Chrysobalanaceae, Melastomataceae, Burseraceae, Apocynaceae, Euphorbiaceae, Violaceae, Annonaceae, and Tiliaceae.

The low local diversity of seedlings establishing in regrowth areas when compared to forest reinforces the idea that natural seedling recruitment in regrowth areas may be restricted by seed dispersal. However, these findings do not exclude the alternative

possibility that the inability of most primary forest species to colonize regrowth areas is due to their low chance of establishment and survival in this habitat type, in spite of successful dispersal.

Seedling Establishment and Survival

The successful establishment of forest species into regrowth areas will depend on their ability to cope with the environment of regrowth habitats. Abiotic factors such as light, air humidity, and soil nutrient concentrations and pH can differ considerably between regrowth and primary forest. I measured the differences between these variables in a series of eight plots that were randomly placed within a 2 ha site located in each habitat. I measured Photosynthetic Active Radiation (PAR, μmol/m²/s) at the soil surface using a Delta T sunflecks ceptometer and air humidity (%) using a whirling hygrometer. The light and air measurements were performed twice, in two consecutive days of clear blue sky, between 1100 h. and 1300 h. Each replicated soil sample was a compound of five random samples collected at 0 to 15 cm depth within a 100 m² area.

In regrowth areas, light penetration at the soil surface can be four times higher and air humidity 20 percent lower than in a forest (table 24.1). These differences are probably due to variations in canopy structure, which range from 30 to 40 m in forest, but only 5 to 7 m in regrowth sites. Soil nutrient conditions can also differ between these two vegetation types, with soil pH and nitrogen concentrations being higher in regrowth soils than in forest soils (table 24.1). Such differences may reflect the land-use history of these sites, because the soil characteristics of regrowth areas suggest that they still maintain soil nutrients from the burning of forest biomass prior to pasture establish-

TABLE 24.1.   Comparisons Between Measurements of Photosynthetically Active Radiation, Air Humidity, Air Temperature, Soil Acidity, and Soil-Nutrient Concentration in Forest and Regrowth Plots

Measurements	Forest		Regrowth		ANOVA	
	Mean	SE	Mean	SE	$F_{1,14}$	P Value
Light (PAR)	20.06	4.01	84.63	18.81	24.64	<0.001
Percentage air humidity	61.63	2.98	48.75	1.91	13.22	<0.01
n Nitrogen (%)	0.12	0.01	0.16	0.01	5.16	<0.05
p Phosphorus ($\mu g.g^{-1}$)	1.88	0.23	2.25	0.16	1.80	ns
k Potassium ($\mu g.g^{-1}$)	22.50	1.68	23.25	3.27	0.05	ns

Note: PAR = photosynthetically active radiation.

ment. Therefore, with respect to primary forest, regrowth areas offer an environment for plant colonizers that has higher light intensity at ground level, drier air conditions, and soils that have lower acidity and are richer in nitrogen.

Theoretically, the first vegetation that establishes itself in an open area can enhance, decrease, or have no effect on the conditions that favor further propagule establishment (Connell and Slatyer 1977). For example, regrowth vegetation can impose high levels of root competition for newly germinated seedlings. Seedling growth in old fields of eastern Amazonia has been shown to be four times lower than seedling growth in forest gaps, and this pattern was largely due to root competition (Nepstad, Uhl, and Serrão 1991). Alternatively, existing patches of shrubs in old fields can favor the establishment of new colonizers by providing habitats suitable for seedling establishment. In eastern Amazonia, the establishment of woody rainforest seedlings under clumps of the shrub *Cordia multispicata* was enhanced eightfold in relation to seedling establishment under pasture grasses. The presence of the shrubs in old-field areas favors the establishment of rainforest species by attracting seed dispersers and providing light and nutrient conditions similar to those found in primary forest (Vieira, Uhl, and Nepstad 1994).

Attack by natural enemies can also influence propagule and seedling survival. For example, regrowth areas may function as shelter for seed predators, pathogens, and herbivores. In eastern Amazonia, seed and seedling attack by leaf-cutter ants and small rodents was an important barrier for forest regeneration in degraded sites (Nepstad, Uhl, and Serrão 1990; Nepstad, Uhl, and Serrão 1991). Alternatively, other evidence suggests that pasture areas sustain lower densities of seed predators than those found in adjacent tropical forest (Osunkoya 1994). In central Amazonia, seed predation by insects and vertebrates does not limit the establishment of woody species in pastures (Miriti 1998) moreover, insect herbivory on woody species was reported to be lower in regrowth areas than in adjacent forest (Benitez-Malvido 1995; Chapter 11).

Establishment and Growth of Planted Species

One way to investigate the overall impact of regrowth areas on colonizing species is to compare the performance of species invading these sites with their performance in forest sites. I conducted such a study by sowing seeds from four tree species in both regrowth and forest sites and measuring their seedling establishment and growth for ten months. The four species of tropical

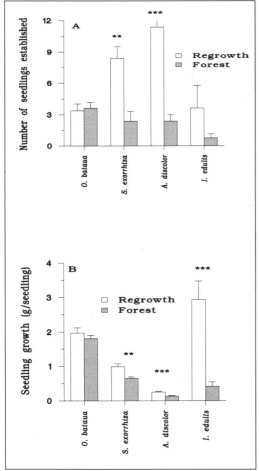

Fig. 24.3. Plant performance of *Oenocarpus bataua, Socratea exorrhiza, Aspidosperma discolor,* and *Inga edulis* in regrowth (unshaded bars) and forest (shaded bars) areas. (A) plots represent the mean (+ 1 SE) of the number of seedlings established, and (B) the mean (+1SE) of growth, measured by seedling dry biomass in g. Significant differences in seedling establishment and seedling growth between regrowth and forest sites, for each studied species, are indicated by * P > 0.05, ** P > 0.01, and *** P > 0.001.

trees used in the experiment were *Aspidosperma discolor* (Apocynaceae); *Inga edulis* (Leguminosae); *Oenocarpus bataua;* and *Socratea exorrhiza* (Arecaceae). *Aspidosperma discolor* is an emergent tree of the

primary forest with round and flat wind-dispersed seeds. *Inga edulis* is a fast-growing species widely cultivated in central Amazonia for the sweet aril that surrounds its large, soft seeds. Fruits of *Inga* species are eaten by large vertebrates and may be dispersed by monkeys. *Oenocarpus bataua* and *S. exorrhiza* are palm trees typically found around freshwater streams in primary forest; they have a thin but very rich aril around their large seeds, and their fruits are eaten and probably dispersed by large birds, tapirs, and primates (including humans).

Seeds of these four species were sown in eight plots placed within a 2 ha regrowth area and in eight plots placed within a 2 ha area of forest. In each plot, one group of sixteen seeds of each study species were sown. Seeds were sown 10 cm apart and marked with wooden sticks. The regrowth and forest areas studied were located 500 m from the edge between both vegetation types to avoid edge affects. Seedling establishment was monitored at monthly intervals for ten months when all seedlings were harvested and weighed to assess their growth as measured by total dry mass. The total number of seedlings established by the end of the experiment and the mean seedling dry mass produced were compared between regrowth and forest areas by one-way analysis of variance (ANOVA).

Establishment and growth of the studied species was, in general, higher in regrowth vegetation compared to forest vegetation (fig. 24.3). Seedling establishment in the regrowth site was four times higher for *S. exorrhiza* and five times higher for *A. discolor* (fig. 24.3a). Once germination occurred, seedlings that established themselves in regrowth sites were able to grow at a faster rate than seedlings established in the forest. Ten months after the beginning of the experiment, the dry biomass of plants in regrowth areas was 50 percent greater for *S. exorrhiza,*

90 percent greater for *A. discolor* and 700 percent greater for *I. edulis* in comparison with the dry biomass of plants established in the forest (fig. 24.3b). The establishment and growth of *O. bataua* seedlings did not differ between the two sites. Thus, regrowth areas represent similar or better environments for colonization, survival, and growth of forest species than do forest areas. Experiments done in similar sites using transplanted seedlings of three species in the Sapotaceae family also found that seedling growth was higher in regrowth areas than in the primary forest (Benitez-Malvido 1995).

Many factors may contribute to the pattern of higher establishment and growth of invading species in regrowth areas. Seed removal and seed attack by herbivores were monitored approximately every two weeks for ten months. The levels of seed predation were in general lower in the regrowth site (fig. 24.4; Ganade 1996). Moreover, the plant litter released by the regrowth vegetation had created moist microsites suitable for germination of primary forest species (Ganade 1996), contrary to some evidence suggesting that plant litter can obstruct seedling establishment (Facceli and Pickett 1991). The presence of litter may have helped forest species to overcome the problems associated with establishing in the drier air conditions experienced in regrowth sites. Finally, high light levels in the regrowth site contribute to the higher growth of colonizing species in these areas. Indeed, it has been suggested that light is the most important resource limiting plant growth in tropical forests (Denslow et al. 1990). It appears that light is the main limiting factor for plant growth in these regrowth areas as well, because the addition of such scarce soil nutrients as phosphorus and potassium did not modify the growth rates of seedlings while an increase in light availability due to the removal of regrowth vegetation enhanced seedling biomass (Ganade 1996).

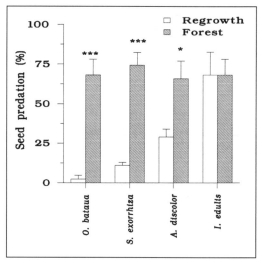

Fig. 24.4. Seed predation of *Oenocarpus bataua*, *Socratea exorrhiza*, *Aspidosperma discolor*, and *Inga edulis* in regrowth (unshaded bars) and forest (shaded bars) areas. Plots represent the mean (+ 1 SE) of the percentage of seeds removed and attacked in groups of sixteen seeds of each studied species. Significant differences in seed predation between regrowth and forest sites, for each studied species, are indicated by * $P > 0.05$, ** $P > 0.01$, and *** $P > 0.001$.

The evidence discussed so far has shown that if seed arrival takes place, plant establishment can promptly occur in regrowth areas of central Amazonia. Furthermore, seedling establishment and growth in regrowth areas is generally more successful than in primary forest. Together, these results strongly suggest that seed arrival is indeed the most important process restricting forest regeneration.

Alternatives for Land Use

Before management plans for regrowth areas can be developed, one has to construct arguments that persuade ranchers to change their present land-use practices. It is naive to believe that ranch owners will abandon their land or convert it back to tropical forest for reasons that are purely based on conservation of natural resources. Areas of abandoned pastures that are within the domain of their land are associated with their livelihood. In many cases, effort is made to restore pasture land by cutting and burning regrowth vegetation. Here, I discuss how partial changes in land-use practices in ranches of central Amazonia can be economically viable, ecologically sound, and in many cases more productive than the practices applied nowadays. These considerations may have significant impact on land-management decisions made by ranchers and governmental agencies.

It is important at the outset to point out that pastures in Amazonia are threatened by decreases in productivity. This is because the fertility of soils can be remarkably low (Vitousek 1984; Chapter 23), and although there is a high input of nutrients, due to the cutting and burning of forest at the time of pasture establishment, the amount of nutrients in the soil decreases with pasture aging (Fearnside 1980b). Five years after pasture establishment, the amounts of available phosphorus in the soil can decrease from 4.18 to 0.46 mg/100g (Fearnside 1980b), and soil compaction and erosion begin due to the torrential tropical rains that fall in the region. Furthermore, pastures are constantly invaded by plants that are inedible to cattle; in some extreme cases pastures are completely overtaken by outbreaks of weeds. For these reasons, the agricultural sustainability of most pastures can be maintained only by means of heavy and sometimes uneconomic inputs (Fearnside 1990a). No wonder many pastures in the Amazonian region have been abandoned or partially abandoned, often for more than ten years (Nepstad, Uhl, and Serrão 1991). This common practice of pasture abandonment clearly points to a need for alternative strategies for land use.

We are living now in a world scenario where there is an increasing interest in the exotic products and exotic species of Amazonia. Within this context, countries in the Amazon basin have the potential to export tropical fruits, oils, and pharmaceutical products to national and international markets. Estimates for the value of currently used nontimber products found in a Peruvian Amazonian forest can reach up to $7,000 (U.S.) per hectare (Peters, Gentry, and Mendelsohn 1989). Some of these products are known but many are still to be discovered from species that may be threatened with extinction. To adopt alternatives for land use that take into account the preservation of both our well-known and still-unknown biological products seems an economically promising strategy.

Alternative strategies of land use such as agroforestry have been traditionally used by indigenous people and local farmers in many countries of the world including those in the Amazon region (Peck 1990; Anderson 1991; Altieri 1995). This technique consists of cultivation of commercially valuable species within a matrix of primary or secondary forests. Because the forest does not need to be completely replaced by cultivation, this practice preserves the diversity and integrity of forests while producing agricultural goods (Alcorn 1990; Fearnside 1990a). Agroforestry techniques have been successfully applied in many tropical countries, such as India, Mexico, Brazil, Guatemala, and Ecuador. In a single field, farmers may grow more than one hundred commercially valuable products, such as wood, fiber, rubber, medicinal and ornamental plants, as well as vegetables, crops, spices, fruit trees, and livestock food (Anderson 1990; Altieri 1995).

Some agroforestry practices are, however, labor intensive and may not be easily embraced by ranchers, who frequently have a very restricted number of personnel. A less labor-intensive alternative would be the practice of forest enrichment using commercially valuable species followed by "extractivism," or the harvesting of forest products from the modified forest. This practice has been successfully applied by rural inhabitants in the floodplains of the Amazon estuary (Anderson 1991). In the following sections I address in more detail how the results of this research can be directly applied to this alternative and how commercial land use and the conservation of fragmented forests may be reconciled by considering the landscape in which forest fragments are embedded as potential income-generating areas.

Management of Regrowth Areas

The results reported here open a window on alternatives for land use in regrowth areas and convey important information on the steps that need to be taken to develop plans for management. An interesting and unexpected outcome of these studies was that the higher light intensity combined with the lower chance of being attacked by herbivores when compared to the forest makes regrowth areas a more suitable place for seedling establishment than primary forest. It follows from this that ranch owners have the opportunity to enrich these areas with commercially valuable species and sell local products to regional and international markets. Moreover, by spatially aggregating target species, ranchers could improve product accessibility and therefore their harvesting capabilities.

Programs to enrich species composition in regrowth areas may be achieved by techniques as simple as throwing seeds in desirable localities within regrowth sites. After seedlings become established, the canopy of pioneer trees could be partially removed to increase light penetration, while protection against desiccation, wind, and other cli-

matic hazards would be maintained by trees and plant litter retained at the site. The choice of species to be introduced is another important consideration, and decisions need to be made on the basis of future market forecasts. Products such as wood, tropical fruits, palm heart, palm oil, and nuts are some of the many that can be produced in the region.

A Landscape Perspective for Management

Some arrangements of the landscape mosaic are more effective than others in preserving wildlife and biodiversity. Species migration, population survival, and the dynamics and structure of plant and animal communities can be altered depending on the composition and spatial arrangements of the landscape (Wiens 1995). A good example of how the landscape mosaic may alter species dynamics within remnants comes from studies of the fragmented forests at the BDFFP experimental site in central Amazonia. In these areas, the type of matrix of regrowth vegetation surrounding fragments was shown to be more important than fragment size in defining the species composition of small mammals and birds in forest fragments (Malcolm 1991; Stouffer and Bierregaard 1995a, 1995b; Bierregaard and Stouffer 1997). This suggests that the way in which managed areas are distributed in the landscape may be critical for the long-term survival of species that inhabit forest fragments.

Management of regrowth areas could be strategically designed in two ways. First, regrowth areas can be maintained around forest fragments as belts for forest restoration that function as buffer zones and reduce edge effects within fragments (Kapos et al. 1993, 1997). As quite wide areas of regrowth vegetation could be maintained, they may function as extremely efficient buffer zones, given that Rankin–de Mérona and Hutchings (Chapter 9) have shown that buffer zones will need to be on the order of a kilometer to guarantee that forest fragments would be maintained structurally intact. Second, the spatial position of regrowth areas could be planned in a way that allows the regrowth matrix to link forest fragments with continuous forest. Thus, if ranchers decide to maintain areas of pastures between fragments and continuous forest, forest fragments could be linked by corridors of regrowth vegetation enriched with commercially valuable species. As a consequence, the size of pasture clearings may be reduced, which, in terms of conservation of natural resources, would be of great benefit for the region given that the speed of natural forest invasion would be increased if pastures are eventually abandoned (Silva, Uhl, and Murray 1996). Linking forest remnants by corridors of regrowth vegetation would also favor the transit of primary forest animals and plant species between patches of forest. In this way, populations confined to forest fragments would no longer be isolated in an ocean of green pastures but instead be connected by favorable sites for their establishment and reproduction.

What are the economic benefits of adopting a landscape perspective when planning programs to manage regrowth areas? One possible economic benefit for ranchers is that direct contact with an edge of forest can enhance the migration of forest species to regrowth areas, and these species can be useful for the management of commercially valuable species. The presence of pollinators supported by adjacent primary forest can increase the harvesting capabilities of ranchers by increasing overall fruit production (Buchmann and Nabhan 1996). Furthermore, the colonization of regrowth areas

by soil-dwelling microorganisms such as mycorrhizal fungi and insect decomposers may enhance nutrient cycling in the managed system.

Clearly, the invasion of regrowth areas by beneficial pollinators and decomposers of organic matter can potentially be accompanied by the invasion of undesirable seed predators and herbivores. The extent to which these invasions will be harmful for commercially valuable species remains to be tested, and the subject could be an interesting focus of future research. Nevertheless, the invasion of undesirable species in managed areas should not be seen as an impediment for the establishment of species enrichment practices in this region. Rather, it represents a part of the production cost of agronomic practices that preserve biodiversity. Application of ecological principles will improve the management of these areas and help minimize this cost. Alternatively, this is an area where creative insurance plans and fiscal incentive policies can play an important role.

Conservation Lessons

1. Studies of the patterns and processes related to forest regeneration in abandoned pastures of central Amazonia suggest alternatives for land use that decrease the degree to which species are isolated within forest fragments.

2. Areas of abandoned pastures that were naturally invaded by pioneer trees, here called regrowth areas, can be enriched with commercially valuable species that provide an income stream for the landowner and can also serve as corridors of vegetation linking forest fragments with larger patches of forest, increasing the chances of species dispersal between forest remnants.

3. Because seed arrival is an important constraint for species invasion and because seed predation is lower, and seedling establishment and growth of different tree species are higher under regrowth vegetation than under forest vegetation, regrowth areas are a useful matrix for the management of commercially valuable plant species.

4. There is an alternative to managing abandoned pasture lands that preserves Amazonia's biological diversity while reaping a profit from the land. The proactive management of regrowth areas aims to maximize the long-term gain of preserving natural products (species)—both known and yet to be discovered in these tropical forests.

CHAPTER 25

Effects of Fire on Rainforest Regeneration in the Amazon Basin

G. BRUCE WILLIAMSON AND RITA C. G. MESQUITA

In the rainforests of the Amazon, fire is used extensively to create pastures following timber extraction and to establish farms following deforestation. Fire is the principal tool for removing slash after the forest has been cut and dried. Without fires, removing slash would be both impractical and uneconomical. With fire, vast areas of slash can be eliminated in a few days with minimal effort, after the forest has been felled.

Natural fires in primary rainforest of the Amazon appear to be rare today, except in seasonal localities on the edge of the Basin that experience an extended dry season (Uhl, Kaufman, and Cummings 1988) or during extremely severe El Niño droughts. The presence of charcoal in soil cores in many Amazonian localities (Sanford et al. 1985; Saldarriaga and West 1986; Bassini and Becker 1990) reflects drier conditions in the geological past, perhaps mega El Niño events (Absy 1982; Markgraf and Bradbury 1982; Meggers 1994a, 1994b) or anthropogenic burning by Native Americans in pre-Columbian times. Where evidence for fire exists, the fire return intervals appear to be on the order of 400–1,000 years, longer than the life expectancy of most trees (but see Chambers, Higuchi, and Schimel [1998]

and Chapter 7 for very old trees). In more recent history, it is doubtful that today's intact rainforests have been subjected to frequent fires (Kaufman and Uhl 1990). In fact, rainforest trees do not exhibit adaptations to fire as do trees in areas known to burn naturally (Williamson, Schatz, et al. 1986; Kaufman and Uhl 1990; Rebertus, Williamson, and Platt 1993).

In primary forest of the wet tropics the moisture content of the forest litter's fine fuels generally remains above the level below which combustion can be maintained (Kaufman and Uhl 1990). This "moisture of extinction," derived for temperate hardwoods as 25 percent (Albini 1976), is assumed to apply to tropical hardwood forests as well (Kaufman and Uhl 1990). Furthermore, woody fuel is almost always above its "fuel ignition threshold" of 15 percent (Kaufman and Uhl 1990). However, after the rainforest is cut, the tropical sun in the Amazon can dry the slash before it is burned. A general rule of thumb is that after ten consecutive days of sunshine without rain and cloud cover, the moisture content of the slash drops and the litter becomes combustible (Uhl and Kaufman 1990). The ignition source is always anthropogenic.

Where the rainforest has been disturbed

This is publication number 292 in the BDFFP Technical Series.

by selective timber extraction or by edge effects around forest fragments, humidity is reduced and sufficient fuel may be present to allow fires to penetrate the forest without clear-cutting (Uhl 1987; Kaufman 1991; Nepstad, Veríssimo, et al. 1999). Fire frequently spreads through secondary forest and selectively cut forest without penetrating primary forest (Kaufman and Uhl 1990). The differential susceptibility to fire is especially evident during the El Niño Southern Oscillation (ENSO)-related drought of 1982–83 in Borneo, where fires spread rapidly through secondary and high-graded rainforests but burned little primary rainforest (Leighton and Wiraman 1986; Woods 1989). Preliminary data suggest similar differences between primary and secondary Amazonian forests in the 1997–98 drought (Hammond and Steege 1998).

Responses of Rainforest Trees to Fire

In ecosystems where fires occur naturally, there are three basic mechanisms for genetic survival of woody plants (Carpenter and Recher 1979; Williamson, Schatz, et al. 1986; Rebertus, Williamson, and Platt 1993). Where fire intensity is low (surface fires), fire frequency is high (many times in the lifetime of a tree) and fires are relatively periodic (low variance about the mean frequency), trees often survive fires by having thick bark to protect the lateral cambium and trunks tall enough that apical buds survive above the heat (table 25.1). This "adult stem lives" strategy, which is common among tree species of the *cerrado* (woody savannah) vegetation in Brazil (Coutinho 1990) and the sub-tropical savannas of the United States (Rebertus, Williamson, and Platt 1993), is unknown among Amazon rainforest trees. In the rainforest many trees have long boles, but they lack thick bark. Uhl and Kaufmann (1990) measured bark

thickness in trees in undisturbed rainforest in the eastern Amazon and estimated that 98 percent of all stems greater than 1 cm DBH would be killed in the event of a surface fire.

Beyond the absence of thick bark, there are no rainforest trees known to exhibit the temporal survival mode of the "adult dies" strategy (table 25.1), in which fire kills the adult trees but breaks the dormancy of seeds stored in the soil or in serotinous fruits or cones on the tree, as is the case for some North American pines (Richardson 1988), Australian *Eucalyptus* (Gill 1981), chaparral shrubs (Horton and Kraebel 1955), Mediterranean shrubs (Le Houreau 1973), Australian shrubs (Lamont and Barker 1988), and tropical montane shrubs (Williamson, Schatz, et al. 1986). With seeds tolerant of fires and seed dormancy broken by fire, this mode of genetic survival often results in a cohort that germinates, grows, and reproduces the next generation of seeds that remain dormant until the next fire. However, most species of old-growth rainforests do not share these key adaptations; to the contrary, their seeds are not dormant (Garwood 1989, Thompson 1992), and even when buried in the soil they seem to be particularly susceptible to fire (Brinkmann and Vieira 1971).

Where fires are intense, but infrequent and aperiodic (table 25.1), woody plants exhibit a "resprout" strategy of which the key adaptations are subterranean buds at the base of the stem or on lateral roots and a highly distorted root:shoot ratio (below-ground:above-ground distribution of biomass). Among fire-adapted resprouters, these root:shoot ratios are distorted as high as 6:1 in favor of roots and may reach 10:1, even in controlled circumstances where the above-ground tissue is protected from fires (Rebertus, Williamson, and Moser 1989). Most fire-adapted resprouters can be burned in successive years, losing only a fraction of

TABLE 25.1.   Responses of Adult Trees to Natural Fires, Their Mode of Genetic Survival and Their Key Adaptations for Survival, All Associated with Fire Regimes Varying in Freqency, Intensity, and Periodicity

Adult plant response	Survival mode	Key adaptations for survival	Fire regime		
			Frequency	Intensity	Period
Adult stem lives	Spatial, above ground	Thick bark Tall bole	high	low	high
Adult dies	Temporal Fire-tolerant seeds	Fire-dormant seeds	low	high	high
Adult resprouts	Spatial, below ground	Subterranean buds High roots:shoot ratio	high	high	low

their biomass in each fire; for example, a resprouter losing 10 percent in five successive fires would retain 59 percent (0.90^5) of its biomass, even assuming no gain from growth between fires. The evolutionary cost of maintaining such a high proportion of biomass underground is, of course, slow growth. Annual biomass accumulation in classical resprouting species may be less than 20 percent even in the first few years of growth, a time when many woody plants double in size annually.

At first appearance some rainforest trees might be expected to resprout after fires because they have demonstrated abilities to resprout along the main stem when it is broken in treefalls (Putz and Brokaw 1989), snapped by hurricanes (Vandermeer et al. 1995), cut (Hartshorn 1989; Gorchov et al. 1993), or uprooted (Negrelle 1995). Even so, few of them have dormant buds or regenerative capacity underground where it would be protected from fire. Consequently, burning that girdles trees close to the ground causes higher mortality than cutting or snapping, which occurs higher on the stem well above the root collar. Kaufman (1991) reported that prescribed fire at two selectively logged forests in the state of Pará, Brazil, killed 94 percent of the tree crowns and 66 percent of the individuals.

The fate of rainforest individuals that re-sprout after fire—for example, the 34 percent of individuals showing basal or epicormic sprouting (along the main stem) in the study cited above (Kaufman 1991)—is fairly predictable in anthropogenic systems. Successive fires in pastures exhaust the reserves of resprouting plants and cause death before the pasture is abandoned (five to fifteen years after clear-cutting) for most species because rainforest trees do not maintain a large portion of their biomass underground. Where pastures are managed less intensively by burning every two or three years, resprouting rainforest trees with subterranean buds may survive as new shoots. Stems resprouting on slash-and-burn agricultural sites often face only one fire but successive weedings that exhaust their reserves as well. For example, Uhl (1987) reported the density of woody resprouts dropped from $0.43/m^2$ six months after an initial slash-and-burn prior to weeding and then dropped to $0.07/m^2$ after several weedings.

Responses of Rainforest Pioneers to Fire

Species in the two main genera of pioneers of the Brazilian Amazon (Albuquerque 1980), *Vismia* and *Cecropia,* appear sensitive to hot surface fires, with most individuals suffering crown mortality. For example,

80 km north of Manaus around the Colosso
Reserves (1202 and 1104; see fig. 4.1, table
4.3, and plate 1) of the BDFFP, following a
fire in 6 yr old second growth we surveyed
trees with a DBH of 5 cm or more in 5 ha and
noted 100 percent mortality in *C. purpura-
scens* and *C. sciadophylla* and 99 percent
mortality in *V. cayennensis, V. guianensis,*
and *V. japurensis.* Individual stems sometimes survived the fire in incompletely
burned patches such as low pockets and
some protected slopes.

Although cecropia and *Vismia* spp. share
sensitivity to fire, they differ in their ability
to resprout. Cecropia species appear able to
resprout only epicormically but not from the
stem base, nor from lateral roots. In general,
cut cecropia trees resprout, but surface fires
kill the epicormic buds and the tree. While
there may be variation in sensitivity to fire
among the 100 species of *Cecropia* (Berg
1978; Gentry 1993), none have been recorded with subterranean buds that would
allow the trees to regenerate following fires.
Therefore, most cecropia individuals are
killed in surface fires, and certainly all are
eliminated where repeated burning occurs,
as in heavily used pastures. Kaufman (1991)
reported mortality of 85 percent in cecropia
individuals in fires that were ignited in secondary forests without clear-cutting fires
that were of lower intensity than those that
follow clear-cutting.

In contrast, of the thirty *Vismia* species in
the Neotropics (Ewan 1962; Gentry 1993),
the few that have been investigated appear
capable of resprouting from the base of the
stem (Uhl 1987; Dias-Filho 1990), and some
species resprout as well from lateral roots
(Williamson, Mesquita, et al. 1998). For example, along a four-year-old road cut, adjacent to the BDFFP primary forest at Reserve
Km 41, 75 percent of the adult stems of *Vismia cayennensis* were ramets—stems connected to other adjacent stems via lateral

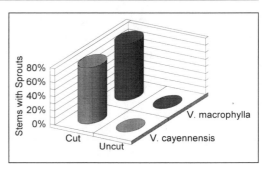

Fig. 25.1. Percentage of stems that produced
lateral, basal, or epicormic resprouts for stems
that were cut and those that were uncut for
Vismia cayennensis and *V. macrophylla.*

roots (table 25.2). However, another species,
V. japurensis, did not show lateral root connections between adjacent adult stems, although this species does resprout basally
when the main stem is cut or burned. In a
nearby forest light gap a small sample of the
same species exhibited no lateral root connections between adjacent adults (table
25.2).

In the primary forest, *Vismia* species are
present as pioneers in light gaps, often co-
occurring with *Cecropia* species, although
their leaves are inconspicuous and their
presence has been overlooked (Lucas et al.
1998). *Vismia* is not clonal in these forest
openings, perhaps because the stems are
never cut or burned. In fact, *Vismia* in forest
lightgaps rarely exhibits even basal sprouts.
To test the resprouting ability of *Vismia* in
the primary forest at Km 41 we cut stems at
heights of 5 to 150 cm above the base and
compared their responses to those of uncut
stems one month later. Cut stems showed
basal, lateral, or epicormic sprouts in 80 percent of the cases for both *V. cayennensis* and
V. macrophylla, whereas uncut stems never
produced a resprout (fig. 25.1). For both
species, half the stems producing sprouts included lateral root sprouts, up to 50 cm
away from the main stem. Some stems produced lateral, basal, and epicormic sprouts,

TABLE 25.2. Percentage of Stems that Were Ramets for Two *Vismia* Species

| Species | Percent ramets (n) | |
	Along road cut	Forest lightgap
Vismia cayennensis (12)	75% (12)	0%
Vismia japurensis	0% (27)	0% (4)

Notes: Ramets are stems connected to adjacent stems via lateral roots. Stems were sampled along a road cut and in an adjacent forest lightgap, 80 km north of Manaus. Total number of stems sampled is shown in parentheses (n).

although higher-cut stems produced more epicormic sprouts and fewer basal and lateral sprouts. Thus, *Vismia* in the primary forest has the capacity to resprout, although, as far as we know, resprouting has no role in its natural regeneration and may simply be derived phylogenetically. Root sprouting occurs in other Guttiferae genera, such as *Garcinia* and *Mammea,* as well as in the closely related African genus *Psorosperum,* which readily resprouts after fires (P. F. Stevens, pers. comm.).

In contrast to the lack of resprouts by *Vismia* species in forest light gaps, the fire and disturbance associated with pastures seem to stimulate massive resprouting. Following a pasture burn in October 1997 at the BDFFP's Colosso site, we carefully excavated all young shoots 1.0 cm or more in height along a 5 m transect from the base of stem-killed *Vismia* spp. adults in mid-December 1997 to determine whether the

shoots were basal resprouts from a previously existing stem, lateral resprouts from existing lateral roots, or newly germinated seedlings. The density of shoots varied by two orders of magnitude, from 125/m² for *V. cayennensis* to 16/m² for *V. guianensis* to 1.3/m² for *V. japurensis* (table 25.3). For *V. cayennensis* and *V. guianensis,* more than 99 percent of the shoots in the total area sampled (10 m²) were lateral resprouts, but *V. japurensis* produced as many basal sprouts as lateral sprouts (table 25.3).

Whereas *Vismia* spp.'s ability to resprout suggests adaptation to fire or disturbance, its root:shoot ratio is not one of a fire-adapted resprouter (see table 25.1). Root:shoot ratios of young *Vismia guianensis* seedlings were 1:2, a normal ratio for non-fire-adapted tree seedlings (Uhl 1987; Dias-Filho 1995). Under an annual burning regime, a seedling which would lose two-thirds of its biomass to fire would then have to triple in biomass, a growth increment of 200 percent, simply to regain its previous size. Such recovery is not impossible (Uhl 1987) but is unlikely on a sustained basis of annual burning, assuming that each year involves additional mortality risk from fire and reduced nutrient levels. However, in pastures under moderate use that are burned irregularly (for example, every second or third year), seedlings have ample time to recover and grow between fires. Pastures under light and moderate use frequently contain *Vismia* spp. clones, whereas heavily used pastures are often

TABLE 25.3. Number of Basal Sprouts, Lateral Sprouts, and Seedlings Two Months After a Pasture Fire

Species and tree number	Basal sprouts	Lateral sprouts	Seedlings
Vismia cayennensis	6	1,238	6
Vismia guianensis	0	154	2
Vismia japurensis #1	12	0	0
Vismia japurensis #2	2	17	0
Vismia japurensis #3	4	5	0

Note: Trees measured along a 5 m transect (10m²) from the base of stem-killed *Vismia* species.

void of *Vismia* spp. (Uhl, Buschbacher, and Serrão 1988; Nepstad, Uhl, et al. 1996).

The differential resprouting capacity of *Vismia* spp. and cecropia has led to the perception by some authors that *Vismia* species "thrive" on intensively managed land (Lucas et al. 1998). Actually *Vismia* spp., rather than thriving, are, at best, surviving relative to *Cecropia* spp., because both genera, along with other woody plants from the rainforest, are eliminated by repeated annual fires if the period of use is long enough (Fearnside 1990c; Nepstad, Uhl, and Serrão 1990). Thus, pastures subjected to heavy use, such as eight to ten years of frequent burning, as well as some weeding and mechanical scraping or discing, are nearly free of arboreal vegetation, *Vismia* spp. notwithstanding (Uhl, Buschbacher, and Serrão 1988; Nepstad, Uhl, and Serrão 1990).

There are no data on how many successive fires *Vismia* species can tolerate, but obviously many pastures are abandoned before all *Vismia* spp. individuals are eliminated. Throughout Amazonas, pastures abandoned after six to twelve years of moderate use (without mechanized disturbance or herbicides) have patches of *Vismia* clones (Uhl, Buschbacher, and Serrão 1988; Saldarriaga and Uhl 1991; Williamson, Mesquita, et al. 1998), leading to the misconception that *Vismia* spp. may be "fire-adapted." Indeed, *Vismia* spp.'s apparent "success" in pastures results from pastures being degraded and abandoned faster than *Vismia* spp. are eliminated.

As active pastures age, they degrade through soil compaction, nutrient depletion, and invasion by weeds—all factors that contribute to the demise of arboreal vegetation. However, we believe that fire is the ultimate culprit and that cecropia, as well as other genera, are absent from pastures where they otherwise could readily flourish (Pereira and Uhl 1998). To test whether cecropia could

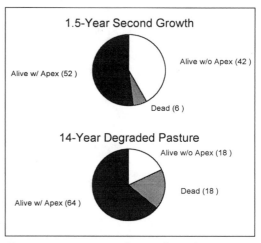

Fig. 25.2. Percentage of transplanted *Cecropia sciadophylla* that after two months were (1) alive with apical bud intact, (2) alive without the apex, or (3) dead, for two habitats, 1.5-year-old second growth and 14-year-old degraded pasture.

survive soil conditions in a moderately used pasture, we transplanted *C. sciadophylla* saplings, 1 to 3 m tall, from an area of 1.5 yr old second growth into a 7.5 yr old pasture where cecropia had been eliminated. For comparison, control saplings were dug up in the 1.5 yr old second growth and transplanted back into it. After two months the saplings were classified as alive with the apical bud intact, alive by epicormic or basal sprouting after death of the apical bud, or dead. The large apical bud, important in cecropia's rapid growth, wilts when the saplings are transplanted. The bud's survival is a measure of the sapling's subsequent recovery. There were no significant differences in the responses of the transplanted cecropia saplings between the pasture and the second growth (fig. 25.2, Chi-square test of independence, 3×2, P = 0.06).

To test the impact of fire on *C. sciadophylla*, the same pasture was burned two months after the saplings were transplanted, with the result that ten of the eighteen cecropia saplings that had survived the transplant were killed. The fire, early in the dry season, was very light and resulted in an incomplete burn of the grass but still killed 55 percent of the cecropia saplings.

Another fact pointing to the importance of fire in the elimination of cecropia is the occasional cecropia individual in moderately and intensively used pastures. On close inspection, these individuals are invariably associated with rocky outcrops where grass will not grow or moist low pockets where fires are extinguished. These edaphic refugia are regular features of moderately used pastures, although they rarely fall within study plots that are usually located within the main extent of the grassland (Nepstad, Uhl, et al. 1996).

Dominated by woody vegetation, these edaphic pockets and other "tree islands" (*sensu* Nepstad, Uhl, et al. 1996) may play a crucial role as perches or feeding roosts for birds and bats that otherwise might not venture far into abandoned pastures (Guevara, Purata, and van der Maarel 1986; Gorchov et al. 1993; Vieira, Uhl, and Nepstad 1994; Miriti 1998). The vertebrates, of course, are illustrious dispersal agents of seeds of second-growth trees, especially cecropia and *Vismia* spp. (Vázquez-Yanes et al. 1975; Fleming and Heithaus 1981; Estrada, Coates-Estrada, and Vásquez-Yanes et al. 1984; Gorchov et al. 1993). Their importance cannot be overestimated because bird- and bat-dispersed trees make up about 80 percent of natural second growth in the Amazon rainforest (Charles-Dominique 1986).

Seed rain, via dispersers, may be crucial in the colonization of abandoned pastures because dormant seeds in the soil will have already germinated after successive years of exposure to full sunlight (Nepstad, Uhl, et al. 1996; Chapter 24). Of course, most of these seedlings will have perished if the pastures have suffered annual burns. If so, re-cruitment of seedlings following the third or fourth fire may be limited to the annual seed input via the seed rain from the adjacent forest. Disperser movements into pastures will have a dramatic effect on the rate of recovery of woody vegetation (Charles-Dominique 1986; Gorchov et al. 1993), and tree islands will have a dramatic effect on disperser movements.

Alternative Successional Sequences for the Central Amazon

The pattern of succession on abandoned agricultural land of terra firme in the central Amazon will depend on the history of land use (see Chapter 25). Where the forest is clear-cut and abandoned with little management for agriculture, pasture, or timber extraction, a high density of second-growth species, dominated by cecropia but including *Vismia* and *Bellucia* species, will regenerate from the forest seed bank to combine with a diverse recruitment of stump sprouts of primary forest species (Williamson, Schatz, et al. 1998). The forest will recover rapidly and maintain much of its vegetative species richness throughout the period of regeneration, as if a hurricane had simply snapped most of the trees (e.g., Vandermeer et al. 1995). However, clearing of old-growth forest is almost always followed by anthropogenic fire.

Where land management is very intensive and includes the use of fire, dormant seeds will be exhausted from the seed bank, and stump sprouts from most primary forest trees will be killed. Even a single fire to clear the slash will eliminate most primary forest sprouts and their nondormant seeds (Brinkman and Vieira 1971; Garwood 1989). The dormant portion of the seed bank is somewhat more resilient, as the seeds are drier and deeper in the soil where they are better protected from the heat of the fire. The demise of the dormant seeds is brought about by the post-clearing exposure to full sunlight, which stimulates germination (Schafer and Chilcote 1970; Dias-Filho 1998) followed by subsequent fires that kill the emerging seedlings. Only such resprouting genera as *Vismia* will be maintained. Moderately used pastures after five to ten years of biennial burning contain an ample supply of resprouts of *Vismia* spp. The second-growth forests that dominate abandoned pastures are almost exclusively *Vismia* spp. and *Bellucia* spp. (Saldarriaga and Uhl 1991; Williamson, Mesquita, et al. 1998; Chapter 24). The latter genus is not an active resprouter but rapidly colonizes the clones of *Vismia* spp. sprouts, as seeds of both genera are dispersed by bats (Saldarriaga and Uhl 1991). *Bellucia* spp. gain some protection from fire by casting deep shade that hinders grass growth, thereby leaving the area around its trunk free of fine fuel from grasses. Presumably, fires burn cooler near *Bellucia* spp. trunks. For example, the primary forest at the Colosso Reserves of BDFFP was clear-cut in 1980 for pasture, used moderately for several years, and then abandoned in 1989. By 1997 it produced a second-growth forest dominated 92 percent by *Vismia* and *Bellucia* species (fig. 25.3). For trees larger than 5 cm DBH, we encountered only nine species outside these two dominant genera, in 676 m².

There is one report that the *Vismia-Bellucia* canopy and the historical conditions that created it may impede succession relative to the cecropia-dominated canopy and the conditions that created it; plant species richness recruited under cecropia canopies after nearly a decade was double that under *Vismia* canopies (Williamson, Mesquita, et al. 1998). *Vismia* species are extremely persistent as well, in other parts of the Amazon Basin (Saldarriaga 1985; Saldarriaga and Uhl 1991; Williamson, Mesquita,

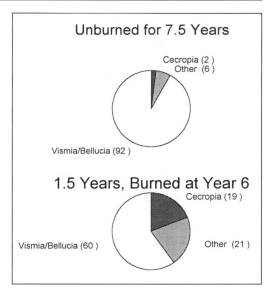

Fig. 25.3. Relative density in percent of *Cecropia* versus *Vismia* and *Bellucia* versus other genera in second growth derived from pasture abandoned for 7.5 years. In plots unburned for the entire 7.5 years, relative density is based on stems at least 5 cm DBH. In plots burned after 6 years, relative density in the 1.5-year-old regrowth is based on stems at least 1.5 m tall.

et al. 1998), and in French Guiana cecropia litter has been shown to be more effective than *Vismia* spp. litter in restoring soil conditions that facilitate the return of the mature forest species (Maury-Lechon 1991). If *Vismia* spp. and *Bellucia* spp. inhibit succession relative to cecropia, then land management of abandoned pastures may harbor some surprising techniques to diversify the pioneer community (see also Chapter 24).

Using Fire to Increase Biodiversity of Abandoned Pastures

A portion of the *Vismia-Bellucia* dominated pasture at the Colosso Reserves was burned at the end of 1995 with some interesting results. The fire killed nearly all the aboveground stems, although some stems resprouted from the base. However, resprouting was much more frequent by smaller individuals than by larger individuals, so the pre-fire dominant individuals were either killed by the fire or resprouted from their smaller lateral roots. This pattern of decreased probability of resprouting by larger individuals following crown death is common among fire-adapted resprouters (Rebertus, Williamson, and Moser 1989), although it has never been examined in rainforest trees. With the prior canopy stems dead, resprouts were small and unable to completely dominate the recovery, so many seeds germinated from the then 6 yr old seed bank to provide a more diverse community of pioneers. Some of these pioneers may have entered from seed rain after the fire, and we do not know the relative contributions of resprouts, the 6 yr old seed bank, and the post-fire seed rain, but the density of seedlings early after the fire is indicative of a significant contribution by the seed bank. A comparison of the 7.5 yr old pioneer community (larger than 5 cm DBH) to the 1.5 yr old pioneer community (stems taller than 1.5 m) indicated a dramatic increase in cecropia individuals and other non-*Vismia*, non-*Bellucia* pioneer genera (e.g., *Solanum, Trema, Laetia, Goupia, Miconia*). On average, dominance of *Vismia* spp. and *Bellucia* spp. dropped from 92 percent to 60 percent as a result of the fire (fig. 25.3).

The result that burning a young *Vismia-Bellucia* stand increased species richness is somewhat counterintuitive because annual or biennial fires were the cause of the loss of diversity when the forest was cleared. Time since abandonment and the last fire, in this case six years, had two important effects. First, with time, the stand developed a canopy that probably increased visitation by forest birds and bats dispersing seeds (e.g., Chapter 20), thereby restoring a diverse soil seed bank. Second, the dominant trees be-

came large enough that they were unable to produce large resprouts following crown death in the fire, thereby leaving regeneration to the seed bank and resprouts of smaller individuals.

Whether that diversity will be maintained through the early years of succession remains unknown. It is notable that some of the new pioneer genera (*Cecropia, Solanum, Trema*) were among the tallest saplings in the study plots, so they are not likely to be suppressed by *Vismia* and *Bellucia* species. It is also unknown whether the net gain in biodiversity will compensate for the lost time—for example, ten years into the future will today's *Vismia-Bellucia* dominated stand at age 17.5 yr still be less diverse than today's burned stand at age 11.5?

Conservation Lessons

1. Clear-cutting of the rainforest, followed by moderate and heavy use as pasture, will leave a depauperate woody flora with few resprouts from the original forest and a seed bank exhausted of forest species; succession to a primary forest will be very slow and perhaps diverted unless steps are taken to increase the diversity of the pioneer and early secondary species. How this enrichment is attempted will vary among sites, land use histories, and status of other forests in the surrounding landscape.

2. Increasing the number of woody trees, which provide perch sites or feeding roosts, prior to abandonment or in the first one to two years of succession may have a dramatic effect of increasing the rate of seed rain from forest trees and the rate of forest regeneration.

3. Artificial seeding of diversifying pioneer genera (*Cecropia, Trema, Solanum, Stryphnodendron*, etc.) will facilitate succes-

sion, although care must be taken to create favorable conditions for seed germination and seedling establishment; for example, seeding at the beginning of the rainy season following the last dry season fire when graminoid cover is reduced (Miriti 1998).

4. The two options above require land owners or land managers to have the foresight to take action around the time that land is abandoned and to have the necessity, perhaps by law (Uhl, Barreto, et al. 1998), to implement steps to facilitate forest regeneration. Where three to ten years have elapsed following abandonment and the regeneration is completely dominated by *Vismia* and *Bellucia,* another option may be necessary to foster biodiversity.

5. Prescribed burning may produce a more diverse flora, but only under certain conditions. The fire must be intense enough to kill most of the dominant canopy trees, *Vismia* and *Bellucia,* but not sufficiently intense to kill seeds in the seed bank and most surface roots. The fire must be prescribed for small to medium parcels so that other remaining wooded parcels can maintain the community of seed dispersers. And finally, the fire can be controlled to preclude damage to adjacent secondary forests and edges of primary forests.

6. Where fire is applied outside the above conditions, it will have a negative effect of setting succession back in the burned plot if the seed bank is destroyed or of reducing the diversity of the landscape if the fire permeates adjacent, mature forests. We are reluctant to recommend the use of prescribed burns because fire, improperly applied, is extremely destructive. However, the apparent alternative of cutting the existing dominant trees will impede, not facilitate, succession as they will resprout with a fury only to yield the same *Vismia-Bellucia* community. Certainly more experimental research into prescribed fire is essential before any broad scale application is implemented.

Acknowledgments

We gratefully acknowledge the support of the BDFFP, both for funding our research and for permission to conduct research in their reserves. The U.S. National Aeronautics and Space Administration provided funding through its Large Scale Biosphere-Atmosphere Experiment in Amazonia. G. Bruce Williamson was supported by grants from the Brazilian Fulbright Commission and the Fundação Coordenação de Aperfeiçoamento de Pessoal de Nível Superior. Rita C. G. Mesquita was supported by the Centro Nacional de Desenvolvimento Científico e Tecnológico. D. A. Clark provided exceptionally useful comments on the manuscript.

CHAPTER 26

Selective Logging in the Brazilian Amazon

Its Relationship to Deforestation and the International Tropical Hardwood Market

NIRO HIGUCHI

Discussions of biodiversity, global climate change, and the supply of tropical hardwood on Earth invariably involve the Amazon. This region is the largest continuous rainforest area in the world, covering approximately 5 million km², or 60 percent of the Brazilian territory. Through 1989, the accumulated deforestation, including areas of the wooded savannas known as *cerrado* in the southern reaches of Brazil's legally defined Amazon, was approximately 500,000 km² (table 26.1).

The Brazilian Amazon forest area is 3.65 million km², or 364.8 million ha (table 26.2). Dense terra firme forests are predominant despite *várzea* forests (seasonal flooded forests on white-water river drainages) playing an important role in the economy of many Amazonian states. The dominant botanical families are Leguminosae, Lecythidaceae, and Sapotaceae on terra firme and Myristicaceae in *várzea*. Mahogany (*Swietenia macrophylla*) is the most important wood species from a commercial standpoint. The Brazil-nut tree (*Bertholletia excelsa*) and the rubber tree (*Hevea* sp.) are also important, but for the products they produce rather than as timber. The total timber volume in the Amazon is estimated at 50 billion m³, of which 10 percent

have the potential to be used by the timber industry.

The Amazon region awakens a great deal of curiosity, passion, covetousness, respect, and legitimate interest in protecting it. For some time now the forest has no longer been an obstacle to regional economical development. On the contrary, available timber has substituted for the government's fiscal incentives, such as the free-trade zone in Manaus, as a motor to economic development. Furthermore, the international tropical hardwood market has begun to move from southeastern Asia to the Amazon region.

Deforestation in the Brazilian Amazon

As of 1989, according to Fearnside, Tardin, and Meira Filho (1990), 478,882 km² (47,888,200 ha) of native forests had been transformed into other types of land use in the "Legal" Amazon—all in the name of developing the region. The principal "development" projects driving this habitat transformation and normally receiving incentives (subsidies) from the federal government, with resources raised from the international financial community, used to be farming and cattle raising, mining, and hydroelectricity.

This is publication number 293 in the BDFFP Technical Series.

TABLE 26.1.   Land Cover Types and Deforestation (km²) for the Brazilian States Part of Legal Amazonia

State	Original vegetation		Deforestation		Percentage deforested
	Forest	Cerrado	Forest	Cerrado	
Acre	152,589	—	8,836	—	5.8
Amapá	99,525	42,834	1,016	—	0.7
Amazonas	1,562,488	5,465	21,551	—	1.4
Maranhão	139,215	121,017	88,664	20,664	42.0
Mato Grosso	572,669	308,332	79,549	25,568	10.0
Pará	1,180,004	66,829	139,605	1,722	7.3
Rondônia	215,259	27,785	31,476	169	13.0
Roraima	173,282	51,735	3,621	—	1.6
Tocantins/Goiás	100,629	169,282	22,327	34,114	20.9
TOTALS	4,195,660	793,279	396,645	82,237	

Source: Fearnside et al. 1990.

During U.N. Conference on the Environment held in Rio de Janeiro in 1992, the National Institute of Space Research (INPE) presented, at that time, the most recent statistics of deforestation in the region, which were as follows: 21,130 km² per year in the 1978–88 period; 17,860 km² in 1989; 13,810 km² in 1990; and 11,130 km² in 1991 (INPE 1992). The most recent deforestation data, according to INPE, show a slight increase from 1992 to 1994 (average of 15,000 km²/year) and a much larger increase in 1995 to 29,000 km². The 1996 data show a decrease of the annual deforested area down to 18,000 km². Also, deforestation in individual Amazonian states varies (table 26.3). There has been a tendency for deforestation rates to fall since 1988, becoming stable in 1990 in the vicinity of 12,000 km² per year. The reason for the decrease from 21,130 km²/year (1978–88) to 11,130 km²/year (1991) was mainly the political climate in Brazil and the lack of financial subsidies for development projects in the Amazon.

Despite fluctuations in deforestation rates, in absolute values deforestation in the Amazon is still enormous. In this context, a new factor begins to play an important role in the maintenance of the status quo in deforestation–selective logging. As long as there were plentiful resources for financing farming and cattle-raising projects (until 1989 and 1990), timber was always considered a subproduct of these activities. Today, due to the lack of financing, along with a slight increase in the value of Amazonian hardwood on the national and international markets, timber now subsidizes the farming and cattle-raising projects, principally in the southern region of Pará, in southeastern Amazonia. Nepstad, Veríssimo, et al. (1999) showed that in 1996–97 selective logging in Amazonia damaged between 10,000 and 15,000 km² of primary forest, which was not included in official (INPE) estimates of deforestation.

Production of Tropical Hardwood

In Brazil, until the middle 1980s timber companies and the forestry sector were able to dissociate themselves from the responsibility for any inappropriate use of the Amazonian landscape, because timber was considered a by-product of other development projects. In the beginning of the 1990s, however, the situation changed completely as timber replaced official fiscal incentives, which had by then disappeared, as the motivating force behind much of the deforestation taking place. In addition, timber prod-

TABLE 26.2. Forest and Nonforest Cover Area in the Amazon Basin

Forest and Non-forest Type	Area (km²)
Terra firme forests	
Dense forest	3,303,000
Dense liana forest	100,000
Open bamboo forest	85,000
Coastal forest	10,000
High *campina*, or *campinarana*[1]	30,000
Dry forest	15,000
Várzea forest[2]	55,000
Igapó forest[3]	15,000
Mangrove forests	1,000
Campinas	34,000
Forest subtotal	3,648,000
Várzea grasslands	15,000
Terra firme savannas	150,000
Serrana vegetation	26,000
Restinga vegetation	1,000
Water	100,000
Nonforest subtotal	292,000
Total Amazon Basin	3,940,000

Notes: [1]*Campina* and *campinarana* refer to the unique vegetation that occurs on areas of white sand found predominantly north of the Amazon. [2]Varzea forests are annually flooded by white-water rivers. [3]*Igapó* forests are annually flooded by black-water rivers, such as the Rio Negro. *Source:* Braga 1979.

ucts began to have a greater liquidity on the national as well as the international market.

Despite all the difficulties encountered in obtaining trustworthy statistics on timber production in logs for industrial use in the Amazon region, there is a clear increase during the period of 1975–85, mainly for the national market. There was a jump from 4,512,000 m³ in 1975 to 19,538,000 m³ in 1985, with dramatic increases in every state but the remote Acre (table 26.4). The International Tropical Timber Organization (ITTO) statistics show that in 1989 the Brazilian Amazon production was 27,200,000 m³ of timber.

The International Tropical Hardwood Market

The ITTO annually compiles from its signatory countries statistics on production, consumption, exportation, and market values of hardwood timber. The statistics on the volume of timber are furnished in cubic meters equivalent in logs and are calculated for sawnwood, plywood, and veneer, using standard conversion factors (see table 26.5.)

Producers

The statistics reviewed in this chapter were compiled from the ITTO's *Annual Review and Assessment of the World Tropical Timber Situation* for 1990–91, 1992, 1993–94, and 1995. (Each report covers a period of four years, with the overlapping years updating or correcting figures from a previous report.) Three sets of figures for timber production (equivalent in logs) between 1988 and 1995 are presented: all countries affiliated with the ITTO; large tropical regions; and largest individual producers.

In this period, production has remained stable, with an average of 136 million m³ of logs per year (table 26.5). The Asia/Pacific region, despite a 12 percent decrease in production between 1989 and 1995, is still the largest tropical hardwood producer. Production in the two other regions, Latin America/Caribbean and Africa, has remained stable but at a much lower level than the Asia/Pacific region.

TABLE 26.3. Yearly Gross Deforestation Rate (km²/year) for Legal Amazonia, from 1979 to 1996

State	1979/1989	'87–88/89	'89/90	'90/91	'92	'93–94	'95	'96
Acre	620	540	550	380	400	482	1,208	433
Amapá	60	130	250	410	36	—	9	—
Amazonas	1,510	1,180	520	980	799	370	2,114	1,023
Maranhão	2,450	1,420	1,100	670	1,135	372	1,745	1,061
Mato Grosso	5,140	5,960	4,020	2,840	4,674	6,220	10,391	6,543
Pará	6,990	5,750	4,890	3,780	3,787	4,284	7,845	6,135
Rondônia	2,340	1,430	1,670	1,110	2,265	2,595	4,730	2,432
Roraima	290	630	150	420	281	240	220	214
Tocantins	1,650	730	580	440	409	333	797	320
Totals	21,130	17,860	13,810	11,130	13,786	14,896	29,059	18,161

Source: INPE 1992, 1998.

TABLE 26.4. Tropical Timber Production by State from Native Amazonian Forests, from 1975 to 1985

State	1975	1980	1985
Acre	31	87	23
Amapá	330	400	413
Amazonas	135	325	1,382
Pará	3,942	10,283	16,361
Rondônia	60	307	1,320
Roraima	14	72	39
Tocantins	—	—	—
Totals	4,512	11,474	19,538

Note: Production expressed in 1,000 m³ of logs.
Source: IBGE 1992, and Deusdará Filho 1996.

During these years, the Asia/Pacific region produced 71 percent of the world's tropical hardwood (including export and internal consumption), Latin America and the Caribbean produced 22 percent, and Africa accounted for 7 percent of world production (table 26.5b). The three largest individual producers were Malaysia with 38,359,375 m³/year (28 percent), Indonesia with 33,250,000 m³/year (24 percent), and Brazil with 23,512,500 m³/year (17 percent). These three countries represent 70 percent of the world's tropical hardwood production.

EXPORTERS AND CONSUMERS

Countries export nearly 50 percent of their production in the form of lumber, plywood, and veneer, and, occasionally, unprocessed logs (table 26.6). Some countries like Indonesia and Brazil export practically no wood in the form of logs. Overall, the international tropical hardwood market consumes an average of 65,806,250 m³/year.

During these years, the Asia and Pacific region shipped 84 percent of worldwide exports, Latin America and the Caribbean exported 5 percent of the global total, and Africa accounted for 9 percent (table 26.6b). The two largest individual exporters were Malaysia with 29,633,625 m³/year and Indonesia with 21,007,625 m³/year; the two countries alone contributed 80 percent of the total exports by tropical countries (table 26.6c).

The largest tropical hardwood consumers in 1988–95 include none of the producing countries (table 26.7). Based on annual averages for the period, the principal consumers were Japan (39 percent of worldwide consumption), European Union (22.7 percent), Taiwan (11.5 percent), South Korea (11.1 percent), China (9.8 percent), and the United States (5 percent) (table 26.7).

In 1993, exports of timber from tropical countries reached US$12 billion. Malaysia and Indonesia exported US$4.45 billion (37 percent) and US$4.59 billion (38 percent), respectively. Brazil exported only US$560 million (4.5 percent).

TABLE 26.5.   Tropical Hardwood Timber Production from 1988 to 1995 (in 1,000 m^3)

A. All countries affiliated to the ITTO

Year	Product			
	Logs[1]	Sawnwood[2]	Plywood[3]	Veneers[4]
1988	130,736	40,439	11,248	1,455
1989	144,079	42,389	12,113	1,214
1990	138,625	40,207	12,762	1,332
1991	134,114	43,282	23,314	2,371
1992	139,804	44,726	23,853	2,819
1993	136,481	42,847	23,557	3,520
1994	135,917	43,127	23,161	3,491
1995	133,649	42,849	23,411	3,225

Notes: [1]Sum of the volume of sawnwood, plywood, and veneers converted to log-equivalents. [2]Conversion factor for lumber (sawnwood) = 1.82. The volume of lumber that each country produces is multiplied by this factor to estimate the equivalent in logs removed from the forest; or, 55 percent of the volume of one log, world average, is transformed into boards or other lumber products. [3]Conversion factor for plywood = 2.3; 43 percent of the volume of one log, world average, is made into plywood. [4]Conversion factor for veneers = 1.9; 53 percent of the volume of one log, world average, is transformed into veneers.

B. By tropical region

Region	Product	1989	1991	1993	1995	Annual average
Africa	Logs	9,387	8,842	9,097	9,660	
	Sawnwood	2,236	1,781	1,849	2,330	
	Plywood	276	162	158	167	
	Veneers	321	302	340	365	
	Total					9,431 (7%)
Latin America and Caribbean	Logs	33,050	24,948	30,204	33,509	
	Sawnwood	10,998	10,196	12,074	13,498	
	Plywood	1,432	1,502	1,967	2,330	
	Veneers	225	232	328	329	
	Total					30,148 (22%)
Asia and Pacific	Logs	101,642	99,785	96,259	89,640	
	Sawnwood	29,155	27,655	26,007	24,183	
	Plywood	10,405	12,340	14,372	14,838	
	Veneers	638	952	2,323	1,952	
	Total					96,615 (71%)

C. Largest individual producers—only considering log-equivalent volumes

Country	Year			
	1989	1991	1993	1995
Malaysia	38,900	39,840	37,260	35,000
Indonesia	29,500	37,000	37,000	34,000
Brazil	27,200	18,500	23,000	26,000
India	18,350	15,812	15,812	15,000
Papua–New Guinea	1,700	2,200	3,050	3,250
Philippines	2,773	1,919	1,022	865
Cameroon	2,121	2,290	2,815	2,995
Congo	808	572	511	600
Ivory Coast	2,491	2,447	1,961	2,220
Gabon	1,322	1,300	1,815	1,990
Ghana	1,200	1,229	1,682	1,500

TABLE 26.6.   Tropical Hardwood Timber Exports from 1988 to 1995 (in 1000 m³)

A. All tropical countries affiliated to the ITTO

Year	Product			
	Logs[1]	Sawnwood[2]	Plywood[3]	Veneers[4]
1988	63,381	9,228	8,652	557
1989	68,489	10,575	9,557	578
1990	63,445	7,447	10,049	674
1991	67,933	8,003	11,220	818
1992	70,996	8,559	12,601	1,212
1993	65,257	8,468	13,350	1,195
1994	63,834	7,736	13,169	1,061
1995	63,115	7,144	13,833	1,071

B. By tropical region

Region	Product	1989	1991	1993	1995	Annual average
Africa	Logs	5,978	5,763	5,668	6,218	
	Sawnwood	960	1,020	1,028	1,135	
	Plywood	75	53	70	91	
	Veneers	197	171	203	246	
	Total					5,978 (9%)
Latin America and Caribbean	Logs	1,747	1,875	4,079	6,118	
	Sawnwood	444	440	992	1,405	
	Plywood	372	381	714	1,000	
	Veneers	41	46	200	111	
	Total					3,288 (5%)
Asia and Pacific	Logs	60,764	58,298	54,052	49,402	
	Sawnwood	9,171	6,187	6,181	4,369	
	Plywood	9,110	10,295	12,220	12,437	
	Veneers	340	543	745	651	
	Total					55,529 (84%)

C. Largest individual producers—only considering log-equivalent volumes

Country	Year			
	1989	1991	1993	1995
Malaysia	33,021	31,930	26,093	25,337
Indonesia	23,446	22,393	23,553	20,115
Brazil	1,509	1,345	3,207	5,080
India	15	54	75	75
Papua–New Guinea	1,265	1,505	2,872	2,907
Philippines	1,309	436	248	62
Cameroon	1,300	1,404	1,447	1,395
Congo	542	429	345	431
Ivory Coast	1,613	1,508	1,400	1,541
Gabon	1,094	1,245	1,799	1,910
Ghana	512	586	985	754

TABLE 26.7. Largest Individual Consumers—Considering Only Log-Equivalent Volumes (in 1,000 m³)

Country	1989	1991	1993	1995
Japan	23,398	19,485	20,950	18,901
European Union	14,111	12,582	10,727	10,280
Taiwan	5,157	6,452	6,294	6,056
Korea	5,841	6,612	5,808	5,588
China	2,346	4,865	6,573	5,379
USA	3,184	2,549	2,466	2,565

TABLE 26.8. Contribution of Brazil to the International Tropical Timber Market from 1988–1995 (in 1,000 m³)

Year	Product			
	Logs[1]	Sawnwood	Plywood	Veneers
1988	1,960	533	364	56
1989	1,509	345	350	40
1990	1,481	445	248	53
1991	1,345	200	350	40
1992	2,330	484	509	109
1993	3,207	627	656	188
1994	4,549	850	800	85
1995	5,080	1,000	900	100

Note: [1]Sum of the volume of sawnwood, plywood, and veneers converted to log-equivalents.

The contribution of the Amazon to the international market has been modest even though Brazil, on the whole, produced 23 million m³ annually (table 26.8). In an international market of approximately 65 million m³ of wood (equivalent in logs), the Amazon contributes less than 10 percent of the international market. The reasons for this are various: better access and infrastructure of the southeastern Asian countries, predominance of a few tree families (e.g., Dipterocarpaceae) of great commercial value in the Asian forests, and, principally, the low quality of wood produced in the Amazon; the exportation of logs is prohibited in Brazil, and the technology used to transform them does not meet the quality standards demanded by the international market, namely, Europe, the United States, and Japan.

One Amazonian species alone, mahogany (*Swietenia macrophylla*), contributes 10 percent of the total wood volume commercialized abroad (Carvalho 1995), with two-thirds of the country's exports coming from the state of Pará, in southeastern Amazonia. Regionwide, the contribution is not larger only because export quotas were established in 1990. The export quota of mahogany has been gradually decreasing, from 150,000 m³ in 1990, to 90,000 m³ in 1995. More recently, a moratorium was decreed for mahogany extraction from native Brazilian forests.

TENDENCIES OF THE INTERNATIONAL TROPICAL HARDWOOD MARKET

Grainger (1987) predicted the following shifts in the tropical hardwood sector: a peak in southeast Asian production in the middle 1990s, at which time there would be a shift to production from Latin America, especially the Amazon, to supply the markets of Europe, Japan, and North America. Indeed, this scenario seems to be playing itself out as predicted. Examining the dynamics of hardwood exports for the period of 1989–95 by the largest world producers (see table 26.6) we see that the wood supply by the Asian countries has decreased in this period, while Brazil's participation has been growing on the international market.

Moreover, combining the forest area estimates furnished by FAO (Schmidt 1991) and the production level averages obtained for the 1988–95 period, it is reasonable to expect that the timber stocks of Malaysia, Indonesia, and Brazil will be depleted in less than 10, 40, and 350 years, respectively.

At this rate, early in the twenty-first century the Amazon will figure on the list of the biggest tropical hardwood producers and quickly top the list. The increasing presence of timber companies in the Amazon, mainly from southeastern Asia, trying to purchase land in this region lends further credence to this prediction. Because of its extensive unexploited territory, the state of Amazonas is the region most coveted by timber companies; international as well as Brazilian companies are beginning to abandon other regions in the Amazon, especially southern Pará, with an eye to the unexploited expanses of the state of Amazonas.

Forest Management in the Brazilian Amazon: Evaluations

Despite Brazil's modern environmental legislation and Article 15 of the Brazilian Forest Code, which covers forest management in the Amazon, finally regulated in 1994, it is hard to find a sustainable, forest-management plan in operation in the region. Plans are approved by the state agencies of IBAMA, Brazil's Institute for the Environment and Renewable Natural Resources, but few projects, or plans, are duly inspected because of the lack of resources and personnel in these government agencies.

THE STATE OF PARÁ

The Paragominas micro-region in southern Pará represents 40.3 percent of the forest management projects out of the 576 approved by IBAMA, between 1981 and July 1995. A recent evaluation of forest management projects approved by IBAMA for this region (PAP, coordinated by EMBRAPA-CPATU [EMBRAPA is the Brazilian Center for Agriculture and Ranching Research]) reached a very clear conclusion: the situation is simply chaotic. Timber volume estimates of the projects do not agree with estimates from the field and not even with the volumes actually removed from the project area; the projects are poorly formulated; the technical teams of the companies are not duly trained for practicing tropical silviculture; and no group evaluated fulfilled the demands of Brazilian Decree 1282 (Sustained Forest Management for the Amazon), nor the demands of ITTO–2000 (a set of guidelines for producers and consumers of tropical hardwoods produced by the International Tropical Hardwood Organization) that will commercialize only timber resulting from a sustainable management plan. This preliminary report was discussed with all the sectors involved in Paragominas in March 1996 (Silva, Uhl, and Murray 1996).

Few differences in relation to the execution of the forest management plans will be found in other micro-regions of Pará, or even in other Amazonian states. Perhaps only the intensity and the duration of the intervention will change, depending on the region. In *várzea* forests of the state of Amazonas where the main source of raw timber for the state is concentrated, timber extraction is much more selective, and, as a result, extracted timber volume per unit area is lower

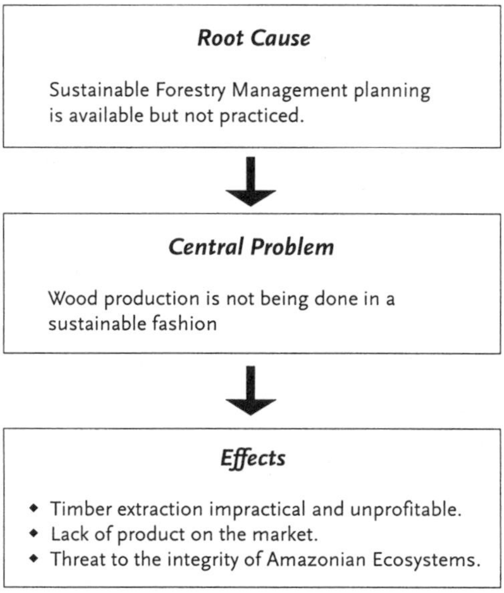

Root Cause

Sustainable Forestry Management planning is available but not practiced.

↓

Central Problem

Wood production is not being done in a sustainable fashion

↓

Effects

- Timber extraction impractical and unprofitable.
- Lack of product on the market.
- Threat to the integrity of Amazonian Ecosystems.

Fig. 26.1. Summary of the main problems of forest management in the Amazon.

than that for Paragominas, but the technical and legal questions involved in management plans are similar to those of Paragominas.

The State of Amazonas

In April 1996, at a workshop promoted by the BIONTE Project (INPA/ODA) in Manaus, researchers, professors, public authorities, businessmen, and environmental nongovernmental organizations debated the existing problems of the forest sector of the state of Amazonas.

The main conclusion from this meeting was that the central problem of forest management in the state of Amazonas is that timber production is not being carried out in a sustainable way (fig. 26.1). As a consequence, there is the risk of timber products being in short supply on the market at the same time that current practices threaten the integrity of Amazonian forest ecosystems. The principal cause is that the sustainable forest management plans are not duly implemented. The main reasons for the non-

implementation of the management plans lie within three domains: implementation, knowledge, and sectorial politics.

Implementation Domain

Forest exploitation is poorly performed because of a lack of qualified personnel at all levels, low pay, inadequate equipment, and lack of technical assistance. Investors have a one-harvest-only mentality owing to a culture of immediate gain, the size of the available forest reserve, natural difficulties, untrustworthy statistics, deficient technological knowledge, lack of knowledge concerning the market, nonexistence of cost-benefit analysis, and lack of incentives for this sector. Finally, there is no inspection or monitoring because of a lack of resources, equipment, orientation, and personnel.

Knowledge Domain

Several problems of an informational nature confound the problems of developing and maintaining sustainable management plans. The curricula of forest engineering schools are inadequate. Communication between the teaching-and-research and production sectors is faulty; lines of research are defined by researchers at universities and research institutes, researchers publish for colleagues rather than an applied audience, there are no extension programs, and businessmen invest very little in research. Lastly, there is a lack of both basic and applied knowledge in such crucial areas as tropical silviculture, wood technology, ergonomics, economy, commercialization and marketing, and environmental, social, and cultural impacts.

Sectorial Political Domain

Fragmented sectorial politics also hinder attempts at sustainable management. Some of the manifestations include a lack of ecological-economical zoning, conflicts among the different regulations of land use, and untrustworthy statistics. And even when legislation is passed it often is unfulfilled or unenforced. Political norms easily change, and there is a lack of validation and inspection systems for established and agreed-upon norms.

Conclusions

The cliché "The natural resources of the Amazon are overexploited and under-utilized" is still sadly appropriate for this region. Ecological impacts are imposed on several hectares of primary forests in the process of extracting a single tree for wood production; extensive areas are clear-cut for farming and cattle raising projects of low productivity (see Chapter 23); vast areas of forest are flooded to form lakes for inefficient hydroelectric projects; and forest soils are totally laid bare for mineral production with minimal improvement. Few of these types of land use would pass a cost-benefit analysis. The deforested area in the Amazon —more than 50 million hectares—has impoverished its wealth and diminished this vast storehouse of natural resources.

The environmental impacts of existing land-use practices are well known and have worried society as a whole. The most important impacts include greenhouse gas emissions into the atmosphere, principally from land clearing and burning of dead-wood as well as biomass collapse in fragmented forest mosaics (W. Laurance, Laurance, et al. 1997), and the decomposition of standing trees in lakes on hydroelectric projects; potential alteration in the hydrological cycle through the removal of the forest cover; genetic erosion by clear-cutting as well as by selective timber extraction; loss of biodiversity; and sedimentation and pollution of rivers and streams.

The social and cultural impacts are also equally important. There are cases of imposed compensations and translocation (evacuation) of indigenous people in the name of development projects. Diseases and their dissemination are serious problems in the region. Common illnesses such as influenza and measles are devastating to the indigenous people. Likewise, endemic diseases, such as malaria and leishmaniasis, cause serious problems for colonizers. Land problems (and the landless) in the Amazon, despite its large territorial size, have also increased and have international repercussions.

With the drastic reduction of fiscal incentives in 1990, principally for farming and cattle-raising projects, the expectation was that this main driver of deforestation in the Amazon would be eliminated. Without subsidies, farming and cattle raising in the region would become less attractive and competitive with other regions of Brazil. This situation was, however, short lived, because as the subsidies were reduced, timber extraction became operationally and economically viable, becoming the new driver to deforestation and acting as a subsidy for subsequent farming and cattle-raising projects. The minimum infrastructure investments made for timber extraction were quickly covered by increasing timber values. Timber companies, through the establishment of roads and the reduction of wood density and volume in primary forests, are facilitating the subsequent clear-cutting and land preparation for cattle ranching and farming and contributing substantially to the fragmentation of the landscape.

Timber production, however, may not necessarily become the next big villain of Amazonian land use. During the last decades, while Brazilian timber did not find an international market, Brazil prepared itself relatively well for receiving new investors in this sector. The principal accomplishments include the regulation of Article 15 of the Brazilian Forest Code in 1994 (which treats questions of forest management in the Amazon) of Decree 1282; passage by all Amazonian states of their own modern environmental legislation to monitor the use of the Amazonian forest; and the development of two important research projects on forest management in the Amazon —one in the Tapajós National Forest (EMBRAPA-CPATU) and the other in Manaus (INPA), since 1980.

Preliminary results of these two studies indicate that it is possible to combine timber production with ecosystem conservation. Specifically, results show that (1) the remaining forest responds positively to the opening of the canopy—the injuries are rapidly healed (but see Chapter 27); (2) the increment in volume is compatible with the commercial cutting cycle; (3) it is possible to orient the felling of trees and thus control the size of the clearing, thereby protecting and stimulating existing natural regeneration, and also control the microclimatic changes, forest succession, seed bank, and seed dispersal; (4) it is possible to minimize nutrient exportation out of the system; and (5) it is possible to adequately plan the forest harvest, minimizing impacts on the soil (principally compaction), the nutrient and water cycles of the soil, and the soil's meso- and microfauna.

Finally, we should not lose sight of the pressing need to conclude the ecological-economical zoning for the region, with areas especially designated for various ends (such as timber production). Beyond this, it is necessary to rethink and consolidate the policies of Amazonian land use as a whole and the exploitation of mineral resources. Likewise, it is necessary to rethink in the short term the question of liquidity and aggregate values for many Amazonian products.

Conservation Lessons

1. The principles and concepts that would permit sustainable forestry management in the tropics have been known for more than 150 years.

2. However, sustainable forestry management has been used throughout the tropical world as a pretext to remove primary forest cover for other ends than investments, strictly speaking. If, in fact, the concept and principles of sustainable forestry management had been appropriately used, the great majority of the world's demand for tropical hardwoods would be coming from secondary, and not primary, forests.

3. Ultimately, the most significant negative impact of selective logging is that it facilitates the establishment of other projects, usually cattle ranching or farming.

4. In a more immediate sense, selective logging has a number of noteworthy effects on the remaining ecosystem:

a) Microclimatic changes are expected in clearings and at forest edges.

b) After removing leaves and branches from the trunks to be removed, an uneven distribution of nutrients will be left on the forest floor.

c) Physical and chemical properties of the soil are changed under logging roads and clearings.

d) Changes are likely in the soil's mesofauna.

e) Overall levels of biodiversity will decrease as species strictly evolved for living in primary forest find the disturbed ecosystem inhospitable.

f) The chance for forest fires increases because of the unusual amount of tinder on the forest floor.

Extending Models of Edge Effects to Diverse Landscape Configurations, with a Test Case from the Neotropics

JAY R. MALCOLM

The extent, timing, and spatial configuration of deforestation in the tropics
varies greatly from one region and
development activity to another. Typically,
forest loss proceeds through "perforation" or
"dissection," in which the original forest
dominates and connectivity is high, to fragmentation, where a shift in dominant cover
types occurs and secondary habitats come to
dominate the landscape, to shrinkage and
attrition, where individual primary forest
patches are eroded, become progressively
isolated, and may eventually disappear altogether (Forman 1995). Superimposed on
this variation in primary forest cover and
connectivity is variation due to the spatial
grain of deforestation (Forman 1995). Average clearing sizes may vary by three or four
orders of magnitude among regions, even
though the total magnitude of canopy loss
may be similar. For example, at one extreme,
large-scale industrial agriculture such as in
northern Mato Grosso, Brazil, can result in
coarse-grained landscapes, where clearings
(and fragments) may be kilometers or tens of
kilometers across. At the other extreme, selective logging results in a fine-grained mosaic of small clearings, treefall gaps, and
roads that are tens of meters across or less
(White 1994b). Although canopy loss in

both coarse-grained and fine-grained landscapes is known to have important implications for the flora and fauna of the remaining
forest (e.g., Laurance and Bierregaard 1997a;
Struhsaker 1997), possible parallels between
the two have rarely been investigated. The
identification of these links is of great importance for conservation. In light of accelerating loss and disturbance of tropical
forests, and urgent needs for effective conservation guidelines and management techniques, the ability to apply knowledge and
understanding from one type of landscape
transformation to another is welcomed.

Unfortunately, it is not yet clear what generalizations can be extended from one landscape configuration to another, nor is there a
general body of theory that applies to the
full range of landscape configurations. The
major body of theory available for fragmented landscapes, namely island biogeography (MacArthur and Wilson 1967) and
metapopulation dynamics (Levins 1969,
1970; Hanski 1991), is not directly applicable in perforated or "shredded" forests
(*sensu* Feinsinger 1997) where the remaining primary forest is still broadly contiguous
and interconnected. For example, in an ide-
alized checkerboard landscape, where clearings alternate with forest patches, island

This is publication number 294 in the BDFFP Technical Series.

biogeography offers few predictions with respect to fragmentation effects, because each patch is connected at its four corners to a neighboring patch and hence none is a true "island." At the same time, checkerboard landscapes can be expected to result in highly modified forest communities. Many studies in tropical forests have shown that a key effect of deforestation is alteration of the forest habitat close to newly created edges (e.g., Kapos 1989; Malcolm 1994; Ferreira and Laurance 1997; Kapos et al. 1997; Turton and Freiburger 1997; Didham 1997b). Thus, deforestation not only affects the spatial configuration of remnant patches, and hence probabilities of colonization and local extinction, but also results in changes within the forest itself under the influence of altered moisture, light, and wind regimes close to edges.

Because both edge and island processes may be operating simultaneously, the study of human-induced landscape change in the tropics becomes considerably more complicated. The two sets of processes are potentially confounded, for example, because both predict that fragment communities will vary as a function of patch size; in one case because smaller patches have greater perimeter:area ratios than do larger patches and hence have proportionally more edge-modified habitat (Levenson 1981); in the other because smaller populations may be subject to more frequent extinction than larger populations (MacArthur and Wilson 1967). In some cases, one of the two processes may be unimportant or absent. For example, island effects may be irrelevant where connectivity is high or where remnant patches are still close to large contiguous areas of rainforest and populations are periodically rescued by immigration (Brown and Kodric-Brown 1977). As an example, I was able to predict vegetation structure, in-

sect biomass, and the abundance of small mammal species in recently created forest fragments in the central Amazon based solely on information from the edge of continuous forest; effects due to insularization per se were not discernible (Malcolm 1991, 1995, 1997a, 1997b). Similarly, where edge effects are weak or extend only a relatively short distance into the forest, metapopulation dynamics may be of overriding importance in driving community change. Evidently, a comprehensive and quantitative theory of landscape transformation in the tropics must include both edge and island effects.

In this chapter I pursue the overall objective of developing landscape models that can be applied to a broad range of landscape configurations, and hence can be used to provide a firm basis for interlandscape comparisons and for management decisions. I focus on edge effects because of their potential importance in both perforated and fragmented landscapes. First, I examine several edge models that provide an increasingly realistic description of the additivity inherent in edge effects (Malcolm 1994). Second, I apply a recently developed model to the problem of grain size and model a set of hypothetical landscapes where grain size is systematically varied across several orders of magnitude, from scales similar to those in selective logging to those observed during large-scale agricultural development. I was particularly interested in testing agreement between this relatively sophisticated model and a simple nonadditive edge model based only on perimeter:area ratios. Finally, as a possible case study in the transfer of knowledge from one landscape configuration to another, I compared avian communities between two Neotropical sites where the potential importance of edge effects was high, but where the grain size of habitat disturbance differed markedly (a selectively logged landscape versus a fragmented landscape).

Incorporating Additivity into Edge Models

Perhaps the simplest edge model is to imagine a strip of edge-modified habitat parallel to an edge, wherein the edge effect declines with increasing distance from the edge, perhaps in a linear or curvilinear fashion (e.g., Laurance and Yensen 1991). These models are useful in coarse-grained landscapes but are unable to treat the additive effects that must occur in small fragments or fine-grain landscapes where edge effects from neighboring edges overlap extensively (e.g., Kapos 1989; Struhsaker 1997). Thus, they are of little use in many fine-grained tropical landscapes that result from such activities as selective logging and small-scale shifting agriculture. In an earlier paper I developed a more realistic model that incorporated additivity from all nearby edges (Malcolm 1994). I assumed that the total edge effect at a location in the interior of a patch was the sum of "point" edge effects along the edges of the patch, weighted by the distance of the edge points to the location. Symbolically, we denote a fragment as a region Ω on the x-y plane with boundary δ. Defining the vector $\mathbf{b}$ as a point on the boundary, and $\mathbf{i}$ as a point in the interior, the function $g(\mathbf{b}, \mathbf{i})$ gives the point edge effect on whatever community characteristic is being considered at $\mathbf{i}$ due to an edge at $\mathbf{b}$. The total edge effect (E) at $\mathbf{i}$ due to the edge effects is obtained by summing $g(\mathbf{b}, \mathbf{i})$ along the boundary δ, which is accomplished by using line integrals:

$$E = \int_{\delta} g(\mathbf{b}, \mathbf{i}) \, ds$$

where ds is the differential of arc length along the boundary δ.

This model is one-dimensional in that the point edge function is evaluated solely over the fragment perimeter; clearings that abut edges are not directly incorporated. Thus, although a significant improvement over non-additive strip models (Malcolm 1994), the model does not describe edge effects in the most general sense because it fails to directly incorporate variation in clearing widths. A logical extension therefore is to extend the model to two dimensions and to imagine that each point in the surrounding clearings (instead of just the points on boundaries) exerts a "point" edge effect. Thus, a narrow road will have less of an edge effect than a large pasture, because the summation will be over fewer points. For this case, we define a region Σ of clearings in the landscape surrounding Ω, $\mathbf{c}$ as a point in Σ and $h(\mathbf{c}, \mathbf{i})$ as the point edge effect. To accomplish the summation we use a surface integral:

$$E = \iint_{\Sigma} h(\mathbf{c}, \mathbf{i}) \, dS$$

where dS is the differential of surface area over the surface Σ. For an application of this model, see Struhsaker (1997).

An alternate approach to these empirical ones is to directly model the abiotic and biotic processes that are altered at edges (which may not result in simple gradients [Chen, Franklin, and Lowe 1996; Kapos et al. 1997]). Heat conduction is a useful physical process to model because it is a simple, diffusion-like process, incorporates the two-dimensional additivity that realistic edge models must incorporate, and is an important component of a forest's thermal balance (other main components are latent heat and convection). Perhaps equally important, process-based models can serve as potential proxy measures of other processes. From a biodiversity standpoint, heat flow models might be useful because they may be able to capture the approximate spatial behavior of several other types of edge-induced habitat changes, such as changes in understory density and insect biomass. They may prove useful as statistical models with parameters that can be estimated in one landscape and subsequently used to predict edge effects in others. Thus, rather than being restricted to the study of the forest temperature or its role in ecosystem processes, heat conduction may serve as a useful proxy measure of edge-induced disturbance gradients and associated community change.

Briefly, in a model of heat conduction, a landscape is envisioned as a two-dimensional grid composed of cells with various thermal properties (Malcolm 1998). If we consider heat flow across a single cell, the rate of heat transfer, Q, from the hotter side of the cell to the cooler side is directly proportional to: (1) the surface area of the cell, A (which is normal to the direction of heat flow); (2) the temperature difference across the cell, $(T_i - T_0)$; and (3) the inverse of the wall thickness, L (Karlekar and Desmond 1977). That is

$$Q \alpha \frac{A(T_i - T_0)}{L} \quad \text{or} \quad Q = \frac{kA(T_i - T_0)}{L}$$

where k is a proportionality constant equal to the thermal conductivity of the cell. Heat flow through a composite of two cells with differing conductivity is obtained in a straightforward way (Malcolm 1998).

To model two-dimensional unsteady heat flow in the full landscape, a finite difference grid is established for time as well as for space, and the rate of heat flow during a small interval of time (Δt) is modeled among cells that have centers spaced at regular intervals of Δx. According to the principle of energy conservation, the net rate of energy arriving at a cell center during Δt is equal to the rate of change of internal energy of the cell, which is a function of the temperature of the center at time $t+\Delta t$, the temperature at time t, and the mass density and specific

heat of the cell. Therefore, the temperature of a cell at time $t+\Delta t$ can be expressed as a function of: (1) Δx; (2) temperature differences with the surrounding cells at time t; (3) the thermal conductivities of the surrounding cells; and (4) the mass density, specific heat, and thermal conductivity of the cell (see Malcolm [1998] for details). Thus, the temperature in the landscape can be followed over time.

Edge Effects as a Function of Landscape Grain

METHODS

To design hypothetical landscapes, I used cells that were 10 m on a side (Δx) within grids that encompassed 10,000 ha (1,000 × 1,000 cells). Three grain sizes were simulated by varying the size of the clearings in a landscape. To approximate individual treefall gaps, the finest-grained landscapes had clearings that were 10 × 20 m (i.e., two adjacent grid cells). Clearing sizes in the other two grain sizes were 100 × 100 m and 1000 × 1000 m. Clearings were oriented either north-south or east-west and were created at randomly determined locations on the grid until the total area of cleared forest comprised 10, 20, 30, 40, or 50 percent of the whole grid. Fifteen landscape grids were created in total (three clearing sizes for each of five levels of deforestation). Because of the relatively large size of the grids, one per combination of clearing size and deforestation extent was sufficient to provide a precise description of temperature changes.

Cells in the clearings were assumed to have the thermal diffusivity and conductivity of air (α = 0.2 m²/s, k = 0.02 W/°C [Karlekar and Desmond 1977]). Thermal diffusivity and conductivity in forested cells were set to 0.053 m²/s and 0.075 W/°C, respectively (see Malcolm 1998). Over the course of a day (see below), these parameters resulted in pronounced edge effects up to 100 m deep along the borders of large gaps, in general agreement with experiments in the central Amazon (see Malcolm 1994).

Heat conduction was simulated during the course of a single day from 0600 to 1800 h. Shading served to control radiative input and was incorporated by assuming that the sun rose at 0600 h in the east and set at 1800 h in the west and moved only in the east-west dimension (zero declination). Thus, shading occurred only in the east-west dimension. The forest was assumed to be 30 m high and to act as a complete barrier to radiative heat input. At the start of each clock interval, clearing cells that were at least partly in the sun were "turned on" to 35° C —i.e., they acted as heat sources. All cells were set to 20° C at the start of the simulation, and except for unshaded cells that were reset to 35° C at the start of a clock interval, cell temperatures were allowed to freely vary. To calculate shading along the periphery of the landscape grid, I assumed that the grid was surrounded by forest. For heat conduction calculations, forested cells outside the grid were neutral—i.e., they had the same temperature as adjacent grid cells. In all simulations, the time interval (Δt) was 120 s. For analysis, cell temperatures at the start of the time intervals were averaged over the 360 intervals in the day.

RESULTS

Edge effects in the simulations varied both quantitatively and qualitatively as a function of the extent and grain of deforestation. Average forest temperature (and hence the proportion of edge-modified habitat) increased with the extent of deforestation, but decreased with increasing grain size (fig. 27.1a). Deforestation of only 10 percent of the landscape using 1 ha clearings resulted in average forest temperatures equivalent to

50 percent deforestation using 100 ha clearings. Similarly, 10 percent deforestation using 0.02 ha clearings resulted in average forest temperatures equivalent to 40 percent deforestation using 1 ha clearings. In addition to these relative effects, grain size also had a strong effect on the absolute amount of edge-modified habitat in the remaining forest. For example, holding deforestation constant at 10 percent, the amount of forest that remained at less than 21° C ("interior" habitat) varied from approximately 8,400 ha under coarse-grained deforestation, to approximately 7,000 ha under medium-grained deforestation, to less than 3,700 ha under fine-grained deforestation (fig. 27.2). In other words, at least 3,000 ha of interior habitat could be maintained under approximately 50 percent deforestation in the coarse-grained landscape, 30 percent deforestation in the medium-grained landscape, and 10 percent deforestation in the fine-grained habitat. These results make intuitive sense because, given the same amount of deforestation, finer-grain landscapes will have more forest edges and hence more pervasive edge effects.

The opposite patterns held true for temperatures in forest clearings (figs. 27.1b, 27.3). Relative to the other landscapes, clearings in the fine-grained landscape on average had cooler conditions more typical of the closed forest environment. Thus, the fine-grained landscape on average had highly edge-modified (warm) forest, but relatively moderated (cool) clearings due to shading. In an absolute sense, average forest temperatures in the fine-grained landscape closely approximated average temperatures in the clearings; hence in comparison to the other landscapes, its average temperature profile was the most homogeneous. However, this homogeneity over the whole landscape belied significant heterogeneity within the forest and clearings themselves (figs. 27.2 and

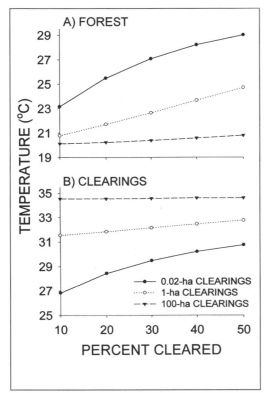

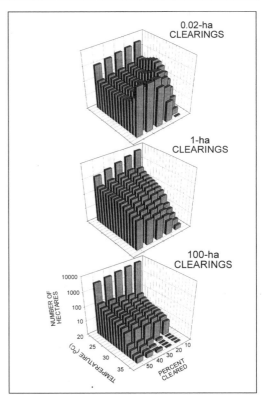

Fig. 27.1. Average temperatures of grid cells in forest (A) and clearings (B) in simulated tropical landscapes. Clearings of one of three sizes were used to deforest 10, 20, 30, 40, or 50 percent of the landscape. Heat conduction was simulated during the course of a 12 h day, with unshaded grid cells in clearings acting as heat sources. Cell temperatures were recorded at 120 s intervals, and were averaged over the 360 intervals in the day.

Fig. 27.2. Frequency distribution of temperatures of forested land in simulated landscape grids. Temperature values were truncated to create the plot intervals (e.g., 20.9 = 20° C). For each clearing size, deforestation varied from 10 to 50 percent. Each grid consisted of 10,000 ha in total.

27.3). At medium (20 to 30 percent) levels of deforestation in the fine-grained landscape, primary forest was composed of approximately equal areas of highly edge-modified forest and interior forest, whereas at these levels of deforestation in the coarser-grained landscape, the primary forest was mostly interior forest. Similarly, cells in forest clearings in the fine-grained landscape were about evenly split between hot and cold conditions, whereas clearings in the coarser-grain landscapes were characterized by hot conditions. Thus, the variance among temperatures within either forest or clearings in the fine-grained landscape was greater than in the other landscapes.

The effects of additivity were especially evident in the fine-grained landscape. In the coarser-grained landscapes, average forest temperate was a linear function of gap perimeter:forest area ratios, indicating that simple strip-based edge models were sufficient to describe the overall patterns (fig. 27.4). In contrast, in the fine-grain landscape, where edge effects from neighboring gaps often overlapped, the relationship be-

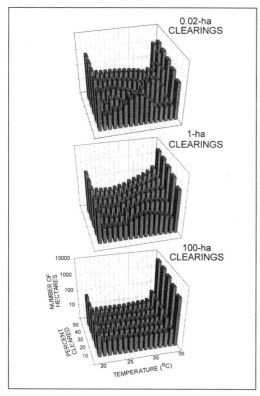

Fig. 27.3. As fig. 27.2, except that temperatures are of clearings.

tween forest temperature and perimeter:area ratios was curvilinear. In addition, the curve for the fine-grained landscape was below an extrapolation of the coarser-grained curves, indicating that for equal perimeter:area ratios, the fine-grained landscape had more extensive heating (i.e., more severe edge effects). The shape of the curve (concave downward) indicated that the increase in edge effects with increasing deforestation was disproportionately high when the absolute amount of deforestation was low. I have noted the serious conservation implications of this relationship before (Malcolm 1998); it suggests, for example, that the most rapid changes in the forest environment as a function of increased logging intensity will be at low harvest intensities.

Forest Bird Communities in Selectively Logged and Fragmented Landscapes

METHODS

The dimensions of the forest clearings in Thiollay (1992) and Bierregaard and Lovejoy (1989) differed by nearly two orders of magnitude. In the selectively logged landscape studied by Thiollay, gaps were usually 25 m across or less, although they were often long (access roads) (see fig. 1 in Thiollay [1992]). At the BDFFP in the agricultural landscape of the central Amazon, clearings were 2,000 m or more across, and the 1 and 10 ha fragments censused were at least 100–800 m from other forest (see fig. 4.1; also see fig. 2 in Malcolm [1994], which included many of the same sites that Bierregaard and Lovejoy [1989] censused).

Despite this difference in the grain size of canopy openings, plots in both studies were in areas that were presumably strongly influenced by edge effects. In the BDFFP study, obvious biotic and abiotic edge effects extended at least 100 m into fragments, and the centers of 1 ha fragments were influenced by additivity from all four edges (see fig. 29.1; Kapos 1989; Malcolm 1994; Laurance, Bierregaard, et. al 1997). Assuming a 100 m wide strip of edge-modified forest, 10 ha fragments contained only 1.35 ha of "interior" forest. I have argued that changes in small mammal communities in these fragments in the first ten years after deforestation were entirely attributed to edge effects (Malcolm 1997a). Although no information on the magnitude of edge effects is available in Thiollay's (1992) study, all of the forest in his illustration was within 100 m of a road, log-gathering area, or treefall gap. Although the logging was highly selective, it resulted in considerable damage (Thiollay 1992). On average, 3.04 trees were cut per hectare, but at one to two years later, an additional 10 to 43/ha had fallen (versus 0.7–1.2 treefalls/ha/yr in nearby primary

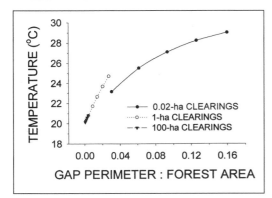

Fig. 27.4. Average temperatures in forested grid cells as a function of the gap perimeter/forest area ratios in the landscape. Results are shown for fifteen landscape grids (five levels of deforestation for each of three clearing sizes).

forest). At ten years after logging, transects in logged forest consisted of 41.9 percent gaps (versus 12.8 percent in primary forest). Thus, the potential for additive edge effects was high.

I examined two aspects of the bird community: bird species richness (as reported by the authors) and abundances of the fifty-three species in common to the two studies (see appendices in Thiollay [1992] and Bierregaard and Lovejoy [1989]). I classified each species as showing positive, negative, or no change in abundance in response to the landscape modification, and the agreement in classification between the two studies was compared with a null model based on chance agreement. A positive response was assigned if a species was more abundant in both of the disturbed habitats in a study (1 and 10 ha fragments in the case of Bierregaard and Lovejoy [1989] and 1- and 10-year-old forest in the case of Thiollay [1992]) than in undisturbed primary forest. A negative response was assigned if the species was less abundant in both disturbed habitats than in continuous forest. A species was judged to have shown no response if abundance in undisturbed forest was inter-

mediate between that in the two disturbed habitats.

RESULTS

In both studies, the landscape transformation resulted in a marked decline in species richness (fig. 27.5). Moreover, individual species tended to respond in the same way in the two sites (fig. 27.6). Most species (44 of 53) showed a negative response to the landscape transformation in at least one of the studies (hence the decline in species richness), but more important, of the 24 species affected by the landscape transformation in both studies, the sign of the impact (either positive or negative) was the same for 17 (P = 0.06, two-tailed binomial test).

DISCUSSION

The additive edge models described here are an improvement over previous models because they can be more easily applied to a broader range of landscape configurations. Simple strip-based models are inaccurate whenever edge effects overlap, which is a common occurrence where forest clearings are relatively close to one another (as investigated here) or where edge effects extend far into the forest (as investigated by Malcolm [1998]). These results suggest that edge effects may provide a useful metric of grain size. A relatively coarse-grained landscape under one edge environment will be considered finer-grained when edge effects are stronger, and vice versa. In turn, a given landscape configuration may have different grain sizes from the perspective of different processes and organisms, depending on the depths of edge effects.

The similar responses of the bird community under two different grains of deforestation support the possibility that edge effects in some cases may provide a common

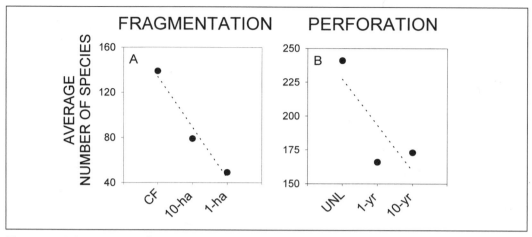

Fig. 27.5. Avian species richness in continuous forest (CF) and 10 and 1 ha forest fragments in the central Amazon (Bierregaard and Lovejoy 1989) and in French Guiana (UNL = unlogged; 1 yr = 1 year after logging; 10 yr = 10 years after logging [Thiollay 1992]). In each graph, edge-modified sites are to the right of undisturbed forest sites.

currency for understanding the impacts of diverse landscape change. Thus, edge models may provide a quantitative framework for transferring understanding and empirical results from one landscape to another. I did not undertake a quantitative test here because I did not have information on the spatial context of all of the individual plots. However, as a rough approximation, fragments in the large agricultural clearings at the BDFFP site in the central Amazon probably corresponded in size to residual forest in the simulated medium-grained, high deforestation landscape, whereas edge effects at the selectively logged sites probably corresponded to the fine-grained, low deforestation simulation. In my simulations, average forest temperatures in these two situations were approximately equal (see fig. 27.1), suggesting that if census sites randomly sampled the forest, similar magnitudes of community change would be observed in the two landscapes. More detailed predictions would require detailed information on census locations.

The above very approximate analysis begs the question: What sort of information is required in order to develop more quantitative

predictions of the extent of edge-induced habitat change? Previously, I used one landscape configuration to provide the parameters for an edge model that was subsequently applied to other landscape configurations (Malcolm 1994). In that study, I used the linear edges of continuous forest to provide the parameters for the model and then used the resulting parameter estimates to predict edge effects in 1 and 10 ha fragments. To apply this approach to the two-dimensional model described here, information on a more diverse array of clearing sizes is desired. An ideal experimental design would be one in which censuses are undertaken in forest strips of various widths isolated by clearings of various widths (including narrow roads and large clearings). A further refinement would be to extend the univariate approach in Malcolm (1994), where I examined understory and overstory vegetation density as a function of the edge gradient, to a multivariate community approach.

Detailed predictions of edge effects provide a baseline from which to investigate other processes acting in the landscape. Up to approximately 10 years after deforestation, and perhaps beyond, small mammal

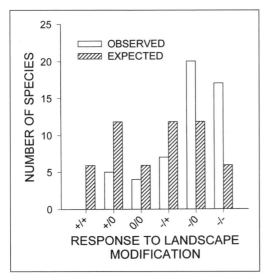

Fig. 27.6. Frequency of fifty-three bird species in common to Bierregaard and Lovejoy (1989) and Thiollay (1992) that showed a positive response to landscape modification in both studies (+/+), one study but not the other (+/0), neither study (0/0), a negative response in one study but a positive response in the other (+/-), etc. Shaded histograms are null expectations under the assumption that any response is equally likely.

populations at the BDFFP site appear to be at levels expected solely from edge-induced habitat change, presumably because the fragments periodically received immigrants from nearby continuous forest (Malcolm 1991). With time, as the fragments become further isolated, their communities can be expected to evolve not only as a function of direct and indirect edge-based change within the fragments, but also because of decreased possibilities for colonization from source populations in primary forest. Thus, species that avoid secondary and edge habitats (such as the wooly opossum [*Caluromys philander*] and the rice rat [*Oryzomys macconnelli*]) can be expected to eventually disappear from the fragments (Malcolm 1991). Simplification and loss of secondary forests in the matrix can be expected to lead to simplification in the matrix small-mammal

community (Malcolm 1995) and to further changes in the pool of immigrants. This deterioration in the original small-mammal community may have already taken place in the Atlantic rainforest of Brazil, an ecosystem that has been highly modified by human activities (see Fonseca 1988). A predominance of marsupials, even in large primary forest tracts, may be indicative of these changes on a landscape level (Fonseca 1988).

Similar expectations with regard to changes in fragment communities may apply in Africa as well. Indications are that African rainforests are strongly convergent with Neotropical ones, both in the importance of vegetation structure as a correlate of small mammal community structure and with respect to the apparent paucity of small mammal interior-forest specialists (Malcolm 1997a; Malcolm and Ray, in press). In addition to edge-induced habitat changes and altered immigration and extinction probabilities comes the possibility of invasion by savanna species and other exotics as modern human-created "savannas" interconnect naturally occurring ones. The extent to which these invasions will occur in the Neotropics is unknown, but in Africa degradation of primary forest habitats is frequently accompanied by a decrease in the primary-forest rodent fauna and increases in taxa associated with anthropogenic and savanna habitats (e.g., Rahm 1972; Happold 1975, 1977; Adam 1977; Iyawe 1989; Okia 1992; Struhsaker 1997; Malcolm and Ray, in press).

The simulations presented here show that the grain size of clearings (as defined relative to the depth that edge effects extend) has important implications for the resultant spatial distribution of edge-modified habitats, both in forest and in clearings. In fine-grained landscapes, such as those resulting from selective logging and small-scale shifting agriculture, the overall shift in the forest environment is to conditions more typical of clearings. At the same time, despite the overall convergence in forest and gap environments, both cover types come to contain habitat extremes that are rare in coarser-grained landscapes. On one hand, the relatively benign conditions in some clearings hold out hope for the regeneration of shade-tolerant forest tree species. However, because of additive neighborhood effects, areas of extremely edge-modified forest are also common.

The implications of this intimate juxtaposition of extremes in fine-grained habitats is unknown, but because of the potential for changes in species interactions at small spatial scales, it does not bode well for many forest species. The counterintuitive loss from selectively logged forests of species that specialize on treefall gaps in primary forest (e.g., Iyawe 1989; Thiollay 1992) may be a result of this juxtaposition—although suitable gaps are present, suitable nearby forest is not. The nonlinear nature of the change in habitat characteristics with increasing deforestation moreover suggests an especially rapid change in interior (cool) conditions at low absolute quantities of deforestation (*contra* Johns 1997). Highly edge-modified forest as a result becomes widespread throughout the landscape. Struhsaker (1997) suggested that these areas of intense edge effects may act as foci for further deterioration in the forest environment. In even moderately logged African forests, elephants and rodents were attracted to areas where gap densities were high, which may have resulted in a positive feedback that led to further deterioration of the forest environment. Thus, although small clearings may seem benign (e.g., Dale et al. 1994), the edge effects that they create can quickly sum up to create disproportionate change.

Although the simulations here seem to suggest that with respect to edge effects in the remaining primary forest, coarse-grained deforestation is preferable to fine-grained

deforestation, two critical factors must be kept in mind. First, although the idea of large forest remnants is an attractive one, rarely do they remain large for long. Instead, a frequent progression in coarse-grained landscapes is shrinkage and attrition of the remaining fragments until only heavily edge-affected, isolated fragments remain— a worst-case scenario from a conservation viewpoint. Second, compared to coarse-grained landscapes, conditions in the clearings of fine-grained landscapes are benign. These clearings will be more easily colonized by forest-dwelling species than the radically altered habitats created during large-scale clear-cutting. The conservation message of this paper instead is to point out the potential dangers of edge effects, even in fine-grained landscapes. A key conservation and management tool is to reduce edge effects whenever and wherever possible, by maintaining large reserves and corridors (on the order of tens or hundreds of kilometers in width), and by reducing canopy damage during selective logging, through low harvesting volumes and careful logging techniques (Johns, Barreto, and Uhl 1996).

The model presented here provides a starting point for simulations of edge effects under a broad range of landscape configurations. It is an improvement over existing models (e.g., Chen, Franklin, and Spies 1993; Chen, Franklin, and Lowe 1996) because it incorporates the additivity that is critical in fine-grained landscapes. My comparisons of avian communities between the two Neotropical sites suggest that edge models such as this one will be useful in extending knowledge from one landscape configuration to others. These models provide an important addition to metapopulation models, which for convenience often break landscapes into "suitable" or "unsuitable" patches (e.g., Bascompte and Solé 1996). In reality, processes such as edge effects create a gradient of habitat conditions and a complex mosaic of abiotic and biotic change (Chen, Franklin, and Lowe 1996). Models based solely on the configuration of canopy cover that do not incorporate abundances of certain habitat types such as interior habitat may lead to incorrect estimates of extinction thresholds.

Conservation Lessons

1. An important conservation strategy appears to be to reduce edge effects wherever and whenever possible. This can be accomplished by establishing reserves and corridors that are at least tens of kilometers across. In fine-grained landscapes (many small clearings and fragments), the number and sizes of canopy gaps should be minimized; for example, by harvesting small timber volumes and by practicing careful logging techniques.

2. Edge effects are additive and can be particularly strong near corners of fragments or in thin corridors. Additivity is especially important in fine-grained landscapes, such as in selectively logged forests where clearing and fragment sizes are small, but where edge effects from neighboring gaps overlap. In these landscapes, simple edge models based on perimeter:area ratios are inaccurate — they underestimate the magnitude of edge effects.

3. A model of heat conduction can be used as a proxy to investigate additive edge-based disturbances in tropical landscapes. Edge effects in simulated landscapes varied both quantitatively and qualitatively as a function of the extent and grain of deforestation. The proportion of edge-modified forest habitat increased with the extent of deforestation, but decreased with increasing grain size. However, relative to other landscapes, clearings in the fine-grained landscapes on average had conditions more typical of the closed forest environment. In comparison to the other landscapes, average conditions in forest and clearings in fine-grained landscapes were more similar.

4. This homogeneity over the whole fine-grained landscape belied significant heterogeneity within the forest and clearings themselves. The variance among conditions within either forest or clearings was greater in the fine-grained landscapes than in the other landscapes. The results of this intimate juxtaposition of extremes are unknown but may have important conservation implications.

5. In fine-grained landscapes, the relationship between the average magnitude of edge effects and the magnitude of deforestation indicated that the increase in edge effects with increasing deforestation was disproportionately high when the absolute amount of deforestation was low. Thus, the most rapid changes in the forest environment as a function of increased logging intensity may be at low harvest intensities. Edge effects from neighboring gaps may quickly sum up to create disproportionate change.

6. Additive edge models provide an important addition to metapopulation models. Estimates of habitat connectivity based solely on the configuration of canopy cover may seriously overestimate connectivity for species that rely on interior forest conditions.

Acknowledgments

I am grateful to R. Bierregaard, W. Laurance, J. Ray, and an anonymous reviewer for helpful comments on an earlier draft of this manuscript.

CHAPTER 28

Characterizing the Changing Spatial Structure of the Landscape

MILES G. LOGSDON, VALERIE KAPOS, AND JOHN B. ADAMS

The concern about tropical deforestation first voiced in the 1970s and 1980s has led to concerted efforts to monitor changes in forest cover in tropical regions and, more recently, worldwide. Advances in remote sensing and improvements in the availability of remotely sensed data, along with increasingly sophisticated tools for spatial data analysis, have greatly facilitated this monitoring. However, the rate of deforestation and amounts of remaining forest arising from changes in forest cover are not the only issues of concern to ecologists. Skole and Tucker (1993) highlighted the importance of the forest fragmentation that accompanies deforestation. They reported that in 1988, the total area of forest potentially affected by fragmentation and edge influences was 2.5 times the area actually deforested.

The ecological implications of these effects are being evaluated through such experimental studies as the BDFFP and other field research. Much evidence suggests that the value of a habitat patch for any given species depends not only on the size and configuration of that patch but also on the nature and spatial characteristics of the matrix (surrounding vegetation) in which it occurs (see, e.g., Bierregaard and Stouffer 1997; Malcolm 1997a). Therefore, the implications of landscape dynamics for land use planing and conservation management can be significant.

All landscapes are composed of recognizable elements that change over time. The spatial scale at which those elements are recognized and defined, and the temporal scale that corresponds to measured change, are central issues in a line of research that now integrates remote sensing (RS) and geographic information systems (GIS) analysis. The emerging disciplines of landscape ecology and conservation biology have highlighted the importance of viewing landscapes in their entirety and considering landscape elements and their relationships to each other in both space and time. Simplistically, it is important not only to consider how much forest area remains but also to examine the sizes and shapes of the elements and to identify what surrounds them. For example, other chapters in this volume document the dynamic nature of the forest fragments of the BDFFP and the plant and animal communities within them, as well as the effects of changes in the surrounding vegetation on these dynamics (e.g., Chapters 11–13 and 19–20). In this chapter, our purpose is to demonstrate how an integrated approach combining remote sensing and

This is publication number 295 in the BDFFP Technical Series.

spatial analysis is used to describe land-scape pattern and how it changes over time. This quantitative description of the changing landscape provides the context for many of the issues of concern to landscape ecologists and conservation biologists.

The landscape we characterize is the fragmented tropical forest landscape that includes the BDFFP sites (see fig. 4.1). This region is predominantly covered in mature terra firme, evergreen tropical moist forest on clayey and sandy yellow latosols (for more details, see Lovejoy and Bierregaard 1990; Chapter 23). Some clearing for cultivation and pasture has occurred since the late 1970s; some cattle ranches have maintained pasture, while other areas are now dominated by woody regrowth (Bierregaard et al. 1992). The easily identifiable rectilinear pattern of ranches, fields, and research forest fragments creates a strong contrast to the surrounding forest.

In this landscape, the mature tropical forest forms a landscape matrix, or backdrop, on which patches of distinct land-cover types reside. The pattern of these different landscape elements and their changes through time are products of both cultural and natural processes. In this chapter we will first characterize the spatial pattern of that well-understood fragmented landscape and then discuss the extent to which this same approach can be used to inform us about less studied areas.

Methods

THE DATA

The remotely sensed data interpreted here are measurements by sensors on satellite platforms of electromagnetic radiation reflected from the earth's surface. These sensors record spatial variations in reflectance at various wavelengths of the electromagnetic spectrum repeatedly through time. A single remotely sensed image is, therefore, a digital record of the reflectance properties of the material covering earth's surface at one point in time.

Because the electromagnetic radiation passes through the earth's atmosphere twice before being recorded by the satellite, and because transient atmospheric conditions can influence the signal recorded, the first step in the image processing procedure is to eliminate the influence of atmospheric conditions. The remaining signal is a spectral signature of the land's reflective properties. Different land covers reflect the incoming radiation with unique characteristics that allow us to classify or name the land-cover type within broad classes of surface material. This spatially distributed information is recorded in individual picture elements (pixels) and is "geo-referenced," or linked to real-world map coordinates. This information is then maintained in a geographic information system where relationships may be mapped, queried, or modeled for any number of purposes.

The choice of which satellite sensor to use in a particular investigation depends on the spatial and temporal resolution and the wavelength sensitivity of the sensors. A variety of remote sensing investigations have focused on identifying land-cover types and rates of deforestation in the Amazon River Basin using various sensors (Tucker, Holben, and Goff 1984; Nelson and Holben 1986; Townshend, Justic, and Kalb 1987; Malingreau and Tucker 1988; Setzer and Pereira 1991; Skole and Tucker 1993; Stone et al. 1994).

In this investigation, our interest was to describe interannual change at a grain (resolution) applicable to remotely sensed data, and at the spatial extent of the BDFFP site. Therefore, we began with four land-cover-classified Landsat Thematic Mapper (TM) images (August 1988, '89, '90 and '91) of a 443 km² area encompassing the BDFFP site in northern Brazil (3° S, 60° W).

Adams, Sabol, et al. (1995) described the classified land-cover data set. In this remotely sensed data set, each pixel is assigned to a land-cover class through application of a spectral mixing model (Adams, Kapos et al. 1990). This procedure uses data from six Landsat spectral bands to compare the spectral response of a pixel with spectra from four of its probable components (green vegetation, nonphotosynthetic vegetation [e.g., wood, dry grass], soil, and shade), which are known as "endmembers." The modeling procedure provides best-fit estimates of the composition of each pixel in terms of the chosen endmembers.

Based on an understanding of the physical characteristics of different vegetation types, the classification step concludes by assigning one of eight recognizable land-cover labels to each pixel. The eight land-cover classes comprised three wooded classes (forest, closed-canopy regrowth, and open-canopy regrowth), two types of pasture (pasture and pasture/crops), and finally classes of slash and exposed soil. No attempt was made to assess the quantitative accuracy of the fractional composition of the endmembers, but rather the classification was tested using field observations, ground photographs, and aerial photographs where the relative proportions of the four endmembers were verified and the classification was found to be qualitatively correct (Adams, Kapos, et al. 1990). Each annual classified image was converted into a raster GIS data model at a grid cell size of 30 m × 30 m (fig. 28.1). In this way the value of each cell in each annual land-cover data set represents that cell's land-cover type.

THE ANALYSIS

Discrete spatial groupings of cells that shared a common land-cover assignment were defined as landscape patches. At this stage it is possible that individual isolated cells or small groups of cells were defined as unique patches. These small patches have the spectral properties consistent with general land-cover labels but may be spatially inconsistent with the meaning of those labels. As an example, a single 30 m × 30 m cell may indeed have the spectral properties of a pasture, yet such a small area would not normally be labeled a pasture.

A classification system has an implied spatial scale. The intent of this modeling design is to allow analysts to match the "scale" (or resolution) of a landscape patch model with the process under investigation. The goal was to provide a description of general land-cover change from a landscape perspective, therefore a spatial filter was applied to deal with patches whose size fell below an appropriate scale for their land-cover classification. In the context of this study, the 1 ha size of the smallest experimental forest fragment in the BDFFP study was deemed to be an appropriate minimum patch size for all the land-cover classes. Therefore, individual isolated cells or groups of cells with a size that was below our value for the allowable minimum size of a patch were assigned to the adjoining patch sharing the longest border.

Once patches were defined, the number and type of cells from the original land-cover data set were then used to determine the spatial composition and configuration of each patch. In this way a measure of internal patch heterogeneity for each of the modeled landscape patches was derived. Patches were recognized as landscape elements containing heterogeneous land-cover types. For example, a pasture may represent a single

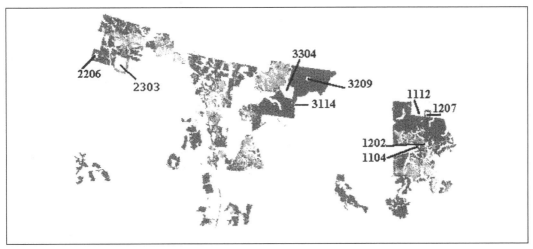

Fig. 28.1. False-color image of the BDFFP study site derived from Landsat Thematic Mapper data. Primary forest (including isolated reserves) are white. Extensive dark gray areas are old cecropia-dominated second growth (see plate 9).

landscape element but be recognized as a patch containing small groupings of non-pasture land cover. The land-cover label of individual cells in patches that fall below the allowable minimum patch size are replaced with the land-cover label of the largest neighboring landscape patch.

Various measures of spatial structure can be obtained from commercially available GIS or public-domain, spatial-analysis software. The public-domain spatial pattern analysis program, Fragstats version 2.0. (McGarigal and Marks 1994), was used in this investigation. The aim of quantifying the landscape is to understand not only the composition but also the configuration of the landscape elements. Describing the landscape's spatial structure in this way is sometimes referred to as landscape physiognomy or landscape pattern analysis (M. Turner 1989; Dunning, Danielson, and Pulliam 1992).

The distinction between measures of landscape composition and landscape configuration is sometimes unclear. In general, landscape indices describe either the variety or abundance of patches (composition) or the shape and relative placement of those patches (configuration) for the landscape as a whole. These indices together are descriptive of the spatial structure of the landscape at a point in time and are not inferential statistics for spatial process through time (see Ripley 1981 or Cressie 1991). It is only through constructing a time series of these indices that possible linkages between pattern and process can be inferred.

For this investigation, we chose simple measures of landscape composition and configuration to compare over a four-year period. The choice of these metrics was influenced by our interest in describing the fragmentation of the landscape. Individually, the measures provide little information about the spatial patterns within the landscape. Collectively, they provide a basis for a quantitative analysis of landscape change.

Landscape composition may be characterized using measures of the variety and abundance of landscape patches. The number of patches (NP) and mean patch size (MPS), reported in hectares, are influenced by the minimum patch size limitation established by the landscape model. In this investigation we were interested in monitoring and characterizing patches larger than 1 ha. Total core area (TCA) and mean core area (MCA) describe the degree to which locations within every patch are distanced from the edge of the patch. These indices summarize the amount of edge environment within the landscape as a whole. We have simplified this analysis of edge environment by reporting on a fixed edge length of 30 m, which is a conservative measurement of the edge influence assessed within the BDFFP reserves (Kapos et al. 1997). A Shannon Diversity index (H'; M. Turner, O'Neill et al. 1989) is used to report the proportional distribution of area among patch types.

The relative placement and adjacency of patches defines the landscape's configuration. Total edge (TE) and edge density (ED) in meters and meters per hectare report the sum of all edge segments in the landscape. As more unique patches appear in the landscape, total edge increases. The shape of each patch, described by dividing the patch perimeter by the square root of patch area, also reflects the amount of reported edge environment. For this measure the single term, mean shape index (MSI), describes the average shape index for all landscape patches. A measure of interspersion and juxtaposition (IJI) of patches reports on the percentage of all possible shared edge environments that are present in the landscape. In other words, with a given number of patch types, a limited number of edge environments are possible. The value for the term IJI reports the percentage of those edges that were observed. Finally, the mean patch fractal dimension (MPFD) term describes the shape complexity of the average landscape patch. For a two-dimensional landscape, MPFD will approach 1.0 for landscapes with simple circular or square patches, and increase to 2.0 when patches take on more convo-

luted, plane-filling shapes (O'Neill et al. 1988).

To evaluate changes in individual patches, measures of area, core area, perimeter, shape index, and fractal dimension were calculated for the patches of interest. The measure of nearest neighbor distance (NND) is a measure of how isolated these individual patches are from the nearest patches with the same land-cover assignment. This distance is a straight-line distance and does not reflect the topography of the site. The original classified remote-sensing data provided land-cover assignments for individual 30 m cells that were then modeled to the coarser resolution of these patches. In this multi-scaled perspective, each individual cell potential may be modeled as a "patch within a patch." Applying this form of spatial hierarchy to our data allowed us to investigate the structure of individual patches and attributes of internal patch heterogeneity. The count of unique occurrences in land-cover classes that are present in each patch defines the index of patch variety (VAR). While we expect patches to be heavily dominated by one land-cover type, we recognize that at the resolution of the individual cells, patches may have some variation in composition.

Results

A number of conditions are explicit in our characterization of this landscape and its changes through time. The land surface was classified into discrete homogeneous units (patches) and labeled with understandable and descriptive names that only suggest ecological function and character (plates 12 and 13). This landscape description is limited to patches no finer than 1 ha in resolution; however, for selected patches, the finer resolution of the classified, remotely sensed data cells provided information concerning

TABLE 28.1. Change in Descriptive Measures of Landscape Composition and Configuration for Four Years

A. Indices of landscape composition and configuration

	NP	MPS	TCA	NCA	H'	TE	ED	MSI	IJI	MPFD
1988	655	67.6	39456	1241	0.74	1240560	28.02	2.31	76.68	1.14
1989	608	72.8	39496	1177	0.71	1211040	27.35	2.25	65.44	1.14
1990	615.00	72.00	39018	1417	0.71	1386240	31.31	2.47	58.12	1.15
1991	474.00	93.41	39296	1192	0.64	1308600	29.55	2.65	48.19	1.16

B. Relative change to 1988 indices

	NP	MPS	TCA	NCA	H'	TE	ED	MSI	IJI	MPFD
1988	1.00	1.00	1.00	1.00	1.00	1.00	1.00	1.00	1.00	1.00
1989	0.93	1.08	1.00	0.95	0.96	0.98	0.98	0.97	0.85	1.00
1990	0.94	1.07	0.99	1.14	0.96	1.12	1.12	1.07	0.76	1.01
1991	0.72	1.38	1.00	0.96	0.86	1.05	1.05	1.15	0.63	1.02

Notes: NP = number or patches   TE = Total edge (m)
MPS = mean patch size (ha)   ED = edge density (m/ha)
TCA = total core area (ha)   MSI = mean shape index
NCA = number of core area   IJI = interspersion of juxtaposition index (%)
H' = Shannon's diversity   MPFD = mean patch fractal dimension

internal patch heterogeneity. The temporal resolution was annual, so processes occurring over shorter time periods may have been missed entirely, though they can often be inferred from the changes observed. It is important to note that the landscape changes we report here are changes in the composition and configuration of the classified land-cover data and therefore represent actual change.

CHARACTERIZING THE LANDSCAPE AS A WHOLE

The assignment of grid cells to landscape patches of 1 ha or larger produced landscape configurations that were relatively simpler than the classified satellite images because of the effectively coarser grain (plates 12 and 13). The relatively simple measures presented in table 28.1 provide a quantitative means of describing a changing landscape that is difficult to understand through visual inspection alone. Researchers whose investigations ignore the changing spatial context of this landscape and the dynamics of the

individual patch ignore an important and dynamic process.

Various changes in the landscape as a whole occurred over the four-year period covered by this investigation. There was a general reduction in the patchiness of the landscape—the number of individual patches was reduced by 28 percent and an increase in mean patch size from 67.6 ha to 93.41 ha, which, in combination with the fall in number of patches, suggests that some of the original smaller patches were either coalescing or being subsumed into larger ones. The amount of total edge environment increased 5 percent, yet the total area of core environment remained unchanged, despite a small decrease in the total number of core areas. The mean shape index increased by 15 percent, suggesting that patch shapes became increasingly complex over the study period, but mean patch fractal dimension remained unchanged. There was a 37 percent decrease in the measure of interspersion and juxtaposition (IJI), indicating that the landscape moved toward a more uneven distribution of adjacent land-cover types, and the

TABLE 28.2.   Relative Changes for Indcies of Land-Cover Class Composition and
Configuration over four years

A. 1988 indices of land-cover class composition and configuration

	CA	NP	MPS	MSI	MPFD	TE	ED	TCA	MCAI	IJI
Forest	35732	43	830.98	1.99	1.11	665430	15.03	34105	36.76	70.26
Closed canopy	4015	181	22.18	2.24	1.14	585420	13.22	2932	28.96	60.24
Opened canopy	860	144	5.97	2.45	1.16	341040	7.70	307	22.91	71.74
Pasture/crops	481	103	4.67	2.12	1.14	184380	4.16	172	23.19	78.13
Pasture	2713	80	33.91	2.68	1.16	515280	11.64	1772	27.04	92.03
Slash	372	66	5.64	2.30	1.15	136020	3.07	147	19.82	66.77
Exposed soil	105	38	2.75	2.22	1.15	53550	1.21	21	12.85	47.82

B. 1991 indices of land-cover class composition and configuration

	CA	NP	MPS	MSI	MPFD	TE	ED	TCA	MCAI	IJI
Forest	35917	67	536.08	2.01	1.11	810420	18.3	34055	32.29	38.55
Closed canopy	6129	158	38.79	2.74	1.16	966420	21.83	4375	31.65	40.61
Opened canopy	1420	129	11	3.04	1.19	537180	12.13	556	20.16	55.26
Pasture/crops	127	28	4.52	2.28	1.15	54720	1.24	37	21.06	75.82
Pasture	584	79	7.39	2.57	1.16	218160	4.93	225	18.84	66.16
Slash	93	9	10.3	2.29	1.14	24840	0.56	48	35.83	67.58
Exposed soil	6	3	2.1	2.41	1.18	4200	0.09	0.27	3.58	47.79

C. Relative change in indices over the complete time series (1988–1991)

	CA	NP	MPS	MSI	MPFD	TE	ED	TCA	MCAI	IJI
Forest	1.01	1.56	0.65	1.01	1.00	1.22	1.22	1.00	0.88	0.55
Closed canopy	1.53	0.87	1.75	1.22	1.02	1.65	1.65	1.49	1.09	0.67
Opened canopy	1.65	0.90	1.84	1.24	1.03	1.58	1.58	1.81	0.88	0.77
Pasture/crops	0.26	0.27	0.97	1.08	1.01	0.30	0.30	0.22	0.91	0.97
Pasture	0.22	0.99	0.22	0.96	1.00	0.42	0.42	0.13	0.70	0.72
Slash	0.25	0.14	1.83	1.00	0.99	0.18	0.18	0.33	1.81	1.01
Exposed soil	0.06	0.08	0.76	1.09	1.03	0.08	0.07	0.01	0.28	1.00

Notes: CA = sum of areas of all patches of corresponding land-cover class; MCAI = sum of core areas of
each patch of corresponding land-cover class divided by number of patches of that same land-cover class.

overall diversity of the landscape, as represented by Shannon's Index, decreased by
more than 10 percent over the study period.

CHARACTERIZING LANDSCAPE CLASSES

Additional information on landscape
processes is obtained by characterizing the
changes in the spatial configuration of each
class, or land-cover type, present in the
landscape (table 28.2). In 1988, the landscape was composed of approximately 80
percent forest, 9 percent closed canopy regrowth, 6 percent pasture, 2 percent open
canopy regrowth, and smaller proportions
in slash and exposed soil. By 1991 forest and

both classes of woody regrowth had increased in area, while all other classes had
decreased, especially pasture. The 1991
landscape composition was 81 percent forest, 14 percent closed canopy regrowth, 3
percent open canopy regrowth, and 1 percent pasture.

Primary forest was the only land-cover
type to increase in the number of patches,
and the mean size of forest patches decreased concomitantly (table 28.2). This suggests that forest fragmentation continued,
despite the lack of maintenance of pasture
areas that had already been cleared. The two
classes of woody regrowth increased in area
over the study period through increases (of

about 80 percent) in mean patch size, which accompanied a slight (c. 10 percent) decrease in number of patches. This pattern is consistent with the expansion and coalescence of patches of colonizing vegetation within pastures that is observed on the ground. The increase in amount and density of edge seen for all three woody vegetation types and the rising shape index for regrowth patches are also compatible with increasing fragmentation of forest and irregular expansion of the regrowth patches.

As observed for the landscape as a whole, the mixing of the different land-cover types decreased over time. In 1988, of all eight classes, pasture was the most evenly interspersed with other land-covers (IJI = 92.03), but by 1991 its index of interspersion and juxtaposition had fallen by 28 percent of its original value. Mature forest and both regrowth classes also showed marked reductions in IJI over the study period.

The relatively large shifts in the indices of configuration by class, compared with the relatively small changes in measures of landscape composition, suggest that the combined strength of these descriptive indices may provide an important tool for detecting and describing key landscape changes.

CHARACTERIZING SITES

Due to our interest in changes to BDFFP research sites (see fig. 28.1), each site that could be distinguished in all four years was monitored as an individual landscape patch. At least two of the smallest project reserves disappeared from the landscape map (i.e., fell below 1 ha in size) during the study period, and one large reserve (2303; see table 4.3) became a discrete patch part way through the study. All of the reserves showed small year-to-year fluctuations in size and shape, which can be explained by changes in the management of the surrounding areas and processes of succession and regrowth. Of the eight reserves that were followed through the full four-year study period (table 28.3), half decreased in apparent area—two appreciably—and the others grew slightly. Core area of three of the shrinking reserves declined substantially, and one reserve (2206) which increased slightly in size, suffered a significant reduction in core area from 1988 to 1991. This latter reserve became more complex in shape, as reflected by the increasing shape index, and also became less isolated from other patches of forest.

Because we defined a landscape patch as a discrete, homogeneous grouping of cells sharing a single common land-cover class, we have ignored, until this point, the variability in the composition of a patch at any finer spatial data resolution. However, using the land-cover classes for individual 30 m cells within these patches from the original remote sensing data we have assessed internal patch heterogeneity of the BDFFP reserves. The index of patch variety, which is a count of the unique occurrences of land-cover classes that are present in each patch (table 28.3), varied from 1 to 8. Not surprisingly, the greatest heterogeneity was found in the largest reserve (3304; see plate 2) which contained all eight land-cover classes throughout the time series. Three sites became less heterogeneous through time, and three of the four sites containing a single land-cover class in 1988 remained unchanged at the end of the time series despite some interim fluctuation. Increases in heterogeneity were usually associated with increases in apparent patch area, suggesting that the incorporation of additional cells into the patch was responsible for the change rather than changes in identity of individual cells.

Of the eight individual isolated BDFFP patch sites monitored in this investigation,

TABLE 28.3.   Changes in Patch Indices for BDFF Project Sites

	Site	A	CAP	P	SI	D	NND	VAR
1988	2206	13.77	9.54	2160	1.455	1.063	90	5
	3304	137.25	113.94	11400	2.433	1.126	30	8
	3209	11.52	7.83	1650	1.237	1.037	760	1
	3114	1.44	0.36	600	1.250	1.047	150	1
	1207	11.7	7.92	1680	1.228	1.035	30	1
	1112	2.16	0.81	660	1.123	1.023	240	1
	1202	14.13	10.44	1920	1.277	1.041	85	5
	1104	3.33	1.62	960	1.315	1.053	85	4
1989	2206	19.89	12.51	3720	2.085	1.12	30	6
	3304	134.55	114.12	9780	2.108	1.106	42	8
	3209	12.06	8.46	1620	1.166	1.026	511	1
	3114	1.44	0.36	600	1.250	1.047	162	1
	1207	11.97	8.10	1800	1.301	1.045	30	2
	1112	1.98	0.72	780	1.386	1.066	210	1
	1202	44.64	23.67	10980	4.108	1.217	30	7
	1104	3.24	1.62	900	1.250	1.043	109	1
1990	2206	15.57	10.08	2760	1.749	1.093	85	5
	3304	133.38	111.42	10500	2.273	1.116	60	8
	3209	11.61	8.01	1740	1.277	1.042	511	2
	3114	1.08	0.27	480	1.555	1.031	175	1
	1207	11.34	7.65	1740	1.292	1.044	30	2
	1112	1.62	0.54	600	1.179	1.034	270	1
	1202	15.3	11.43	2040	1.304	1.044	42	6
	1104	2.97	1.35	840	1.219	1.038	124	3
1991	2206	14.04	8.46	3000	2.002	1.117	60	2
	3304	136.44	115.02	10260	2.196	1.111	60	8
	3209	12.06	8.37	1920	1.382	1.055	750	2
	3114	1.26	0.27	540	1.203	1.039	153	1
	1207	11.88	8.19	1800	1.306	1.046	30	1
	1112	1.71	0.54	600	1.147	1.028	240	1
	1202	13.95	10.17	1920	1.285	1.042	30	2
	1104	3.42	1.62	960	1.298	1.050	85	2

Reserve 2206 experienced the most change. The site's shape index and fractal dimension changed, revealing a shift toward a more complex shape, while core area was lost and the patch became more isolated. The patch lost three types of land cover as elements of internal patch heterogeneity and the proportion of area among intra-patch land-cover types became more equitable.

Discussion

The landscape changes documented in this analysis correspond well with the perceptions of researchers working in the field. The abandonment of many cattle pastures and the sporadic management of others resulted in a mosaic of woody regeneration and forest patches fluctuating in size, with an overall increase in woody vegetation. From a quantitative perspective, the landscape in which the BDFFP research sites reside changed between 1988 and 1991 into a somewhat less diverse landscape of fewer and larger elements when modeled at a 1 ha patch scale. However, these elements became more complex in shape, creating a small increase in the amount, yet a decrease in the diversity, of edge environments (table 28.4).

TABLE 28.4.   Relative Change in BDFF Project Sites Patch Indices Between 1988 and 1991

Site ID	A	CAP	P	SI	D	NND	VAR
2206	1.02	0.89	1.39	1.38	1.05	0.67	−3
3304	0.99	1.01	0.90	0.90	0.99	2.00	0
3209	1.05	1.07	1.16	1.12	1.02	0.99	1
3114	0.88	0.75	0.90	0.96	0.99	1.02	0
1207	1.02	1.03	1.07	1.06	1.01	1.00	0
1112	0.79	0.67	0.91	1.02	1.00	1.00	0
1202	0.99	0.97	1.00	1.01	1.00	0.35	−3
1104	1.03	1.00	1.00	0.99	1.00	1.00	−2

Throughout the entire landscape, patches classified as both pasture and pasture/crops decreased in total area, but the configuration of this change was quite different. While the pasture land-cover class retained similar numbers of patches with similar shape characteristics, these patches were much smaller with less core area. Patches of pasture/crop remained approximately the same size but were reduced in number.

Closed and open canopy regrowth patches increased in size and core area while decreasing in number throughout the entire landscape. This increased the total amount of edge environment surrounding these land-cover types, yet decreased the variety of those environments. Some of this change accounts for the increase in the number of forest patches, as closed canopy patches changed enough spectrally to be classified as forest patches in the remotely sensed images. However, there was little change in the total cumulative area of all forest patches. While forest patches became smaller in size, they remained unchanged in the complexity of their shape. The increase in amount of edge environment is therefore due to the increase in the number of patches.

Such changes are potentially very important for the forest fauna and flora that such reserves are designed to preserve. Edges have been clearly shown to affect microclimate and forest structure (Kapos 1989; Kapos et al. 1997; Laurance 1997) and some

forest animals (Quintela 1985; Didham 1997a, 1997b; Laurance 1997; Chapter 20). The composition of the fauna remaining in isolated forest reserves many be strongly dependent on the structure and composition of the surrounding matrix; development of woody regrowth may permit movement of mature forest species between fragments (Bierregaard and Stouffer 1997; Malcolm 1997a; Tocher, Gascon, and Zimmerman 1997).

Characterization of changes in space and time is a complex problem. Even reporting simple descriptive measures about the landscape involves numerous issues of data quality, data management, model assumptions, statistical reliability, and data visualization. Few terms share universal agreement in definition and application for discussing issues of "scale," "edge," "composition," and "change" in the context of landscape analysis. We found, as should all ecosystem research, that an important step in research involving a landscape perspective is reaching early agreement on the definition of such terms.

The application of this approach to the relatively well-known landscape of the BDFFP research sites has shown that it successfully quantifies landscape processes that are observed by researchers on the ground, and may highlight processes that are not immediately observable. Thus, there is potential for applying this approach to areas that are being less actively studied on the ground as a way of identifying areas that are undergoing change and may require further investigation or management intervention. The approach may also be used to formulate hypotheses for testing within ongoing research frameworks and to aid researchers and managers in explaining patterns they encounter in the field.

Conservation Lessons

1. Given the vast area of Amazonia (and other tropical forests around the world) and the limited amount of time, resources, and personnel to monitor the effects of human development in the region, remote sensing provides an efficient and accurate method to track landscape changes at a regional scale.

2. Describing changes in spatial pattern includes both qualitative and quantitative terms. While the visual analysis of images is often informative for the conservation biologist, simple pattern metrics offer opportunities to link landscape pattern to measured changes in species patterns or landscape level processes.

3. The resolution of a landscape "patch" model sets an important parameter in interpreting the results and implications of any landscape-level analysis. Not only do the remote-sensing data impose a minimum resolution for detecting change, but the land-cover labels likewise impose an implied resolution for meaningful interpretation of results. Conservation plans that are designed in response to change measured at a single spatial resolution may fail to incorporate

useful information concerning changes in the finer resolutions. Internal patch heterogeneity may influence patch stability over time.

4. From the landscape perspective, a conservation goal to decrease edge environments by creating fewer and larger patches may be thwarted if the patches take on more complex shapes. This trend was evident in this study as the landscape changed to fewer patches with more edge environment.

5. The use of these and similar landscape metrics along with long-term species inventories and observational measurements of landscape processes can assist the land manager in assessing the impact of landscape change on landscape function. As the land manager selects an appropriate goal (such as to increase edge environments or to maintain current conditions) these metrics will prove helpful in assessing the success of their actions. Likewise, monitoring the changing spatial pattern of the landscape with these simple metrics can offer the land manager a useful tool for identifying trends that may adversely affect management goals.

Acknowledgments

This work has been supported through a grant from NASA Mission to Planet Earth Program (NAGW-2652) and the cooperation of the World Conservation Monitoring Center. We would also like to acknowledge the tremendous editorial assistance provided by the reviewers for this chapter.

PART V
Synthesis

CHAPTER 29

Principles of Forest Fragmentation and Conservation in the Amazon

RICHARD O. BIERREGAARD, JR., WILLIAM F. LAURANCE,
CLAUDE GASCON, JULIETA BENITEZ-MALVIDO, PHILIP M.
FEARNSIDE, CARLOS R. FONSECA, GISLENE GANADE, JAY R.
MALCOLM, MARLÚCIA B. MARTINS, SCOTT MORI, MÁRCIO
OLIVEIRA, J. RANKIN–DE MÉRONA, ALDICIR SCARIOT, WILSON
SPIRONELLO, AND BRUCE WILLIAMSON

Scientists recognize that habitat loss and its inevitably associated fragmentation collectively pose the single greatest threat to earth's biological diversity (Laurance and Bierregaard 1997b). While humans have been altering forests for millennia, in this century changes have occurred in tropical forests that are unprecedented in their pace and scale. These changes are not qualitatively different from the encroachment of humankind into the world's temperate forests and other ecosystems, which are today broadly reduced to small relics across much of their former extent. As history once again repeats itself, at least in the case of Amazonia, which remains largely covered by structurally intact primary forest, we have the opportunity to direct the process of development in the region with the benefit of an awareness of some of the ecological consequences of our actions.

In the tropics, in the absence of a major reorientation of policies affecting forests, deforestation will continue apace for the foreseeable future (see Chapters 3, 23, and 26). At present rates, most of the world's remaining tropical forests will disappear in the span of a single human lifetime (Whitmore 1997). Their loss is driven by a burgeoning population with rising aspirations, a rush toward industrialization, the establishment of cattle and extensive monocultural agriculture in developing nations, misguided government policies, and the demand for forest products (e.g., timber and ornamental flora and fauna) by developing countries.

The Amazon is undeniably the world's greatest biological frontier on land. Unfortunately, despite initiatives in Brazil to slow forest loss, deforestation rates in Brazil's Legal Amazonia, which includes the savannas and deciduous forests north and south of the rainforest, have actually accelerated during the 1990s (see Chapter 3). To date, large-scale deforestation has been concentrated along the margins of the Amazon Basin—in the east, ramifying out from the Belém-Brasília Highway in southern Pará and northern Mato Grosso (Skole and Tucker 1993). In southwestern Amazonia in the states of Acre and Rondônia, deforestation spreads in a fishbone pattern off government-sponsored roads cut to facilitate colonization, cattle ranching, logging, and

This is publication number 296 in the BDFFP Technical Series.

land speculation (Fearnside 1987, 1997f; fig. 3.2).

Now, for the first time, the heart of the Amazon is being opened up for large-scale development. Major highways are being cut into some of the remotest parts of the basin, providing access to vast new areas for development, exploitation, and colonization. Logging is a major driver of this expansion (see Chapter 26). In recent years, aggressive, multinational timber companies from Asia have moved rapidly into the Amazon, by either buying large forest tracts or purchasing major interests in Brazilian timber firms (Viana 1997).

In this chapter we distill some of the most vital lessons for rainforest conservation arising from the research carried out at the BDFFP, reported both in this volume and elsewhere (e.g., Laurance and Bierregaard 1997a; see also the BDFFP Technical Series in the end materials of this volume) interpreted in the context of regional social and political trends in Amazonia. What have we really learned over the past two decades, and how can we use these principles to help lessen the ecological impacts of rapid Amazonian development?

What Is Happening to the Amazon?

PRINCIPLE 1: CURRENT TRENDS HERALD A DECADE OF DECISION

Many elements have been converging in the last several years to create a volatile situation in the heart of the Amazon (Chapter 3). New roads are appearing; stocks of tropical hardwood timber outside of the Amazon are dwindling (thereby driving up the prices of Amazonian species [see Chapter 26]); there is a rapid influx of migrants to the city of Manaus, which harbors 2.2 of the 3 million inhabitants of the state of Amazonas; employment opportunities in urban centers are

decreasing; and national and international investments in land and infrastructure for timber extraction are dramatically increasing. Meanwhile, state and municipal governments promote "development" of the Amazon through extensive plantings of soybeans and the issuance of timber concessions. Under such circumstances it is no exaggeration to say that the next decade or two will decide the future development trajectories of the Amazon.

As in other regions of the tropics, this increasing pressure will lead to the deforestation and transformation of primary forest into a landscape dominated by human activities. In such a landscape, by either accidents of topography or the design of landowners or policy makers, remnant patches of the native habitat are inevitably left. A burgeoning body of research, including the results of the BDFFP summarized in this volume, has documented the deleterious effects of such habitat fragmentation and points clearly to the need for informed planning if we are to be effective stewards of the richest reservoir of biological diversity on the planet's land surface.

Where Is the Forest Disappearing?

PRINCIPLE 2: DEFORESTATION IS SPATIALLY HIGHLY NONRANDOM

In the Amazon, as elsewhere, patterns of forest clearing and fragmentation are not randomly distributed but are strongly influenced by factors such as physical access, topography, and soil fertility (although soil fertility is far too rarely a consideration—see Chapter 23). Historically, development in the Amazon has been concentrated along rivers—traditionally the highways of the region. More recently roads are providing overland access to virgin forest. For this reason, major new highways such as the

Manaus–Boa Vista Highway (BR-174), which bisects north-central Amazonia, will have a dramatic impact on formerly undisturbed forests, leading to the fishbone pattern of deforestation discussed in Chapter 3 (fig. 3.2).

How Do the Flora and Fauna Respond to Forest Fragmentation?

PRINCIPLE 3: THE AMAZON HAS MANY FRAGMENTATION-SENSITIVE SPECIES

As illustrated by many of the chapters in this book and other publications of the BDFFP, numerous rainforest species respond negatively to forest fragmentation. Groups demonstrated to be most vulnerable to fragmentation effects include some understory birds (Stouffer and Bierregaard 1996; Bierregaard and Stouffer 1997; Chapter 20), primates (Rylands and Keuroghlian 1988; Schwarzkopf and Rylands 1989; Chapters 21 and 22), shade-loving butterflies (Brown and Hutchings 1997), solitary wasps, carpenter and leaf-cutting bees (Morato 1993), euglossine bees (Powell and Powell 1987; Morato 1994; Chapter 17), coprophagous beetles (Klein 1989), forest-interior beetles (Didham 1997b), termites (Souza and Brown 1994), ants (Vasconcelos and Cherrett 1995; Carvalho 1998; Chapter 16), flying insects (Fowler, Silva, and Venticinque 1993), drosophilid fruit flies (Chapter 14), forest interior seedlings (Benitez-Malvido 1998; Chapter 11), palms (Scariot 1999; Chapter 10), and existing canopy and supercanopy rainforest trees (Ferreira and Laurance 1997; Laurance, Ferreira, et al. 1998a; Chapter 9). This list is not exhaustive—other groups that have not been studied will certainly prove to be equally at risk in fragmented landscapes. Most obvious among these are the large terrestrial mammals, such as deer, tapirs, and big cats, which are especially vulnerable to hunting

and likely to disappear or decline in fragmented landscapes, even if suitable habitat remains (Woodroffe and Ginsburg 1998).

At least four general traits increase the vulnerability of Amazonian forest plants and wildlife: rarity, heterogeneous species distributions, a preponderance of co-evolved pairs or suites of species (including "keystone" species [see Howe 1977, 1984; Mills, Soulé, and Doak 1993]), and a reluctance of many wildlife species to leave forest cover. Similarly, many species of plants have a very low likelihood of establishing themselves or reproducing in the absence of adequate forest cover or the things forest cover guarantees, ranging from adequate groundwater to pollinators. This, coupled with the reluctance of seed dispersers to cross open areas, greatly reduces the ability of primary forest plants to recolonize pasture or isolated fragments (Nepstad, Uhl, and Serrão 1990; Nepstad, Uhl, et al. 1996). Many species occur at very low densities (see Chapters 5 and 7). As a result, their populations in fragments are likely to be small and thus prone to deleterious random effects—genetic drift, inbreeding, and stochastic demographic changes—which collectively can be a powerful driving force of extinction (Gilpin and Soulé 1986; Gaston 1994; Alvarez-Buylla et al. 1996).

Populations of many rare tree and palm species (see Chapters 8 and 9 for examples in the Lecythidaceae), for instance, are unlikely to be viable in small fragments (Alvarez-Buylla et al. 1996; Scariot 1996). This holds true especially for dioecious species. For example, 50 percent of the Sapotaceae species in the area are dioecious, and 75 percent of them are rare (less than 1 tree/ha, Pennington, pers. comm.). Also, 40 percent of the individuals in the Sapotaceae population with a DBH of at least 10 cm are not yet reproductive (Pennington, pers. comm.). These basic patterns of rarity and reproductive condition hold true across a broad range of the tree species (Rankin–de Mérona and Ackerly 1987; Chapter 5). In addition, sex ratios may be unequal. For dioecious Myristicaceae, for example, the male-biased sex ratios are between two and three (Ackerly, Rankin–de Mérona, and Rodrigues 1990). Therefore, the number of reproductive females in the population is likely to be very low and the effects of forest fragmentation especially strong for such rare, dioecious species.

Much heterogeneity exists in the distribution patterns of many species in most taxonomic groups that have been studied to date (Rankin–de Mérona, Prance, et al. 1992; Milliken 1998; Chapters 5, 6, 16, and 17). This heterogeneity clearly renders many species vulnerable to extinction, especially for species that are specialists on microhabitats that are preferentially deforested, such as forest plateaus.

Two factors seem to have contributed to the high levels of co-evolved plant-animal interactions in tropical forests. The low density of each tree species renders wind pollination ineffective, so plants must rely on animal vectors to distribute their pollen (Bawa 1990). Secondly, most plants rely on animals, rather than wind, to disperse their seeds (Gilbert and Raven 1975).

Finally, many species are vulnerable to fragmentation because they are strongly dependent on forest cover. When forest patches are surrounded by cattle pastures, crops, or other nonforest habitats, populations of forest-dependent species will be isolated and thus prone to deleterious effects. However, not all species are forest-dependent, and a surprising number of forest species tolerate open, or more commonly, second-growth habitats (see Principle 6 below, and Chapters 11, 16, 19, 20, and 24).

A crucial finding of the BDFFP is that many forest-dependent animals avoid even narrow (less than 100 m wide) clearings (Lovejoy, Bierregaard, et al. 1986; Bierregaard et al. 1992), especially if the opening is maintained (Malcolm 1991; Stouffer and Bierregaard 1996; Bierregaard and Stouffer 1997). Studies in other rainforest ecosystems mirror these results; certain terrestrial and arboreal vertebrates avoid even narrower (less than 50 m wide) forest clearings (Laurance 1990; Goosem 1997). This finding suggests that highways or high-voltage transmission lines, for example, could create significant barriers to movement for some fauna, and affect forest dynamics by reducing the likelihood of colonization for the many forest trees that depend on animal dispersers.

Do Continuous and Fragmented Forests Function in the Same Way?

PRINCIPLE 4: ECOLOGICAL PROCESSES IN FRAGMENTED FORESTS ARE SUBSTANTIALLY ALTERED

Because forest fragments are small samples of a much bigger landscape, it is evident that they contain only a subset of the species occurring in continuous forest. Logically, therefore, fragments experience only a subset of the ecological interactions and processes that occur in the pristine environment. When birds and monkeys are absent from fragments, some remnant tree species will lack their mutualistic seed dispersers, therefore jeopardizing the regeneration process (e.g., Santos and Tellería 1994). When shade-loving butterflies and euglossine bees are missing, many flowers will remain unfertilized, threatening their reproduction (e.g., Powell and Powell 1987; Aizen and Feinsinger 1994a; Alvarez-Buylla et al. 1996; Benitez-Malvido 1998; Chapter 17). Hence, the recruitment of new plant individuals into the fragments is biased away from mature forest species (Alvarez-Buylla et al. 1996; Benitez-Malvido 1998; Laurance, Ferreira, et al. 1998a).

Potentially equally important are second-order effects. For example, changes in rodent and marsupial populations may result in changes in patterns of seed dispersal and predation on seeds and seedlings. Whereas rodents, including spiny rats (*Proechymis* spp.), acouchis (*Mioprocta agouti*), and agoutis (*Dasyprocta leporina*), are important seed predators in the area, they may also play a role as a secondary seed disperser because they bury large seeds, such as those of *Pouteria laevigata* (W. Spironello, pers. comm.). Changes in the populations of these small mammals, for whatever reason, are likely to change the population structure of the plant community.

When top-order predators such as jaguars and harpy eagles are lost, their role as suppressors of prey populations will be unfilled, which can lead to a cascade of effects through the food chain (Pimm and Laughton 1980; Terborgh 1992) and eventually affect plant population structure. With decreased predation on the small mammals and subsequent increases in their populations, successful establishment of trees whose seeds are eaten by these species may be reduced when rodent populations are high because their predators have been reduced or are absent (Leigh et al. 1993).

Malcolm (1997a) has shown that the creation of edges can lead to higher seed and seedling predation rates due to superabundant small mammal populations in the dense understory of forest edges. This, in turn, has the potential to suppress regeneration of large-seeded tree species. Similar effects in intensively logged rainforest in Africa are thought to be contributing to outright suppression of tree regeneration (Struhsaker 1997).

Alternatively, some mammal populations might decrease with a reduction of their food resources or with increased human hunting pressure. In Amazonian forests, male euglossine bees are known to pollinate certain species of orchids, while both male and female euglossines pollinate species of the Brazil nut family. The disappearance of orchids for one reason or another could lead to the loss of the bees needed for seed set in some species of the Brazil nut family, which are, in turn, important sources of food for many species of mammals. Thus, the loss of orchids could ripple through the system and wind up affecting seedling recruitment in the forest. In this scenario, seed predation rates would decline for smaller-seeded species, and the dispersal of larger-seeded species would be diminished.

Primates are important primary seed-dispersal agents for trees in the BDFFP area, especially species in the Sapotaceae (Chapter 21; Spironello, unpubl. data). Decreases in primate populations may influence seedling recruitment of many plant species. For example, twenty forest fragments in which primate populations had been reduced had lower seedling density and fewer species in comparison to an intact continuous forest at Kibale National Park, Uganda (Chapman and Onderdonk 1998). In can be expected, therefore, that the effects of local extinctions will be felt in the community long after the most sensitive species have disappeared.

When forests are fragmented, many physical processes that affect ecological functions at the ecosystem level can also be modified. Processes such as decomposition of organic matter, water retention, and rates of carbon fixation and emission may be affected by species extinction or by microclimatic and structural changes caused by edge effects as discussed in the next principle (Laurance, Bierregaard, et al. 1997). In the BDFFP study areas, Sizer (1992) discovered that rates of litter decomposition decreased in smaller fragments. Moreover, W. Laurance,

Laurance, et al. (1997) have shown that tree mortality within 100 m of fragment edges caused a loss of up to 36 percent of the biomass in the first ten years after fragmentation. This structural change can affect water retention by soil and vegetation (Camargo and Kapos 1995) and may also influence the balance between carbon emission due to decomposition of dead trees and carbon fixation by regrowth vegetation. In this way, habitat fragmentation in the Amazon may influence cycles of water, nutrients, and atmospheric gases at local and potentially global scales (W. Laurance, Laurance, et al. 1997; Laurance, Gascon, and Delamônica 1998).

PRINCIPLE 5: EDGE EFFECTS HAVE A MAJOR IMPACT ON FRAGMENT BIOTAS

One of the important conclusions of the BDFFP is that edge effects—the diverse physical and biotic changes associated with the abrupt, artificial margins of forest fragment—can have a dramatic impact on fragment ecology and dynamics (Lovejoy, Bierregaard, et al. 1986; Bierregaard et al. 1992; Chapters 9, 13, and 27). BDFFP researchers (and others) have discovered that edge effects influence many aspects of the microclimate, structure, composition, and functioning of the 1 to 100 ha BDFFP remnants (Kapos 1989; Wandelli 1991; Camargo and Kapos 1995; Malcolm 1994; Didham 1977; Ferreira and Laurance 1997; Kapos et al. 1997; Laurance, Ferreria, et al. 1998a, 1998b; Benitez-Malvido 1998). Even much larger (1,000 ha) fragments may be markedly influenced by edge effects, especially if irregularly shaped, because fragment shape itself is important (Malcolm 1994; Benitez-Malvido 1998). Angular edges appear to be poor habitats for the establishment of tropical rainforest seedlings (Chapter 11). Accelerated changes in the structure, floristic composition, and microclimate of forests are likely to exacerbate effects of fragmentation on disturbance-sensitive species (Scariot 1996, 1999; Laurance, Gascon, and Rankin–de Mérona 1999).

Habitat Fragments Are Not Islands

PRINCIPLE 6: THE SURROUNDING HABITAT DRAMATICALLY INFLUENCES FRAGMENTED COMMUNITIES

This might be rephrased in contrast to Principle 3: Many apparently primary-forest species persist in second-growth forests.

Researchers are becoming increasingly aware of the key influence of the vegetation in the "matrix" surrounding fragment biotas and dynamics (Fahrig and Merriam 1994; Laurance, Bierregaard, et al. 1997). Fragments surrounded by an inhospitable matrix are likely to support lower numbers of vulnerable forest-dependent species than are those encircled by a matrix (such as regrowth forest) which more closely approximates the vegetation structure of the original habitat (Laurance 1990; Malcolm 1991; Stouffer and Bierregaard 1996). At the BDFFP study sites, a higher mortality of trees was reported in fragments that were surrounded by a matrix of pasture than were those surrounded by secondary forest (Mesquita, Delamônica, and Laurance 1999). Laurance (1991b) demonstrated that for Australian rainforest mammals, species that tolerated the matrix of cattle pastures and regrowth forest were far more likely to persist in fragments than were those that rarely or never used the matrix. In the BDFFP study area, a number of studies have shown similar patterns, strongly supporting the generality of this relationship.

Benitez-Malvido (1995) and Ganade (Chapter 24) found seeds and seedlings of primary forest tree species performing better under a second-growth canopy than in primary forest. Seedlings of the canopy tree *Chrysophyllum pomiferum* grew ten times faster in a secondary regrowth patch, with high light levels, than in the shaded forest understory (Chapter 11 and Benitez-Malvido, unpubl. data). The causes of higher survival and growth of seeds and seedlings under second-growth canopy may be related to reduced numbers of natural enemies and high light availability.

Obligate army-ant-following birds, because they were the first to disappear from recently isolated 1 and 10 ha fragments, were thought to be primary-forest specialists and very sensitive to fragmentation (Bierregaard and Lovejoy 1988). Further studies, after the matrix around the isolates shifted to a second-growth forest less than ten years old, showed these species to be among the first to move readily through the second growth, although not revisiting the small fragments (Stouffer and Bierregaard 1995a; Chapter 20). Studies of snakes in the landscape of primary forest and much older (c. 100 yrs) second growth around Belém showed only four of about eighty species of snakes in the region were restricted to primary forest (Cunha and Nascimento 1978). Vasconcelos and his colleagues have found ground-dwelling, "primary-forest" ant species in pasture before second-growth vegetation had even gotten established (Chapter 16). Malcolm (1991, 1994, 1995) was able to predict the composition of the small nonvolant mammal community in 1 and 10 ha fragments based solely on the nature of the edge of the fragments and the matrix habitat around them. He found more species and higher biomasses in small fragments surrounded by second growth than in those surrounded by pasture. And Tocher and her colleagues found a number of primary-forest frogs breeding more successfully in ponds in second growth than in primary forest (Tocher, Gascon, and Zimmerman 1997;

Chapter 19). Successful establishment of some animal species in second-growth areas may be due to the loss of natural enemies that do not survive in the altered habitat or to an increased abundance of food.

Other studies suggest that multispecies interactions could be at risk of being lost in small fragments and successional vegetation (Benitez-Malvido, Garcia-Guzman, and Kossmann-Ferraz 1999). Leaf-fungal infections on primary forest seedlings in continuous forest were found to be mostly associated with injuries caused by herbivorous insects. In contrast, for the same seedling species growing in fragments and second-growth forest, fungal infection associated with herbivory was reduced, probably as a result of reduced attacks by herbivores. Many insect fungal vectors may be absent in small fragments and second-growth forest.

While we must emphasize that these results come from a mosaic of fragments and second-growth forest that, in the case of the BDFFP study area, are themselves surrounded by vast expanses of primary forest, it appears that a landscape mosaic of forest fragments on the scale of hundreds of hectares in a matrix of second-growth forest and agriculture will support, at least for a number of decades, many species of the original regional flora and fauna, albeit with a loss of some of the ecological relationships and processes of intact forest. But, depending on the quantity of intact primary forest in the landscape mosaic, it is equally clear that significant components of the rainforest biota may be lost, even in a primarily forested landscape (Chapter 27). For example, in a post-logging matrix of secondary and edge-modified primary forest in French Guiana, a large proportion of the original bird community had declined in abundance, an effect that apparently persisted for several decades (Thiollay 1992).

Another change associated with the appearance of the matrix habitat is the invasion of matrix species and their effect on fragment dynamics (Janzen 1983). In the BDFFP landscape, we are now finding many such cases of species associated with disturbed areas having a strong influence on fragment dynamics (Borges 1995, Hutchings 1991; Brown and Hutchings 1997; Chapters 14 and 19). In some cases, species originally present in primary forest may become superabundant in the matrix and thus invade fragments near their edges (Malcolm 1991, 1997a). Invasive species may influence fragment dynamics in different ways. They may invade fragment areas, thereby increasing species richness in post-isolation communities. This is characteristic of communities of amphibians (Tocher, Gascon, and Zimmerman 1997), small mammals (Malcolm 1995), and palms (Scariot 1996) in the BDFFP. Despite the isolation of the mosaic of fragments, pastures, and second-growth forest amid an ocean of primary forest and the short time since isolation, an invasive palm species typical of disturbed areas is becoming the most abundant palm in the 1 and 10 ha isolates (Chapter 10). It is too soon too tell whether this species will increase as typical forest species disappear or add to the overall species richness of the group, as with amphibians.

Alternatively, invasive species may displace species present in fragments prior to isolation, resulting in significant turnover of the original forest community. This was observed for butterflies (Hutchings 1991; Brown and Hutchings 1997), although evidence that the shift in species composition is related to competition rather than microhabitat changes is lacking. In some cases, the success of the invasion was associated with a significant alteration of the composition of the original fauna, which in turn could be a response to other perturbations, such as hunting, harvesting, and other human activities (Chapter 14; see principle 8).

Human-altered habitats are homogeneous due to uniformity in treatment. Thus, pastures, woodlots, and farms lose environmental heterogeneity and patchiness characteristic of old-growth forests. Even secondary forests appear uniform, often even aged and dominated by a single genus (plate 8; Chapters 24 and 25). In this context, the matrix habitat can be viewed as a filter of animal use, dispersal, and movement. The effectiveness of the "filter" will be determined by the nature, extent, and configuration of its vegetation. Both forest-dependent species and those adapted to disturbed areas can respond to the filter and occupy, or at least move through, the matrix habitat. Their probability of moving through the matrix will depend on species-specific requirements (Gascon et al. 1999).

In the terms of the island biogeography model, a more hospitable matrix is equivalent to the island moving closer to the mainland—colonization rates will increase and consequently increase the possibility of periodic "rescues" of fragment populations from nearby continuous habitat (Brown and Kodric-Brown 1977) or from other fragments in the landscape (Levins' [1970] concept of the "metapopulation"). An added complication, not present in models of true islands, is that species living in (rather than just those passing over) the "ocean" around habitat fragments can invade the forest remnants. All of these results suggest that the matrix can be managed in ways that will strongly influence the dynamics of communities inside forest fragments.

PRINCIPLE 7: FOREST FRAGMENTS ARE STRUCTURALLY AND BIOLOGICALLY DYNAMIC

The creation of abrupt edges during deforestation has powerful effects on the remaining rainforest fragments, especially near the edges. As noted above, changes in the wind and light environment can lead to fundamental changes in ecosystem functioning and the avoidance of the edge by many forest species. Evidently the zone of influence of the edge will evolve as the edge evolves. Additionally, as the surrounding habitat changes around the fragments (see Principle 6), invasive species are likely to play a role in ecological processes, while isolation-sensitive species disappear, perhaps leaving other ecological roles (e.g., pollination or seed dispersal) unfulfilled. Studies on forest fragmentation need to be long term because of this dynamic nature; conditions that we find at any given time will change as characteristics of the edge change and patterns of species loss and invasion change.

PRINCIPLE 8: HUNTING AND HARVESTING IN FORESTS CAN DRASTICALLY EXACERBATE FRAGMENTATION EFFECTS

In developing nations, hunting and harvesting are often integral to the fragmentation process. Forest clearing, fragmentation, and hunting and harvesting are all direct consequences of increased human access. In the BDFFP study area, for example, improved access to forest following the completion of the Manaus–Boa Vista highway has resulted in a drastic increase in hunting pressure. Hunting is likely to be greatest near towns and settlements (Smith 1997). In tropical Mexico many palm species of the ornamental genus *Chamaedorea* have been harvested close to human settlements, driving many palm populations to the verge of extinction (J. Rodriguez, pers. comm.). In Costa Rica the edible palm heart of *Iriartea deltoidea* has led to its extinction near human habitation (D. A. Clark, Clark, et al. 1995).

For species such as jaguars, pumas, deer, peccaries, tapirs, and larger monkeys and birds, hunting can drastically increase the deleterious effects of habitat fragmentation. This can occur because hunting pressure is so great that fragmented landscapes become population sinks for target species (Pulliam and Danielson 1991; Woodroffe and Ginsberg 1998), or because animals learn to avoid areas frequented by hunters, rendering large areas of the fragmented landscape ineffective in conserving the regional fauna.

For conservation purposes, it may be fruitful to think of hunting as an edge effect (Woodroffe and Ginsberg 1998), albeit one that can extend deeply into the forest. This model assumes that hunters originate from the modified matrix but penetrate into forest tracts while hunting. Unfortunately, there are few empirical estimates of the distances to which hunters penetrate into forest tracts (Laurance, Bierregaard, et al. 1997), although in French Guiana Thiollay (1992) has shown measurable changes in the community structure of both game and nongame species in the face of remarkably low levels of human hunting pressure. Redford (1992) suggested that vast, structurally intact areas in the Amazon forest have already suffered from heavy hunting pressure. More data on this will be invaluable for establishing effective buffer zones around nature reserves and for other types of conservation planning.

Forest Reserves: How Big Is Big Enough?

PRINCIPLE 9: THERE IS MORE THAN ONE MINIMUM CRITICAL SIZE OF AN ECOSYSTEM

Nearly two decades ago, the BDFFP was launched under another name with a very specific goal—to determine the area required to maintain the species and ecological processes found in continuous primary forest (Lovejoy and Oren 1981). The proj-

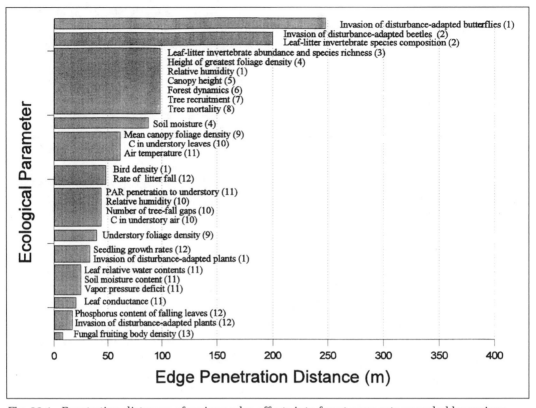

Fig. 29.1. Penetration distances of various edge effects into forest remnants recorded by various BDFFP studies. Sources: 1 = Lovejoy, Bierregaard, et al. 1986; 2 = Didham 1997a; 3 = Didham 1997b; 4 = Camargo and Kapos 1995; 5 = Camargo 1993; 6 = Malcolm 1994; 7 = Kapos et al. 1993; 8 = Kapos 1989; 9 = Bierregaard et al. 1992; 10 = R. K. Didham, pers. obs. PAR = Photosynthetically Active Radiation. (Reproduced from W. Laurance, Bierregaard et al. [1997] with permission of University of Chicago Press.)

ect's first name, the Minimum Critical Size of Ecosystems, reflected this goal. We now feel that the admittedly unsatisfactory answer to this seminal question is, "It depends."

At a basinwide scale, the answer is probably on the order of a million or more hectares, based on territory sizes of large, territory-demanding predators such as harpy eagles (*Harpia harpyja*) and standard estimates of minimum viable population sizes of around 500 pairs (Thiollay 1989). By necessity, the BDFFP experimental design left us with a suite of experimental forest reserves orders of magnitude smaller than

this absolute "all-species, all-processes" minimum critical size. Consequently we cannot attempt an answer at this scale. Nevertheless, while recognizing that such speculation is fraught with uncertainties, we can at least offer some insight into other minima, recognizing that the estimates will be site specific. As reported by Spironello (Chapter 22), the home ranges used by brown capuchin monkeys in the BDFFP area are an order of magnitude larger than those reported elsewhere in the basin. Territory size of mixed species flocks of birds north of Manaus are three times larger than those reported for the same flocks in eastern Peru

(Powell 1985). Such differences imply that the answer to the minimum critical size question depends not only on how one defines the question but also on where it is being asked.

While applauding Brazil's recent announcement of an ambitious plan to preserve vast tracts of the region in virgin forest, we must recognize that even if the government's plan is brought to fruition, most of the Amazon will soon be part of the mosaic of developed land and forest fragments outside of national parks. It is therefore relevant to ask, recognizing the caveats mentioned above, What is the minimum size of a fragment that can make a meaningful contribution to conservation in the vast landscape that will not be under federal protection? We define "meaningful" here as the size at which a fragment can support a *population* of a given species, albeit one that might be too small on its own to be viable in the long run, but which, with genetic rescue in the form of occasional immigration, might be part of a viable local metapopulation.

To estimate this minimal size we might look first at edge effects, which should vary noticeably across regions in the basin. What size reserve at least protects a core area wherein species will be unaffected by edge effects? Given that measurable changes have been detected 250 m into continuous forest (fig. 29.1), a 100 ha reserve will preserve only a small portion of its core unaffected (Laurance, Bierregaard, et al. 1997; Chapter 9). However, Didham (1997b) has shown that the insect fauna of the forest floor is recognizably altered in a 100 ha patch when compared to continuous forest. We would conclude therefore that several hundred hectares would be a minimal size for forest fragments to even contain the ecological habitat conditions characteristic of undisturbed forest and populations of the vast ma-

jority of species. Our data, obviously, are still based on a time frame too short to detect how long-lived sessile organisms will behave. Trees and palms, for instance, may take decades to respond to fragmentation. It is not unlikely that many populations may slowly decrease to levels where recovery would be impossible, or may even persist, but not reproduce, for a long time in a sort of "living-dead" population (*sensu* Janzen 1986b).

Minimum reserve size estimates necessarily depend on spatial and temporal context. If the fragment is one of many with larger tracts of extensive forest nearby and the time frame is on the order of decades, very roughly speaking, the answer seems to be "hundreds of hectares." If we envision a landscape with widely separated forest remnants far from large tracts and extend our time frame to a century from now, the minimum would have to be thousands or tens of thousands of hectares. There are obviously some species that are very territory demanding, such as the large predators (jaguars, pumas, harpy eagles) and some primates (capuchins and spider monkeys [see Chapter 21], and bearded sakis [Frazo 1992]) that will require many tens or hundreds of thousands of hectares to maintain a viable population, at least in the central Amazon, but for many species, an area of some thousands of hectares will probably maintain substantial populations (especially when one considers that most animal species are insects).

Where Should Conservation Efforts Be Focused?

PRINCIPLE 10: THE CENTRAL AMAZON IS SPECIES RICH

Basinwide patterns of species richness and diversity are remarkably complex and beyond the scope of the BDFFP's investigations. However, they are important considerations in reserve planning, and our data supply some important insights.

Reserve design includes the size effects (stochastic extinction, edge effects) and connectivity (immigration) but must also consider representation (gap analyses). If diversity tends to be concentrated in hot spots, then we can conserve more species by focusing on those areas. As regional diversity goes up, we need to increase the number of reserves.

The central Amazon has often been considered depauperate in species relative to the western Amazon that adjoins the Andean foothills. For example, Gentry (1988) found 283 and 275 trees species (with a DBH of at least 10 cm) in each of two hectares near Iquitos, Peru, and prematurely concluded that the western Amazon harbored the most species-rich rainforests in the world. A similar theme was echoed by Phillips et al. (1994), who argued that in tropical forests of the central Amazon, lower species richness resulted from lower productivity. While the central Amazonian soils are highly mineralized and productivity is undoubtedly lower than at the foothills of the Andes, species richness is nearly identical in the two regions. Recent work at the BDFFP control plots in continuous forest adjacent to the fragments revealed 285, 280, and 280 tree species in each of three hectares (Oliveira and Mori 1999). In fact, based on a larger survey, the forests north of Manaus are the most species-rich forests in the world with well over a thousand species known or expected to be identified from the 66 hectares of permanent plots at the BDFFP (Rankin–de Mérona et al. 1992; Nee 1995; Chapter 5).

Another myth of higher biodiversity in the western Amazon has been debunked for birds by BDFFP researchers (Cohn-Haft, Whittaker, and Stouffer 1997), who showed that bird species richness for terra firme forests at BDFFP (394 species) is nearly identical to western Amazonian bird diversity in the same habitat. Earlier comparisons showing higher diversity in the western Amazon had included other habitat types, such as bamboo and flooded forests.

Biogeographic patterns are poorly understood for most Amazonian taxa, greatly complicating efforts to delineate effective networks of nature reserves (Patton et al. 1997). While many Amazonian species have large geographic ranges (Fjeldså and Rahbek 1997), others exhibit strong patterns of local endemism that have resulted from a complex variety of ecological and historical factors. For example, Milliken (1998) found that the most abundant tree species at the BDFFP site is not even registered in Maré, only 200 km north of Manaus and on the same side of the Rio Negro. Mori (1990) and Oliveira (1997) argued that species richness of Amazonian trees can be partially attributed to the presence of endemics as well as species that have migrated into central Amazonia following the receding of Pleistocene floods. Central Amazonia is a crossroads for species migration from other areas in an area rich with endemics (Chapter 6) and thus an important target area for habitat preservation. The rapidly expanding deforestation radiating out from Manaus lends the utmost urgency to conservation planning in this area.

How Can We Reduce the Effects of Fragmentation?

PRINCIPLE 11: ROADS ARE AN ENEMY

The Amazon is a region characterized by extensive encroachment and poaching from illegal gold miners, loggers, and settlers. The region also has little policing of the environmental legislation that exists. Agencies such as IBAMA, Brazil's bureau for natural resource enforcement, are hard pressed to meet existing demands to regulate the dramatically expanding logging and extractive industries. According to IBAMA's former chief, Eduardo Martins, "Multi-million dollar investments in the Amazonian logging industry would spell disaster as things stand. We don't want that kind of investment" (Laurance et al. 1997b).

The absence of anything approaching an effective regulatory system for resource extraction or land use means that chaotic, unregulated development is the rule, rather than the exception, in the Amazon (Fearnside 1990b; Chapter 26). Under these conditions, the only effective means to prevent overexploitation of resources is to limit human access, which can be accomplished only by stopping initiative for major new roads (Riswan and Hartanti 1995). Roads are the great threat to the Amazon, for they bring influxes of people, who inevitably clear, log, burn, fragment, overhunt, and otherwise degrade the natural values of the rainforest. Consider the example of the Belém-Brasília highway, which runs from the national capital north to the mouth of the Amazon River. Completed in 1960, the highway has functioned as a massive artery for colonization. Dozens of roads now radiate out from the highway, creating a 400 km network of deforestation in the eastern Amazon. In a few decades the Manaus–Boa Vista highway may look much the same. Stopping new highways, especially those penetrating deep into forest tracts, is clearly the most effective way to limit forest destruction.

PRINCIPLE 12: SIMPLE LAND-USE GUIDELINES COULD PROFOUNDLY INFLUENCE THE DEFORESTATION PROCESS

In the Amazon there is enormous scope to reduce the effects of ongoing deforestation and habitat fragmentation on wildlife communities, but such efforts are far more likely to be effective before or during—rather than after—the deforestation process.

For example, in regions such as Pará, Acre, and Rondônia, which are experiencing rapid forest conversion, although Malcolm (Chapter 27) calls for corridors of a kilometer or more in width, the retention of a bare minimum of 300 m wide strips of primary forest along rivers and major streams would help maintain ecosystem connectivity for forest-dependent species, thereby limiting the deleterious effects of habitat fragmentation (Laurance and Gascon 1997). In the BDFFP landscape, linear forest remnants have been shown to harbor a large number of small mammals and litter frogs (Lima 1998). Because habitat corridors should be functional pieces of the local ecosystem, the 300 m width is probably close to a minimum. Given that the deepest penetration of physical changes associated with edges is 100 m for relative humidity (see fig. 29.1), a 300 m wide corridor would maintain only 100 m of unaffected forest. However, the presence of a stream or river would clearly mitigate this effect, so such a corridor may well be effective. However, faunal changes have been reported as far as 300 m from a forest edge (Chapter 18), so to the extent that these species are responding to physical changes that we have not been able to detect, 300 m may not be a maximally effective corridor width.

Proactive land-management guidelines could also include such considerations as the protection of rare vegetation types, protection of forests on steep slopes, minimum viable sizes for forest remnants, maximum sizes for clearings, and buffer zones for nature reserves.

Soil fertility also has a major effect on development activities and needs to be considered in the planning process. For example, despite ambitious government-sponsored plans, large-scale agricultural development has been hindered across much of the north-central Amazon basin, which is an area of ancient, nutrient-poor soils. It is clear from the soil data presented by Fearnside (Chapter 23) that large areas of Brazilian Amazonia are not suitable for agriculture and that industrial agriculture should not be considered in these areas. Fortunately, because this region is rich in species (see Principle 10), biodiversity can be protected with no lost development opportunity costs.

PRINCIPLE 13: CONSERVATION IS BEST ACHIEVED THROUGH PEOPLE

Conservation through people can be attained via different avenues. Recently, for instance, community-based conservation models such as the Jaú National Park northwest of Manaus have been extremely successful at using local people as guardians of natural resources while increasing the standard of living (housing, education, health, etc.) of local communities (see also Vivian and Ghai 1992 and chapters in Schelhas and Greenberg 1996).

Ecotourism, while not without its potential negative effects, provides a mechanism whereby the preservation of native habitat provides a livelihood for locals who might otherwise be forced to clear the forest to grow crops.

Large-scale and long-term research projects can likewise provide similar local employment opportunities. Additionally, they can enlighten landowners to the value of the natural landcover, educate students, and, of course, provide essential data relevant to conservation issues. South of the Amazon, the Golden-Lion Tamarin project in the Atlantic forests of southern Brazil is another example of how local involvement can lead to effective conservation. This project, fo-

cused on the reintroduction of an endangered primate species, has turned local landowners who once hunted the tamarins into tamarin protectors who proudly boast to others when their land supports a troop of the monkeys. The reintroduction of tamarins on private property has increased by nearly 25 percent the amount of remaining forest under protection. Also, some of the ranch owners are turning their property into private reserves formally and are hoping to attract ecotourism, thus contributing to the economic base of the region (D. Kleiman, pers. comm.).

When we conceived of the BDFFP we had no idea how important education would be to the success of the project. We knew we would need student interns and graduate students to carry out much of our research but did not know what an important niche our project would fill in providing a site and infrastructure for the training of young conservation professionals for careers in management as well as research. So far, several hundred young South Americans have been exposed to the conservation issues relevant to regional development issues through internships, graduate fellowships, and courses at the BDFFP. Today, many of these trainees are in position to apply their knowledge to the research and management of natural resources in tropical countries.

Concluding Remarks

Preserving the myriad ecosystems that comprise the Amazonian rainforests in the face of relentless pressures to develop the region will require first and foremost a driving desire to do so and second at least a rudimentary understanding of the complex issues involved in managing the change from a forest-dominated to a human-dominated landscape. All the data in the world will be of no use if the development juggernaut proceeds unhindered across the basin.

Over the two decades that the BDFFP has been operating in Brazil, we have witnessed the maturing of a real conservation ethic in the country. There is clearly grassroots support for preserving the forest and an understanding at the highest levels of the government that it simply does not make sense to sit back and watch uninhibited development destroy the resources that lie beneath the vast forest canopy that covers so much of the country. Brazil's ambitious plan to preserve substantial tracts of rainforest demonstrates this understanding. However, in the face of economic shockwaves reverberating from collapsing Asian stock markets, the Brazilian government chose to redirect funds that were scheduled for conservation and research to other areas of the economy and to provide additional incentives to exploiting timber and other resources. We can only hope that the lost ground is regained. We can also hope that the results of our research will provide a sound basis for decision making, and that the students trained on our study sites will be among those making the decisions.

Theses and Publications of the Biological Dynamics of Forest Fragments Project

Arthropods

THESES

Didham, R. 1997. The litter beetle fauna and habitat fragmentation. Ph.D. thesis, University of London.

Hutchings, R. 1991. The dynamics of three communities of Papilionoidea (Lepidoptera: Insecta) in forest fragments in central Amazonia. M.Sc. thesis, INPA.

Klein, B. C. 1987. The effects of forest fragmentation on dung and carrion beetle (Scarabaeinae) communities in central Amazonia. M.Sc. thesis, University of Florida.

Lemes, M. R. 1991. Agregação de ovos como estratégia reprodutiva de *Hypothirys euclea barii* (Lepidoptera, Nymphalisae: Ithomiinae) na região de Manaus, Amazonas. M.Sc. thesis, UNICAMP.

Martins, M. 1985. Influência da modificação do habitat sobre a diversidade e abundância de espécies de *Drosophila* (Diptera, Drosophilidae) em uma floresta tropical da Amazônia Central. M.Sc. thesis, INPA/FUA.

Morato, E. 1993. Effects of forest fragmentation on solitary bees and wasps in central Amazonia. M.Sc. thesis, Universidade Federal de Viçosa.

Oliveira, M. L. 1994. A fauna de abelhas Euglossinae em florestas contínuas de terra firme na Amazonia central. M.Sc. thesis, INPA.

Ribeiro, J. 1997. The ecology of termites in tropical terra firme forest. Ph.D. thesis, INPA.

Silva, R. O. 1992. Observations on the community structure and ecology of Odonata (Insecta, Hemimetabola) in terra firme streams and igapo forests. M.Sc. thesis, INPA/FUA.

Souza, O. F. 1988. Levantamento da composição em espécies de térmitas (Insecta, Isoptera) em fragmentos florestais na Amazônia Central. M.Sc. thesis, Universidade Federal de Viçosa.

Vasconcelos, H. L. 1987. Atividade forrageira, fundação e distribuição de colônias de saúvas (*Atta* spp.) em uma floresta da Amazônia Central. M.Sc. thesis, INPA/FUA.

Venticinque, E. M. 1995. Architecture, parisitism, foundation, and mortality of social spider (*Anelosimus eximius*, Araneae, Theridiidae) colonies in edges and interior of continuous forest and forest fragments in the central Amazon. M.Sc. thesis, UNESP.

Venticinque, E. M. 1999. A estrutura de metapopulação de aranhas sociais. Ph.D. thesis, UNESP.

Viera, R. S. 1995. The effect of army ant foraging on spider communities in tropical rainforest. M.Sc. thesis, INPA/FUA.

PAPERS

Becker, P., J. S. Moure, and F. J. A. Peralta. 1991. More about euglossine bees in Amazonian forest fragments. Biotropica 23: 586–91.

Benson, W. W., and A. Y. Harada. 1988. Local diversity of tropical and temperate ant faunas (Hymenoptera, Formicidae). Acta Amazónica 18: 275–89.

Brown, K. S., Jr., and R. W. Hutchings. 1997. Disturbance, fragmentation, and the dynamics of diversity in Amazonian forest butterflies. In *Tropical Forest Remnants: Ecology, Management, and Conservation of Fragmented Communities,* ed. W. F. Laurance and R. O. Bierregaard, Jr., pp. 91–110. University of Chicago Press, Chicago.

Cruz Neto, A. P., and M. Gordo. 1996. Temperatura corpórea e comportamento termoregulativo de *Ameiva ameiva.* Studies on Neotropical Fauna and Environmental 31: 11–16.

Darlington, J. P. E. C., E. M. Cancello, and O. F. F. de Souza. 1992. Ergatoid reproductives in termites of the genus *Dolichorhinotermes* (Isoptera, Rhinotermatidae). Sociobiology 20: 41–47.

De Marco, P. 1998. The Amazonian campina dragonfly assemblage: Patterns in microhabitat use and behaviour in a foraging habitat. Odonatologica 27: 239–48.

DeSouza, O. F. F., and V. K. Brown. 1994. Effects of habitat fragmentation on Amazonian termite communities. J. Tropical Ecology 10: 197–206.

Didham, R. K. 1997. An overview of invertebrate responses to forest fragmentation. In *Forest and Insects,* ed. A. Watt, N. E. Stork, and M. Hunter, pp. 303–20. Chapman and Hall, London.

Didham, R. K. 1998. Altered leaf-litter decomposition rates in tropical forest fragments. Oecologia 116: 397–406.

Didham, R. K., J. Ghazoul, N. E. Stork, and A. J. Davis. 1996. Insects in fragmented forests: A functional approach. Trends Ecol. Evol. 11: 255–60.

Didham, R., P. M. Hammond, J. H Lawton, P. Eggleton, and N. Stork. 1998. Beetle species responses to tropical forest frag-

mentation. Ecological Monographs 68: 295–323.

Didham, R. K., J. H. Lawton, P. M. Hammond, and P. Eggleton. 1998. Trophic structure stability and extinction dynamics of beetles (Coleoptera) in tropical forest fragments. Phil. Trans. Royal Soc., Series B, 353: 437–51.

Fonseca, C. R. 2000. Cooperação, conflitos e razão sexual em himenópteros sociais: a perspectiva de uma formiga Amazónica. In R. P. Martins, T. M. Lewinsohn, and M. S. Barbeitos (eds). Ecologia e comportamento de insetos. Série Oecologia Brasiliensis, vol. 8. PPGE-UFRJ. Rio de Janeiro, pp. 131–48.

Fowler, H. G. 1994. *Agelaia brevistigma* (Hymenoptera: Vespidae) in central Amazonia. Revista de Biologia Tropical 47: 387–88.

Fowler, H. G. 1994. Interference competition between ants (Hymenoptera: Formicidae) in Amazonian clearings. Ecologia Austral. 4: 35–39.

Fowler, H. G. In press. The colonial spider, *Parawixia bistriata* (Rengger, 1835) (Araneae: Arandidae), in the central Amazon. Boletim de Zoologia da Universidade de São Paulo.

Fowler, H. G., C. A. Silva, and E. Venticinque. 1993. Size, taxonomic, and biomass distributions of flying insects in central Amazonia: Forest edge vs. understory. Revista de Biologia Tropical 41: 755–60.

Fowler, H. G., and E. M. Venticinque. 1996. Interference competition and scavenging by *Crematogaster* ants associated with the webs of the social spider *Anelosimus eximius* in the central Amazon. J. Kansas Entomological Society 69: 267–69.

Hofer, H., A. D. Brescovit, and T. Gasnier. 1994. The wandering spiders of the genus *Ctenus* of Reserva Ducke, a rainforest reserve in central Amazonia. Andrias 13: 81–98.

Klein, B. C. 1989. The effects of forest fragmentation on dung and carrion beetle (Scarabaeinae) communities in central Amazonia. Ecology 70: 1715–25.

389

Malcolm, J. R. 1997. Insect biomass in Amazonian forest fragments. In *Canopy Arthropods,* ed. N. E. Stork, J. Adis, and R. K. Didham, pp. 510–33. Chapman and Hall, London.

Martins, M. B. 1987. Variação espacial e temporal de algumas espécies de *Drosophila* (Diptera) em duas reservas de matas isoladas, nas vizinhanças de Manaus (Amazonas, Brazil). Bol. Museu Paraense Emílio Goeldi, ser. Zool. 3: 195–218.

Martins, M. B. 1989. Invasão de fragmentos florestais por espécies oportunistas de *Drosophila* (Diptera, Drosophilidae). Acta Amazónica 19: 265–71.

Morato, E. 1994. Abundância e riqueza de machos de Euglossini em mata de terra firme e areas de derrubadas nas vizinhanças de Manaus. Bol. Museu Paraense Emílio Goeldi, ser. Zool. 10: 95–105.

Morato, E. 1999. Biologia de *Centris fabricius* (Hymenoptera, Anthophoridae, Centridini) em matas contínuas e fragmentos na Amazônia Central. Rev. bras. Zool. 16: 1213–22.

Morato, E. F., L. A. de O. Campos. 1994. Aspetos da biologia de *Pisoxylon xanthosa* na Amazônia Central. Rev. bras. Entomol. 38: 585–94.

Morato, E. F., L. A. de O. Campos, and J. S. Moure. 1992. Abelhas Euglossini (Hymenoptera, Apidae) coletadas na Amazônia Central. Rev. bras. Entomol. 36: 767–71.

Oliveira, M. L. 1999. Sazonalidade e horário de atividade de abelhas Euglossinae (Hymenoptera, Apidae), em florestas de terra-firme na Amazônia Central. Rev. bras. Zool. 16: 83–90.

Oliveira, M. L. In press. Stingless bees (Hymenoptera: Apidae) feeding on stinkhorns spores (Fungi: Phallales): Robbery or dispersal? Rev. bras. Zool.

Oliveira, M. L., and L. A. O. Campos. 1995. Abundância, riqueza, e diversidade de abelhas Euglossinae em florestas contínuas de terra firme na Amazônia Central. Rev. bras. Zool. 12: 547–56.

Oliveira, M. L., and L. A. O. Campos. 1996. Preferência por estratos florestais e por substâncias odoríferas em abelhas euglossinae. Rev. bras. Zool. 13: 1075–85.

Oliveira, M. L., E. F. Morato, and M. V. B. Garcia. 1995. Diversidade de espécies e densidade de ninhos de abelhas sociais sem ferrão (Hymenoptera: Apidae, Meliponinae) em floresta de terra firme na Amazônia Central. Rev. bras. Zool. 12: 13–24.

Powell, A. H., and G. V. N. Powell. 1987. Population dynamics of euglossine bees in Amazonian forest fragments. Biotropica 19: 176–79.

Vasconcelos, H. L. 1988. Distribution of *Atta* (Hymenoptera-Formicidae) in "terra-firme" rain forest of central Amazonia: Density, species composition, and preliminary results on effects of forest fragmentation. Acta Amazónica 18: 309–15.

Vasconcelos, H. L. 1990. Habitat selection by the queens of the leaf-cutting ant *Atta sexdens* in Brazil. J. Tropical Ecology 6: 249–52.

Venticinque, E. M., and H. G. Fowler. 1998. Sheet-web regularity: Fixed allometric relationship in the social spider *Anelosimus eximius*. Ciência e Cultura 50: 371–73.

Venticinque, E. M., H. G. Fowler, and C. A. Silva. 1993. Modes and frequencies of colonization and its relation to extinctions, habitat, and seasonality in the social spider *Anelosimus eximius* in the Amazon (Araneidae: Theridiidae). Pysche 100: 35–41.

Viera, R. S., and H. Hofer. 1994. Prey spectrum of two army ants in central Amazonia, with special attention on their effect on spider populations. Andrias 13: 189–98.

Viera, R. S., and H. Hofer. 1998. Efeito de forrageamento de *Echiton burchelli* sobre a aranofauna de liteira em uma floresta tropical de terra firme na Amazônia central. Acta Amazónica 28: 345–51.

Frogs

THESES

Acuna, F. In progress. Reproductive Ecology of *Colostethus stepheni*. Ph.D. thesis, University of São Paulo.

Acuna, F. J. 1994. Ecologia e biologia reprodutiva de duas espécies de *Colostethus* da região de Manaus, Amazônia Central. M.Sc. thesis, USP.

Barreto, L. 1999. The effects of forest fragmentation on the structure and dynamics of tadpole communities in central Amazonia. Ph.D. thesis, INPA.

Buchacher, C. In progress. Reproductive biology of *Pipa pipa* (Amphibia, Anura). M.Sc. thesis, University of Vienna.

Gascon, C. 1990. The relative importance of habitat characteristics in the maintenance of a species assemblage of tropical forest-breeding frogs. Ph.D. thesis, Florida State University.

Gordo, M. 1998. Factors that affect habitat use of adult frogs. M.Sc. thesis, INPA.

Neckel, S. 1996. Factors structuring tadpole distributions. M.Sc. thesis, INPA.

Oliveira, S. N. In progress. Ecologia de *Phyllomedusa tomopterna* e *Phyllomedusa tarsius* (Amphibia-Anura) em áreas de floresta primária e em habitats matrizes. Ph.D. thesis, INPA.

Tocher, M. 1996. The effects of forest disturbance on habitat use and abundance of frogs in central Amazonia. Ph.D. thesis, University of Canterbury.

Zimmerman, B. L. 1983. A survey of forest frog species in the central Amazon and an analysis of vocal behavior. M.Sc. thesis, University of Guelph.

Zimmerman, B. 1991. Distribution and abundances of forest frogs at a site in the central Amazon. Ph.D. thesis, Florida State University.

PAPERS

Allmon, W. 1991. A plot study of forest floor litter frogs, central Amazon, Brazil. J. Tropical Ecology 7: 503–22.

Buchacher, C. O. 1993. Field studies on the small Surinam toad *Pipa arrabali* (Pipidae, Anura) near Manaus, Brazil. Amphibia-Reptilia 14: 59–69.

Gascon, C. 1989. Predator-prey size interaction in tropical ponds. Rev. bras. Zool. 6: 701–6.

Gascon, C. 1989. The tadpole of *Atelopus pulcher* Boulenger (Annura, Bufonidae) from Manaus, Amazonas. Rev. bras. Zool. 6: 235–39.

Gascon, C. 1991. Breeding of *Leptodactylus knudseni:* Responses to rainfall variation. Copeia 1: 248–52.

Gascon, C. 1991. Population- and community-level analyses of species occurrences of central Amazonian rainforest tadpoles. Ecology 72: 1731–46.

Gascon, C. 1992. Aquatic predators and tadpole prey at a central Amazonian site: Field data and experimental manipulations. Ecology 73: 971–80.

Gascon, C. 1992. The effects of reproductive phenology on larval performance traits in a three-species assemblage of central Amazonian tadpoles. Oikos 65: 307–13.

Gascon, C. 1992. Spatial distribution of *Osteocephalus taurinus* and *Pipa arrabali* in a central Amazonian forest. Copeia 1992: 894–97.

Gascon, C. 1993. Breeding-habitat use by Amazonian primary-forest frogs species at forest edge. Biodiversity and Conservation 2: 438–44.

Gascon, C. 1994. Bottom-nets as a new method to quantitatively sample tadpole populations. Rev. bras. Zool. 11: 355–59.

Gascon, C. 1994. Sampling with artificial pools. In *Measuring and Monitoring Biological Diversity: Standard Methods for Amphibians,* ed. W. R. Heyer, M. A. Donnelly, R. McDiarmid, L. C. Hayek, and M. S. Foster, pp. 144–45. Smithsonian Institution Press, Washington, D.C.

Gascon, C. 1995. Natural history notes on frogs from the Manaus region. Rev. bras. Zool. 12: 9–12.

Gascon, C. 1995. Tropica larval anuran fitness in the absence of direct effects of predation and competition. Ecology 76: 2222–29.

Gascon, C., and J. Meyer. 1999. Sapos resistem na floresta fragmentada. Ciência Hoje 26: 65–68.

Hero, J. M., C. Gascon, and W. E. Magnusson. 1998. Direct and indirect effects of predation on tadpole community structure in the Amazon rainforest. Australian J. Ecology 23: 474–82.

Juncá, F. A. 1996. Parental care and egg mortality in *Colostethus stepheni*. J. Herpetology 30.

Juncá, F. A., R. Altig, and C. Gascon. 1994. Breeding biology of *Colostethus stepheni:* A dendrobatid with a non-transported nidicolous tadpole. Copeia 1994: 747–50.

Schiesari, L. C., and G. Moreira. 1996. The tadpole of *Phrynohyas coriacea* with comments on the species reproduction. J. Herpetology 30: 404–7.

Schmidt, B. R., and A. Amézquita. In press. Predator-induced behavioural responses: Tadpoles of the Neotropical frog Phyllomedusa tarsius do not respond to all predators. Herpetological Journal.

Shaffer, H. B., R. A. Alford, B. G. Woodward, S. J. Richards, R. A. Altig, and C. Gascon. 1994. Quantitative sampling of amphibian larvae. In *Measuring and Monitoring Biological Diversity: Standard Methods for Amphibians,* ed. W. R. Heyer, M. A. Donnelly, R. McDiarmid, L. C. Hayek, and M. S. Foster, pp. 130–41. Smithsonian Institution Press, Washington, D.C.

Tocher, M. 1998. A comunidade de anfíbios da Amazônia central: Diferenças na composição específica entre a mata primária e pastagens. In *Floresta Amazónica: Dinâmica, Regeneração e Manejo,* ed. C. Gascon and P. Moutinho, pp. 219–32. Instituto Nacional de Pesquisas da Amazônia, Manaus.

Tocher, M., C. Gascon, and B. Zimmerman. 1997. Fragmentation effects on a central Amazonian frog community: A ten-year study. In *Tropical Forest Remnants: Ecology, Management, and Conservation of Fragmented Communities,* ed. W. F. Lau-

rance and R. O. Bierregaard, Jr., pp. 124–37. University of Chicago Press, Chicago.

Zimmerman, B. L. 1983. A comparison of structural features of calls of open and forest habitat frog species in the central Amazon. Herpetologica 39: 235–46.

Zimmerman, B. L. 1994. Audio strip transects. In *Measuring and Monitoring Biological Diversity: Standard Methods for Amphibians,* ed. W. R. Heyer, M. A. Donnelly, R. McDiarmid, L. C. Hayek, and M. S. Foster, pp. 92–96. Smithsonian Institution Press, Washington D.C.

Zimmerman, B. L., and J. P. Bogart. 1986. Vocalizations of primary forest frog species in the central Amazon. Acta Amazónica 14: 473–519.

Zimmerman, B. L., and J. P. Bogart. 1988. Ecology and calls of four little-known central Amazonian forest species of frogs. Herpetology 22: 97–108.

Zimmerman, B. L., and M. T. Rodrigues. 1990. Frogs, snakes, and lizards of the INPA-WWF reserves near Manaus, Brazil. In *Four Neotropical Rainforests,* ed. A. Gentry, pp. 426–54. Yale University Press, New Haven.

Zimmerman, B., and D. Simberloff. 1996. An historical interpretation of habitat use by frogs in a central Amazonian forest. J. Biogeography 23: 27–46.

Birds

THESES

Borges, S. 1995. Comunidade de aves em dois tipos de vegetação secundária da Amazônia central. M.Sc. thesis, INPA.

Cohn-Haft, M. 1995. Dietary specialization by lowland tropical rainforest birds: Forest interior versus canopy and edge habitats. M.Sc. thesis, Tulane University.

Harper, L. H. 1987. The conservation of ant-following birds in small Amazonian forest fragments. Ph.D. thesis, State University of New York.

Quintela, C. E. 1985. Forest fragmentation

and differential use of natural and man-made edges by understory birds in central Amazonia. M.Sc. thesis, University of Illinois.

Stratford, J. 1999. Effects of forest fragmentation on terrestrial insectivorous birds. M.Sc. thesis, Southeastern Louisiana University.

PAPERS

Bierregaard, R. O., Jr. 1982. Levantamentos ornitológicos no dossel da mata pluvial de terra firme. Acta Amazónica 12: 107–11.

Bierregaard, R. O., Jr. 1984. Observations on the nesting biology of the Guiana Crested Eagle (*Morphnus guianensis*). Wilson Bull. 96: 1–5.

Bierregaard, R. O., Jr. 1988. Morphological data from understory birds in terra firme forest in the central Amazonian basin. Rev. bras. Biol. 48: 169–78.

Bierregaard, R. O., Jr. 1990. Avian communities in the understory of Amazonian forest fragments. In *Biogeography and Ecology of Forest Bird Communities,* ed. A. Keast and J. Kikkawa, pp. 333–43. SPB Academic Publishing, The Hague.

Bierregaard, R. O., Jr. 1990. Species composition and trophic organization of the understory bird community in a central Amazonian terra firme forest. In *Four Neotropical Rainforests,* ed. A. Gentry, pp. 217–36. Yale University Press, New Haven.

Bierregaard, R. O., Jr., and T. E. Lovejoy. 1988. Birds in Amazonian forest fragments: Effects of insularization. In Acta XIX Cong. Int. Ornith, vol. 2, ed. H. Ouellet, pp. 1564–79. University of Ottawa Press, Ottawa.

Bierregaard, R. O., Jr., and T. E. Lovejoy. 1989. Effects of forest fragmentation on Amazonian understory bird communities. Acta Amazónica 19: 215–41.

Bierregaard, R. O., Jr., D. F. Stotz, L. H. Harper, and G. V. N. Powell. 1987. Observations on the occurrence and behavior of the Crimson Fruit Crow (*Haematoderus militaris*), in central Amazonia. Bull. Brit. Ornith. Club 107: 134–37.

Bierregaard, R. O., Jr., and P. Stouffer. 1997. Birds in forest fragments. In *Tropical Forest Remnants: Ecology, Management, and Conservation of Fragmented Communities,* ed. W. F. Laurance and R. O. Bierregaard, Jr. University of Chicago Press, Chicago.

Borges, S. 1995. Ninhos e ovos de *Caryothraustes candensis*. Ararajuba 3: 76.

Borges, S. 1999. Relative use of secondary forests by cracids in central Amazonia. Ornith. Neotropical 10: 77–80.

Borges, S. H., and P. C. Stouffer. 1999. Bird communities in two types of anthropogenic successional vegetation in central Amazonia. Condor 101: 529–36.

Borges, S., P. Stouffer, and A. Carvalhães. In press. Comportamento de "Lek" em *Topazza pella* na Amazônia central (Aves, Trochilidae). In *História Natural da Biota Amazónica,* ed. R. Cintra. INPA, Manaus.

Karr, J. R., J. Blake, S. Robinson, and R. O. Bierregaard, Jr. 1990. Birds of four Neotropical forests. In *Four Neotropical Rainforests,* ed. A. Gentry, pp. 237–69. Yale University Press, New Haven.

Klein, B. C., and R. O. Bierregaard, Jr. 1988. Capture and telemetry techniques for the Lined Forest-falcon, *Micrastur gilvicollis*. Raptor Research 22: 29.

Klein, B. C., and R. O. Bierregaard, Jr. 1988. Movement and calling behavior of the Lined Forest-falcon (*Micrastur gilvicollis*) in the Brazilian Amazon. Condor 90: 497–99.

Klein, B. C., L. H. Harper, R. O. Bierregaard, Jr., and G. V. N. Powell. 1988. Nesting and feeding behavior of the Ornate Hawk-eagle, *Spizaetus ornatus*. Condor 90: 239–41.

Powell, G. V. N. 1989. On the possible contribution of mixed species flocks to species richness in Neotropical avifaunas. Behavioral Ecology and Sociobiology 24: 387–93.

Quintela, C. E. 1987. First report of the nest

and young of the Variegated Antpitta (*Grallaria varia*). Wilson Bull. 99: 499–500.

Stouffer, P. 1997. Interspecific aggression in Amazonian antthrushes? The view from central Amazonian Brazil. Auk 114: 780–85.

Stouffer, P. C., and R. O. Bierregaard, Jr. 1993. Seasonal rainfall patterns and the abundance of Ruddy Quail-doves (*Geotrygon montana*) near Manaus, Brazil. Condor 95: 896–903.

Stouffer, P., and R. O. Bierregaard, Jr. 1995. Effects of forest fragmentation on understory hummingbirds in Amazonia, Brazil. Conserv. Biol. 9: 1085–94.

Stouffer, P., and R. O. Bierregaard, Jr. 1995. Use of Amazonian forest fragments by understory insectivorous birds. Ecology 76: 2429–43.

Stouffer, P., and R. O. Bierregaard, Jr. 1996. Forest fragmentation and seasonal patterns of hummingbird abundance in Amazonian Brazil. Ararajuba 4: 9–15.

Stratford, J. A. In press. A ferruginous-backed antbird, Myrmeciza ferruginea, nest from central Amazonas, Brazil. Ararajuba.

Stratford, J. A., and P. C. Stouffer. 1999. Local extinctions of terrestrial insectivorous birds in a fragmented landscape near Manaus, Brazil. Conserv. Biol. 13: 1416–23.

Whittaker, A. 1993. Notes on the behavior of the Crimson Fruitcrow *Haematoderus militaris* near Manaus, Brazil, with the first nesting record for this species. Bull. Brit. Ornith. Club 113: 93–96.

Whittaker, A. 1995. Bird Observations. Bull. Brit. Ornith. Club 115: 45–48.

Whittaker, A. 1995. Notes of feeding behavior, diet, and anting of some cotingas. Bull. Brit. Ornith. Club 116: 58–62.

Whittaker, A. 1998. Observation on the vocalization behavior and distribution of the glossy-backed becard (*Pachyramphus surinamis*), a poorly known canopy inhabitant of the Amazonian rainforest. Ararajuba 6: 37–41.

Mammals

THESES

Bernard, E. 1997. Estrutura vertical de uma comunidade de morcegos na floresta primária no PDBFF, Manaus. M.Sc. thesis, INPA.

Frazão, E. 1992. Diet and foraging strategy of *Chiropotes satanas chiropotes* (Cebidae: Primates) in central Brazilian Amazonia. M.Sc. thesis, INPA.

Gilbert, K. 1994. Prevalence and abundance of endoparisitic infection in *Alouatta seniculus* in forest fragments and continuous forest in central Amazonia. Ph.D. thesis, Rutgers University.

Gomez, M. S. 1999. Padrões de atividade, hábitos alimentares, e dispersão de sementes de *Alouatta seniculus* em floresta de terra firme na Amazônia central. M.Sc. thesis, UFMG.

Malcolm, J. 1991. The small mammals of Amazonian forest fragments: Pattern and process. Ph.D. thesis, University of Florida.

Sampaio, E. In progress. The effects of fragmentation on structure and diversity of bat communities in Amazonian tropical rain forest. Ph.D. thesis, University of Tuebingen.

Setz, E. 1993. Feeding ecology of *Pithecia pithecia:* A comparison between groups in continuous forest and an isolated forest fragment. Ph.D. thesis, UNICAMP.

Waldick, R. In progress. Ecological and evolutionary implications of forest fragmentation for Neotropical primate populations. Ph.D. thesis, McMaster University.

PAPERS

Bernard, E. In press. Vertical stratification of bat communities in primary forests of Central Amazon, Brazil. J. of Trop. Ecology.

Bernard, E. In press. Structure of bat communities in primary forests of Central Amazon, Brazil.

Bernard, E., A. K. L. M. Albernaz, and W. E. Magnusson. In press. Bat species composition in three localities in the Amazon

394

Basin. Studies on Neotropical Fauna and Environment. Mammalia.

Frazão, E. 1991. Insectivory in free-ranging Bearded Saki (*Chiropotes satanas*). Primates (Japan) 32: 245–47.

Gilbert, K. A. 1997. Red howling monkey use of specific defecation sites as a parasite avoidance strategy. Animal Behaviour 54: 451–55.

Gilbert, K. A. In press. Predation on a white-faced saki in the central Amazon. Neotropical Primates.

Kalko, E. K. V. 1998. Organization and diversity of tropical bat communities through space and time. Zoology (Analysis of Complex Systems) 101: 281–97.

Malcolm, J. R. 1988. Small mammal abundances in isolated and non-isolated primary forest reserves near Manaus, Brazil. Acta Amazónica 18: 67–83.

Malcolm, J. R. 1990. Estimation of mammalian densities in continuous forest north of Manaus. In *Four Neotropical Rainforests,* ed. A. Gentry, pp. 339–57. Yale University Press, New Haven.

Malcolm, J. R. 1991. Comparative abundances of Neotropical small mammals by trap height. J. Mammalogy 72: 188–92.

Malcolm, J. R. 1992. Use of tooth impressions to identify and age live *Proechimys guyannensis* and *P. cuvieiri* (Rodentia: Echimyidae). J. Zoology 227: 537–46.

Malcolm, J. R. 1995. Forest structure and the abundance and diversity of Neotropical small mammals. In *Forest Canopies: Ecology, Biodiversity, and Conservation,* ed. M. D. Lowman and N. M. Nadkarni, pp. 179–97. Academic Press, New York.

Malcolm, J. 1997. Biomass and diversity of small mammals in forest fragments. In *Tropical Forest Remnants: Ecology, Management, and Conservation of Fragmented Communities,* ed. W. F. Laurance and R. O. Bierregaard, Jr., pp. 207–21. University of Chicago Press, Chicago.

Rylands, A. B., and A. Keuroghlian. 1988. Primate populations in continuous forest and forest fragments in central Amazonia. Acta Amazónica 18: 291–307.

Rylands, A. B., and A. S. Neves. 1992. Diet of a group of howler monkeys (*Alouatta seniculus*) in an isolated forest patch in central Amazonia. Primatologia no Brasil 3.

Schwarzkopf, L., and A. B. Rylands. 1989. Primate species richness in relation to habitat structure in Amazonian rainforest fragments. Biol. Conserv. 48: 1–12.

Setz, E. 1992. Comportamento de alimentação de *Pithecia pithecia* (Cebidae, Primatas) em um fragmento florestal. Primatologia no Brasil 3: 327–30.

Setz, E., Z. Freira, D. de Alemar. In press. Scent-marking in free-ranging golden-faced saki monkeys, *Pithecia pithecia chrysocephala:* Sex differences and context. J. Zoology.

Setz, E., Z. Freira, J. Enzweiler, V. N. Solferini, M. P. Amêndola, and R. S. Berton. In press. Geography of the golden-faced saki monkey in the central Amazon. J. Zoology (London).

Spironello, W. R. 1987. Range size of a group of *Cebus apella* in central Amazonia. American J. Primatology 8: 522.

Forest Ecology

THESES

Almeida, S. S. 1989. Clareiras naturais na Amazônia central: Distribuição, estrutura e aspectos da regeneração vegetal. M.Sc. thesis, INPA.

Benitez-Malvido, J. 1995. Studies of seedling communities in Amazonian forest fragments with special reference to Sapotaceae. Ph.D. thesis, Cambridge University.

Bruna, E. In progress. The effects of forest fragmentation on mating patterns and population structure of tropical plants. Ph.D. thesis, University of California.

Castilho, C. V. d. 2000. Regeneração natural de duas espécies de palmeiras em floresta de terra firme na Amazônia Central: Efeito de fatores bióticos e abióticos na sobrevivência de sementes e estabelecimento de plântulas. M.Sc. thesis, INPA.

Chambers, J. Q. 1998. The role of large wood

in the carbon cycle of central Amazon rain forest. Ph.D. thesis, University of California.

Elias, M. E. d. A. 1997. The influence of fragmentation and succession on the growth and survival of seedlings of *Copaifera multijuga*. M.Sc. thesis, INPA.

Imakawa, A. 1996. Establishment and distribution pattern of *Lecythis barnebyi* and *L. prancei*. M.Sc. thesis, INPA.

Nascimento, H. E. M. In progress. Efeitos da fragmentação florestal sobre a biomassa vegetal do sub-bosque em floresta de terra-firme na Amazônia central. Ph.D. thesis, INPA.

Oliveira, A. A. 1997. Composition, structure, and phenology of a terra firme forest in central Amazonia. Ph.D. thesis, USP.

Pacheco, M. 1999. Mechanisms of seed and seedling mortality controlling the distribution of two tropical trees. Ph.D. thesis, University of Illinois.

Scariot, A. 1996. The ecological consequences of fragmentation on a tropical palm community. Ph.D. thesis, University of California.

Spironello, W. 1999. Sapotaceae community ecology in a central Amazon forest. Ph.D. thesis, Cambridge University.

PAPERS

Ackerly, D. D., J. M. Rankin–de Mérona, and W. A. Rodrigues. 1990. Tree densities and sex ratios in breeding populations of dioecious central Amazonian Myristicaceae. J. Tropical Ecology 6: 239–48.

Bassini, F., and P. Becker. 1990. Charcoal's occurrence in soil depends on topography in terra firme forest near Manaus, Brazil. Biotropica 22: 420–22.

Benitez-Malvido, J. 1998. Impact of forest fragmentation on seedling abundance in a tropical rain forest. Conserv. Biol. 12: 380–89.

Benitez-Malvido, J., and I. D. Kossman-Ferraz. 1999. Litter cover variability affects seedling performance and herbivory. Biotropica 31: 598–606.

Boom, B., and M. T. V. A. Campos. 1991. A preliminary account of the Rubiaceae of a central Amazonian terra firme forest. Bol. Museu Paraense Emílio Goeldi, ser. Bot 7: 223–47.

Bowen, W. T., P. Becker, and F. Bassini. 1990. Spatial variability of extractable phosphorus in an Amazon forest. In Proceedings of the SCOPE workshop on phosphorus cycles in terrestrial and aquatic ecosystems in Latin America with emphasis on the Amazon Basin, Venezuela.

Chambers, J. Q., N. Higuchi, J. P. Schimel, L. V. Ferreira, and J. Melack. 2000. Decomposition and carbon cycling of dead trees in tropical forests of central Amazon. Oecologia 261: 1–9.

Chambers, J. Q., J. P. Schimel, and A. D. Nobre. In press. Respiration from coarse wood litter in central Amazon forests. Biogeochemistry.

Ferreira, L. V., and J. Rankin–de Mérona. 1997. Floristic composition and structure of a one-hectare plot in terra firme forest in central Amazonia. In *Forest Biodiversity in North, Central and South America and the Caribbean: Research and monitoring,* ed. F. Dallmeier and J. A. Comiskey, Man and Biosphere series, vol. 22, chapter 34. UNESCO and Parthenon, Carnforth, Lancashire, U.K., pp. 655–68.

Henderson, A., B. A. Fischer, M. P. Scariot, and R. Pardini. In press. Flowering phenology of a palm community in a central Amazon forest. Brittonia.

Laurance, W. F., P. M. Fearnside, S. G. Laurance, P. Delamônica, T. E. Lovejoy, J. M. Rankin–de Mérona, J., J. Chambers, and C. Gascon. 1999. Relationships between soils and Amazon forest biomass: Landscape-scale study. Forest Ecology and Management 118: 127–38.

Laurance, W. F., L. V. Ferreira, C. Gascon, and T. E. Lovejoy. 1998. Biomass loss in forest fragments. Science 282: 1611a.

Laurance, W. F., L. V. Ferreira, J. Rankin–de Mérona, and R. Hutchings. 1998. Influence of plot shape on estimates of tree diversity and community composition in central Amazon. Biotropica 30: 662–65.

Laurance, W. F., L. V. Ferreira, J. M. Rankin–de Mérona, and S. G. Laurance. 1998. Rain-forest fragmentation and the dynamics of Amazonian tree communities. Ecology 79: 2032–40.

Laurance, W. F., L. V. Ferreira, J. M. Rankin–de Mérona, S. G. Laurance, R. W. Hutchings, and T. E. Lovejoy. 1998. Effects of forest fragmentation on recruitment patterns in central Amazonia. Conserv. Biol. 12: 460–64.

Laurance, W. F., C. Gascon, and J. M. Rankin–de Mérona. 1999. Predicting effects on habitat destruction on plant communities; a test of a model using Amazonian trees. Ecological Applications 9: 548–54.

Lepsch-Cunha, N., and S. Mori. 1999. Reproductive phenology and mating potential in a low density population of *Courataria multiflora* (Lecythidaceae) in central Amazonia. J. Tropical Ecology 15: 97–121.

Lewis, S., and E. V. J. Tanner. 2000. Effects of above- and below-ground competition on growth and survival of tree seedlings in Amazonian rain forest. Ecology 81: 2525–38.

Moreira, F. W., L. A. de Oliveira, and P. Becker. 1997. Ausência de micorrizas vesículo-arbusculares efetivas em Lecythidaceaes numa área de floresta primária da Amazônia central. Acta Amazónica 27.

Mori, S., and P. Becker. 1991. Flooding affects survival of Lecythidaceae in terra firme forest near Manaus, Brazil. Biotropica 23: 87–90.

Mori, S., and N. Lepsch-Cunha. 1995. The Lecythidaceae of a central Amazonian moist forest. Memoirs of the NYBG 75.

Oliveira, A. A., and D. C. Daly. 1999. Geographic distribution of tree species occurring in the region of Manaus, Brazil: Implications for regional diversity and conservation. Biodiversity and Conservation 8: 1245–59.

Oliveira, A. A., and S. Mori. 1999. Central Amazonian terra firme forests: High tree species richness. Biodiversity and Conservation 8: 1219–44.

Pacheco, M., and A. Henderson. 1996. Testing association between species abundance and a continuous variable with Kolmogorov-Smirnov statistics. Vegetatio 124: 95–99.

Phillips, O. L., Y. Mahli, N. Higuchi, W. F. Laurance, P. Nunez, V. R. Vasquez, S. G. Laurance, L. V. Ferreira, M. Stern, S. Brown, and J. Grace. 1998. Changes in the carbon balance of tropical forests: Evidence from long-term plots. Science 282: 439–41.

Piperno, D. R., and P. Becker. 1996. Vegetational history of a site in the central Amazon derived from phytolith and charcoal records from natural soils. Quaternary Research 45: 202–9.

Rankin, J. M. 1980. Diversidade e dispersão de Lecythidaceae na floresta de terra firme da Amazônia Central. Abstract from the 32 Reunião da Sociedade Brasileira para o Progresso da Ciência (SBPC). Ciência e Cultura, Suplemento, 17.

Rankin–de Mérona, J. M., R. W. Hutchings, and T. E. Lovejoy. 1990. Tree mortality and recruitment over a five-year period in undisturbed upland rain forest of the central Amazon. In *Four Neotropical Rainforests,* ed. A. Gentry, pp. 573–84. Yale University Press, New Haven.

Rankin–de Mérona, J. M., G. T. Prance, R. W. Hutchings, F. M. Silva, W. A. Rodrigues, M. E. Uehling. 1992. Preliminary results of large-scale tree inventory of upland rain forest in the central Amazon. Acta Amazónica 22: 493–534.

Rebelo, F. C., and G. B. Williamson. 1996. Driptips vis-à-vis soil types from Central Amazônia. Biotropica 28: 159–63.

Scariot, A. 1999. Forest fragmentation: Effects on palm diversity in central Amazonia. J. Ecology 87: 66–76.

Scariot, A. In press. Seedling mortality by litterfall in Amazonian forest fragments. Biotropica.

Williamson, G. B., and F. Costa. In press. Dispersal of Amazonian trees: Hydrochory in *Pentaclethra macroloba.* Biotropica.

Williamson, G. B., F. Costa, and C. V. Minte Vera. 1999. Dispersal of Amazonian trees:

Hydrochory in *Swartzea polyphela*. Biotropica 31: 460–65.

Williamson, G. B., W. F. Laurance, A. A. Oliveira, P. Delamônica, C. Gascon, T. E. Lovejoy, and L. Pohl. In press. Amazonian tree mortality during the 1997 El Niño drought. Conserv. Biol.

Williamson, G. B., T. Van Eldik, P. Delamônica, and W. F. Laurance. 1999. How many millenarians in Amazonia: Sizing the ages of large trees. Trends in Plant Science. 4: 397.

Edge Effects

THESES

Camargo, J. L. 1993. Variation in soil moisture and air vapour pressure deficit relative to tropical rain forest edges near Manaus, Brazil. M.Sc. thesis, Cambridge University.

Carvalho, K. S. 1998. Efeitos de borda e do isolamento florestal sobre a comunidade de formigas associadas a pequenos troncos na liteira. M.Sc. thesis, INPA.

Sizer, N. 1992. The impact of edge formation on regeneration and litterfall in a tropical rain forest fragment in Amazonia. Ph.D. thesis, Cambridge University.

Tavares, L. N. J. 1998. Efeitos de borda e do crescimento secundário sobre pequenos mamíferos da Amazônia. M.Sc. thesis, INPA.

Wandelli, E. 1991. Eco-physiological response of the understory palm *Astrocaryum sociale* to environmental changes resulting from the forest edge effect. M.Sc. thesis, INPA.

PAPERS

Camargo, J. L. C., and V. Kapos. 1995. Complex edge effects on soil moisture and microclimate in central Amazonian forest. J. Tropical Ecology 11: 205–21.

Didham, R. 1997. The influence of edge effects and forest fragmentation on leaf litter invertebrates in central Amazonia. In *Tropical Forest Remnants: Ecology, Management, and Conservation of Fragmented Communities,* ed. W. F. Laurance and R. O. Bierregaard, Jr., pp. 55–70. University of Chicago Press, Chicago.

Didham, R. K., and J. H. Lawton. 1999. Edge structure determines the magnitude of changes in microclimate and vegetation structure in tropical forest fragments. Biotropica 31: 17–30.

Gascon, C., B. G. Williamson, and G. A. B. Fonseca. 2000. Receding forest edges and vanishing reserves. Science 288: 1356–58.

Kapos, V. 1989. Effects of isolation on the water status of forest patches in the Brazilian Amazon. J. Tropical Ecology 5: 173–85.

Kapos, V., G. M. Ganade, E. Matsui, and R. L. Victoria. 1993. $\delta^{13}C$ as an indicator of edge effects in tropical rain forest reserves. J. Ecology 81: 425–32.

Kapos, V., E. Wandelli, J. L. C. Camargo, and G. Ganade. 1997. Edge-related changes in environment and plant responses due to forest fragmentation in central Amazonia. In *Tropical Forest Remnants: Ecology, Management, and Conservation of Fragmented Communities,* ed. W. F. Laurance and R. O. Bierregaard, Jr., pp. 33–44. University of Chicago Press, Chicago.

Laurance, W. F. 2000. Edge effects and large-scale ecological processes: Reply to Ickes and Williamson. Trends Ecol. Evol. 15: 373.

Lovejoy, T. E., R. O. Bierregaard, Jr., A. B. Rylands, J. R. Malcolm, C. E. Quintela, L. H. Harper, K. S. Brown, Jr., A. H. Powell, G. V. N. Powell, H. O. R. Schubart, and M. Hays. 1986. Edge and other effects of isolation on Amazon forest fragments. In *Conservation Biology,* ed. M. Soulé, pp. 257–85. Sinauer, Sunderland, Mass.

Malcolm, J. R. 1994. Edge effects in central Amazonian forest fragments. Ecology 75: 2438–45.

Malcolm, J. 1998. A model of conductive heat flow in forest edges and fragmented landscapes. Climatic Change 39: 487–502.

Mesquita, R. C. G., P. Delamônica, and W. F. Laurance. 1999. Effect of surrounding vegetation on edge-related tree mortality in Amazonian forest fragments. Biol. Conserv. 91: 129–34.

Sizer, N., and E. V. J. Tanner. 1999. Responses of woody plant seedlings to edge formation in a lowland tropical rainforest, Amazonia. Biol. Conserv. 91: 135–42.

Plant-Animal Interactions

THESES

Andersen, E. 2000. Effect of dung beetles and rodents on the post-dispersal fate of different-sized seeds in central Amazonia. Ph.D. thesis, University of Florida.

Fáveri, S. B. 2000. Riqueza e composição da comunidade de artrópodes associados a *Cecropia*. M.Sc. thesis, INPA.

Fonseca, C. R. S. 1991. Interaction between *Tachigalia myrmecophila* Ducke (Caesalpinaceae) and associated ants. M.Sc. thesis, UNICAMP.

Fonseca, C. R. S. 1995. The growth and reproduction of ant colonies living in sympatric Amazonian myrmecophytes. Ph.D. thesis, Oxford University.

Mesquita, R. 1989. Aspectos da biologia reprodutiva e remoção de sementes de *Clusia grandiflora* numa reserva florestal na Amazonia central. M.Sc. thesis, INPA.

Meyer, J. 1999. Efeitos da fragmentação da floresta sobre predação de ovos de aves e remoção de sementes. M.Sc. thesis, INPA.

Miriti, M. 1999. Promotion of seed dispersal as a technique for restoration of tropical rain forests. Ph.D. thesis, University of Illinois.

Vasconcelos, H. L. 1994. Interactions between leafcutter ants and forest regeneration in Amazonia. Ph.D. thesis, University of Wales, Bangor.

PAPERS

Andersen, E. In press. Effects of dung presence, dung amount, and secondary dispersal by dung beetles on the fate of Micropholis guyanensis (Sapotaceae) seeds in Central Amazonia. J. Trop. Ecology.

Benitez-Malvido, J., G. Garcia-Guzman, and I. D. Kossmann-Ferraz. 1999. Leaf-fungal incidence and herbivory on tree seedlings in tropical rainforest fragments: An experimental study. Biol. Conserv. 91: 143–50.

Fonseca, C. R. S. 1993. Nesting space limits colony size of the plant-ant *Pseudomyrmex concolor*. Oikos 67: 473–82.

Fonseca, C. R. S. 1994. Herbivory and long-lived leaves of an Amazonian ant-tree. J. Ecology 82: 833–42.

Fonseca, C. R. S. 2000. Amazonian ant-plant interactions and the nesting space limitation hypothesis. J. Tropical Ecology 15: 807–25.

Fonseca, C. R. S., and G. Ganade. 1996. Asymmetry, compartments and null interactions in an Amazonian ant-plant community. J. Animal Ecology 65: 339–47.

Fonseca, C. R. S., and J. L. John. 1996. Connectance: A role for community allometry. Oikos 77: 353–58.

Fowler, H. G. 1993. Herbivory and assemblage structure of myrmecophytous understory plants and their associated ants in the central Amazon. Insectes Sociaux 40: 137–45.

Fowler, H. G. In press. Apparency and accumulated herbivory in understory myrmecophytes of the central Amazon. Ciência e Cultura.

Fowler, H. G., and E. M. Venticinque. 1996. Spiders and understory myrmecophytes of the central Amazon, Brazil. Rev. bras. Entomol. 40: 71–73.

Garcia, M. V. B., M. L. Oliveira, and L. A. O. Campos. 1992. Use of seeds of *Coussapoa asperifolia magnifolia* (Cecropiaceae) by stingless bees in the central Amazonian forest (Apidae: Meliponinae). Entomologia Generalis 17: 255–58.

Harper, L. H. 1989. Birds and army ants (*Eciton burchelli*), observations on their ecology in undisturbed forest and isolated reserves. Acta Amazónica 19: 249–63.

Henderson, A. R., J. F. S. Rebelo, S. Vanin, and D. Almeida. In press. Pollination of *Bactris* (Palmae) in an Amazon Forest. Brittonia.

Mesquita, R., and C. H. Franciscon. 1995. Flower visitors of *Clusia nemorosa* G. F. W. Meyer (Clusiaceae) in an Amazonian white-sand campina. Biotropica 27: 254–57.

Spironelo, W. 1991. A importância de frutos de palmeiras (Palmae) na dieta de um grupo de *Cebus apella* (Cebidae: Primates) na Amazônia central. Primatologia no Brasil 3: 285–96.

Vasconcelos, H. L. 1990. Effects of litter collection by understory palms on the associated macroinvertebrate fauna in central Amazonia. Pedobiologia 34: 157–60.

Vasconcelos, H. L. 1990. Foraging activity of two species of leaf-cutting ants (*Atta*) in a primary forest of central Amazon. Insectes Sociaux 37: 131–45.

Vasconcelos, H. L. 1991. Mutualism between *Maieta guianensis,* a myrmecophytic melastome, and one of its ant inhabitants: Ant protection against insect herbivores. Oecologia 81: 295–98.

Vasconcelos, H. L. 1993. Ant colonization of *Maieta guianensis* seedlings, an Amazon ant-plant. Oecologia 95: 439–43.

Vasconcelos, H. L. 1997. Foraging activity of an Amazonian leaf-cutting ant: Responses to changes in the availability of woody plants and to previous plant damage. Oecologia 112: 370–78.

Vasconcelos, H. L. 1999. Effects of forest disturbance on the structure of ground-foraging ant communities in central Amazonia. Biodiversity and Conservation 8: 409–20.

Vasconcelos, H. L. 1999. Levels of leaf herbivory in Amazonian trees from different stages in forest regeneration. Acta Amazónica 29: 615–23.

Vasconcelos, H. L., and A. B. Casimiro. 1997. Influence of *Azteca alfari* ants on the exploitation of *Cecropia* trees by a leaf-cutting ant. Biotropica 29: 84–92.

Vasconcelos, H. L., and J. M. Cherrett. 1995. Colonization changes in leaf-cutting ant populations after the clearing of mature forest in Amazonia. Studies on Neotropical Fauna and Environment 30: 107–13.

Vasconcelos, H. L., and J. M. Cherrett. 1996. The effect of wilting in the selection of leaves by the leaf-cutting and *Atta laevigata*. Entomologia Experimentalis et Applicata 78: 215–20.

Vasconcelos, H. L., and J. M. Cherrett. 1997. Leaf-cutting ants and early forest regeneration in central Amazonia: Effects of herbivory on tree seedling establishment. J. Tropical Ecology 13: 357–70.

Vasconcelos, H. L., and J. M. Cherrett. 1998. Efeitos da herbivoria por saúvas (*Atta laevigata*) sobre a regeneração florestal em uma área agrícola abandonada da Amazônia central. In *Floresta Amazónica: Dinâmica, Regeneração e Manejo,* ed. C. Gascon and P. Moutinho, pp. 171–78. Instituto Nacional de Pesquisas da Amazônia, Manaus.

Vasconcelos, H. L., and D. W. Davidson. 2000. Relationship between plant size and ant associates in two Amazonian ant-plants. Biotropica 32: 100–11.

Forest Regeneration

Theses

Andrade, A. C. S. 2000. Variação na área foliar específica de espécies pioneiras da Amazônia Central. M.Sc. thesis, INPA.

Freitas, M. 1998. Factors affecting colonization and establishment of arboreal species in forest clearings. M.Sc. thesis, INPA.

Ganade, G. 1995. On the survival of forest seeds and seedlings in abandoned fields of central Amazonia. Ph.D. thesis, Imperial College, London.

Lewis, S. 1998. Treefall gaps and regeneration: A comparison of continuous forest and fragmented forest in central Amazônia. Ph.D. thesis, University of Cambridge.

Mesquita, R. C. G. 1995. Utilization of *Cecropia*-dominated secondary forest for establishment and growth of primary forest seedlings in the Brazilian Amazon. Ph.D. thesis, University of Georgia.

Monaco, L. 1998. O uso do solo e a regeneração florestal em área abandonadas. M.Sc. thesis, INPA.

Puerta, R. 2000. Aspectos da regeneração florestal em pasto abandonado da região de Manaus. M.Sc. thesis, INPA.

PAPERS

Mesquita, R. C. 1998. O impacto da remoção do dossel em florestas secundárias sobre o crescimento de duas espécies de árvores de importância econômica. *In Floresta Amazónica: Dinâmica, Regeneração e Manejo,* ed. C. Gascon and P. Moutinho, pp. 261–75. Instituto Nacional de Pesquisas da Amazônia, Manaus.

Mesquita, R. C. G. 2000. Management of advanced regeneration in secondary forests of the Brazilian Amazon. Forest Ecology and Management 130: 131–40.

Mesquita, R. C. G., S. W. Workman, and C. L Neely. 1998. Slow decomposition in a *Cecropia*-dominated secondary forest of central Amazonia. Soil Biology and Biochemistry 30: 167–75.

Miriti, M. 1998. Regeneração florestal em pastagens abandonadas na Amazônia central: Competição, predação, e dispersão de sementes. In *Floresta Amazónica: Dinâmica, Regeneração e Manejo,* ed. C. Gascon and P. Moutinho. Instituto Nacional de Pesquisas da Amazônia, Manaus.

Mônaco, L., R. C. G. Mesquita, G. B. Williamson. In press. O banco de sementes de uma floresta secundária Amazônica dominada por Vismia spp. Acta Amazónica.

Nelson, B. W. 1994. Natural forest disturbance and change in the Brazilian Amazon. Remote Sensing Reviews 10: 105–25.

Nelson, B., V. Kapos, J. B. Adams, W. J. Oliveira, O. P Braun, and I. do Amaral. 1994. Forest disturbance by large blowdowns in the Brazilian Amazon. Ecology 75: 853–58.

Nelson, B. W., R. Mesquita, J. L. G. Pereira, S. G. A, Souza, G. T. Batista, and L. B. Couto. 1999. Allometric regressions for improved estimate of secondary forest biomass in the central Amazon. Forest Ecology and Management 117: 149–67.

Williamson, B. G., R. C. Mesquita, K. Ickes, and G. Ganade. 1998. Estratégias de pioneiras nos trópicos. In *Floresta Amazónica: Dinâmica, Regeneração e Manejo,* ed. C. Gascon and P. Moutinho. Instituto Nacional de Pesquisas da Amazônia, Manaus.

Population Genetics

THESES

Dick, C. 1999. The effect of fragmentation on the genetic structure of tropical forest trees. Ph.D. thesis, Harvard University.

Lepsch-Cunha, N. 1996. Genetic variation of a rare species in central Amazonia with regards to its conservation. M.Sc. thesis, ESALQ.

Lepsch-Cunha, N. In progress. Efeito da fragmentação na estrutura genética e no sucesso reprodutivo de *Oenocarpus bacaba*. Ph.D. thesis, INPA.

PAPERS

Hamilton, M. B. 1999. Four primer pairs for the amplification of chloroplast intergenic regions with intraspecific variation. Molecular Ecology 8: 513–25.

Hamilton, M. B. 1999. Tropical gene flow and seed dispersal. Nature 401: 129–30.

Lepsch-Cunha, N., P. Y. Kageyama, and B. R. Vemovsky. 1999. Genetic diversity of *Couratari multiflora* and *Couratari guianensis* (Lecythidaceae): Consequences of two types of rarity in central Amazonia. Biodiversity and Conservation 8: 1205–18.

Corridors

THESES

Laurance, S. In progress. The influence of linear corridors and edge effects on understory rainforest birds in central Amazônia. Ph.D. thesis, University of New England.

Lima, M. G. 1998. Uso de matas associadas à igarapés como áreas de dispersão de pequenos mamíferos terrestres e sapos no PDBFF. M.Sc. thesis, INPA.

PAPERS

Laurance, S. G., and W. F. Laurance. 1999. Tropical corridors: Use of linear rainforest remnants by arboreal mammals. Biol. Conserv. 91: 231–40.

Laurance, S. G. W., and W. F. Laurance. In press. Bandages for wounded landscapes: Faunal corridors and their roles in wildlife conservation in the Americas. In *Disruptions and Variability: The Dynamics of Climate, Human Disturbance, and Ecosystem in the Americas,* ed. G. A. Bradshaw, P. A. Marquet, and H. A. Mooney. Columbia University Press, New York.

Lima, M. G. d., and C. Gascon. 1999. The conservation value of linear forest remnants in central Amazonia. Biol. Conserv. 91: 241–47.

Aquatic Ecology

THESES

Bührnheim, C. M. 1998. Composição, riqueza e comportamento de peixes de igarapés de floresta de terra firme na Amazônia central. M.Sc. thesis, INPA.

Martins, C. S. In progress. Estrutura de comunidades de peixes em igarapés sob influência do desmatamento na reserva Porto Alegre. M.Sc. thesis, INPA.

Pascoaloto, D. 1999. Levantamento taxonômico, variação sazonal e padrões de distribuição de macroalgas lóticos na região de Manaus e arredores. Ph.D. thesis, INPA.

PAPERS

Bührnheim, C. M. 1998. Habitat abundance patterns of fish communities in three Amazonia rainforest streams. In *Biology of Tropical Fishes,* ed. A. L. Val and V. M. F. Almeida-Val, pp. 65–77. INPA, Manaus.

Geographical Information Systems and Remote Sensing

THESES

Garcia, J. P. M. In progress. Análise geomorfológica como instrumento de identificação de padrões de vegetação na Amazônia central. Ph.D. thesis, USP.

PAPERS

Adams, J. B., V. Kapos, M. O. Smith, R. Almeida Filho, A. R. Gillespie, and D. A. Roberts. 1990. A new Landsat view of land use in Amazonia. Proc. International Symp. on Primary Data Acquisition. International Archives of Photogrammetry and Remote Sensing 28: 177–85.

Deforestation and Forest Fragmentation

PAPERS

Bierregaard, R. O., Jr. 1989. Conservation of tropical rainforests: Facing a fragmented future. In Proc. Regional Meeting American Association of Zoological Parks and Aquariums, pp. 4–12.

Bierregaard, R. O., Jr., and V. H. Dale. 1996. Islands in an ever-changing sea: The ecological and socioeconomic dynamics of Amazonian rainforest fragments. In *Forest Patches in Tropical Landscapes,* ed. J. Schelhas and R. Greenberg. Island Press, Washington D.C., pp. 187–204.

Bierregaard, R. O. Jr., W. F. Laurance, J. W. Sites, A .J. Lynam, R. K. Didham, M. Andersen, C. Gascon, M. D. Tocher, A. P. Smith, V. M. Viana, T. E. Lovejoy, K. E. Sieving, E. A. Kramer, C. Restrepo, and C. Moritz.

1997. Key priorities for the study of fragmented ecosystems. In *Tropical Forest Remnants: Ecology, Management, and Conservation of Fragmented Communities,* ed. W. F. Laurance and R. O. Bierregaard, Jr., pp. 515–25. University of Chicago Press, Chicago.

Bierregaard, R. O., Jr., T. E. Lovejoy, V. Kapos, A. A. dos Santos, and R. W. Hutchings. 1992. The biological dynamics of tropical rainforest fragments. BioScience 42: 859–66.

Bruna, E. M. 1999. Seed germination in rainforest fragments. Nature 402: 139.

Carvalho, K. S., and H. L. Vasconcelos. 1999. Forest fragmentation in central Amazonia and its effects on litter-dwelling ants. Biol. Conserv. 91: 151–58.

Delamônica, P., W. F. Laurance, and S. G. Laurance. In press. A fragmentação de floresta e as estratégias para conservação. In *As Florestas do Rio Negro,* ed. D. Daly and A. A. Oliveira. Universidade Paulista, São Paulo.

Fearnside, P. M. In press. Can pasture intensification discourage deforestation in the Amazon and Pantanal regions of Brazil? In *Patterns and Processes of Land Use and Forest Change in the Amazon,* ed. C. H. Wood. University Presses of Florida, Gainesville.

Fearnside, P. M., P. M. L. A. Graça, and F. J. A. Rodriguez. In press. Burning efficiency and charcoal formation in forest cleared for cattle pasture near Manaus, Brazil. Forest Ecology and Management.

Fearnside, P. M., N. Leal Filho, and F. M. Fernandes. 1993. Rainforest burning and the global carbon budget: Biomass, combustion efficiency, and charcoal formation in the Brazilian Amazon. J. Geophysical Research 98: 16733–43.

Ferreira, L. V., and W. F. Laurance. 1997. Effects of forest fragmentation on mortality and damage of selected trees in central Amazonia. Conserv. Biol. 11: 797–801.

Fowler, H. G. 1991. A teoria de biogeografia de ilhas e a preservação: Um paradigma

que atrapalha? Revista de Geografia, São Paulo 10: 39–49.

Gascon, C., W. F. Laurance, and T. E. Lovejoy. In press. Fragmentação Florestal e Biodiversidade na Amazônia central. In *Conservação da Biodiversidade em Ecossistemas Tropicais: Avanços Conceituais e Revisão de Novas Metodologias de Avaliação e Monitoramento,* ed. B. F. de Souza and I. Garay. Editora Vozes, São Paulo.

Gascon, C., and T. E. Lovejoy. 1998. Ecological impacts of forest fragmentation in central Amazonia. Zoology, Analysis of Complex Systems 101: 273–80.

Gascon, C., T. E. Lovejoy, R. O. Bierregaard, Jr., J. R Malcolm, P. C. Stouffer, H. Vasconcelos, W. F. Laurance, B. Zimmerman, M. Tocher, and S. Borges. 1999. Matrix habitat and species persistence in tropical forest remnants. Biol. Conserv. 91: 223–30.

Gascon, C., R. Mesquita, and N. Higuchi. 1998. Tropical logging and the World Bank. Science (Science Letters) 281: 1453.

Harrison, S., and E. Bruna. 1999. Habitat fragmentation and large-scale conservation: What do we know for sure. Ecography 22: 225–32.

Laurance, W. F. 1996. Tropical forest remnants: Ecology, management, and conservation of fragmented communities (symposium report). Environmental Conservation 23: 89–90.

Laurance, W. F. 1997. Effects of logging on wildlife in the tropics. Conserv. Biol. 11: 311–12.

Laurance, W. F. 1997. Landscape alteration in the Americas. Trends Ecol. Evol. 12: 253–54.

Laurance, W. F. 1997. Logging and wildlife in tropical forests. Environmental Conservation 23: 365–75.

Laurance, W. F. 1998. A crisis in the making: Responses of Amazonian forests to land use and climate change. Trends Ecol. Evol. 13: 411–15.

Laurance, W. F. 1998. Dynamics and biomass of Amazonian forest fragments. ITTO Tropical Forest Update 8: 12–13.

Laurance, W. F. 1998. Forest fragmentation: Another perspective. Trends Ecol. Evol. 11: 75.

Laurance, W. F. 1998. A long-term study of Amazonian forest fragments. CTFS Summer 1998: 14.

Laurance, W. F. 1999. Introduction and synthesis. Biol. Conserv. 91: 101–8.

Laurance, W. F. 1999. Reflections on the tropical deforestation crisis. Biol. Conserv. 91: 109–18.

Laurance, W. F. 2000. Mega-development trends in the Amazon: Implications for global change. Environmental Monitoring and Assessment 61: 113–22.

Laurance, W. F. In press. Cut and run: The dramatic rise of transnational logging in the tropics. Trends in Ecology and Evolution.

Laurance, W. F. In press. Tropical logging and human invasion: An insoluble dilemma. Conserv. Biol.

Laurance, W. F., and R. O. Bierregaard, Jr. 1996. Fragmented tropical forests (report on symposium). Bull. Ecol. Soc. America 77: 34–36.

Laurance, W. F., R. O. Bierregaard, Jr., C. Gascon, R. K. Didham, A. P. Smith, A. J. Lynam, V. M. Viana, T. E. Lovejoy, K. E. Sieveing, J. W. Sites, M. Andersen, M. D. Tocher, E. A. Kramer, C. Restrepo, and C. Moritz. 1997. Tropical forest fragmentation: Synthesis of a diverse and dynamic discipline. In *Tropical Forest Remnants: Ecology, Management, and Conservation of Fragmented Communities,* ed. W. F. Laurance and R. O. Bierregaard, Jr., pp. 502–14. University of Chicago Press, Chicago.

Laurance, W. F., M. Cochrane, S. Bergen, P. M. Fearnside, P. Delamônica, S. Agra, and C. Barber. In press. The future of the Amazon. In E. Bermingham, C. Dick, and C. Moritz (eds.), *Tropical Rainforests: Past, Present, and Future.* University of Chicago Press, Chicago.

Laurance, W. F., and P. Delamônica. 1998. Ilhas de sobrevivência. Ciência Hoje 24: 26–31.

Laurance, W. F., P. Delamônica, S. G. Laurance, H. Vasconcelos, and T. E. Lovejoy. 2000. Rainforest fragmentation kills big trees. Nature 404: 836.

Laurance, W. F., and P. M. Fearnside. 1999. Amazon burning. Trends Ecol. Evol. 14: 457.

Laurance, W. F., and C. Gascon. 1997. How to creatively fragment a landscape. Conserv. Biol. 11: 577–79.

Laurance, W. F., S. G. Laurance, and P. Delamônica. 1998. Tropical forest fragmentation and greenhouse gas emissions. Forest Ecology and Management 110: 173–80.

Laurance, W. F., S. G. Laurance, L. V. Ferreira, J. M. Rankin–de Mérona, C. Gascon, and T. E. Lovejoy. 1997. Biomass collapse in Amazonian forest fragments. Science 278: 1117–18.

Laurance, W. F., D. Perez-Salicrup, P. Delamônica, P. M. Fearnside, S. Agra, A. Jerozolinski, L. Pohl, and T. E. Lovejoy. In press. Rain forest fragmentation and the structure of Amazonian liana communities. Ecology.

Laurance, W. F., and H. L. Vasconcelos. 2000. Amazônia: Consequências das mudanças climáticas e de uso da terra para a floresta. Ciência Hoje 160: 59–62.

Laurance, W. F., H. L. Vasconcelos, and T. E. Lovejoy. 2000. Forest loss and fragmentation in the Amazon: Implications for wildlife conservation. Oryx 34: 39–45.

Lovejoy, T. E. 1981. Discontinuous wilderness: Minimum area for conservation. Parks 5: 13–15.

Lovejoy, T. E. 1983. Biosphere reserves: The size question. In 1st International Biosphere Reserve Congress. UNESCO, Man and the Biosphere Program, Minsk, USSR.

Lovejoy, T. E. 1984. Application of ecological theory to conservation planning. In *Ecology in Practics,* ed. F. di Castri, F. W. G. Bakeri, and M. Hadley, pp. 402–13. UNESCO, Paris.

Lovejoy, T. E. 1985. Forest fragmentation in the Amazon: A case study. In *The Study of*

Populations, ed. H. Messel, pp. 243–51. Pergamon, New York.

Lovejoy, T. E. 1985. Minimum size for bird species and avian habitats. In Acta XVIII Cong. Inter. Orni, ed. V. D. Ilyichev and V. M. Gavrilov, pp. 324–27. Nauka, Moscow.

Lovejoy, T. E. 1986. Conservation planning in a checkerboard world: The problem of size of natural areas. In *Land and Its Uses, Actual and Potential: An Environmental Appraisal,* ed. F. T. Last, M. C. B. Hotz, and B. G. Bell. Plenum, New York.

Lovejoy, T. E. In press. Amazon forest degradation and fragmentation: Implications for biodiversity. In *Amazonia 2000,* ed. A. L. Hall. Institute of Latin American Studies, University of London, and the Brookings Institution, Washington, D.C.

Lovejoy, T. E., and R. O. Bierregaard, Jr. 1990. Central Amazonian forests and the Minimum Critical Size of Ecosystems Project. In *Four Neotropical Rainforests,* ed. A. Gentry, pp. 60–74. Yale University Press, New Haven.

Lovejoy, T. E., R. O. Bierregaard, Jr., J. M. Rankin, and H. O. R. Schubart. 1983. Ecological dynamics of forest fragments. In *Tropical Rain Forest: Ecology and Management,* ed. S. L. Sutton, T. C. Whitmore, and A. C. Chaddwick, pp. 377–84. Blackwell Scientific Publications, Oxford.

Lovejoy, T. E., and D. C. Oren. 1981. Minimum critical size of ecosystems. In *Forest Island Dynamics in Man-Dominated Landscapes,* ed. R. L. Burgess and D. M. Sharp, pp. 7–12. Springer-Verlag, New York.

Lovejoy, T. E., and J. M. Rankin. 1981. Uma fisionomia ameaçada: As implicações da dinâmica de parcelas florestais no planejamento silvicultural e de reservas. Bol. Fundação Brasileira para a Conservação da Natureza 16: 136–39.

Lovejoy, T. E., J. M. Rankin, R. O. Bierregaard, Jr., K. S. Brown, Jr., L. H. Emmons, and M. van der Voort. 1984. Ecosystem decay of Amazon forest remnants. In *Extinctions,* ed. M. H. Nitecki, pp. 295–325. University of Chicago Press, Chicago.

Morato, E. F., and L. A. O. Campos. 2000. Efeitos da fragmentação florestal sobre vespas e abelhas solitárias em uma área da amazônia central. Rev. bras. Zool. 17: 429–44.

Offerman, H., V. H. Dale, S. M. Pearson, R. O. Bierregaard, Jr., and R. V. O'Neill. 1995. Effects of forest fragmentation on Neotropical fauna: Current research and data availability. Environmental Reviews 3: 191–211.

Sizer, N. C., E. V. J. Tanner, and I. D. K Ferraz. In press. Edge effects on litterfall mass and nutrient concentrations in forest fragments in central Amazonia. J. Trop. Ecology.

Stratford, J. A., and P. C. Stouffer. In press. Reduced feather growth rates of two common birds inhabiting central Amazonian forest fragments. Conservation Biology.

Vasconcelos, H. L., and J. H. C. Delabie. 2000. Ground ant communities from central Amazonia forest fragments. In D. Agosti, J. D. Majer, L. Alonso, and T. Schultz (eds). Sampling Ground-Dwelling Ants: Case Studies from the World's Rain Forests. Curtin School of Environmental Biology. Bulletin no. 18. Perth, Australia, 59–69.

Vasconcelos, H. L., J. M. S. Vilhena, and G. J. A Caliri. 2000. Responses of ants to selective logging in a central Amazonian forest. J. App. Ecology 37: 508–14.

Zimmerman, B. L., and R. O. Bierregaard, Jr. 1986. Relevance of the equilibrium theory of island biogeography with an example from Amazonia. J. Biogeography 13: 133–43.

Taxonomy and Distribution

PLANTS

Mori, S. 1992. *Eschweilera pseudodecolorans* (Lecythidaceae), a new species from Amazonian Brazil. Brittonia 44: 244–46.

Prance, G. T. 1991. Five new species of Neotropical Chrysobalanaceae. Kew Bulletin 47: 247–56.

Rodrigues, W. A. 1982. Duas novas espécies da flora Amazónica. Acta Amazónica 12: 295–300.

INSECTS

Brandão, C. F. R., J. L. M. Diniz, D. Agosti, and J. Delabie. 1999. Revision of the Neotropical ant subfamily Leptanilloidinae. Systematic Entomology 24: 17–36.

Fowler, H. G. In press. A remarkable record of *Cylindromyrmex brasiliensis* Emery, 1901 (Hymenoptera: Formicidae: Ponerinae: Cylindromyrmecini) in central Amazonia, with notes of behavior. Bol. Museu Paraense Emílio Goeldi.

Hrabovsky, M. 1987. On *Anelaphus* Linsley 1936—One new species and two new combinations (Coleoptera, Cerambycidae, Elaphidionini). Rev. bras. Entomol. 31: 135–37.

Rafael, J. A., and R. Ale-Rocha. 1990. Primeiro registro do gênero *Ocydromia* Meigen na região neotropical e descrição de *O. amazonica* sp. n. (Diptera, Empididae, Ocydromiinae). Rev. bras. Entomol. 34: 739–41.

REPTILES AND AMPHIBIANS

Gascon, C., and O. de Souza Pereira. 1993. Preliminary checklist of the herpetofauna of the upper Rio Urucú, Amazonas, Brazil. Rev. bras. Zool. 10: 179–83.

Hero, J. M. 1990. An illustrated key to tadpoles occurring in the central Amazon rainforest, Manaus, Amazonas, Brazil. Amazoniana 11: 201–62.

Heyer, W. R., and L. M. Hardy 1991. A new species of frog of the *Eleutherodactylus lacrimosus* assembly from Amazonia, South America (Amphibia: Anura: Leptodactylidae). Proc. Biol. Soc. Washington 104: 436–47.

Rankin–de Mérona, J. M., and D. D. Ackerly. 1987. Estudos populacionais de árvores em florestas fragmentadas e as implicações para conservação in situ das mesmas na floresta tropical da Amazônia central. Instituto de Pesquisas e Estudos Florestais da Escola Superior de Agricultura "Luiz de Queiroz" 35: 47–59.

Rodrigues, M. T. 1988. A new anole of the *Punctatus* group from central Amazonia (Sauria, Iguanidae). Papeis Avulsos de Zoologia, São Paulo 36: 333–36.

Vanzolini, P. E. 1985. *Micrurus averyi* Schmidt, 1939, in central Amazônia (Serpentes, Elapidae). Papeis Avulsos de Zoologia 36: 77–85.

Zimmerman, B. L., and W. Hodl. 1983. Distinctions of *Phyrnohyas resinfictrix* (Goeldi, 1907) from *Phyrnohyas venulosa* (Laurenti, 1768) based on acoustical and behavioural parameters. Zoology Anzeiger 211: 341–52.

MAMMALS

Gribel, R., and V. A. Tadei. 1989. Notes on the distribution of *Tonatia schulzi* and *Tonatia carrikeri* in the Brazilian Amazon. J. Mammalogy 70.

Rodriguez, B. S., E. M. Sampaio, and C. O. Handley, Jr. 1999. *Thyroptera discifera* (Lichtenstein and Peters) 1855 in the central Amazon. Bat Research News 40: 73.

BIRDS

Bierregaard, R. O., Jr., M. Cohn-Haft, and D. F. Stotz. 1997. Cryptic biodiversity: An overlooked species and new subspecies of antbird (Aves; Formicariidae) with a revision of *Cercomacra tyrannina* in northeastern South America. Ornithological Monographs 48: 111–28.

Cohn-Haft, M. 1993. Rediscovery of the White-winged Potoo. Auk 110.

Cohn-Haft, M. A. W., and P. C. Stouffer. 1997. A new look at "species-poor" central Amazon: The avifauna north of Manaus, Brazil. Ornithological Monographs 48: 205–35.

Stotz, D. F., and R. O. Bierregaard, Jr. 1989. The birds of the fazendas Porto Alegre, Dimona, and Esteio north of Manaus, Amazonas, Brazil. Rev. bras. Biol. 49: 861–72.

Stotz, D. F., R. O. Bierregaard, Jr., M. Cohn-Haft, P. Peterman, J. Smith, A. Whittaker, and S. V. Wilson. 1992. The status of North American migrants in the central Amazonian Brazil. Condor 94: 608–21.

Bibliography

Aberg, J., G. Jansson, J. E. Swenson, and P. Angelstam. 1995. The effect of matrix on the occurrence of hazel grouse (*Bonasa bonasia*) in isolated habitat fragments. Oecologia 103: 265–69.

Abrantes, C. V. M. 1990. Amostragem de Euglossini (Hymenoptera, Apidae) em Viçosa, MG, com o uso de armadilhas. Unpublished Monograph, Departamento de Biologia Geral, Universidade Federal de Viçosa, Minas Gerais, Brazil.

Absy, M. L. 1982. Quaternary palynological studies in the Amazon Basin. In G. T. Prance (ed.), *Biological Diversification in the Tropics.* Columbia University Press, New York, pp. 63–73.

Ackerly, D. D., J. M. Rankin–de Mérona, and W. A. Rodrigues. 1990. Tree densities and sex ratios in breeding populations of dioecious central Amazonian Myristicaceae. J. Trop. Ecol. 6: 239–48.

Ackerman, J. D. 1985. Euglossine bees and their nectar hosts. In W. G. D'Arcy and M. D. Correa (eds.), *The Botany and Natural History of Panama.* Missouri Botanical Garden, St. Louis, Mo., pp. 225–33.

Ackerman, J. D. 1989. Geographic and seasonal variation in fragrance choices and preferences of male euglossine bees. Biotropica 21: 340–47.

Adam, F. 1977. Données préliminaires sur l'habitat et la stratification des ronguers en forêt de Basse Côte d'Ivoire. Mammalia 41: 283–90.

Adams, J. B., V. Kapos, M. O. Smith, R. A. Filho, A. R. Gillespie, and D. A. Roberts. 1990. A new landsat view of land use in Amazonia. International Archive of Photogrammetry and Remote Sensing 28: 177–85.

Adams, J. B., D. Sabol, V. Kapos, D. A. Roberts, R. A. Filho, M. O. Smith, and A. R. Gillespie. 1995. Classification of multispectral images based on fractions of endmembers: Application to land-cover change in the Brazilian Amazon. Remote Sensing of Environment 52: 137–54.

Agosti, D., J. Majer, T. Schultz, and L. Tennant (eds.). In press. *Measuring and Monitoring Biodiversity: Standard Methods for Ground-dwelling Ants,* Smithsonian Institution Press, Washington, D.C.

Aguillar, J. B. V. 1990. Contribuição ao conhecimento dos Euglossini (Hymenoptera, Apidae) do estado da Bahia, Brasil. M. Sc. Thesis, Departamento de Biociências, Universidade de São Paulo, São Paulo.

Aide, T. M., and J. Cavelier. 1994. Barriers to lowland tropical forest restoration in the Sierra Nevada de Santa Marta, Colombia. Restoration Ecol. 2: 219–29.

Aide, T. M., J. K. Zimmerman, L. Herrera, M. Rosário, and M. Serrano. 1995. Forest recovery in abandoned pastures in Puerto Rico. For. Ecol. & Manag. 77: 77–86.

Aizen, M. A., and P. Feinsinger. 1994a. Forest fragmentation, pollination, and plant reproduction in a Chaco dry forest, Argentina. Ecology 75: 330–51.

Aizen, M. A., and P. Feinsinger. 1994b. Habitat fragmentation, native insect pollinators, and feral honey bees in Argentine "Chaco Serrano." Ecological Applications 4: 378–92.

Albini, F. A. 1976. Estimating wildfire behavior and effects. USDA Forest Service, Gen. Tech. Rep. INT–30. Intermountain Forest Research Station, Odgen, Utah.

Albuquerque, J. M. 1980. Identificação de plantas invasores de cultura da região de Manaus. Acta Amazónica 10: 47–95.

Alcorn, J. B. 1990. Indigenous agroforestry strategies meeting farmers' needs. In A. B. Anderson (ed.), *Alternatives to Deforestation: Steps Toward Sustainable Use of the Amazon Rainforest*. Columbia University Press, New York, pp. 139–48.

Aldrich, P. R., and J. L. Hamrick. 1998. Reproductive dominance of pasture trees in a fragmented tropical forest mosaic. Science 281:103–5.

Alencar, J. C. 1998. Fenologia de espécies arbóreas tropicais na Amazônia central. In C. Gascon and P. Moutinho (eds.), *Floresta Amazônica: Dinâmica, Regeneração e Manejo*. Instituto Nacional de Pesquisas da Amazônia (INPA), Manaus, Amazonas, Brazil, pp. 25–40.

Alencar, J. C., R. A. Almeida, and N. P. Fernandes. 1979. Fenologia de espécies florestais em floresta tropical úmida de terra firme na Amazônia Central. Acta Amazónica 9: 163–98.

Alfenas, A. C., I. Peters, W. Brune, and G. C. Passador. 1991. *Eletroforese de Proteínas e Isoenzimas de Fungos e Essências Florestais*. Imprensa Universitária, Universidade Federal de Viçosa. Viçosa, MG, Brazil.

Altieri, A. M. 1995. *Agroecology: The Science of Sustainable Agriculture*. Westview, Boulder, Colo.

Altmann, J. 1974. Observational study of behaviour: Sampling methods. Behaviour 49: 227–67.

Alvarez-Buylla, E. R., and A. A. Garay. 1994. Population genetic structure of *Cecropia obtusifolia*, a tropical pioneer tree species. Evolution 48: 437–53.

Alvarez-Buylla, E. R., R. García-Barrios, C. Lara-Boreno, and M. Martínea-Bamos. 1996. Demographic and genetic models in conservation biology: Applications and perspectives for tropical rain forest tree species. Ann. Rev. Ecol. Syst. 27: 387–91.

Alvim, P. de T. 1977. Cacao. In P. de T. Alvim and T. T. Kozlowski (eds.), *Ecophysiology of Tropical Crops*. Academic Press, New York, pp. 279–313.

Anderson, A. B. 1990. *Alternatives to Deforestation: Steps Toward Sustainable Use of the Amazon Rainforest*. Columbia University Press, New York.

Anderson, A. B. 1991. Forest management strategies by rural inhabitants in the Amazon estuary. In A. Gomez-Pompa, T. C. Whitmore, and M. Hadley (eds.), *Rain Forest Regeneration and Management. Man and the Biosphere Series*. Parthenon, Paris, pp. 351–60.

Andrén, H. 1994. Effects of habitat fragmentation on birds and mammals in landscapes with different proportions of suitable habitat: A review. Oikos 71: 355–66.

Andrén, H., and P. Angelstam. 1988. Elevated predation rates as an edge effect in habitat islands: Experimental evidence. Ecology 69: 544–7.

Angermeier, P. L. 1995. Ecological attributes of extinction-prone species: Loss of freshwater fishes of Virginia. Conserv. Biol. 9: 143–58.

Anonymous. 1980. Projeto RADAMBRASIL. *Levantamento de Recursos Naturais*. Ministério das Minas e Energia, Departamento de Produção Mineral (DNPM), Rio de Janeiro. Vol 18: p. 261.

Armbruster, W. S. 1993. Within-habitat heterogeneity in baiting samples of male Euglossine bees: Possible causes and implications. Biotropica 25: 122–28.

Armstrong, J. E., and A. K. Irvine. 1989. Floral biology of *Myristica insipida* (Myristicaceae), a distinctive beetle pollination system. Am. J. Bot. 76: 86–94.

Arnold, G. W., J. R. Weeldenburg, and V. M. Ng. 1995. Factors affecting the distribution of Western grey kangaroos (*Macropus fuliginosus*) and euros (*M. robustus*) in a fragmented landscape. Landscape Ecol. 10: 65–74.

Arrenhius, O. 1921. Species and area. J. Ecology 9: 95–99.

Ash, J. 1988. Demography and production of *Balaka microcarpa* Burret (Arecacea), a tropical understory palm in Fiji. Aust. J. Bot. 36: 67–80.

Ashton, P. S. 1969. Speciation among tropical forest trees: Some deductions in light of recent evidence. Biological J. Linnean Soc. 1: 155–96.

Ashton, P. S. 1981. The need for information regarding tree age and growth in tropical forests. In F. Bormann and G. Berlyn (eds.), *Age and Growth Rate of Tropical Trees: New Directions for Research*. Yale University Press, New Haven, pp. 3–6.

Ashton, P. S. 1984. Biosystematics of tropical forest plants: A problem of rare species. In Grant, W. F. (ed.), *Plant Biosystematics*. Academic Press, London, pp. 497–518.

Ashton, P. S., and P. Hall. 1992. Comparisons of structure among mixed dipterocarp forests of northwestern Borneo. J. Ecology 80: 459–81.

Askins, R. A., M. J. Philbrick, and D. S. Sugeno. 1987. Relationship between the regional abundance of forest and the composition of forest bird communities. Biological Conservation 39: 129–52.

Augspurger, C. K. 1984. Seedling survival of tropical tree species: Interactions of dispersal distance, light gaps, and pathogens. Ecology 65: 1705–12.

Aviles, L. 1986. Sex ratio bias and possible group selection in the social spider *Anelosimus eximius*. Am. Nat. 128: 1–12.

Ayres, J. M. 1981. Observações sobre a ecologia e o comportamento dos cuxiús (*Chiropotes albinasus* e *Chiropotes satanas,* Cebidae: Primates). M. Sc. diss., Instituto Nacional de Pesquisas da Amazônia e Fundação Universidade do Amazonas, Manaus, Brasil.

Ayres, J. M. 1989. Comparative feeding ecology of the uakari and bearded saki, *Caca-jao* and *Chiropotes*. J. History of Evolution 18: 697–716.

Baker, H. G. 1959. Reproductive methods as factors in speciation in flowering plants. Cold Spring Harbor Symposium on Quantitative Biology 24: 177–91.

Balslev, H., J. Luteyn, B. Øllgard, and L. B. Holm-Nielsen. 1987. Composition and structure of adjacent unflooded and floodplain forest in Amazonian Ecuador. Opera Botanica 92: 37–57.

Barbosa, R. I. 1990. Analysis of the timber sector in the State of Roraima. Acta Amazónica 20: 193–209.

Barbosa, R. I., and P. M. Fearnside. 2000. Incêndios em Roraima: Implicações Ecológicas e Lições ao Desenvolvimento na Amazônia. *Ciência Hoje.*

Barinaga, M. 1997. Making plants aluminum tolerant. Science 276: 1497.

Barrett, S. C. H., and J. R. Kohn. 1991. Genetic and evolutionary consequences of small population size in plants: Implication for conservation. In D. A. Falk and K. E. Holsinger (eds.), *Genetics and Conservation of Rare Plants*. Center for Plant Conservation, Oxford University Press, pp. 3–30.

Bascompte, J., and R. V. Solé. 1996. Habitat fragmentation and extinction thresholds in spatially explicit models. J. Anim. Ecol. 65: 465–73.

Bascompte, J., and R. V. Solé. 1998. Models of habitat fragmentation. In J. Bascompte and R. V. Solé (eds.), *Modeling Spatiotemporal Dynamics in Ecology*. Springer-Verlag, Berlin, Germany, pp. 127–49.

Bassini, F., and P. Becker. 1990. Charcoal's occurrence in soil depends on topography in terra firme forest near Manaus, Brazil. Biotropica 22: 420–22.

Bawa, K. S. 1974. Breeding system of tree species of a lowland tropical community. Evolution 28: 85–92.

Bawa, K. S. 1990. Plant-pollinator interactions in tropical rain forests. Ann. Rev. Ecol. Syst. 21: 399–422.

Bawa, K. S., and P. S. Asthon. 1991. Conservation of rare trees in tropical rain forests:

a genetic perspective. In D. A. Falk and K. E. Holsinger (eds.), *Genetics and Conservation of Rare Plants*. Center for plant conservation. Oxford University Press, pp. 62–74.

Bawa, K. S., S. H. Bullock, D. R. Perry, R. E. Colville and M. H. Grayum. 1985. Reproductive biology of tropical lowland rain forest trees. II. Pollination systems. Amer. J. Bot. 72: 346–56.

Bawa, K. S., and P. A. Opler. 1975. Dioecism in tropical forest trees. Evolution 29: 167–79.

Bawa, K. S., D. R. Perry, and J. H. Beach. 1985. Reproductive biology of tropical lowland rain forest trees. Sexual systems and incompatibility mechanisms. Amer. J. Bot. 72: 331–45.

Bazzaz, F. A. 1984. Dynamics of wet tropical forests and their species strategies. In E. Medina, H. A. Mooney, and C. Vázquez-Yanes (eds.), *Physiological Ecology of Plants in the Wet Tropics*. Junk Publishers, The Hague, Netherlands, pp. 233–43.

Bazzaz, F. A. 1991. Regeneration of tropical forests: Physiological responses of pioneer and secondary species. In A. Gomez-Pompa, T. C. Whitmore, and M. Hadley (eds.), *Rain Forest Regeneration and Management*. UNESCO, Paris, pp. 91–118.

Becker, P., J. S. Moure, and F. J. A. Peralta. 1991. More about Euglossine bees in Amazonian forest fragments. Biotropica 23: 586–91.

Beinroth, F. H. 1975. Relationships between U.S. soil taxonomy, the Brazilian system, and FAO/UNESCO soil units. In E. Bornemisza and A. Alvarado (eds.), *Soil Management in Tropical America: Proceedings of a Seminar held at CIAT, Cali, Colombia, February 10–14, 1974*. North Carolina State University, Soil Science Department, Raleigh, pp. 97–108.

Beisiegel, W. de R., and W. O. de Souza. 1986. Reservas de fosfatos—Panorama nacional e mundial. In *Instituto Brasileiro de Fosfato (IBRAFOS) III Encontro Nacional de Rocha Fosfática, Brasília, 16–18/06/86*. IBRAFOS, Brasília, Brazil, pp. 55–67.

Belbin, L. 1995. *PATN user's guide*. CSIRO Division of Wildlife and Ecology, Canberra, Australia.

Belshaw, R., and B. Bolton. 1994. A survey of the leaf litter ant fauna in Ghana, West Africa (Hymenoptera: Formicidae). J. Hymenoptera Research 3: 5–16.

Bender, D. J., T. A. Contreras, and L. Fahrig. 1998. Habitat loss and population decline: A meta-analysis of the patch size effect. Ecology 79: 517–33.

Benites, J. R. 1994. Bases de datos de recursos de suelos. In *Tratado de Cooperación Amazonica (TCA). Zonificación Ecológica-Económica: Instrumento para la Conservación y el Desarrollo Sostenible de los Recursos de la Amazonia*. TCA Secretaria Pro Tempore, Lima, Peru, pp. 207–31.

Benitez-Malvido, J. 1995. The ecology of seedlings in central Amazonian forest fragments. Ph.D. diss., University of Cambridge, U.K.

Benitez-Malvido, J. 1998. Impact of forest fragmentation on seedling abundance in a tropical rain forest. Conserv. Biol. 12: 380–89.

Benitez-Malvido, J., G. Garcia-Guzmán, and I. D. Kossmann-Ferraz. 1999. Leaf-fungal incidence and herbivory on tree seedlings in tropical rain forest fragments: An experimental study. Biological Conservation 91: 143–50.

Benitez-Malvido, J., and I. D. Kossmann-Ferraz. 1999. Litter cover variability affects seedling performance and herbivory. Biotropica 31: 598–606.

Bennema, J. 1977. Soils. In P. de T. Alvim and T. T. Kozlowski (eds.), *Ecophysiology of Tropical Crops*. Academic Press, New York, pp. 29–55.

Bennett, A. F. 1987. Conservation of mammals within a fragmented forest environment: The contributions of insular biogeography and autecology. In D. A. Saunders, G. W. Arnold, A. A. Burbidge, and A. J. M. Hopkins (eds.), *Nature Conservation: The Role of Remnants of Native Vegetation*. Surrey Beatty and Sons, Australia, pp. 41–52.

Bennett, A. F. 1990. Habitat corridors and

the conservation of small mammals in a fragmented forest environment. Landscape Ecology 4: 109–22.

Bennett, F. D. 1972. Baited MacPhail fruitfly traps to collect euglossine bees. New York Entomological Society 80: 137–45.

Berg, C. C. 1978. Espécies de *Cecropia* da Amazônia brasileira. Acta Amazónica 8: 149–82.

Bernstein, I. S., P. Balcaen, L. Dresdale, H. Gouzoules, M. Kavanagh, T. Patterson, and P. Neyman-Warner. 1976. Differential effects of forest degradation on primate populations. Primates 17: 401–11.

Berryman, A., and P. Turchin. 1997. Detection of delayed density dependence: Comment. Ecology 78: 318–20.

Besuchet, C., D. H. Burckhardt, and I. Löbl. 1987. The "Winkler/Moczarski" eclector as an efficient extractor for fungus and litter Coleoptera. Coleopterist's Bulletin 41: 392–94.

Bezerra, C. P. 1995. Aspectos ecológicos dos Euglossini (Hymenoptera, Apidae) em áreas de Mata Atlântica da cidade de João Pessoa-PB. M.Sc. Dissertation. Curso de Pós-Graduação em Ciências Biológicas, Univ. Fed. Piracicaba, João Pessoa, Brazil.

Bhat, U. N. (ed.). 1984. *Elements of Applied Stochastic Process*. Wiley, New York.

Bierregaard, R. O., Jr. 1990. Avian communities in the understory of Amazonian forest fragments. In A. Keast (ed.), *Biogeography and Ecology of Forest Bird Communities*. Academic Publishing, The Hague, Netherlands, pp. 333–43.

Bierregaard, R. O., Jr., and V. H. Dale. 1996. Islands in an ever-changing sea: The ecological and socioeconomic dynamics of Amazonian rainforest fragments. In J. Schelhas and R. Greenberg (eds.), *Forest Patches in Tropical Landscapes*. Island, Washington D.C., pp. 187–204.

Bierregaard R. O., Jr., and T. E. Lovejoy. 1988. Birds in Amazonian forest fragments: Effects of insularization. In H. Ouellet (ed.), *Acta XIX Congressus Internationalis Ornitholigici, vol. II*. University of Ottawa Press, Ottawa, Ontario, Canada, pp. 1564–79.

Bierregaard, R. O., Jr., and T. E Lovejoy. 1989. Effects of forest fragmentation on Amazonian understory bird communities. Acta Amazónica 19: 215–41.

Bierregaard R. O., Jr., T. E. Lovejoy, V. Kapos, A. A. dos Santos and R. W. Hutchings. 1992. The biological dynamics of tropical rainforest fragments. Bioscience 42: 859–66.

Bierregaard, R. O., Jr., and P. C. Stouffer. 1997. Understory birds and dynamic habitat mosaics in Amazonian rainforests. In W. F. Laurance and R. O. Bierregaard, Jr. (eds.), *Tropical Forest Remnants: Ecology, Management, and Conservation of Fragmented Communities*. University of Chicago Press, Chicago, pp. 138–55.

Bierzychudek, P. 1981. Pollinator limitation of plant reproductive effort. Am. Nat. 117: 838–40.

Biological Dynamics of Forest Fragments Project. 1990. Eleventh annual report. November 1, 1989–October 31, 1990. The World Wildlife Fund, The Smithsonian Institution, and the Instituto Nacional de Pesquisas da Amazonia.

BIONTE. 1998. Biomass and nutrients in the environment. Final report of the ODA-INPA Collaborative project. Instituto Nacional de Pesquisas da Amazônia, Manaus, Brazil.

Biot, Y., V. Brilhante, J. Veloso, J. Ferraz, N. Leal Filho, N. Higuchi, S. Ferreira and T. Desjardins. 1997. INFORM–O model florestal do INPA. In N. Higuchi, J. B. S. Ferraz, L. Antony, F. Luizão, R. Luizão, Y. Biot, I. Hunter, J. Proctor and S. Ross (eds.), *Bionte: Biomassa e Nutrientes Florestais. Relatório Final*. Instituto Nacional de Pesquisas da Amazônia (INPA), Manaus, Amazonas, Brazil, pp. 271–318.

Black, G. A., T. Dobzhansky, and C. Pavan. 1950. Some attempts to estimate species diversity and population density of trees in Amazonian forests. Botanical Gazette 111: 413–25.

Blake, J. G. 1991. Nested subsets and the

distribution of birds on isolated woodlots. Conserv. Biol. 5: 58–66.

Boecklen, W. J., and D. Simberloff. 1986. Area-based extinction models in conservation. In D. K. Elliot (ed.), *Dynamics of Extinction*. Wiley, New York, pp. 165–80.

Bond, W. J. 1994. Do mutualisms matter? Assessing the impact of pollinator and disperser disruption on plant extinction. Philosophical Transactions of the Royal Society, London B 344: 83–90.

Boom, B. M., and M. T. V. do Amaral Campos. 1991. A preliminary account of the Rubiaceae of a central Amazonian terra firme forest. Bol. Museu Paraense Emílio Goeldi, Nova Série, Botânica 7: 223–47.

Borges, S. H. 1995. Comunidade de aves em dois tipos de vegetação secundária da Amazônia central. M. Sc. thesis, Instituto Nacional de Pesquisas da Amazônia, Manaus, Brazil.

Borges, S. H., and P. C. Stouffer. 1999. Bird communities in two types of anthropogenic successional vegetation in central Amazonia. Condor 101: 529–36.

Boshier, D. H., M. R. Chase, and K. S. Bawa. 1995. Population genetics of *Cordia alliodora* (Boraginaceae), a Neotropical tree. 2. Mating system. Amer. J. Bot. 82: 476–83.

Boyer, J. 1972. Soil potassium. In M. Drosdoff (ed.), *Soils of the Humid Tropics*. National Academy of Sciences, Washington, D.C., pp. 102–35.

Brach, V. 1975. The biology of social spiders *Anelosimus eximius*. Bulletin of Society California of Science 74: 37–41.

Braga, P. I. S. 1976. Atração de abelhas polinizadoras de Orchidaceae com auxílio de iscas-odores na campina, campinarana e floresta tropical úmida da região de Manaus. Ciência e Cultura 28: 767–73.

Braga, P. I. S. 1979. Subdivisão fitogeográfica, tipos de vegetação, conservação e inventário florístico da floresta Amazónica. Acta Amazónica 9: 53–80.

Braga, P. I. S. 1987. Orquídeas, Biologia floral. Ciência Hoje 5: 53–55.

Bravard, S., and D. Righi. 1989. Geochemi-

cal differences in an Oxisol-Spodosol toposequence of Amazonia, Brazil. Geoderma 44: 29–42.

Brazil, SNLCS-EMBRAPA. 1979. *Manual de Métodos de Análise de Solo*. Serviço Nacional de Levantamento e Conservação de Solos-Empresa Brasileira de Pesquisa Agropecuária (SNLCS-EMBRAPA), Rio de Janeiro. Irregular pagination.

Brinkmann, W. L. F., and A. N. Vieira. 1971. The effect of burning on germination of seeds at different soil depths of various tropical tree species. Turrialba 21: 77–82.

Brockwell, P. J., and R. A. Davis. 1991. Time Series: Theory and Methods. Springer-Verlag, New York.

Brothers, T. S., and A. Spingarn. 1992. Forest fragmentation and alien plant invasion of central Indiana old-growth forests. Conserv. Biol. 6: 91–100.

Brown, A. H. D., and B. S. Weir. 1983. Measuring genetic variability in plant populations. In S. D. Tanksley and T. J. Orton (eds.), *Isozymes in Plants Genetics and Breeding*. Part A. Elsevier, Amsterdam, pp. 219–39.

Brown, J. H. 1984. On the relationship between abundance and distribution of species. Am. Nat. 124: 255–79.

Brown, J. H., and A. Kodric-Brown. 1977. Turnover rates in insular biogeography: Effect of immigration on extinction. Ecology 58: 445–49.

Brown, J. H., D. W. Mehlman, and G. C. Stevens. 1995. Spatial variation in abundance. Ecology 76: 2028–43.

Brown, K. S., 1987. Soils and vegetation. In T. C. Whitmore and G. T. Prance (eds.), *Biogeography and Quaternary History in Tropical America*. Oxford Monographs in Biogeography 3, Oxford, U.K, pp. 19–45.

Brown, K. S., Jr., and R. W. Hutchings. 1997. Disturbance, fragmentation, and the dynamics of diversity in Amazonian forest butterflies. In W. F. Laurance and R. O. Bierregaard Jr. (eds.), *Tropical Forest Remnants: Ecology, Management, and Conservation of Fragmented Communities*.

University of Chicago Press, Chicago, pp. 91–110.

Bruna, E. M. 1999. Seed germination in rainforest fragments. Nature 402: 139.

Buchmann, S. L. 1983. Buzz pollination in angiosperms. In C. E. Jones and R. J. Little (eds.), *Handbook of Experimental Pollination Biology*. Van Nostrand Rinehold, New York, pp. 73–113.

Buchmann, S. L., and G. P. Nabhan. 1996. *The Forgotten Pollinators*. Island, Washington, D.C.

Buckley, D. P., D. M. O'Malley, V. Apsit, G. T. Prance, and K. S. Bawa. 1988. Genetics of Brazil nut (*Bertholletia excelsa* Humb. and Bonpl.: Lecythidaceae). 1. Genetic variation in natural populations. Theoretical and Applied Genetics 76: 923–28.

Bull, G. A. D., and E. R. C. Reynolds. 1968. Wind turbulence generated by vegetation and its implications. Forestry (suppl.) 41: 28–37.

Burkey, T. V. 1989. Extinction in nature reserves: the effect of fragmentation and the importance of migration between reserve fragments. Oikos 55: 75–81.

Buschbacher, R. 1986. Tropical deforestation and pasture development. BioScience 36: 22–28.

Butz-Huryn, V. M. 1997. Ecological impacts of introduced honey bees. Quart. Rev. Biol. 72: 275–97.

Byers, D. L., and D. M. Waller. 1999. Do plant populations purge their genetic load? Effects of population size and mating history on inbreeding depression. Ann. Rev. Ecol. Syst. 30: 479–513.

Byrne, M. M. 1994. Ecology of twig-dwelling ants in a wet lowland tropical forest. Biotropica 26: 61–72.

Cain, S. A., G. M. Oliveira Castro, J. Murça-Pires, and N. T. Silva. 1956. Application of some phytosociological techniques to Brazilian rain forest. Amer. J. Bot. 43: 911–41.

Camargo, J. L. C. 1993. Variation in soil moisture and air vapour pressure deficit relative to tropical rain forest edges near Manaus, Brazil. M.Phil. diss., University of Cambridge, U.K.

Camargo, J. L. C., and V. Kapos 1995. Complex edge effects on soil moisture and microclimate in central Amazonian forest. J. Trop. Ecol. 11: 205–21.

Camargo, J. M. F. 1988. Meliponinae (Hymenoptera, Apidae) da coleção do "Istituto di Entomologia Agraria," Portici, Itália. Rev. bras. Entomol. 32: 351–74.

Camargo, J. M. F. 1990. Stingless bees of the Amazon. In G. K. Veeresh, B. Mallik, and C. A. Viraktamath (eds.), *Proc. IUSSI Int. Congr. Social Insects and Environment*. Bangalore, India, pp. 736–38.

Camargo, J. M. F., and S. R. M. Pedro. 1992. Systematics, phylogeny, and biogeography of the Meliponinae (Hymenoptera, Apidae): A mini-review. Apidologie 23: 509–22.

Camargo, P. B. de, R. de P. Salomão, S. Trumbore, and L. A. Martinelli. 1994. How old are large Brazil-nut trees (*Bertholletia excelsa*) in the Amazon? Sc. Agric., Piracicaba 51: 389–91.

Campos, L. A. O., F. A. Silveira, M. L. Oliveira, C. V. M. Abrantes, E. F. Morato, and G. A. R. Melo. 1989. Utilização de armadilhas para a captura de machos de Euglossini (Hymenoptera, Apoidea). Rev. bras. Zool. 6: 621–26.

Canaday, C. 1996. Loss of insectivorous birds along a gradient of human impact in Amazonia. Biological Conservation 77: 63–77.

Carpenter, F. L., and H. F. Recher. 1979. Pollination, reproduction, and fire. Am. Nat. 113: 871–79.

Carroll, C. R., and S. J. Risch. 1984. The dynamics of seed harvesting in early successional communities by a tropical ant, *Solenopsis geminata*. Oecologia 61: 388–92.

Carvalho, G. 1995. Comercialização e exportação de madeiras. *Anais do I Simpósio de Política Florestal no Estado do Amazonas*. UTAM, Manaus, Amazonas, Brazil, pp. 69–73.

Carvalho, G. A., W. E. Kerr, and V. A. Nascimento. 1995. Sex determination in bees. XXXIII. Decrease of xo heteroalleles in a finite population of Melipona scutellaris (Apidae, Meliponini). Rev. bras. Genética 18: 13–16.

Carvalho, K. S. 1998. Efeitos de borda sobre a comunidade de formigas da serrapilheira em florestas da Amazônia central. M. Sc. thesis, Instituto Nacional de Pesquisas da Amazônia and Fundação Universidade do Amazonas, Manaus, Brazil.

Carvalho, K. S., and H. L. Vasconcelos. 1999. Forest fragmentation in central Amazonia and its effects on litter-dwelling ants. Biological Conservation, 151–58.

Case, T. J. 1975. Species numbers, density compensation, and colonizing ability of lizards on islands in the Gulf of California. Ecology 56: 3–18.

Cerri, C. C., B. P. Eduardo and M. C. Piccolo. 1990. *Métodos de Análises em Matéria Orgânica do Solo.* Centro de Energia Nuclear na Agricultura/Universidade de São Paulo (CENA/USP), Piracicaba, São Paulo.

Chambers, J. Q. 1998. The role of large wood in the carbon cycle of central Amazon rain forest. Ecology, Evolution, and Marine Biology. Ph.D. thesis. University of California, Santa Barbara.

Chambers, J. Q., N. Higuchi, and J. P. Schimel. 1998. Ancient trees in Amazonia. Nature 391: 135–36.

Chambers, J. Q., N. Higuchi, J. P. Schimel, L. V. Ferreira, J. M. Melack. In press. Decomposition and carbon cycling of dead trees in tropical evergreen forests of the central Amazon. Oecologia.

Chapman, C. A. 1988. Patterns of foraging and range use by three species of Neotropical primates. Primates 29: 177–94.

Charles-Dominique, P. 1986. Inter-relations between frugivorous vertebrates and pioneer plants: Cecropia, birds and bats in French Guyana. In A. Estrada and T. H. Fleming (eds.), *Frugivores and Seed Dispersal.* Dr. W. Junk Publishers, Dordrecht, Netherlands, pp. 119–35.

Charles-Dominique, P., M. Atramentowicz, M. Charles-Dominique, H. Gérard, A.

Hladik, C. M. Hladik, and M. F. Prevost. 1981. Les mammiferes frugivores arboricoles nocturnes d'une forêt guyanaise: Inter-relations plantes-animaux. Review Ecology (Terre et Vie) 35: 1–335.

Charlesworth, D., and B. Charlesworth. 1987. Inbreeding depression and its evolutionary consequences. Ann. Rev. Ecol. Syst. 18: 237–68.

Charlesworth, D., M. T. Morgan, and B. Charlesworth. 1990. Inbreeding depression, genetic load, and the evolution of outcrossing rates in a multi-locus system with no linkage. Evolution 44: 1469–89.

Chase, M. R., D. H. Boshier, and K. S. Bawa. 1995. Population genetics of *Cordia alliodora* (Boraginaceae), a Neotropical tree. 1. Genetic variation in natural populations. Amer. J. Bot. 82: 468–75.

Chase, M. R., C. Moller, R. Kesseli, and K. S. Bawa. 1996. Distant gene flow in tropical trees. Nature 383: 398–99.

Chauvel, A. 1983. Os latossolos amarelos, álicos, argilosos dentro dos ecosistemas das bacias experimentais do INPA e da região vizinha. Acta Amazónica 12: 38–47.

Chauvel, A., Y. Lucas, and R. Boulet. 1987. On the genesis of the soil mantle of the region of Manaus, Central Amazonia, Brazil. Experientia 43: 234–40.

Chen, J., J. F. Franklin, and J. S. Lowe. 1996. Comparisons of abiotic and structurally defined patch patterns in a hypothetical forest landscape. Conserv. Biol. 10: 854–62.

Chen, J., J. F. Franklin, and T. A. Spies. 1992. Vegetation responses to edge environments in old-growth Douglas-fir forests. Ecological Applications 2: 387–96.

Chen, J., J. F. Franklin, and T. A. Spies. 1993. An empirical model for predicting diurnal air-temperature gradients from edge into old-growth Douglas-fir forest. Ecological Modelling 67: 179–98.

Chew, F. S. 1981. Coexistence and local extinction in two pierid butterflies. Am. Nat. 118: 655–72.

Christenson, T. E. 1984. Behavior of colonial

and solitary spiders of the theridiid species *Anelosimus eximius*. Anim. Behav. 32: 752–734.

Clark, D. A., and D. B. Clark. 1992. Life history diversity of canopy and emergent trees in a Neotropical rain forest. Ecol. Monogr. 62: 315–44.

Clark, D. A., D. B. Clark, M. R. Sandoval, and C.-M. V. Castro. 1995. Edaphic and human effects on landscape-scale distributions of tropical rain forest palms. Ecology 76: 2581–94.

Clark, D. B. 1990. The role of disturbance in the regeneration of Neotropical moist forests. In K. S. Bawa and M. Hadley (eds.), *Reproductive Ecology of Tropical Forest Plants*. UNESCO, Paris, pp. 291–315.

Clark, D. B., and D. A. Clark. 1989. The role of physical damage in the seedling mortality regime of a Neotropical rain forest. Oikos 55: 225–30.

Clark, D. B., and D. A. Clark. 1996. Abundance, growth, and mortality of very large trees in Neotropical lowland rain forest. For. Ecol. & Manag. 80: 235–44.

Clinebell, R. R., O. L. Phillips, A. H. Gentry, N. Stark, and H. Zuuring. 1995. Prediction of Neotropical tree and liana species richness from soil and climatic data. Biodiversity and Conservation 4: 56–90.

Clutton-Brock, T. H. 1977. Some aspects of intraspecific variation in feeding and ranging behaviour in primates. In T. H. Clutton-Brock (ed.), *Primate Ecology*. Academic Press, London, pp. 539–56.

Cochrane, T. T., and P. A. Sanchez. 1982. Land resources, soils, and their management in the Amazon region: A state of knowledge report. In S. B. Hecht (ed.) *Amazonia: Agriculture and Land Use Research*. Centro Internacional de Agricultura Tropical (CIAT), Cali, Colombia, pp. 137–209.

Coelho, F. S., and F. Verlengia. 1972. *Fertilidade do Solo*. Instituto Campineiro de Ensino Agrícola, Campinas, São Paulo. 384 pp.

Cohn-Haft, M., A. Whittaker, and P. C. Stouffer. 1997. A second look at the "species-poor" central Amazon: Updates and corrections to the avifauna north of Manaus, Brazil. In J. V. Remsen, Jr., and J. M. Hagan, III (eds.), *Neotropical Ornithology Honoring Ted Parker*. American Ornithologists' Union, Washington, D.C.

Condit, R., S. P. Hubbell, and R. B. Foster. 1992. Recruitment near conspecific adults and the maintenance of tree and shrub diversity in a Neotropical forest. Am. Nat., 140: 261–86.

Condit, R., S. P. Hubbell, and R. B. Foster. 1994. Density dependence in two understory tree species in a Neotropical forest. Ecology 75: 671–80.

Condit, R., S. P. Hubbell, and R. B. Foster. 1995. Mortality rates of 205 Neotropical tree and shrub species and the impact of severe drought. Ecol. Monogr. 65: 419–39.

Condit, R., S. P. Hubbell, and R. B. Foster. 1996. Changes in tree species abundance in a Neotropical forest: Impact of climate change. J. Trop. Ecol. 12: 231–56.

Condit, R., S. P. Hubbell, J. V. LaFrankie, R. Sukumar, N. Marokaran, R. B. Roster, and P. S. Ashton. 1996. Species-area and species-individual relationships for tropical trees: A comparison of three 50–ha plots. J. Ecology 84: 549–62.

Connell, J. H. 1971. On the role of natural enemies in preventing competitive exclusion in some marine animals and in rain forest trees. In P. J. Den Boer and G. R. Gradwell (eds.), *Dynamics of Numbers in Populations*. Center for Agricultural Publication and Documentation, Wageningen, Netherlands, pp. 298–312.

Connell, J. H. 1978. Diversity in tropical rain forest and coral reefs. Science 199: 1302–9.

Connell, J. H., and R. O. Slatyer. 1977. Mechanisms of succession in natural communities and their role in community stability and organization. Am. Nat. 111: 1119–44.

Connell, J. H., J. G. Tracey, and L. Webb. 1984. Compensatory recruitment, growth, and mortality as factors maintaining rain forest tree diversity. Ecol. Monogr. 54: 141–64.

Corn, P. S., and R. B. Bury. 1989. Logging in Western Oregon: Responses of headwater habitats and stream amphibians. For. Ecol. & Manag. 29: 39–57.

Corner, E. J. H. 1954. The evolution of tropical forest. In J. S. Huxley, A. C. Hardy, and E. B. Ford (eds.), *Evolution as a Process*. Allen and Unwin, London, pp. 34–46.

Coutinho, L. M. 1990. Fire in the ecology of the Brazilian cerrado. In L. G. Goldhammer (ed.), *Fire in the Tropical Biota*. Springer-Verlag, Berlin, Germany, pp. 82–105.

Cox, F. R. 1973a. Micronutrients. In P. A. Sanchez (ed.), *A Review of Soils Research in Tropical Latin America*. North Carolina State University, Soil Science Department, Raleigh, pp. 182–97.

Cox, F. R. 1973b. Potassium. In P. A. Sanchez (ed.), *A Review of Soils Research in Tropical Latin America*. North Carolina State University, Soil Science Department, Raleigh, pp. 162–78.

Crawley, M. 1993. *GLIM for Ecologists*. Blackwell, Cambridge, U.K.

Cressie, N. A. 1991. *Statistics for Spatial Data*. Wiley, New York.

Crow, J. P., and M. Kimura. 1970. *An Introduction to Population Genetics Theory*. Harper and Row, New York.

Cunha, O. V., and F. P. Nascimento, 1978. Ofidios da Amazonia X; As cobras da região leste do Pará. Museu Paraense Emílio Goeldi Publicações Avulsas 31: 1–218.

Curtis, J. T., and G. Cottam. 1962. *Plant Ecology Workbook*. Burgess, Minneapolis, Minn.

D'Andrea, M. 1987. Social behavior in spiders (Arachnida-Araneae). Monitore Zoologico Italiano 3: 1–146.

D'Arrigo, R. D., G. C. Jacoby, and P. G. Krusic. 1994. Progress in dendroclimatic studies in Indonesia. Terrestrial, Atmospheric and Oceanic Sciences 5: 349–63.

D. O. U. 1996. Desmatamento na Amazônia, 1993–94. Diário Oficial da União, 26–07–96, Brasília, Brasil.

Dale, V. H., and S. M. Pearson. 1997. Quantifying habitat fragmentation due to land use change in Amazonia. In W. F. Laurance and R. O. Bierregaard, Jr. (eds.), *Tropical Forest Remnants: Ecology, Management, and Conservation of Fragmented Communities*. University of Chicago Press, Chicago, pp. 400–410.

Dale, V. H., S. M. Pearson, H. L. Offerman, and R. V. O'Neill. 1994. Relating patterns of land-use change to faunal biodiversity in the central Amazon. Conserv. Biol. 8: 1027–36.

Dallmeier, F. (ed.). 1992. *Long-term monitoring of biological diversity in tropical forest areas: Methods for establishment and inventory of permanent plots*. MAP Digest 11, UNESCO, Paris. France.

Darlington, P. J. 1957. Zoogeography: *The Geographical Distribution of Animals*. Wiley, New York.

Davis, A. J., and K. E. Jones. 1994. Drosophila as a indicator of habitat type and habitat disturbance in tropical forest, central Borneo. Drosophila Information Service 75: 150–51.

de Cássia, R. 1997. "BR–174: FHC anuncia abertura de nova fronteira agrícola no Norte." Amazonas em Tempo [Manaus], 25 June 1997, p. A–4.

de la Fuente, J. M., V. Ramírez-Rodríguez, J. L. Cabrera-Ponce, and L. Herrera-Estrella. 1997. Aluminum tolerance in transgenic plants by alteration of citrate synthesis. Science 276: 1566–68.

de Lima, J. M. G. 1976. *Perfil Analítico dos Fertilizantes Fosfatados*. Ministério das Minas e Energia, Departamento Nacional de Produção Mineral (DNPM) Bol. 39. DNPM, Brasília, Brazil.

Defler, T. R. 1979. On the ecology and behavior of *Cebus albifrons* in northern Colombia 1. Primates 20: 475–90.

Demers, M. N., J. W. Simpson, R. J. Boerner, A. Silva, L. Berns, and F. Artigas. 1995. Fencerows, edges, and implications of changing connectivity illustrated by two contiguous Ohio landscapes. Conserv. Biol. 9: 1159–68.

Den Boer, P. J. 1970. On the significance of dispersal power for populations of carabid beetles (Coleoptera, Carabeidae). Oecologia 4: 1–28.

Denslow, J. S. 1980. Gap partitioning among tropical rain forest trees. Biotropica 12 (suppl.): 47–55.

Denslow, J. S. 1987. Tropical rainforest gaps and the tree species diversity. Ann. Rev. Ecol. Syst. 18: 431–51.

Denslow, J. S., J. C. Schultz, P. M. Vitousek, and B. R. Strain. 1990. Growth responses of tropical shrubs to treefall gap environments. Ecology 71: 165–79.

DeSouza, O., and V. K. Brown. 1994. Effects of habitat fragmentation on Amazonian termite communities. J. Trop. Ecol. 10: 197–206.

Deusdará Filho, R. 1996. *Diagnóstico e Avaliação do Setor Florestal Brasileiro–Região Norte*. Relatório Preliminar (Sumário Executivo), Brazilian Forest Service.

Dial, R., and J. Roughgarden. 1995. Experimental removal of insectivores from rain forest canopy: direct and indirect effects. Ecology 76: 1821–34.

Dias-Filho, M. B. 1990. *Plantas invasoras em pastagens cultivadas da Amazônia: Estratégias de manejo e controle*. EMBRAPA-CPATU Documentos 52, Belém, Brazil.

Dias-Filho, M. B. 1995. Root and shoot growth in response to soil drying in seedlings of four Amazonian weedy species. Revista Brasileira de Fisiologia Vegetal 7: 53–59.

Dias-Filho, M. B. 1998. Alguns aspectos da ecologia de sementes de duas espécies de plantas invasoras da Amazônia brasileira: Implicações para o recrutamento de plântulas em áreas manejadas. In C. Gascon and P. Moutinho (eds.), *Floresta Amazônica: Dinâmica, Regeneração e Manejo*. Instituto Nacional de Pesquisas da Amazônia (INPA), Manaus, Amazonas, Brazil, pp. 233–48.

Dick, C. W. 1999. Effect of habitat fragmentation on the breeding structure of rain forest trees. Ph.D. diss., Harvard University, Cambridge, Mass.

Dick, C., and M. Hamilton. 1999. Microsatellites for the Amazonian tree *Dinizia excelsia* (Fabaceae). Molecular Ecology 8: 1765–66.

Didham, R. K. 1997a. An overview of invertebrate responses to forest fragmentation. In A. Watt, N. E. Stork, and M. Hunter (eds.), *Forests and Insects*. Chapman and Hall, London, pp. 303–20.

Didham, R. K. 1997b. The influence of edge effects and forest fragmentation on leaf litter invertebrates in Central Amazonia. In W. F. Laurance, and R. O. Bierregaard, Jr. (eds.), *Tropical Forest Remnants: Ecology, Management, and Conservation of Fragmented Communities*. University of Chicago Press, Chicago, pp. 55–70.

Didham, R. K., J. Ghazoul, N. E. Stork, and A. Davis. 1996. Insects in fragmented forests: A functional approach. Trends Ecol. Evol. 11: 255–60.

Didham, R. K., P. M. Hammond, J. H. Lawton, P. Eggleton, and N. E. Stork. 1998. Beetle species responses to tropical forest fragmentation. Ecol. Monogr. 68: 295–323.

Didham, R. K., J. H. Lawton, P. M. Hammond, and P. Eggleton. 1998. Trophic structure stability and extinction dynamics of beetles (Coleoptera) in tropical forest fragments. Phil. Trans. Royal Soc. London, Series B 353: 437–51.

Diffendorfer, J. E., M. S. Gaines, and R. D. Holt. 1995. Habitat fragmentation and movements of three small mammals (*Sigmodon, Microtus,* and *Peromyscus*). Ecology 76: 827–39.

Diniz, J. A. F., and O. Malaspina. 1996. Geographic variation of Africanized honey bees (*Apis mellifera*) in Brazil—multivariate morphometrics and racial admixture. Braz. J. Genetics 19: 217–24.

Dirzo, R., and A. Miranda. 1991. Altered patterns of herbivory and diversity in the forest understory: A case of study of the possible consequences of contemporary defaunation. In P. W. Price, T. M. Lewinson, G. W. Fernandes, and W. Benson (eds.), *Plant-Animal Interactions: Evolutionary*

Ecology in Tropical and Temperate Regions. Wiley, New York, pp. 273–87.

Döbereiner, J. 1992. Recent changes in concepts of plant bacteria interactions: Endophytic N2 fixing bacteria. Ciência e Cultura 44: 310–13.

Dobzhansky, T. 1950. Evolution in the tropics. American Scientist 32: 209–21.

Dobzhansky, T., and C. Pavan. 1950. Local and seasonal variations in relative frequencies of species of *Drosophila* in Brazil. J. Animal Ecology 19: 1–14.

Dodd, C. K., Jr. 1990. Effects of habitat fragmentation on a stream-dwelling species, the flattened muck turtle *Stenothrus depressus.* Biological Conservation 54: 3–45.

Dodson, C. H., R. L. Dressler, H. G. Hills, R. M. Adams and N. H. Williams. 1969. Biologically active compounds in orchid fragrances. Science 164: 1243–49.

Domínguez, C. A., R. Dirzo, and S. H. Bullock. 1989. On the function of floral nectar in *Croton suberosus* (Euphorbiaceae). Oikos 56: 109–14.

Dressler, R. L. 1982. Biology of the orchid bees (Euglossini). Ann Rev. Ecol. Syst. 13: 373–94.

Ducke, A. 1922. Plantes nouvelles ou peu connues de la région amazonienne. Archivos do Jardim Botanico do Rio de Janeiro 3: 2–269.

Ducke, A., and G. Black. 1954. Notas sobre a fitogeografia da Amazônia Brasileira. Bol. Técnico do Instituto Agronômico do Norte 29: 3–62.

Duivenvoorden, J. F. 1996. Patterns of tree species richness in rain forests of the Middle Caqueta area, Columbia, NW Amazonia. Biotropica 28: 142–58.

Dunning, J. B., Jr., R. Borella Jr., K. Clements, and G. F. Meffe. 1995. Patch isolation, corridor effects, and colonization by a resident sparrow in managed pine woodland. Conserv. Biol. 9: 542–50.

Dunning, J. B., Jr., B. J. Danielson, and H. R. Pulliam. 1992. Processes that effect populations in complex landscapes. Oikos 65: 169–75.

Dupuis, L. A., J. N. M. Smith, and F. Bunnell. 1995. Relation of terrestrial-breeding amphibian abundance to tree-stand age. Conserv. Biol. 9: 645–53.

Durant, S. M., and G. M. Mace. 1994. Species differences and population structures in population viability analysis. In P. J. S. Olney, G. M. Mace, and A. T. C. Feister (eds.), *Creative Conservation: Interactive Management of Wild and Captive Animals.* Chapman and Hall, London, pp. 67–91.

Eggleton, P., D. E. Bignell, W. A. Sands, N. A. Mawdsley, J. H. Lawton, T. G. Wood, and N. C. Bignell. 1996. The diversity, abundance, and biomass of termites under differing levels of disturbance in the Mbalmayo Forest Reserve, southern Cameroon. Phil. Trans. Royal Soc. London, Series B 351: 51–68.

Eguiarte, L. E., N. Perez-Nasser, and D. Piñero. 1992. Genetic structure, outcrossing rate, and heterosis in Astrocaryum mexicanum (tropical palm): Implications for evolution and conservation. Heredity 69: 217–28.

Ehrlich, P. R. 1988. The loss of diversity: Causes and consequences. In E. O. Wilson (ed.), *Biodiversity.* National Academy Press, Washington, D.C., pp. 29–35.

Ehrlich, P. R., D. E. Breedlove, P. F. Brussard, and M. A. Sharp. 1972. Weather and the "regulation" of subalpine populations. Ecology 53: 243–47.

Ehrlich, P. R., D. D. Murphy, M. C. Singer, C. B. Sherwood, R. R. White, and I. L. Brown. 1980. Extinction, reduction, stability, and increase: The response of checkerspot butterfly (*Euphydryas*) populations to the California drought. Oecologia 46: 101–5.

Ellstrand, N. C., and D. R. Elam. 1993. Population genetic consequences of small population size: Implications for plant conservation. Ann. Rev. Ecol. Syst. 24: 217–42.

Emmons, L. H. 1984. Geographic variation in densities and diversities of non-flying mammals in Amazonia. Biotropica 16: 210–22.

Ennos, A. R. 1997. Wind as an ecological factor. Trends Ecol. Evol. 12: 108–11.

Erwin, T. L. 1982. Tropical forests: Their richness in Coleoptera and other arthropods species. Coleopterists Bull. 36: 74–75.

Erwin, T. L. 1983. Beetles and other insects in tropical forest canopies at Manaus, Brazil, sampled by insecticidal fogging. In S. L. Sutton, T. C. Whitmore, and A. C. Chadwick (eds.), *Tropical Rain Forest: Ecology and Management*. Blackwell, Oxford, U.K., pp. 59–75.

Estrada, A., R. Coates-Estrada, and C. Vázquez-Yanes. 1984. Observations on fruiting and dispersers of *Cecropia obtusifolia* at Los Tuxtlas, Mexico. Biotropica 16: 315–18.

Ewan, J. 1962. Synopsis of the South American species of *Vismia* (Guttiferae). Contributions of the U.S. National Herbarium 35: 293–377.

Faccelli, J. M., and S. T. A. Pickett. 1991. Plant litter: Its dynamics and effects on plant community structure. Botanical Rev. 57: 1–32.

Fahn, A., J. Burley, K. Longman, and A. Maraux. 1981. Possible contributions of wood anatomy to the determination of the age of tropical trees. In F. Bormann and G. Berlyn (eds.), *Age and Growth Rate of Tropical Trees: New Directions for Research*. Yale University Press, New Haven, Conn., pp. 31–54.

Fahrig, L. 1997. Relative effects of habitat loss and fragmentation on population extinction. J. Wildl. Man. 61: 603–10.

Fahrig, L., and G. Merriam. 1994. Conservation of fragmented populations. Conserv. Biol. 8: 50–59.

Fahrig, L., and J. Paloheimo. 1988. Determinants of local populations size in patch habitats. Theor. Pop. Bio. 34: 194–213.

Falesi, I. C. 1974. Soils of the Brazilian Amazon. In C. Wagley (ed.), *Man in the Amazon*. University Presses of Florida, Gainesville, pp. 201–29.

Fearnside, P. M. 1980a. A previsão de perdas através de erosão do solo sob vários usos de terra na área de colonização da Rodovia Transamazônica. Acta Amazónica 10: 505–11.

Fearnside, P. M. 1980b. The effects of cattle pasture on soil fertility in the Brazilian Amazon: Consequences for beef production sustainability. Tropical Ecology 21: 125–37.

Fearnside, P. M. 1984. Initial soil quality conditions on the Transamazon Highway of Brazil and their simulation in models for estimating human carrying capacity. Tropical Ecology 25: 1–21.

Fearnside, P. M. 1986a. Settlement in Rondônia and the token role of science and technology in Brazil's Amazonian development planning. Interciencia 11: 229–36.

Fearnside, P. M. 1986b. *Human Carrying Capacity of the Brazilian Rainforest*. Columbia University Press, New York.

Fearnside, P. M. 1987. Rethinking continuous cultivation in Amazonia. BioScience 37: 209–14.

Fearnside, P. M. 1988. Yurimaguas reply. BioScience 38: 525–27.

Fearnside, P. M. 1989a. Deforestation in Brazilian Amazon: The rates and causes of forest destruction. The Ecologist 19: 214–18.

Fearnside, P. M. 1989b. Deforestation in Amazonia. Environment 31: 16–20.

Fearnside, P. M. 1990a. Predominant land uses in Brazilian Amazonia. In A. B. Anderson (ed.), *Alternatives to Deforestation: Steps Toward Sustainable Use of the Amazon Rain Forest*. Columbia University Press, New York, pp. 233–51.

Fearnside, P. M. 1990b. The rate and extent of deforestation in Brazilian Amazonia. Environmental Conservation 17: 213–26.

Fearnside, P. M. 1990c. Fire in the tropical rain forest of the Amazon Basin. In J. G. Goldhammer (ed.), *Fire in the Tropical Biota*. Springer-Verlag, Berlin, Germany, pp. 106–16.

Fearnside, P. M. 1995. Potential impacts of climatic change on natural forests and forestry in Brazilian Amazonia. For. Ecol. & Manag. 78: 51–70.

Fearnside, P. M. 1996a. Amazonian deforestation and global warming: Carbon stocks in vegetation replacing Brazil's Amazon forest. For. Ecol. & Manag. 80: 21–34.

Fearnside, P. M. 1996b. Amazonia and global warming: Annual balance of greenhouse gas emissions from land-use change in Brazil's Amazon region. In J. Levine (ed.), *Biomass Burning and Global Change. Vol. 2, Biomass Burning in South America, Southeast Asia and Temperate and Boreal Ecosystems and the Oil Fires of Kuwait.* MIT Press, Cambridge, Mass., 606–17.

Fearnside, P. M. 1997a. Limiting factors for development of agriculture and ranching in Brazilian Amazonia. Rev. bras. Biol. 57: 531–49.

Fearnside, P. M. 1997b. Human carrying capacity estimation in Brazilian Amazonia as a basis for sustainable development. Environmental Conservation 24: 271–82.

Fearnside, P. M. 1997c. Environmental services as a strategy for sustainable development in rural Amazonia. Ecological Economics 20: 53–70.

Fearnside, P. M. 1997d. Greenhouse gases from deforestation in Brazilian Amazonia: Net committed emissions. Climatic Change 35: 321–60.

Fearnside, P. M. 1997e. Protection of mahogany: A catalytic species in the destruction of rain forests in the American tropics. Environmental Conservation 24: 303–6.

Fearnside, P. M. 1997f. Monitoring needs to transform Amazonian forest maintenance into a global warming mitigation option. Mitigation and Adaptation Strategies for Global Change 2: 285–302.

Fearnside, P. M. 1998a. Phosphorus and Human Carrying Capacity in Brazilian Amazonia. In J. P. Lynch and J. Deikman (eds.) *Phosphorus in Plant Biology: Regulatory Roles in Molecular, Cellular, Organismic, and Ecosystem Processes.* American Society of Plant Physiologists, Rockville, Maryland, U.S.A, pp. 94–108.

Fearnside, P. M. 1998. Sistemas agroflorestais na política de desenvolvimento na Amazônia brasileira: Papel e limites como uso para áreas degradadas. In C. Gascon and

P. Moutinho (eds.), *Floresta Amazônica: Dinâmica, Regeneração e Manejo.* Instituto Nacional de Pesquisas da Amazônia (INPA), Manaus, Amazonas, Brazil, pp. 293–312.

Fearnside, P. M. In press. Can pasture intensification discourage deforestation in the Amazon and Pantanal regions of Brazil? In C. H. Wood and R. Porro (eds.), *Land Use and Deforestation in the Amazon.* University Press of Florida, Gainesville.

Fearnside, P. M., and R. I. Barbosa. 1996. The Cotingo Dam as a test of Brazil's system for evaluating proposed developments in Amazonia. Envir. Manag. 20: 631–48.

Fearnside, P. M., and R. I. Barbosa. 1998. Soil carbon changes from conversion of forest to pasture in Brazilian Amazonia. For. Ecol. & Manag. 108: 147–66.

Fearnside, P. M., and W. M. Guimarães. 1996. Carbon uptake by secondary forests in Brazilian Amazonia. For. Ecol. & Manag. 80: 35–46.

Fearnside, P. M., A. T. Tardin, and L. G. Meira Filho. 1990. *Deforestation Rate in Brazilian Amazon,* INPE, São José dos Campos, São Paulo.

Federal House of Representatives Report. 1997. *The Timber Situation in Amazônia.* Brasília, DF, Brazil.

Fedorov, A. A. 1966. The structure of the tropical rain forest and speciation in the humid tropics. J. Ecology 54: 1–11.

Feinsinger, P. 1997. Habitat "shredding." In G. K. Meffe, C. R. Carroll, and contributors (eds.), *Principles of Conservation Biology.* Sinauer, Sunderland, Mass., pp. 270–72.

Fenster, W. E., and L. A. León. 1979. Management of phosphorus fertilization in establishing and maintaining improved pastures on acid, infertile soils of tropical America. In P. A. Sanchez and L. E. Tergas (eds.), *Pasture Production in Acid Soils of the Tropics.* Proceedings of a Seminar held at CIAT, Cali, Colombia, April 17–21, 1978, CIAT Series 03 EG-05. Centro Internacional de Agricultura Tropical (CIAT), Cali, Colombia, pp. 109–22.

Ferrari, S. F., and V. H. Diego. 1995. Habitat fragmentation and primate conservation in the Atlantic Forest of eastern Minas Gerais, Brazil. Oryx 29: 192–96.

Ferraz, J., N. Higuchi, J. dos Santos, Y. Biot, F. Marques, K. Baker, R. Baker, I. Hunter, and J. Proctor. 1997. Distribuição de nutrientes nas árvores e exportação de nutrientes pela exploração seletiva de madeira. In N. Higuchi, J. B. S. Ferraz, L. Antony, F. Luizão, R. Luizão, Y. Biot, I. Hunter, J. Proctor and S. Ross (eds.), *Bionte: Biomassa e Nutrientes Florestais.* Relatório Final. Instituto Nacional de Pesquisas da Amazônia (INPA), Manaus, Amazonas, Brazil, pp. 133–49.

Ferreira, L. V., and W. F. Laurance. 1997. Effects of forest fragmentation on mortality and damage of selected trees in central Amazonia. Conserv. Biol. 11: 797–801.

Ferreira, L. V., and J. Rankin–de Mérona. 1997. Floristic composition and structure of a one-hectare plot in terra firme forest in Central Amazonia. In F. Dallmeier and J. A. Comiskey (eds.), *Forest Biodiversity in North, Central and South America and the Caribbean: Research and monitoring,* Man and Biosphere series, Vol 22, Chapter 34. UNESCO and Parthenon, Carnforth, Lancashire, U.K, pp. 655–68.

Ferreira, S. J. F. 1997. Efeitos da exploração seletiva de madeira sobre a hidrologia e hidroquímica do solo. In N. Higuchi, J. B. S. Ferraz, L. Antony, F. Luizão, R. Luizão, Y. Biot, I. Hunter, J. Proctor and S. Ross (eds.), *Bionte: Biomassa e Nutrientes Florestais, Relatório Final.* Instituto Nacional de Pesquisas da Amazônia (INPA), Manaus, Amazonas, Brazil, pp. 151–69.

Fisher, R. A., A. S. Corbet, and C. B. Williams. 1943. The relation between the number of species and the number of individuals in a random sample of an animal population. J. Anim. Ecol. 12: 42–58.

Fittkau, E. J., and H. Klinge. 1973. On biomass and trophic structure of the Central Amazonian rain forest ecosystem. Biotropica 5: 2–14.

Fjeldså, J., and C. Rahbek. 1997. Species richness and endemism in South American birds: Implications for the design of networks of nature reserves. In W. F. Laurance and R. O. Bierregaard, Jr. (eds.), *Tropical Forest Remnants: Ecology, Management, and Conservation of Fragmented Communities.* University of Chicago Press, Chicago, pp. 466–82.

Flather, C. H., and J. R. Sauer. 1996. Using landscape ecology to test hypotheses about large-scale abundance patterns in migratory birds. Ecology 77: 28–35.

Fleming, T. H., and E. R. Heithaus. 1981. Frugivorous bats, seed shadows, and the structure of tropical forests. Biotropica (suppl. Reproductive Botany) 13: 4–53.

Folsom, J. P. 1985. Dos nuevas tecnicas para capturar y marcar abejas machos de la tribu Euglossini (Hymenoptera, Apidae). Actualidades Biologicas 14: 20–25.

Fonseca, G. A. B. 1988. Patterns of small mammal species diversity in the Brazilian Atlantic Forest. Ph.D. diss., University of Florida, Gainesville.

Fonseca, G. A. B., and J. G. Robinson. 1990. Forest size and structure: Competitive and predatory effects on small mammal communities. Biological Conservation 53: 265–94.

Forman, R. T. T. 1995. *Land mosaics: The Ecology of Landscapes and Regions.* Cambridge University Press, Cambridge, U.K.

Forman, R. T. T., A. E. Galli, and C. F. Leck. 1976. Forest size and avian diversity in New Jersey woodlots with some land use implications. Oecologia 26: 1–8.

Forman, R. T. T., and M. Godron. 1986. *Landscape Ecology.* Wiley, New York.

Foster, R. B. 1980. Heterogeneity and disturbance in tropical vegetation. In M. E. Soulé and B. A. Wilcox (eds.), *Conservation Biology: An Evolutionary-Ecological Perspective.* Sinauer, Sunderland, Mass., pp. 75–92.

Foster, R. B. 1990. The floristic composition of Rio Manu floodplain forest. In A. H. Gentry (ed.) *Four Tropical Rainforests.* Yale University Press, New Haven, Conn., pp. 99–111.

Fowler H. G., and H. W. Levi. 1979. A new quasi social *Anelosimus* spider (Araneae, Theridiidae) from Paraguay. Psyche 86: 11–18.

Fowler, H. G., C. A. Silva, and E. M. Venticinque. 1993. Size, taxonomic, and biomass distributions of flying insects in Central Amazonia: Forest edge vs. understory. Rev. Biologia Tropical 41: 755–60.

Fowler, H. G., and E. M. Venticinque. 1996. Interference competition and scavenging by *Crematogaster* ants (Hymenoptera:Formicidae) associated with the webs of the social spider Anelosimus eximius (Araneae:Theridiidae) in Central Amazon. J. Kansas Entomological Society 69: 267–69.

Frailey, C. D., E. L. Lavina, A. Rancy, and J. P. de Souza Filho. 1988. A proposed Pleistocene/Holocene lake in the Amazon basin and its significance to Amazonian geology and biogeography. Acta Amazónica 18: 119–43.

Frankie, G. W., W. A. Haber. 1983. Why bees move among mass-flowering Neotropical trees. In C. E. Jones and R. J Little (eds.), *Handbook of Experimental Pollination Biology*. Van Rostrand Rinehold, New York, pp. 360–72.

Frankie, G. W., S. B. Vinson, and J. F. Barthell. 1988. Nest site and habitat preferences of *Centris* bees in the Costa Rican dry forest. Biotropica 20: 301–10.

Franklin, I. R. 1980. Evolutionary change in small populations. In M. E. Soulé, and B. A. Wilcox (eds.), *Conservation Biology: An Evolutionary-Ecological Perspective*. Sinauer, Sunderland, Mass., pp. 135–49.

Frazão, E. 1991. Insectivory in free-ranging bearded sakis, *Chiropotes satanas*. Primates 32: 245–47.

Frazão, E. da R. 1992. Dieta e estrategia de forragear de *Chirpotes satanas chiropotes* (Cebidae: Primates) na Amazônia central Brasileira. M. Sc. thesis, Instituto Nacional de Pesquisas da Amazônia, Manaus, Brazil.

Freese, C. H. 1976. Censusing *Alouatta palliata, Ateles geofroyi,* and *Cebus capucinus* in the Costa Rican dry forest. *Neotropical Primates: Field Studies and Conservation*. In R. W. Thorington, Jr., and P. G. Heltne (eds.), National Academy of Science, Washington, D.C., pp. 4–9.

Freese, C. H., and J. R. Oppenheimer. 1981. The capuchin monkeys, Genus *Cebus*. In A. F. Coimbra-Filho and R. A. Mittermeier (eds.), *Ecology and Behavior of Neotropical Primates*. Academia Brasileira de Ciências, Rio de Janeiro, pp. 331–90.

Freitas, B. M., and R. J. Paxton. 1996. The role of wind and insects in cashew (*Anacardium occidentale*) pollination in NE Brazil. J. Agricultural Science 126: 319–26.

Fritts, H. 1991. *Reconstructing large-scale climatic patterns from tree-ring data*. University of Arizona Press, Tucson.

Fritz, R. S. 1979. Consequences of insular population structure: Distribution and extinction of spruce grouse populations. Oecologia 42: 57–65.

Galetti, M. 1996. Fruits and frugivores in a Brazilian Atlantic Forest. Ph.D. thesis, University of Cambridge, Cambridge, U.K.

Galil, J., and D. Eisikowitch. 1971. Studies on mutualistic symbiosis between syconia and sycophilous wasps in monoeicious figs. New Phytologist 70: 773–87.

Gama, W. N. G. 1997. O Projeto Dinâmica Biológica de Fragmentos Florestais-PDBFF [INPA/Smithsonian]: Uma base científica norte-americana na Amazônia brasileira. M. Sc. Thesis. Universidade Federal Do Pará; Belém, Para, Brazil.

Ganade, G. 1996. Seedling establishment in Amazon rainforest and old fields. Ph.D. diss., University of London, London.

Gandara, F. B. 1996. Diversidade genética, taxa de cruzamento e estrutura espacial dos genótipos em uma população de *Cedrela fissilis* Vell. (Meliaceae). M. Sc. Thesis, Universidade de Campinas (UNICAMP), Campinas, SP, Brasil.

Garcia, J. E., and J. Mba. 1997. Distribution, status and conservation of primates in Monte Alen National Park, Equatorial Guinea. Oryx 31: 67–76.

Garcia, M. V. B., M. L. Oliveira and L. A. O.

Campos. 1992. Use of seeds of *Coussapoa asperifolia magnifolia* (Cecropiaceae) by stingless bees in the Central Amazonian Forest (Hymenoptera: Apidae: Meliponinae). Entomol. Gener. 17: 255–58.

Garwood, N. 1989. Tropical soil seed banks: A review. In M. A. Leck, V. T. Parker, and R. L. Simpson (eds.), *Ecology of Soil Seed Banks*. Academic Press, San Diego.

Gascon, C. 1990. The relative importance of habitat characteristics in the maintenance of a species assemblage of tropical forest-breeding frogs. Florida State University, Tallahassee.

Gascon, C. 1991. Population and community level analyses of species occurrences of central Amazonian rainforest tadpoles. Ecology 72: 731–46.

Gascon, C. 1993. Breeding-habitat use by five Amazonian frogs at forest edge. Biodiversity and Conservation 2: 38–44.

Gascon, C. 1995. Habitat effects on tropical larval anuran fitness in the absence of direct effects of predation and competition. Ecology 76: 2222–29.

Gascon, C., and T. E. Lovejoy. 1998. Ecological impacts of forest fragmentation in central Amazonia. Zoology, Analysis of Complex Systems 101: 273–80.

Gascon, C., T. E. Lovejoy, R. O. Bierregaard, Jr., J. R Malcolm, P. C. Stouffer, H. Vasconcelos, W. F. Laurance, B. Zimmerman, M. Tocher, and S. Borges. 1999. Matrix habitat and species persistence in tropical forest remnants. Biological Conservation 91: 223–30.

Gascon, C., and P. Moutinho (eds.). 1998. *Floresta Amazônica: Dinâmica, Regeneração e Manejo*. Instituto Nacional de Pesquisas da Amazônia (INPA), Manaus, Brazil.

Gascon, C., G. B. Williamson, and G. A. B. Fonseca. 2000. Receding edges and vanishing fragments. Science 288: 1356–58.

Gaston, K. J. 1994. *Rarity*. Chapman & Hall, London.

Gentry, A. H. 1982. Neotropical floristic diversity: Phytogeographical connections between Central and South America, Pleistocene climatic fluctuations or an accident of the Andean orogeny? Ann. Missouri Bot. Gard. 69: 557–93.

Gentry, A. H. 1986. Endemism in tropical vs. temperate plant communities. In M. E. Soulé (ed.), *Conservation Biology: The Science of Scarcity and Diversity*. Sinauer, Sunderland, Mass., pp. 153–81.

Gentry, A. H. 1988. Changes in plant community diversity and floristic composition on environmental geographical gradients. Ann. Missouri Bot. Gard. 75: 1–34.

Gentry, A. H. 1988. Tree species richness of upper Amazonian forests. Proc. Nat. Acad. Sci. 85: 156–59.

Gentry, A. H. 1990. Floristic similarities and differences between southern Central America and upper and central Amazonia. In A. H. Gentry (ed.), *Four Neotropical Rainforests*. Yale University Press, New Haven, Conn., pp. 141–57.

Gentry, A. H. 1992. Tropical forest biodiversity: Distributional patterns and their conservational significance. Oikos 63: 19–28.

Gentry, A. H. 1993. A field guide to the families and genera of woody plants of Northwest South America. Conservation International, Washington, DC.

Gentry, A. H., and C. Dodson. 1987. Diversity and phytogeography of Neotropical epiphytes. Ann. Missouri Bot. Gard. 74: 205–33.

Gentry, A. H., and L. H. Emmons. 1987. Geographical variation in fertility and compositions of the understory of Neotropical forests. Biotropica 19: 216–27.

Gibbs, J. P., and J. Faaborg. 1990. Estimating the viability of Ovenbird and Kentucky Warbler populations in forest fragments. Conserv. Biol. 4: 193–96.

Gilbert, K. A. 1994. Endoparasitic infection in red howling monkeys (*Alouatta seniculus*) in the central Amazonian basin: A cost of sociality? Ph.D. diss., Rutgers University, New Brunswick, N.J.

Gilbert, L. E. 1980. Food web organization and the conservation of Neotropical diversity. In M. E. Soulé and B. A. Wilcox (eds.), *Conservation Biology: An Evolutionary-Ecological Perspective*. Sinauer, Sunderland, Mass., pp. 11–33.

Gilbert, L. E., and P. H. Raven. 1975. *Coevolution of Animals and Plants*. University of Texas Press, Austin.

Gill, A. M. 1981. Coping with fire. In J. S. Pate and A. J. McComb (eds.), *The Biology of Australian Plants*. University of Western Australia Press, Nedlands, pp. 65–87.

Gillman, G. P., and G. G. Murtha. 1983. Effects of sample handling on some chemical properties in soils from high rainfall coastal North Queensland. Australian J. Soil Research 21: 67–72.

Gilpin, M. E. 1988. A comment on Quinn and Hastings: Extinction in subdivided habitats. Conserv. Biol. 2: 290–92.

Gilpin, M. E. 1990. Extinction of finite metapopulations in correlated environments. In B. Shorrocks and I. Swingland (eds.), *Living in a Patchy Environment*. Oxford Scientific Publications, Oxford, pp. 177–86.

Gilpin, M. E., and M. E. Soulé. 1986. Minimum viable population: processes of species extinction. In M. E. Soulé (ed.), *Conserv. Biol.: the Science of Scarcity and Diversity*. Sinauer, Sunderland, Mass., pp. 19–34.

Gleason, H. A. 1922. On the relation between species and area. Ecology 3: 158–62.

Gleason, H. A. 1925. Species and area. Ecology 6: 66–74.

Gomes, L. F. 1991. Diversidade e flutuação de populações de abelhas da tribo Euglossini (Hymenoptera, Apidae) em dois ecossistemas de São Luís-MA: Mata e restinga. Unpublished Monograph, Centro de Ciências da Saúde/UFMA, São Luís, Brazil.

Goosem, M. 1997. Internal fragmentation: Effects of roads, highways, and powerline clearings on movements and mortality of rainforest vertebrates. In W. F. Laurance and R. O. Bierregaard, Jr. (eds.), *Tropical Forest Remnants: Ecology, Management,*

and Conservation of Fragmented Communities. University of Chicago Press, Chicago, pp. 241–55.

Gorchov, D. L., F. Cornejo, C. Ascorra, and M. Jaramillo. 1993. The role of seed dispersal in the natural regeneration of rain forest after strip-cutting in the Peruvian Amazon. Vegetatio 107/108: 339–49.

Gosner, K. L. 1960. A simplified table for staging Anuran embryos and larvae with notes on identification. Herpetologica 16: 83–90.

Gotelli, N. J. 1991. Metapopulation models: The rescue effect, the propagule rain, and the core-satellite hypothesis. Am. Nat. 138: 768–76.

Grainger, A. 1987. Tropform: A Model of Future Tropical Timber Hardwood Supplies. In *CINTRAFOR Symposium in Forest Sector and Trade Models*. University of Washington, Seattle.

Grant, B. W., K. L. Brown, G. W. Ferguson, and J. W. Gibbons. 1994. Changes in amphibian biodiversity associated with 25 years of pine forest regeneration: Implications for biodiversity management. In S. K. Majumdar, F. J. Brenner, J. E. Lovich, J. F. Schalles, and E. W. Miller (eds.), *Biological Diversity: Problems and Challenges*. The Pennsylvania Academy of Sciences, Philadelphia, PA, pp. 354–67.

Green, F. B. M., and C. Payne. 1994. *GLIM 4: The Statistical System for Generalized Linear Interactive Modeling*. Clarendon, Oxford, U.K.

Greenberg, R. 1983. The role of neophobia in determining the degree of foraging specialization in some migrant warblers. Am. Nat. 122: 444–53.

Greenberg, R. 1989. Neophobia, aversion to open space, and ecological plasticity in Song and Swamp sparrows. Canadian J. Zool. 67: 1194–99.

Greenslade, P. J. M., and P. Greenslade. 1977. Some effects of vegetation cover and disturbance on a tropical ant fauna. Insectes Sociaux 24: 163–82.

Greenstone, H. M. 1984. Determinants of

web spider species diversity: Vegetation structural diversity vs. prey availability. Oecologia 62: 299–304.

Greenwood, S. R. 1987. The role of insects in tropical forest food webs. Ambio 16: 267–71.

Guevara, S., and J. Laborde. 1993. Monitoring seed dispersal at isolated standing trees in tropical pastures: Consequences for local species availability. In T. H. Fleming and A. Estrada (eds.), *Frugivory and seed dispersal: Ecological and evolutionary aspects*. Kluwer Academic, Dordrecht, The Netherlands, pp. 319–38.

Guevara, S., S. E. Purata, and E. van der Maarel. 1986. The role of remnant forest trees in tropical secondary succession. Vegetatio 66: 77–84.

Guillaumet, J.–L. 1987. Some structural and floristic aspects of the forest. Experientia 43: 241–51.

Guimarães, G. de A., J. B. Bastos, and E. de C. Lopes. 1970. Métodos de análise física, química e instrumental de solos. Instituto de Pesquisas e Experimentação Agropecuárias do Norte (IPEAN), Série: Química de Solos 1: 1–108.

Gullison, R. E., S. N. Panfil, J. J. Strouse, and S. P. Hubbell. 1996. Ecology and management of mahogany (*Swietenia macrophylla* King) in the Chimanes Forest, Beni, Bolivia. Botanical J. Linnean Soc. 122: 9–34.

Gunnarson, B. 1988. Spruce-living spiders and forest decline; the importance of needle-loss. Biological Conservation 43: 309–19.

Haas, C. A. 1995. Dispersal and use of corridors by birds in wooded patches on an agricultural landscape. Conserv. Biol. 9: 845–54.

Haila, Y., I. K. Hanski, and S. Raivio. 1987. Breeding bird distribution in fragmented coniferous taiga in Southern Finland. Ornis Fennica 64: 90–106.

Hall, P., L. C. Orrell, and K. S. Bawa. 1994. Genetic diversity and mating system in a tropical tree, *Carapa guianensis* (Meliaceae). Amer. J. Bot. 81: 1104–11.

Hall, P., S. Walker, and K. Bawa. 1996. Effect of forest fragmentation on genetic diversity and mating system in a tropical tree *Pithecellobium elegans*. Conserv. Biol. 10: 757–68.

Hamilton, M. 1999. Tropical tree gene flow and seed dispersal. Nature 401: 129–30.

Hammond, D. S., and H. ter Steege. 1998. Propensity for fire in the Guianan highlands. Conserv. Biol. 12: 944–47.

Hammond, P. M. 1990. Insect 97. Tree crown beetles in context: A comparison of canopy and other ecotone assemblages in a lowland tropical forest in Sulawesi. In N. E. Stork, J. Adis, and R. K. Didham (eds.), *Canopy Arthropods*. Chapman and Hall, London, pp. 184–223.

Hammond, P. M., N. E. Stork, and M. J. D. Brendell. 1997. Tree crown beetles in context: A comparison of canopy and other ecotone assemblages in a lowland tropical forest in Sulawesi. In N. E. Stork, J. Adis, and R. K. Didham (eds.), *Canopy Arthropods*. Chapman and Hall, London, pp. 184–223.

Hamrick, J. L., and M. T. Godt. 1990. Allozyme diversity in plant species. In A. H. D. Brown, M. T. Clegg, A. L. Kahler, and B. S. Weir (eds.), *Plant Population Genetics, Breeding and Genetic Resources*. Sinauer, Sunderland, Mass., pp. 43–63.

Hamrick, J. L., and M. D. Loveless. 1986. Isozyme variation in tropical trees: Procedures and preliminary results. Biotropica 18: 201–7.

Hamrick, J. L., and M. D. Loveless. 1989. The genetic structure of tropical tree populations: Association with reproductive biology. In J. H. Bock and Y. B. Linhart (eds.), *The Evolutionary Ecology of Plants*. Westview, Boulder, Colo., pp. 129–46.

Hamrick, J. L., and D. A. Murawski. 1990. The breeding structure of tropical tree populations. Plant Species Biol. 5: 157–65.

Hamrick, J. L., and D. A. Murawski. 1991. Levels of allozyme diversity in populations of uncommon Neotropical tree species. J. Trop. Ecol. 7: 395–99.

Hanski, I. 1991. Single-species metapopulation dynamics: Concepts, models, and observations. Biol. J. Linnean Soc. 42: 17–38.

Hanski, I. 1994. A practical model of metapopulation dynamics. J. Anim. Ecol. 63: 151–62.

Hanski, I and M. E. Gilpin. 1991. Metapopulation dynamics: Brief history and conceptual domain. Biol. J. Linnean Soc. 42: 3–16.

Happold, D. C. D. 1975. The effect of climate and vegetation on the distribution of small rodents in western Nigeria. Zeitschrift Saügetier 40: 221–42.

Happold, D. C. D. 1977. A population study of small rodents in the tropical rain forest of Nigeria. Terre et Vie 31: 385–458.

Hardy, F. 1961. *Manual de Cacao.* Instituto Interamericano de Ciências Agrícolas (IICA), Turrialba, Costa Rica.

Harper, J. L. 1977. *Population biology of plants.* Academic Press, New York.

Harper, J. L. 1981. The concepts of population in modular organisms. In R. M. May (ed.), *Theoretical Ecology: Principles and Applications.* Blackwell, Oxford, U.K, pp. 55–77.

Harper, L. H. 1987. The conservation of ant-following birds in small Amazonian forest fragments. Ph.D. thesis, State University of New York, Albany.

Harper, L. H. 1989. The persistence of ant-following birds in small Amazonian forest fragments. Acta Amazónica 19: 249–63.

Harris, L. D. 1984. *The Fragmented Forest: Island Biogeographic Theory and the Preservation of Biotic Diversity.* University of Chicago Press, Chicago.

Harris, R. J. 1975. *A Primer of Multivariate Statistics.* Academic Press, New York.

Harrison, S. 1991. Local extinction in a metapopulation context: An empirical evaluation. Biol. J. Linnean Soc. 42: 73–88.

Harrison, S. 1994. Metapopulations and Conservation. In P. J. Edwards, N. R. Webb and R. M. May (eds.), *Large-Scale Ecology and Conservation.* Blackwell, Oxford, U.K., pp. 111–28.

Harrison, S., and J. F. Quinn. 1989. Correlated environments and persistence of metapopulations. Oikos 56: 293–98.

Harrison, S., D. D. Murphy, and P. R. Ehrlich. 1988. Distribution of the bay checkerspot butterfly, *Euphydryas editha bayensis:* Evidence for a metapopulation model. Am. Nat. 132: 360–82.

Hartshorn, G. S. 1980. Neotropical forest dynamics. Biotropica 12 (suppl.): 23–30.

Hartshorn, G. S. 1989. Application of gap theory to tropical forest management: Natural regeneration on strip clear-cuts in the Peruvian Amazon. Ecology 70: 567–69.

Hawkins, C. P. S., and J. A. MacMahon. 1989. Guilds: The multiple meanings of a concept. Ann. Rev. Entomology 34: 423–51.

Helle, P., and J. Muona. 1985. Invertebrate numbers in edges between clear-fellings and mature forests in northern Finland. Silva Fennica 19: 281–94.

Henderson, A. 1995. *The Palms of the Amazon.* Oxford University Press, New York.

Henderson, A., G. Galeano, and R. Bernal. 1995. *Field Guide to the Palms of the Americas.* Princeton University Press, Princeton, N.J.

Henry, J. D., and J. M. A. Swan. 1974. Reconstructing forest history from live and dead plant material—an approach to the study of forest succession in southwest New Hampshire. Ecology 55: 772–83.

Herkert, J. R. 1994. The effects of habitat fragmentation on midwestern grassland bird communities. Ecological Applications 4: 461–71.

Herrera, R., T. Merida, N. Stark, C. F. Jordan. 1978. Direct phosphorus transfer from leaf litter to roots. Naturwissenschaften 65: 208–9.

Hietzseifert, U., P. Hietz, and S. Guevara. 1996. Epiphyte vegetation and diversity on remnant trees after forest clearance in southern Veracruz, Mexico. Biological Conservation 75: 103–11.

Hill, R. J., G. T. Prance, S. A Mori, W. C. Steward, D. Shimabukuru, and D., J. Bernardi. 1978. Estudo eletroforético da dinâmica de variação genética em três taxa ribeirinhos ao longo do rio Solimões, América do Sul. Acta Amazónica 8: 183–99.

Hobbs, R. J. 1989. The nature and effects of disturbance relative to invasions. In J. Drake and H. A. Mooney (eds.), *Biological Invasions: A Global Perspective*. Wiley, Chichester, U.K., pp. 389–405.

Hobbs, R. J. 1992. The role of corridors in conservation: Solution or bandwagon? Trends Ecol. Evol. 11: 389–92.

Hobbs, R. J. 1993. Effects of landscape fragmentation on ecosystem processes in the Western Australian wheat belt. Biological Conservation 64: 193–201.

Holdsworth, A. R., and C. Uhl. 1997. Fire in Amazonian selectively logged rain forest and the potential for fire reduction. Ecological Applications 7: 713–25.

Holt, R. D. 1977. Predation, apparent competition, and the structure of prey communities. Theor. Pop. Bio. 12: 197–229.

Holt, R. D. 1993. Ecology at the mesoscale: The influence of regional processes on local communities. In R. E. Ricklefs and D. Schluter (eds.), *Species Diversity in Ecological Communities*. University of Chicago Press, Chicago, pp. 77–88.

Horton, J. S., and C. J. Kraebel. 1955. Development of vegetation after fire in the chaise chaparral of southern California. Ecology 36: 244–62.

Howe, H. F. 1977. Bird activity and seed dispersal of a tropical wet forest tree. Ecology 58: 539–50.

Howe, H. F. 1984. Implications of seed dispersal by animals for the management of tropical reserves. Biological Conservation 30: 261–81.

Howe, H. F., and J. Smallwood. 1982. Ecology of seed dispersal. Ann. Rev. Ecol. Syst. 13: 201–28.

Hsu, M. J., and G. Agoramoorthy. 1996. Conservation status of primates in Trinidad, West Indies. Oryx 30: 285–91.

Hubbell, S. P., and R. B. Foster. 1986. Commonness and rarity in a Neotropical forest: Implications for tropical tree conservation. In M. E. Soulé (ed.), *Conservation Biology: The Science of Scarcity and Diversity*. Sinauer, Sunderland, Mass., pp. 205–32.

Hubbell, S. P., and R. B. Foster. 1990. Structure, dynamics, and equilibrium status of old-growth forest on Barro Colorado Island. In A. H. Gentry (ed.), *Four Neotropical Forests*. Yale University Press, New Haven, Conn., pp. 522–41.

Hubbell, S. P., and L. K. Johnson. 1977. Competition and nest spacing in a tropical stingless bee community. Ecology 58: 949–63.

Hunt, R. 1990. *Basic Growth Analysis*. Unwin Hyman, London.

Husband, B. C., and D. W. Schemske. 1996. Evolution of the magnitude and timing of inbreeding depression in plants. Evolution 50: 54–70.

Huston, M. A. 1994. *Biological Diversity: The Coexistence of Species on Changing Landscapes*. Cambridge University Press, Cambridge, U.K.

Hutchings, R. W. 1991. The Dynamics of Three Communities of Papilionoidea (Lepidoptera: Insecta) in Forest Fragments in Central Amazonia. M.Sc. Thesis, INPA, Manaus, Brazil.

Hutchinson, T. F., and J. L. Vankat. 1997. Invasibility and effects of amur honeysuckle in southwestern Ohio forests. Conserv. Biol. 11: 1117–24.

IBDF (Instituto Brasileiro de Desenvolvimento Florestal). 1965. *Codigo Florestal*. Instituto Brasiliero de Desevolvimento Florestal, Ministério da Agricultura, Brasilia, Brazil.

IBGE (Fundação Instituto Brasileiro de Geografia e Estatística). 1992. Anuário Estatístico, Capítulo 44: Extração Vegetal e Silvicultura.

INPE, 1998. Deforestation 1995–97 Amazônia. Instituto Nacional de Pesquisas Espaciais and Ministério da Ciência e Tecnologia, Brasília DF, Brazil.

Irion, G. 1978. Soil infertility in the Amazonian rain forest. Naturwissenschaften 65: 515–19.

Isham, G., and D. Eisikowitch. 1993. The behaviour of honey bees (*Apis mellifera*) visiting avocado (*Persea americana*) flowers and their contribution to its pollination. J. Apicultural Res. 32: 175–86.

ITTO (International Tropical Timber Organization). Annual Review and Assessment of the World Tropical Timber Situation. Reports from 1990–91, 1992, 1993–94 e 1995, ITTO, Tokyo.

IUCN (International Union for Conservation of Nature and Natural Resources). 1984. Categories, objectives, and criteria for protected areas. In J. A. McNeely and K. R. Miller (eds.), *National Parks, Conservation and Development*. Smithsonian Institution Press, Washington, D.C., pp. 47–53.

IUCN (International Union for Conservation of Nature and Natural Resources). 1988. *1988 IUCN Red List of Threatened Animals*. IUCN, Gland, Switzerland.

Iyawe, J. G. 1989. The ecology of small mammals in Ogba Forest Reserve, Nigeria. J. Trop. Ecol. 5: 51–64.

Izar, P. 1999. Aspectos de ecologia e comportamento de um grupo de macacos-pregos (*Cebus apella*) em área de Mata Atlântica em São Paulo. Ph.D. diss., Universidade de São Paulo, São Paulo.

Izawa, K. 1980. Social behavior of the wild black-capped capuchin (*Cebus apella*). Primates 21: 443–67.

Jaenike, J. 1978. Effect of island area on Drosophila population densities. Oecologia 36: 327–32.

Janson, C. H. 1984. Female choice and mating system of the brown capuchin monkey *Cebus apella* (Primates: Cebidae). Zeitschrift fur Tierpsychologie 65: 177–200.

Janson, C. H. 1986. Capuchin counterpoint. Natural History 2: 44–53.

Janson, C. H., and L. H. Emmons 1990. Ecological structure of the nonflying mammal community at Cocha Cashu Biological Station, Manu National Park, Peru. In A. H. Gentry (ed.), *Four Neotropical Rainforests*. Yale University Press, New Haven, Conn., pp. 314–38.

Janzen, D. H. 1970. Herbivores and the number of tree species in tropical forests. Am. Nat. 104: 501–28.

Janzen, D. H. 1971. Euglossine bees as long distance pollinators of tropical plants. Science 171: 203–5.

Janzen, D. H. 1975. *Ecology of plants in the tropics*. Institute of Biology's Studies in Biology, No. 58, Arnold, London.

Janzen D. H. 1983. No park is an island: Increased interference from outside as park size decreases. Oikos 41: 402–10.

Janzen, D. H. 1986a. The eternal external threat. In M. E. Soulé (ed.), *Conservation Biology: The Science of Scarcity and Diversity*. Sinauer, Sunderland, Mass., pp. 286–303.

Janzen, D. H. 1986b. The future of tropical biology. Ann. Rev. Ecol. Syst. 17: 305–24.

Janzen, D. H., P. J. De Vries, M. L. Higgins, and L. S. Kimsey. 1982. Seasonal and site variation in Costa Rican euglossine bees at chemical baits in lowland deciduous forest and evergreen forests. Ecology 63: 66–74.

Janzen, D. H., and C. Vásquez-Yanes. 1991. Aspects of tropical seed ecology of relevance to management of tropical forest wildlands. In A. Gómez-Pompa, T. C. Whitmore, and M. Hadley (eds.), *Rain Forest Regeneration and Management. Man and the Biosphere Series,* Parthenon, Paris, pp. 137–57.

Jefferies, M. J., and J. H. Lawton. 1984. Enemy-free space and the structure of ecological communities. Biol. J. Linnean Soc. 23: 269–86.

Jenkins, G. M., and D. G. Watts. 1968. Spectral analysis and its applications. Holden-Day, San Francisco.

Jenkins, S. I. 1993. Exotic plants in the rainforest fragments of the Atherton and Evelyn Tablelands, north Queensland. B. Sc.

Honors Thesis, James Cook University, Townsville, Queensland, Australia.

Jennersten, O. 1988. Pollination in *Dianthus deltoides* (Caryophyllaceae): Effects of habitat fragmentation on visitation and seed set. Conserv. Biol. 2: 359–66.

Johns, A. D. 1991. Responses of Amazonian rain forest birds to habitat modification. J. Trop. Ecol. 7: 417–37.

Johns, A. G. 1997. *Timber Production and Biodiversity Conservation in Tropical Rain Forests.* Cambridge University Press, Cambridge, U.K.

Johns, J. S., P. Barreto, and C. Uhl. 1996. Logging damage during planned and unplanned logging operations in the eastern Amazon. For. Ecol. & Manag. 89: 59–77.

Jonkers, W. B. J., and P. Schmidt. 1984. Ecology and timber production in tropical rainforest in Suriname. Interciencia 9: 290–97.

Jordan, C. 1985. Soils of the Amazon rainforest. In G. T. Prance and T. E. Lovejoy (eds.), *Key Environments: Amazonia.* Pergamon, Oxford, U.K, pp. 83–94.

Jorgensen, S. S. 1977. *Guia Analítico: Metodologia Utilizada para Análises Químicas de Rotina.* Centro de Energia Nuclear na Agricultura (CENA), Piracicaba, São Paulo, Brazil.

Jules, E. S. 1998. Habitat fragmentation and demographic change for a common plant: Trillium in old-growth forest. Ecology 79: 1645–56.

Kageyama, P., C. Castro, A. Reis, F. Gandara, N. Lepsch-Cunha, and R. Vencovsky. 1997. In situ genetic conservation of tropical tree species in Brazil. Forgen News (IPGRI) 1: 16–18.

Kahn, F. 1987. The distribution of palms as a function of local topography in Amazonian terra-firme forests. Experimentia 43: 251–59.

Kahn, F., and A. Castro. 1985. The palm community in a forest of Central Amazonia, Brazil. Biotropica 17: 210–16.

Kahn, F., K. Mejia, and A. Castro. 1988. Species richness and diversity of palms in terra firme forests of Amazonia. Biotropica 20: 266-269.

Kahn, J. R., and J. A. McDonald. 1997. The role of economic factors in tropical deforestation. In W. F. Laurance and R. O. Bierregaard, Jr. (eds.), *Tropical Forest Remnants: Ecology, Management, and Conservation of Fragmented Communities.* University of Chicago Press, Chicago, pp. 13–28.

Kalko, E. K. V. 1998. Organization and diversity of tropical bat communities through space and time. Zoology (Analysis of Complex Systems). 101: 281–97.

Kamprath, E. J. 1973a. Phosphorus. In P. A. Sanchez (ed.), *A Review of Soils Research in Tropical Latin America.* North Carolina State University, Soil Science Department, Raleigh, pp. 138–61.

Kamprath, E. J. 1973b. Soil acidity and liming. In P. A. Sanchez (ed.), *A Review of Soils Research in Tropical Latin America.* North Carolina State University, Soil Science Department, Raleigh, pp. 126–37.

Kapos, V. 1989. Effects of isolation on the water status of forest patches in the Brazilian Amazon. J. Trop. Ecol. 5: 173–85.

Kapos, V., G. Ganade, E. Matusi, and R. L. Victoria. 1993. δ ^{13}C as an indicator of edge effects in tropical rainforest reserves. J. Ecology 81: 425–32.

Kapos, V., E. Wandelli, J. L. Camargo, and G. Ganade. 1997. Edge-related changes in environment and plant responses due to forest fragmentation in central Amazonia. In W. F. Lawrence, and R. O. Bierregaard, Jr. (eds.), *Tropical Forest Remnants: Ecology, Management, and Conservation of Fragmented Communities.* University of Chicago Press, Chicago, pp. 33–44.

Karlekar, B. V., and R. M. Desmond. 1977. *Engineering heat transfer.* West Publishing, New York.

Kattenberg, A., F. Giorgi, H. Grassl, G. A. Meehl, J. F. B. Mitchell, J. Stouffer, T. Tokioka, A. J. Weaver, and T. M. L. Wigley. 1996. Climate models projections of future climate. In J. T. Houghton, L. G. Meira Filho, B. A. Callander, N. Harris, A. Kattenberg, and K. Maskell (eds.), *Climate Change 1995: The Science of Climate Change.* Cambridge University Press, Cambridge, U.K., pp. 285–57.

Kaufman, J. B. 1991. Survival by sprouting following fire in tropical forests of the Eastern Amazon. Biotropica 23: 219–24.

Kaufman, J. B., and C. Uhl. 1990. Interactions of anthropogenic activities, fire and rain forests in the Amazon Basin. In J. G. Goldhammer (ed.), *Fire in the Tropical Biota.* Springer-Verlag, Berlin, Germany, pp. 117–34.

Kearns, C. A., and D. W. Inouye. 1997. Pollinators, flowering plants, and Conserv. Biol. BioScience 47: 297–307.

Kerr, J. T. 1997. Species richness, endemism, and the choice of areas for conservation. Conserv. Biol. 11: 1094–1100.

Kerr, W. E. 1971. Contribuição é egogenetica de algumas espécies de abelhas. Cienc. Cult. São Paulo 23 (Suppl.): 89–90.

Kerr, W. E., and R. Vencovsky. 1982. Melhoramento genético em abelhas. I. Efeito do número de colônias sobre o melhoramento. Rev. Brasil. Genet. 5: 279–85.

Kimsey, L. S., and R. L. Dressler. 1986. Synonymic species list of Euglossini. Pan-Pacific Entomologist 62: 229–36.

Klar, A. E. 1984. *Agua no Sistema Solo-Planta-Atmosfera.* Nobel, São Paulo.

Klein, B. C. 1989. Effects of forest fragmentation on dung and carrion beetle communities in Central Amazonia. Ecology 70: 1715–25.

Koelewijn, H. P. 1998. Effects of different levels of inbreeding on progeny fitness in *Plantago coronopus.* Evolution 52: 692–702.

Kohlmaier, G. H., F. W. Badeck, R. D. Otto, S. Häger, S. Dönges, J. Kindermann, G. Würth, T. Lang, U. Jäkel, A. Nadler, P. Ramge, A. Klaudius, S. Habermehl, and M. K. B. Lüdeke. 1997. The Frankfurt biosphere model: A global process-oriented model of seasonal and long-term CO_2 exchange between terrestrial ecosystems and the atmosphere II. Global results for potential vegetation in an assumed equilibrium state. Climate Research 8: 61–87.

Koster, H. W., E. J. A. Khan, and R. P. Bosshart. 1977. *Programa e Resultados Preliminares dos Estudos de Pastagens na Região de Paragominas, Pará, e nordeste de Mato Grosso junho 1975–dezembro 1976.* Superintendência do Desenvolvimento da Amazônia (SUDAM), Convênio SUDAM/Instituto de Pesquisas IRI, Belem, Pará, Brazil.

Krebs, C. J. 1989. *Ecological Methodology.* Harper and Row, New York.

Kunin, W. E., and K. Gaston. 1993. The biology of rarity: Patterns, causes and consequences. Trends Ecol. Evol. 8: 298–301.

Laan, R., and B. Verboom. 1990. Effect of pool size and isolation on amphibian communities. Biological Conservation 54: 251–62.

Lacy, R. C., A. Petric, and M. Warneke. 1993. Inbreeding and outbreeding in captive population of wild animal species. In N. W. Thornhill (ed.), *The Natural History of Inbreeding and Outbreeding: Theoretical and Empirical Perspectives,* University of Chicago Press, Chicago, pp. 352–74.

LaGro, J., Jr. 1991. Assessing patch shape in landscape mosaics. Photogrammetric Engineering and Remote Sensing 57: 285–93.

Lamont and Barker. 1988. Seed bank dynamics of a serotinous fire-sensitive Banksia species. Australian J. Botany 36: 193–203.

Lande, R. 1988. Genetics and demography in biological conservation. Science 241: 1455–61.

Lande, R. 1994. Risk of population extinction from fixation of new deleterious mutations. Evolution 48: 1460–69.

Lande, R. 1998. Anthropogenic, ecological and genetic factors in extinction and conservation. Res. Pop. Ecol. 40: 259–69.

Lande, R., and G. F. Barrowclough. 1987. Effective population size, genetic variation, and their use in population management. In M. L. Soulé (ed.), *Viable Populations for Conservation.* Cambridge University Press, Cambridge, U.K., pp. 87–124.

Lasalle, J., and I. D. Gauld. 1993. Hymenoptera: Their diversity, and their impact on the diversity of the other organisms. In J. Lasalle and I. D. Gauld (eds.), *Hymenoptera and Biodiversity.* C. A. B. International, Wallingford, U.K., pp. 3–26.

Lauga, J., and J. Joachim. 1987. L'échantillonnage des populations d'oiseaux par la méthode de E. F. P.: Intérêt d'une étude mathématique de la courbe de richesse cumulée. Oecologia Generalis 8: 117–24.

Laurance, S. G. W. 1996. Utilization of linear rainforest remnants by arboreal marsupials in North Queensland. M. Sc. Thesis, University of New England, Armidale, New South Wales, Australia.

Laurance, S. G. W., and W. F. Laurance. 1999. Tropical wildlife corridors: Use of linear rainforest remnants by arboreal mammals. Biol. Conserv. 91: 231–39.

Laurance, S. G. W., and W. F. Laurance. In press. Bandages for wounded landscapes: Faunal corridors and their roles in wildlife conservation in the Americas. In G. A. Bradshaw and P. Marquet (eds.), *Landscape Alteration in the Americas.* University of Minnesota Press, Minneapolis.

Laurance, W. F. 1990. Comparative responses of five arboreal marsupials to tropical forest fragmentation. J. Mammalogy 71: 641–53.

Laurance, W. F. 1991a. Edge effects in tropical forest fragments: Application of a model for the design of nature reserves. Biological Conservation 57: 205–19.

Laurance, W. F. 1991b. Ecological correlates of extinction proneness in Australian tropical rain forest mammals. Conserv. Biol. 5: 79–89.

Laurance, W. F. 1994. Rainforest fragmentation and the structure of small mammal communities in tropical Queensland. Biological Conservation 69: 23–32.

Laurance, W. F. 1997. Hyper-disturbed parks: Edge effects and the ecology of isolated rainforest reserves in tropical Australia. In W. F. Laurance and R. O. Bierregaard, Jr. (eds.), *Tropical Forest Remnants: Ecology, Management, and Conservation of Fragmented Communities.* University of Chicago Press, Chicago, pp. 71–83.

Laurance, W. F., and R. O. Bierregaard, Jr. (eds.). 1997a. *Tropical Forest Remnants: Ecology, Management, and Conservation of Fragmented Communities.* University of Chicago Press, Chicago.

Laurance, W. F., and R. O. Bierregaard, Jr. 1997b. A crisis in the making. In W. F. Laurance and R. O. Bierregaard, Jr. (eds.), *Tropical Forest Remnants: Ecology, Management, and Conservation of Fragmented Communities.* University of Chicago Press, Chicago.

Laurance, W. F., R. O. Bierregaard, Jr., C. Gascon, R. K. Didham, A. P. Smith, A. J. Lynam, V. M. Viana, T. E. Lovejoy, K. E. Sieving, J. W. Sites, Jr., M. Andersen, M. D. Tocher, E. A. Kramer, C. Restrepo, and C. Moritz. 1997. Tropical forest fragmentation: Synthesis of a diverse and dynamic discipline. In W. F. Laurance and R. O. Bierregaard, Jr. (eds.), *Tropical Forest Remnants: Ecology, Management, and Conservation of Fragmented Communities.* University of Chicago Press, Chicago, pp. 515–25.

Laurance, W. F., P. M. Fearnside, S. G. W. Laurance, P. Delamônica, T. E. Lovejoy, J. M. Rankin–de Mérona, J. Q. Chambers, and C. Gascon. 1999. Relationship between soils and Amazon forest biomass: A landscape-scale study. For. Ecol. & Manag. 118: 127–38.

Laurance, W. F, L. V. Ferreira, J. M. Rankin–de Mérona, and S. G. W. Laurance. 1998a. Rain forest fragmentation and the dynamics of Amazonian tree communities. Ecology 79: 2032–40.

Laurance, W. F., L. V. Ferreira, J. M. Rankin–de Mérona, S. G. W. Laurance, R. W. Hutchings, and T. E. Lovejoy. 1998b. Effects of forest fragmentation on recruitment patterns in Amazonian tree communities. Conserv. Biol. 12: 460–64.

Laurance, W. F., L. V. Ferreira, J. M. Rankin–de Mérona, and R. Hutchings. 1999. Influence of plot shape on estimates of tree diversity and community composition in central Amazonia. Biotropica 30: 662–65.

Laurance, W. F., and C. Gascon. 1997. How to creatively fragment a landscape. Conserv. Biol. 11: 577–79.

Laurance, W. F., C. Gascon, and J. M. Rankin–de Mérona. 1999. Predicting effects of habitat destruction on plant communities: A test of a model using Amazonian trees. Ecological Applications 9: 548–54.

Laurance, W. F., S. G. Laurance, L. V. Ferreira, J. M. Rankin–de Mérona, C. Gascon, and T. E. Lovejoy. 1997. Biomass collapse in Amazonian forest fragments. Science 278: 1117–18.

Laurance, W. F., S. G. Laurance, and P. Delamônica. 1998. Tropical forest fragmentation and greenhouse gas emissions. For. Ecol. & Manag. 55: 173–80.

Lawton, J. H. 1993. Range, population abundance and conservation. Trends Ecol. Evol. 8: 409–13.

Lawton, J. H., D. E. Bignell, G. F. Bloemers, P. Eggleton, and M. E. Hodda. 1996. Carbon flux and diversity of nematodes and termites in Cameroon forest soils. Biodiversity and Conservation 5: 261–73.

Lawton, J. H., and G. L. Woodroffe. 1991. Habitat and the distribution of water voles: Why are there gaps in a species range? Animal Ecology 60: 79–91.

Le Houerou, H. N. 1973. Fire and vegetation in the Mediterranean basin. Proceedings Tall Timbers Fire Ecology Conference 13: 237–77.

Leck, C. F. 1979. Avian extinctions in an isolated tropical wet-forest preserve, Ecuador. Auk 96: 343–52.

Leigh, E. G. 1999. Tropical forest ecology: A view from Barro Colorado Island. Oxford University Press, Oxford, U.K.

Leigh, E. G., Jr., S. J. Wright, E. A. Herre, and F. E. Putz. 1993. The decline of tree diversity on newly isolated islands: A test of a null hypothesis and some implications. Evolutionary Ecology 7: 76–102.

Leighton, M., and N. Wirawan. 1986. Catastrophic drought and fire in Borneo rain forests associated with the 1982–3 El Nino southern oscillation event. In G. T. Prance (ed.), Tropical Rain Forests and the World Atmosphere. AAAS Symposium 101, Westview, Boulder, Colo., pp. 75–102.

Leite, L. L., and P. A. Furley. 1985. Land development in the Brazilian Amazon with particular reference to Rondônia and the Ouro Preto colonization project. In J. Hemming (ed.), Change in the Amazon Basin. Manchester University Press, Manchester, U.K, pp. 119–40.

Lenthe, H. R. 1991. Methods for monitoring organic matter in soil. In D. M. Burger (ed.), Studies on the Utilization and Conservation of Soil in the Eastern Amazon Region. Deutsche Gesellschaft fr Technische Zusammenarbeit (GTZ), Eschborn, Germany, pp. 119–29.

Lepsch-Cunha, N. 1996. Estrutura Genética e Fenologia de Espécies Raras de Couratari spp. (Lecythidaceae) na Amazônia Central. M. Sc. Thesis, ESALQ/USP (Universidade de São Paulo, Piracicaba), São Paulo, Brasil.

Lepsch-Cunha, N., P. Y Kageyama, and R. Vencovsky. 1999. Genetic diversity of Couratari multiflora and Couratari guianensis (Lecythidaceae): Consequences of two types of rarity. Biodiversity and Conservation 8: 1205–18.

Lepsch-Cunha, N., and S. Mori. 1999. Reproductive phenoer (ed.), Lectures on Mathematics in the Life Sciences, Vol. 2. Providence, R.I., pp. 77–107.

Lescure, J. P., and R. Boulet. 1985. Relationships between soil and vegetation in a tropical rain forest in French Guiana. Biotropica 17: 155–64.

Levenson, J. B. 1981. Woodlots as biogeographic islands in southeastern Wisconsin. In R. L. Burgess and D. M. Sharpe (eds.), Forest Island Dynamics in Man-

Dominated Landscapes. Springer-Verlag, New York, pp. 12–39.

Levings, S. C., and D. M. Windsor. 1983. Seasonal and annual variation in litter arthropod populations. In E. G. Leigh, A. S. Rand, and D. M. Windsor (eds.), *The Ecology of a Tropical Forest. Seasonal Rhythms and Long-Term Changes.* Oxford University Press, Oxford, U.K., pp. 355–87.

Levins, R. 1969. Some demographic and genetic consequences of environmental heterogeneity for biological control. Bull. Entomol. Soc. Amer. 15: 237–40.

Levins, R. 1970. Extinction. In M. Gerstenhaber (ed.), *Lectures on Mathematics in the Live Sciences,* Vol. 2. American Mathematical Society, Providence, R.I., pp. 77–107.

Lieberman, D., and M. Lieberman. 1987. Forest tree growth and dynamics at La Selva, Costa Rica (1969–72). J. Trop. Ecol. 3: 347–58.

Lima, M. G. 1998. Uso de matas associadas é igarapés como áreas de dispersão de pequenos mamíferos terrestres e sapos no PDBFF. M. Sc. thesis, INPA.

Lima, M., and C. Gascon. 1999. The conservation value of linear forest remnants in central amazonia. Biological Conservation 91: 241–47.

List, R. J. 1950. *Smithsonian Meteorological Tables, 6th ed.* Smithsonian Institution, Washington, D.C.

Loiselle, B. A. and J. G. Blake. 1990. Diets of understory fruit-eating birds in Costa Rica: Seasonality and resource abundance. Studies in Avian Biol. 13: 91–103.

Loiselle, B. A., and J. G. Blake. 1994. Annual variation in birds and plants of a tropical second-growth woodland. Condor 96: 368–80.

Lovejoy, T. E., and R. O. Bierregaard, Jr. 1990. Central Amazonian forests and the Minimum Critical Size of Ecosystems Project. In A. H. Gentry (ed.), *Four Neotropical Forests.* Yale University Press, New Haven, Conn., pp. 60–71.

Lovejoy, T. E., R. O. Bierregaard, Jr., J. M. Rankin, and H. O. R. Schubart. 1983. Ecological dynamics of tropical forest fragments. In S. L. Sutton, T. C. Whitmore and A. C. Chadwick (eds.), *Tropical Rain Forest: Ecology and Management.* Special Publication No. 2, British Ecological Society, Blackwell, Oxford, U.K., pp. 377–84.

Lovejoy, T. E., R. O. Bierregaard, Jr., A. B. Rylands, J. R. Malcolm, C. E. Quintela, L. E. Harper, K. S. Brown Jr., A. H. Powell, H. O. R. Shubart, and M. B. Hays. 1986. Edge and other effects of isolation on Amazon forest fragments. In M. Soulé (ed.), *Conservervation Biology: The Science of Scarcity and Diversity.* Sinauer, Sunderland, Mass., pp. 257–85.

Lovejoy, T. E., and D. C. Oren. 1981. Minimum critical size of ecosystems. In R. L. Burgess and D. M. Sharp (eds.), *Forest Island Dynamics in Man-Dominated Landscapes.* Springer-Verlag, New York, pp. 7–12.

Lovejoy, T. E., and J. M. Rankin. 1981. Uma fisionomia ameaçada. As implicações das parcelas florestais no planejamento silvicultural e de reservas. Bol. FBCN 16: 136–39.

Lovejoy, T. E., J. M. Rankin, R. O. Bierregaard, Jr., K. S. Brown, Jr., L. H. Emmons and M. E. Van der Voort. 1984. Ecosystem decay of Amazon forest remnants. In M. H. Nitecki (ed.), *Extinctions.* University of Chicago Press, Chicago, pp. 295–325.

Loveless, M. D., and J. L. Hamrick. 1987. Distribucion de la variation genetica en espécies arbóreas tropicales. Rev. Biologia Tropical 35: 165–76.

Lubin, Y. D. 1978. Seasonal abundance and diversity of web-building spiders in relation to habitat structure on Barro Colorado Island, Panama. J. Arachnology 6: 31–52.

Lubin, Y. D. 1980. Population studies of two colonial orb-weaving spiders. Zool. J. Linnean Soc. 70: 265–87.

Lubin, Y. D., and M. H. Robinson 1982. Dispersal by swarming in a social spider. Science 216: 319–21.

Lucas, R. M., M. Honzak, I. do Amaral, P. J. Curran, G. M. Foody, and S. Amaral. 1998. Composição florística, biomassa, e estrutura de florestas tropicais em regeneração: Uma avaliação por sensoriamento remoto. In C. Gascon and P. Moutinho (eds.), *Floresta Amazônica: Dinâmica, Regeneração e Manejo.* Instituto Nacional de Pesquisas da Amazônia (INPA), Manaus, Amazonas, Brazil, pp. 61–82.

Ludwig, J. A., and J. F. Reynolds. 1988. *Statistical Ecology: A Primer on Methods and Computing.* Wiley, New York.

Luken, J. O. 1990. *Directing Ecological Succession.* Chapman and Hall, London.

MacArthur, R. H. 1957. Fluctuations of animal populations and a measure of community stability. Ecology 36: 533–36.

MacArthur, R. H., and E. O. Wilson. 1963. An equilibrium theory of insular zoogeography. Evolution 17: 373–87.

MacArthur, R. H., and E. O. Wilson. 1967. *The Theory of Island Biogeography.* Princeton University Press, Princeton, N.J.

McGarigal, K., and B. J. Marks. 1994. *Fragstats: Spatial Pattern Analysis Program for Quantifying Landscape Structure.* Oregon State University, Corvallis.

McGuinness, K. A. 1984. Equations and explanations in the study of species-area curves. Biological Review 59: 423–40.

McNab, B. K. 1963. Bioenergetics and the determination of home range size. Am. Nat. 97: 133–39.

Magurran, A. E. 1988. *Ecological Diversity and Its Measurement.* Princeton University Press, Princeton, N.J.

Majer, J. D. 1983. Ants: Bio-indicators of mine site rehabilitation, land-use, and land conservation. Envir. Manag. 7: 375–83.

Malcolm, J. R. 1988. Small mammal abundances in isolated and non-isolated primary forest reserves near Manaus, Brazil. Acta Amazónica 18: 67–83.

Malcolm, J. R. 1990. Estimation of mammalian densities in continuous forest north of Manaus. In A. H. Gentry (ed.), *Four Neotropical Rainforests.* Yale University Press, New Haven, Conn., pp. 339–57.

Malcolm, J. R. 1991. The small mammals of Amazonian Forest Fragments: Pattern and process. Ph.D. diss., University of Florida, Gainesville.

Malcolm, J. R. 1994. Edge effects in central Amazonian forest fragments. Ecology 75: 2438–45.

Malcolm, J. R. 1995. Forest structure and the abundance and diversity of Neotropical small mammals. In M. D. Lowman and N. M. Nadkarni (eds.), *Forest Canopies.* Academic Press, New York, pp. 179–97.

Malcolm, J. R. 1997a. Biomass and diversity of small mammals in Amazonian forest fragments. In W. F. Laurance and R. O. Bierregaard, Jr. (eds.), *Tropical Forest Remnants: Ecology, Management, and Conservation of Fragmented Communities.* University of Chicago Press, Chicago, pp. 207–21.

Malcolm, J. R. 1997b. Insect biomass in Amazonian forest fragments. In N. E. Stork, J. Adis, and R. K. Didham (eds.), *Canopy Arthropods.* Chapman and Hall, London, pp. 510–33.

Malcolm, J. R. 1998. A model of conductive heat flow in forest edges and fragmented landscapes. Climatic Change 39: 487–502.

Malcolm, J. R., and J. C. Ray. In press. Influence of timber extraction routes on central African small mammal communities, forest structure, and tree diversity. Conserv. Biol.

Malingreau, J. P., and C. J. Tucker. 1988. Large-scale deforestation in the southeastern Amazon basin of Brazil. Ambio 17: 49–55.

Mangel, M., and C. Tier. 1994. Four facts every conservation biologist should know about persistence. Ecology 75: 607–14.

Margules, C. 1999. Conservation planning at the landscape scale. In J. A. Wiens and M. R. Moss (eds.), *Issues in Landscape Ecology.* International Association of Landscape Ecology, Guelph, CANADA.

Markgraf, V., and J. P. Bradbury. 1982. Holocene climatic history of South America. Striae 16: 40–45.

Marquis, R. J., and C. J. Whelan. 1994. Insectivorous birds increase growth of white oaks through consumption of leaf-chewing insects. Ecology 75: 2007–14.

Martins, M. B. 1985. Influência da modificação do habitat sobre a diversidade e abundância de *Drosophila* (Diptera, Drosophilidae) em uma floresta tropical da Amazônia central. M. Sc. diss., INPA/FUA. Manaus, Brasil.

Martins, M. B. 1987. Variação espacial e temporal de algumas espécies e grupos de *Drosophila* (Diptera) em duas reservas de matas isoladas, nas vizinhanças de Manaus (Amazonas, Brasil). Boletim Museu Paraense Emílio Goeldi 3: 195–98.

Martins, M. B. 1989. Invasão de fragmentos florestais por espécies oportunistas de Drosophila (Diptera, Drosophildae). Acta Amazónica 19: 265–71.

Martins, M. B. 1996. Drosófilas e outros insetos associados a frutos de Parahancornia dispersos sobre o solo da floresta. Ph.D. thesis, Universidade Estadual de Campinas, Campinas, Brazil.

Martins, M. B., and F. C. G. Fonseca. 1988. Verificação de atividade diurna de espécies de *Drosophila* (Diptera, Drosophilidae). Resumo do XV Congresso Brasileiro de Zoologia, Curitiba, PR, Brazil.

Martins, M. B., and R. C. Santos. 1988. Observações sobre a utilização de frutos da mata como sítio de reprodução por espécies de *Drosophila* (Diptera, Drosophilidae). Resumo do XV Congresso Brasileiro de Zoologia, Curitiba, PR.

Matlack, G. R. 1994. Vegetation dynamics of the forest edge: Trends in space and successional time. J. Ecol. 82: 113-123.

Maury-Lechon, G. 1991. Comparative dynamics of tropical rain forest regeneration in French Guyana. In A. Gómez-Pompa, T. C. Whitmore, and M. Hadley (eds.), *Rain Forest Regeneration and Management*, pp. 285–93. Man and the Biosphere (MAB) Series vol. 6.

May, R. M. 1975. Patterns of species abundance and diversity. In M. L. Cody and J. M. Diamond (eds.), *Ecology and Evolution of Communities*. Harvard University Press, Cambridge, Mass., pp. 81–120.

Meggers, B. J. 1994a. Biogeographical approaches to reconstructing the prehistory of Amazonia. Biogeographica 70: 97–110.

Meggers, B. J. 1994b. Archeological evidence for the impact of mega-Niño events on Amazonia during the past two millennia. Climatic Change 28: 321–38.

Mellilo, J. M., A. D. McGuire, D. W. Kicklighter, B. Moore III, C. J. Vorosmarty, and A. L. Schloss. 1993. Global climate change and terrestrial net primary productivity. Nature 363: 234–40.

Mendes, S. L. 1989. Estudo ecológico de *Alouatta fusca* (Primates: Cebidae) na Estação Biológica de Caratinga, MG. Rev. Nordestina Biologia 6: 71–104.

Menezes, P., S. de Menezes, and J. M. F. de Camargo. 1991. Interactions on floral resources between the Africanized honey bee *Apis mellifera* L., and the native bee community (Hymenoptera: Apoidea) in a natural "cerrado" ecosystem in southeast Brazil. Apidologie 22: 397–415.

Menges, E. S. 1990. Population viability analysis for an endangered plant. Conserv. Biol. 4: 52–62.

Menges, E. S. 1992. Stochastic modeling of extinction in plant populations. In P. L. Fiedler and S. K. Jain (eds.), *Conservation Biology: The Theory and Practice of Nature Conservation, Preservation and Management*. Chapman and Hall, New York, pp. 253–75.

Merriam, G., and A. Lanoue. 1990. Corridor use by small mammals: Field measurement for three experimental types of *Peromyscus leucopus*. Landscape Ecology 4: 123–31.

Mesquita, R. C. G., P. Delamônica, and W. F. Laurance. 1999. Effects of surrounding vegetation on edge-related tree mortality in Amazonian forest fragments. Biological Conservation 91: 129–34.

Michener, C. D. 1946. Notes on the habits of

some Panamanian stingless bees (Hymenoptera, Apidae). Jour. N. Y. Entomol. Soc. 54: 179–97.

Michener, C. D. 1975. The Brazilian bee problem. Ann. Rev. Entomol. 20: 399–416.

Milliken, W. 1998. Structure and composition of one hectare of central Amazonian terre firme forest. Biotropica 30: 530–37.

Mills, L. S., M. E. Soulé, and D. F. Doak. 1993. The keystone-species concept in ecology and conservation. BioScience 43: 219–24.

Milne, B. T., and R. T. T. Forman. 1986. Peninsulas in Maine: Woody plant diversity, distance, and environmental patterns. Ecology 67: 967–74.

Miriti, M. N. 1998. Regeneração florestal em pastagens abandonadas na Amazônia central: Competição, predação, e dispersão de sementes. In C. Gascon and P. Moutinho (eds.), *Floresta Amazônica: Dinâmica, Regeneração e Manejo*. Instituto Nacional de Pesquisas da Amazônia (INPA), Manaus, Amazonas, Brazil, pp. 179–90.

Mitchell, J. D., and S. A. Mori. 1987. Ecology. In S. A. Mori and collaborators (eds.), *The Lecythidaceae of a Lowland Neotropical Forest: La Fumée Mountain, French Guiana*. Memoirs of the New York Botanical Garden 44, New York Botanical Garden, Bronx, New York, pp. 137–55.

Mittermeier, R. A., and M. G. M. van Roosmalen. 1981. Preliminary observations on habitat utilization and diet in eight Suriname monkeys. Folia Primatologia 36: 1–39.

Monkkonen, M., and P. Reunanen. 1999. On critical thresholds in landscape connectivity: A management perspective. Oikos 84: 302–5.

Mooney, H. A., and J. A. Drake (eds.). 1986. *Ecology of Biological Invasions of North America and Hawaii*. Springer-Verlag, New York.

Moraes, M. L. T. 1992. Variabilidade genética por isoenzimas e caracteres quantitativos em duas populações naturais de aroeira *Myracrodruon urundeuva* M. F. Allemão–

Anacardiaceae. Ph.D. diss., ESALQ/ Universidade de São Paulo, Piracicaba, SP, Brasil.

Moraes, P. B., M. B. Martins, and A. N. Hagler. 1995. Niche partitioning by *Drosophila* feeding on a succession of yeasts in the Amazonian *Parahancornia amapa* fruit. Applied and Environmental Microbiology 61: 4251–57.

Moran, E. F. 1981. *Developing the Amazon*. Indiana University Press, Bloomington.

Moran, E. F., E. Brondizio, P. Mausel, and Y. Wu. 1994. Integrating Amazonian vegetation, land-use, and satellite data. Bioscience 44: 329–38.

Morato, E. F. 1993. Efeitos da fragmentação florestal sobre vespas e abelhas solitárias em uma área da Amazônia central. M. Sc. Thesis, Universidade Federal de Viçosa, Viçosa, Brazil.

Morato, E. F. 1994. Abundância e riqueza de machos de Euglossini (Hymenoptera: Apidae) em mata de terra firme e áreas de derrubada, nas vizinhanças de Manaus (Brasil). Boletim Museu Paraense Emílio Goeldi, sér. Zool. 10: 95–105.

Morato, E. F., L. A. O. Campos, and J. S. Moure. 1992. Abelhas Euglossini (Hymenoptera, Apidae) coletadas na Amazônia Central. Revta. bras. Ent. 36: 767–71.

Mori, S. A. 1990. Diversificação e conservação das Lecythidaceae Neotropicais. Acta Botanica Brasilica 4: 45–68.

Mori, S. A. 1991. The Guayana lowland floristic province. Compte-Rendu des Seances de la Société de Biogéographie 67: 67–75.

Mori, S. A. 1992a. *Eschweilera pseudodecolorans* (Lecythidaceae), a new species from central Amazonian Brazil. Brittonia 44: 244–46.

Mori, S. A. 1992b. Neotropical floristics and inventory: Who will do the work? Brittonia 44: 372–75.

Mori, S. A. 1995. Observações sobre as espécies de Lecythidaceae do leste do Brasil. Boletim de Botânica da Universidade de São Paulo 14: 1–31.

Mori, S. A., and P. Becker. 1991. Flooding affects survival of Lecythidaceae in terra firme forest near Manaus, Brazil. Biotropica 23: 87–90.

Mori, S. A., and collaborators (eds.). 1987. *The Lecythidaceae of a lowland Neotropical forest: La Fumée Mountain, French Guiana.* Memoirs of the New York Botanical Garden 44: 1–190.

Mori, S. A., P. Becker, D. Black, and C. de Zeeuw. 1987. Chapter VII. Habit and bark. In S. A. Mori and collaborators (eds.), *The Lecythidaceae of a lowland Neotropical forest: La Fumée Mountain, French Guiana.* Mem. New York Botanical Garden 44: 86–99.

Mori, S. A., P. Becker, B. V. Rabelo, C.-H. Tsou, and D. Daly. 1989. Composition and structure of an eastern Amazonian forest at Camaipi, Amapá, Brazil. Boletim Museu Paraense de História Natural 5: 3–18.

Mori, S. A., and J. Boeke. 1987. Chapter XII. Pollination. In S. A. Mori and collaborators. *The Lecythidaceae of a lowland Neotropical forest: La Fumée Mountain, French Guiana.* Mem. New York Botanical Garden 44: 137–55.

Mori, S. A., and B. M. Boom. 1987. Chapter II. The forest. In S. A. Mori and collaborators (eds.), *The Lecythidaceae of a lowland Neotropical forest: La Fumée Mountain, French Guiana.* Mem. New York Botanical Garden 44: 9–29.

Mori, S. A and N. Lepsch-Cunha. 1995. Lecythidaceae of a Central Amazonian moist forest. Mem. New York Botanical Garden. 75: 1–55.

Mori, S. A., and G. T. Prance. 1987. Species diversity, phenology, plant-animal interactions, and their correlation with climate, as illustrated by the Brazil nut family (Lecythidaceae). In R. E. Dickinson (ed.), *The Geophysiology of Amazonia.* Wiley, New York, pp. 69–89.

Mori, S. A., and G. T. Prance. 1990. Lecythidaceae–Part II. The zygomorphic-flowered New World Genera (*Couropita, Corythophora, Berthollethia, Couratari, Eschweilera and Lecythis*). Flora Neotropica 21(II): 1–376.

Moutinho, P. R. S. 1998. Impactos do uso da terra sobre a fauna de formigas: Consequências para a recuperação florestal na Amazônia. In C. Gascon and P. Moutinho (eds.), *Floresta Amazônica: Dinâmica, Regeneração e Manejo.* Instituto Nacional de Pesquisas da Amazônia (INPA), Manaus, Amazonas, Brazil, pp. 155–70.

Murawski, D. A. 1995. Reproductive biology and genetics of tropical trees from a canopy perspective. In M. D. Lowman and N. M. Nadkarni (eds.), *Forest Canopies.* Academic Press, London, pp. 457–93.

Murawski, D. A., B. Dayanandan, and K. S. Bawa. 1994. Outcrossing rates of two endemic *Shorea* species from Sri Lankan tropical rain forests. Biotropica 26: 23–29.

Murawski, D. A., and J. L. Hamrick. 1991. The effect of the density of flowering individuals on the mating systems of nine tropical tree species. Heredity 67: 167–74.

Murawski, D. A. and J. L. Hamrick. 1992a. The mating system of *Cavanillesia platanifolia* under extremes of flowering-tree density: A test of predictions. Biotropica 24(1): 99–101.

Murawski, D. A., and J. L. Hamrick. 1992b. Mating system and phenology of *Ceiba pentandra* (Bombacaceae) in Central Panama. J. Heredity 83: 401–4.

Murawski, D. A., J. L. Hamrick, S. P. Hubbell, and R. B. Foster. 1990. Mating system of two bombacaceous trees of a Neotropical moist forest. Oecologia 82: 501–6.

Murawski, D. A., I. A. U. Nimal Gunatilleke, and K. Bawa. 1994. The effects of selective logging on inbreeding in *Shorea megistophylla* (Dipterocarpaceae) from Sri Lanka. Conserv. Biol. 8: 997–1002.

Murcia, C. 1995. Edge effects in fragmented forests: Implications for conservation. Trends Ecol. Evol. 10: 58–62.

Murcia, C. 1996. Forest fragmentation and the pollination of Neotropical plants. In J. Schelhas, and R. Greenberg (eds.), *Forest*

Patches in Tropical Landscapes. Island, Washington, D.C., pp. 19–36.

Murphy, D. D. 1989. Conservation and confusion: Wrong species, wrong scale, wrong conclusion. Conserv. Biol. 3: 82–84.

Myers, N. 1981. Conservation needs and opportunities in tropical moist forests. In H. Synge (ed.), *The Biological Aspects of Rare Plant Conservation*. Wiley, New York, pp. 141–54.

Myers, N. 1988a. Tropical forests and their species: Going, going . . . In E. O. Wilson (ed.), *Biodiversity*. National Academy Press, Washington, D.C., pp. 28–35.

Myers, N. 1988b. Threatened biotas: "Hotspots" in tropical forests. The Environmentalist 8: 1–20.

Nadkarni, N. M., and J. T. Longino. 1990. Invertebrates in canopy and ground organic matter in a Neotropical montane forest, Costa Rica. Biotropica 22: 286–89.

Namba, T., A. Umemoto, and E. Minami. 1999. The effects of habitat fragmentation on persistence of source-sink metapopulations in systems with predators and prey or apparent competition. Theor. Pop. Bio., 56: 123–37.

Nascimento, H. E. M., A. A. J. Tabanez, and V. M. Viana. 1996. Estrutura e dinâmica de dois fragmentos de floresta estacional semidecídua na região de Piracicaba, SP. Resumo do 3º Congresso Nacional de Ecologia do Brasil. Universidade de Brasília, Brasília, DF.

Nason, J. D., P. R. Aldrich, and J. L. Hamrick. 1997. Dispersal and the dynamics of genetic structure in fragmented tropical tree populations. In W. F. Laurance and R. O. Bierregaard, Jr. (eds.), *Tropical Forest Remnants: Ecology, Management, and Conservation of Fragmented Communities*. University of Chicago Press, Chicago, pp. 304–20.

Nason, J. D., and J. L. Hamrick. 1997. Reproductive and genetic consequences of forest fragmentation: Two case studies of Neotropical canopy trees. J. Heredity 88: 264–76.

National Academy of Sciences. 1972. *Soils of the Humid Tropics*. National Academy of Sciences, Washington, D.C.

Nee, M. 1995. *Flora preliminar do Projeto Dinâmica Biológica de Fragmentos Florestais (PDBFF)*. New York Botanical Garden and INPA/Smithsonian Projeto Dinâmica Biológica de Fragmentos Florestais, Manaus, Estado de Amazonas, Brasil.

Negrelle, R. R. B. 1995. Sprouting after uprooting of canopy trees in the Atlantic rain forest of Brazil. Biotropica 27: 448–54.

Nei, M. 1987. *Molecular Evolutionary Genetics*. Columbia University Press, New York.

Nei, M., T. Maruyamo, and R. Chakraborty. 1975. The bottleneck effect and genetic variation in populations. Evolution 29: 1–10.

Nelson, B. W. 1994. Natural forest disturbance and change in Brazilian Amazon. Remote Sensing Reviews 10: 105–25.

Nelson, B. W., M. L. Absy, E. M. Barbosa, and G. T. Prance. 1985. Observations on flower visitors to *Bertholletia excelsa* H. & K., and *Couratari tenuicarpa* A. C. SM. (Lecythidaceae). Acta Amazónica (suppl.) 15: 225–34.

Nelson, B. W, C. A. C. Ferreira, M. F. da Silva, and M. L. Kawasaki. 1990. Endemism centres, refugia, and botanical collection density in Brazilian Amazonia. Nature 345: 714–16.

Nelson, B. W, V. Kapos, J. B. Adams, W. J. Oliveira, O. P. G. Braun, and I. L. do Amaral. 1994. Forest disturbance by large blowdowns in the Brazilian Amazon. Ecology 75: 853–58.

Nelson, R., and B. Holben 1986. Identifying deforestation in Brazil using multiresolution satellite data. International J. Remote Sensing 7: 429–88.

Nentwig, W. 1985. Social spiders catch larger play: A study of *Anelosimus eximius* (Araneae: Theridiidae). Behavior Ecology and Sociobiology 17: 79–85.

Nepstad, D. C., C. R. Carvalho, E. A. Davidson, P. Jipp, P. Lefebvre, G. H. Negreiros, E.

D. Silva, T. A. Stone, S. Trumbore, and S. Vieira. 1994. The role of deep roots in the hydrologic and carbon cycles of Amazonian forest and pastures. Nature 372: 666–69.

Nepstad, D. C, A. G. Moreira and A. A. Alencar. 1999. *Flames in the Rain Forest: Origins, Impacts and Alternatives to Amazonian Fire.* Pilot Program to Conserve the Brazilian Rain Forest, Brasília, Brazil.

Nepstad, D. C., C. Uhl, C. A. Pereira, and J. M. C. da Silva. 1996. A comparative study of tree establishment in abandoned pasture and mature forest of eastern Amazonia. Oikos 76: 25–39.

Nepstad, D. C., C. Uhl, and E. A. S. Serrão. 1990. Surmounting barriers to forest regeneration in abandoned, highly degraded pastures: A case study from Paragominas, Pará, Brasil. In A. Anderson (ed.), *Alternatives to Deforestation, Steps Toward Sustainable Use of the Amazon Rain Forest.* Columbia University Press, New York, pp. 215–29.

Nepstad, D. C., C. Uhl, and E. A. S. Serrão. 1991. Recuperation of a degraded Amazonian landscape: Forest recovery and agricultural restoration. Ambio 20: 248–55.

Nepstad, D. C., A. Veríssimo, A. Alencar, C. Nobre, E. Lima, P. Lefebvre, P. Schlesinger, C. Potter, P. Moutinho, E. Mendoza, M. Cochrane, and V. Brooks. 1999. Large-scale impoverishment of Amazonian forests by logging and fire. Nature 398: 505–8.

Neves, A. M. S. 1985. Alguns aspectos da ecologia de *Alouatta seniculus* em reserva isolada na Amazônia Central. B. S. Monograph, Universidade de São Paulo, Ribeirão Preto, Brasil.

Neves, A. M. S., and A. B. Rylands. 1991. Diet of a group of howling monkeys, *Alouatta seniculus,* in an isolated forest patch in central Amazonia. Primatologia no Brasil 3: 263–74.

Neves, E. L., and B. F. Viana. 1997. Inventário da fauna de Euglossinae (Hymenoptera: Apidae) do baixo sul da Bahia, Brasil. Rev. bras. Zool. 14: 831–37.

Newmark, W. D. 1991. Tropical forest fragmentation and the local extinction of understory birds in the eastern Usambara Mountains, Tanzania. Conserv. Biol. 5: 67–78.

Newstrom, L. E., G. W. Frankie, H. G. Baker, and R. K Colwell. 1994. Diversity of long-term flowering patterns. In L. A. McDade, K. S. Bawa, H. A. Hespenheide, and G. S. Hartshorn (eds.), *La Selva: Ecology and Natural History of a Neotropical Rain Forest.* University of Chicago Press, Chicago, pp. 142–60.

Newton, P. 1992. Feeding and ranging patterns of forest hanuman langurs (*Presbytis entellus*). International J. Primatology 13: 245–85.

Nicholls, N., and 98 others. 1996. Observed climate variability and change. In J. T. Houghton, L. G. Meira Filho, B. A. Callander, N. Harris, A. Kattenberg, and K. Maskell (eds.), *Climate Change 1995: The Science of Climate Change.* Cambridge University Press, Cambridge, U.K., pp. 133–92.

Nichols-Orians, C. M. 1991. Environmentally-induced differences in plant traits: Consequences to a leaf-cutter ant. Ecology 72: 1609–23.

Nobre, C., P. Sellers, and J. Shukla. 1991. Amazonian deforestation and regional climate change. J. Climate 4: 957–88.

Norton, D. A., R. J. Hobbs, and L. Atkins. 1995. Fragmentation, disturbance, and plant distribution: Mistletoes in woodland remnants in the Western Australian wheatbelt. Conserv. Biol. 9: 426–38.

Noss, R. F. 1987. Corridors in real landscapes: A reply to Simberloff and Cox. Conserv. Biol. 1: 159–64.

Nunes, A. 1995. Foraging and ranging patterns in white-bellied spider monkeys. Folia Primatologica 65: 85–99.

Nydal, R., and K. Lövseth. 1970. Tracing bomb C–14 in the atmosphere, 1962–80. J. Geophysical Research 88: 3621–42.

O'Brien, T. G., and M. F. Kinnaird. 1996. Changing populations of birds and mammals in North Sulawesi. Oryx 30: 150–56.

Offerman, H., V. H. Dale, S. M. Pearson, R. O. Bierregaard, Jr., and R. V. O'Neill. 1995. Effects of forest fragmentation on Neotropical fauna: Current research and data availability. Environmental Reviews 3: 191–211.

Okia, N. O. 1992. Aspects of rodent ecology in Lunyo Forest, Uganda. J. Trop. Ecol. 8: 153–67.

Oliveira, A. A. de. 1997. Diversidade, estrutura e dinâmica do componente arbóreo de uma floresta de uma floresta de terra firme de Manaus, Amazonas. Ph.D. diss., Universidade de São Paulo, São Paulo.

Oliveira, A. A. de, and D. C. Daly. 1999. Geographic distribution of tree species occurring in the region of Manaus, Brazil: Implications for regional diversity and conservation. Biodiversity and Conservation 8: 1245–59.

Oliveira, A. A., and S. Mori. 1999. A central Amazonian terra firme forest. I. High tree species richness on poor soils. Biodiversity and Conservation 8: 1219–44.

Oliveira, L. 1993. Flutuação populacional de *D. malerkotliana* nas vizinhanças de Belém, Pará. Undergraduate Monograph. Universidade Federal do Pará, Brasil.

Oliveira, M. L., and L. A. O. Campos. 1995. In press. Abundância, riqueza e diversidade de abelhas Euglossinae (Hymenoptera, Apidae) em florestas de terra firme na Amazônia Central. Rev. bras. Zool 12: 547–56.

Oliveira, M. L., E. F. Morato, and M. V. B. Garcia. 1995. Diversidade de espécies e densidade de ninhos de abelhas sociais sem ferrão (Hymenoptera, Apidae, Meliponinae) em floresta de terra firme na Amazônia Central. Rev. bras. Zool. 12: 13–24.

Oliveira-Filho, A. T., J. M. de Mello, and J. R. S. Scolforo. 1997. Effects of past disturbance and edges on tree community structure and dynamics within a fragment of tropical semideciduous forest in southeastern Brazil over a five-year period (1987–92). Plant Ecology 131: 45–66.

Olson, D. K. 1986. Determining range size for arboreal monkeys: Methods, assump-tions, and accuracy. In D. M. Taub and F. A. King (eds.), *Current Perspectives in Primate Social Dynamics.* van Nostrand Reinhold, New York, pp. 212–27.

Olson, J. S. 1963. Energy storage and the balance of producers and decomposers in ecological systems. Ecology 44: 322–31.

O'Malley, D. M., and K. S. Bawa. 1987. Mating system of a tropical rain forest tree species. Amer. J. Bot. 74: 1143–49.

O'Malley, D. M., D. P. Buckley, G. T. Prance, and K. S. Bawa. 1988. Genetics of Brazil nut (*Bertholletia excelsa* Humb and Bonpl.: Lecythidaceae). 2. Mating system. Theor. Appl. Genet. 76: 929–32.

Onderdonk, D. A., and C. A. Chapman. 1996. The effects of forest fragmentation on primate communities in Kibale National Park, Uganda. Paper presented at 16th Congress of the International Primatological Society, Madison, Wisc.

O'Neill, R. V., J. R. Krummel, R. H. Gardner, G. Sugihara, B. Jackson, D. L. DeAngels, B. T. Milne, M. G. Turner, B. Zygmunt, S. W. Christensen, V. H. Dale, and R. L. Graham. 1988. Indices of landscape pattern. Landscape Ecology 1: 153–62.

Opler, P. A., and K. S. Bawa. 1978. Sex ratio in tropical forest trees. Evolution 32: 812–21.

Osorioberistain, M., C. A. Dominguez, L. E. Eguiarte, and B. Benrey. 1997. Pollination efficiency of native and invading Africanized bees in the tropical dry forest annual plant, *Kallstroemia grandiflora* Torr ex Gray. Apidologie 28: 11–16.

Osunkoya, O. O. 1994. Postdispersal survivorship of north Queensland rainforest seeds and fruits: Effects of forest, habitat and species. Australian J. Ecology 19: 54–64.

Oxley, D. J., M. B. Fenton and G. R. Carmody. 1974. The effects of roads on populations of small mammals. J. Appl. Ecol. 11: 51–59.

Ozanne, C. M., C. Hambler, A. Foggo, and M. R. Speight. 1997. The significance of edge effects in the management of forests for invertebrate biodiversity. In N. E. Stork, J. Adis, and R. K. Didham (eds.), *Canopy Arthropods*. Chapman and Hall, London, pp. 534–50.

Pacífico, C. 1997. "Distrito Agropecuário: ZF–9, o ramal dos iludidos." Amazonas em Tempo [Manaus], 1 June 1997, p. A4.

Paiva, J. R., P. Y. Kageyama, and R. Vencovsky. 1994a. Genetics of rubber tree [*Hevea brasiliensis* (Willd. ex Adr. de Juss.) Müll. Arg.] 2. Mating system. Silvae Genetica 43: 373–76.

Paiva, J. R., P. Y. Kageyama, R. Vencovsky, and E. P. B. Contel. 1994b. Genetics of rubber tree [*Hevea brasiliensis* (Willd. ex Adr. de Juss.) Müll. Arg.] 1. Genetic variation in natural populations. Silvae Genetica 43: 307–12.

Parkinson, J. A., and S. E. Allen. 1975. A wet oxidation procedure suitable for the determination of nitrogen and mineral nutrients in biological material. Communications in Soil Science and Plant Analysis 6: 1–11.

Parsons, P. A. 1991. Biodiversity Conservation under global climatic change: The insect Drosophila as a biological indicator? Global Ecology and Biogeography Letters 1: 77–83.

Paton, D. C. 1993. Honey bees in the Austrailian environment. BioScience 43: 95–103.

Patterson, B. D. 1987. The principle of nested subsets and its implications for biological conservation. Conserv. Biol. 1: 323–34.

Patton, J. L., M. N. F. da Silva, M. C. Lara, and M. A. Mustrangi. 1997. Diversity, differentiation, and the historical biogeography of nonvolant small mammals of the Neotropical forests. In W. F. Laurance and R. O. Bierregaard, Jr. (eds.), *Tropical Forest Remnants: Ecology, Management, and Conservation of Fragmented Communities*. University of Chicago Press, Chicago, pp. 455–65.

Pavan, C. 1959. Relações entre populações naturais de *Drosophila* e o meio ambiente.

Universidade de São Paulo Faculdade de Filosofia, Ciencias e Letras 221: 1–81.

Pearson, D. L., and A. Derr 1986. Seasonal patterns of lowland forest floor arthropod abundance in south-eastern Peru. Biotropica 18: 244–56.

Pearson, D. L., and R. L. Dressler. 1985. Two-year study of male orchid bee (Hymenoptera, Apidae, Euglossini) attraction to chemical baits in lowland south-eastern Peru. J. Trop. Ecol. 1: 37–54.

Pease, C. M., and N. L. Fowler. 1997. A systematic approach to some aspects of Conserv. Biol. Ecology 78: 1321–29.

Peck, R. B. 1990. Promoting agroforestry practices among small producers: The case of the coca agroforestry project in Amazonian Ecuador. In A. B. Anderson (ed.), *Alternatives to Deforestation: Steps Toward Sustainable Use of the Amazon Rain Forest*. Columbia University Press, New York, pp. 167–80.

Peltonen, A., and I. Hanski. 1991. Patterns of island occupancy explained by colonization and extinction rates in shrews. Ecology 72: 1698–708.

Pereira, C. A., and C. Uhl. 1998. Crescimento de árvores de valor econômico em áreas de pastagens abandonadas no nordeste do Estado do Pará. In C. Gascon and P. Moutinho (eds.), *Floresta Amazônica: Dinâmica, Regeneração e Manejo*. Instituto Nacional de Pesquisas da Amazônia (INPA), Manaus, Amazonas, Brazil, pp. 249–60.

Peres, C. A. 1991. Ecology of mixed-species groups of tamarins in Amazonian terra firme forests. Ph.D. diss., University of Cambridge, Cambridge, U.K.

Peres, C. A. 1993. Structure and spatial organisation of an Amazonian terra firme forest primate community. J. Trop. Ecol. 9: 259–76.

Peres, C. A. 1994a. Primate responses to phenological changes in an Amazonian terra firme forest. Biotropica 26: 98–112.

Peres, C. A. 1994b. Diet and feeding ecology of grey woolly monkeys (*Lagothrix lago-*

tricha cana) in central Amazonia: Comparisons with other atelines. International J. Primatology 15: 333–72.

Peres, C. A. 1994c. Composition, density, and fruiting phenology of arborescent palms in an Amazonian terra firme forest. Biotropica 26: 284–94.

Perfecto, I., and R. R. Snelling. 1995. Biodiversity and the transformation of a tropical agroecosystem: Ants in coffee plantations. Ecological Applications 5: 1084–97.

Peters, C. M., A. H. Gentry, and R. O. Mendelsohn. 1989. Valuation of an Amazonian rainforest. Nature 339: 655–56.

Petranka, J. W., M. E. Eldridge, and K. E. Haley. 1993. Effects of timber harvesting on southern Appalachian salamanders. Conserv. Biol. 7: 363–70.

Phelps, J. P., and R. A. Lancia. 1995. Effects of clearcut on the herpetofauna of a South Carolina bottomland swamp. Brimleyana 22: 31–45.

Phillips, O. 1993. The potential for harvesting fruits in tropical rainforests: New data from Amazonian Peru. Biodiversity and Conservation 2: 18–38.

Phillips, O. L., and A. H. Gentry. 1994. Increasing turnover through time in tropical forests. Science 261: 954–58.

Phillips, O. L., P. Hall, A. H. Gentry, R. Vasquez, and S. Sayer. 1994. Dynamics and species richness of tropical rain forests. Proc. Nat. Acad. Sci. 91: 2805–9.

Phillips, O. L., Y. Malhi, N. Higuchi, W. F. Laurance, P. V. Núñez, R. M. Vásquez, S. G. Laurance, L. V. Ferreira, M. Stern, S. Brown, and J. Grace. 1998. Changes in the carbon balance of tropical forests: evidence from long-term plots. Science 282: 439–42.

Pickett, S. T. A., and M. L. Cadenasso. 1995. Landscape ecology: Spatial heterogeneity in ecological systems. Science 269: 331–34.

Pickett, S. T. A., and J. N. Thompson. 1978. Patch dynamics and the design of nature

reserves. Biological Conservation 13: 27–37.

Pielou, E. C. 1984. *The Interpretation of Ecological Data: A Primer on Classification and Ordination.* Wiley,

Pimm, S. L. 1991. *The Balance of Nature?* University of Chicago Press, Chicago.

Pimm, S.L., J. H. Lawton, 1980. Are food webs compartmented? J. Animal Ecology 49: 879–98.

Pinheiro, R. 1997. "Zoneamento econômico-ecológico: Sudeste e nordeste do AM têm maior potencial econômico." Amazonas em Tempo [Manaus], 26 September 1997, p. B–2.

Piperno, D. R., and P. Becker. 1996. Vegetational history of a site in the central Amazon Basin derived from phytolith and charcoal records from natural soils. Quaternary Research 45: 202–9.

Pires, J. M., and G. T. Prance 1985. The vegetation types of the Brazilian Amazon. In G. T. Prance and T. E. Lovejoy (eds.), *Key Environments: Amazonia.* Pergamon Press, Oxford, U. K., pp. 109–145.

Pollard, E., M. L. Hall, and T. J. Bibby. 1986. Monitoring the Abundance of Butterflies, 1976–1985. No. 2. Britain: Nature Conservation Council.

Posey, D. A., and J. M. F. Camargo. 1985. Additional notes on the classification and knowledge of stingless bees (Meliponinae, Apidae, Hymenoptera) by the Kayapo Indians of Gorotire, Pará, Brazil. Annals of the Carnegie Museum 54: 247–74.

Potter, C. S., J. T. Randerson, C. B. Field, P. A. Matson, P. M. Vitousek, H. A. Mooney, and S. A. Klooster. 1993. Terrestrial ecosystem production: A process model based on global satellite and surface data. Global Biogeochemical Cycles 7: 811–41.

Potter, M. A. 1990. Movement of North Island brown kiwi between forest fragments. New Zealand J. Ecology 14: 17–14.

Powell, A. H., and G. V. N. Powell. 1987. Population dynamics of male euglossine bees in Amazonian forest fragments. Biotropica 19: 176–79.

Powell, G. V. N. 1985. Sociobiology and adaptive significance of interspecific foraging flocks in the Neotropics. In P. A. Buckley, M. S. Foster, E. S. Morton, R. S. Ridgely, and F. G. Buckley (eds.), *Neotropical Ornithology*. Ornithological Monograph Number 36. American Ornithologists Union, Washington, D.C., pp. 713–32.

Powell, G. V. N., and R. Bjork. 1995. Implications for intratropical migration on reserve design: A case study using *Pharomachrus mocinno*. Conserv. Biol. 9: 354– 362.

Prance, G. T. 1975. The history of the INPA capoeira based on ecological studies of Lecythidaceae. Acta Amazónica 5: 261– 63.

Prance, G. T. 1976. The pollination and androphore structure of some Amazonian Lecythidaceae. Biotropica 8: 235–41.

Prance, G. T. 1982. A review of the phytogeographic evidences for Pleistocene climate changes in the Neotropics. Ann. Missouri Bot. Gard. 69: 594–624.

Prance, G. T. 1990. The floristic composition of the forests of Central Amazonian Brazil. In A. H. Gentry (ed.), *Four Neotropical Rainforests*. Yale University Press, New Haven, Conn., pp. 112–40.

Prance, G. T., and S. A. Mori. 1979. Lecythidaceae—Part 1. The actinomorphic-flowered New World Lecythidaceae (*Asteranthos, Gustavia, Grias, Allantoma,* and *Cariniana*). Flora Neotropica Monograph 21: 1–270.

Prance, G. T, W. A. Rodrigues, and M. F. da Silva. 1976. Inventário florestal de um hectare de mata de terra firme km 30 da estrada Manaus-Itacoatiara. Acta Amazónica 6: 9–35.

Prendergast, J. R., R. M. Quinn, J. H. Lawton, B. C. Eversham, and D. W. Gibbons. 1993. Rare species, the coincidence of diversity hotspots and conservation strategies. Nature 365: 335–37.

Pressey, R. L., I. R. Johnson, and P. D. Wilson. 1994. Shades of irreplaceability: Towards a measure of the contribution of sites to a reservation goal. Biodiversity and Conservation 3: 242–62.

Preston, F. W. 1948. The commonness, and rarity, of species. Ecology 29: 254–83.

Preston, F. W. 1962. The canonical distribution of commonness and rarity. Ecology 43: 185–95, 410–32.

Primack, R. B. 1992. Dispersal can limit local plant distribution. Conserv. Biol. 6: 513–19.

Primack, R. B. 1993. *Essentials of Conservation Biology*. Sinauer, Sunderland, Mass.

Primavesi, A. 1981. *O Manejo Ecológico do Solo: Agricultura em Regiões Tropicais, 3*[a] ed. Nobel, São Paulo.

Pulliam, H. R. 1988. Sources, sinks, and population regulation. Am. Nat. 132: 652–61.

Pulliam, H. R., and B. J. Danielson. 1991. Sources, sinks and habitat selection: A landscape perspective on population dynamics. Am. Nat. 138 (Suppl.): 50–66.

Putz, F. E., and N. V. L. Brokaw. 1989. Sprouting of broken trees in Barro Colorado Island, Panama. Ecology 70: 508–12.

Quezada-Euan, J. J., and Y. W. De. J. May-Itza. 1996. Características morfométricas, poblacionales y parasitosis de colonias silvestres de *Apis mellifera* (Hymenoptera: Apidae) en Yucatán, Mexico. Folia Entomológica Mexicana 97: 1–19.

Quinn, J. F., and A. Hastings. 1987. Extinctions in subdivided habitats. Conserv. Biol. 1: 198–208.

Quinn, J. F., and A. Hastings. 1988. Extinctions in subdivided habitats: A reply to Gilpin. Conserv. Biol. 2: 293–96.

Quintela, C. E. 1985. Forest fragmentation and differential use of natural and man-made edges by understory birds in central Amazonia. M. Sc. thesis, University of Illinois.

Rabinowitz, D. 1981. Seven forms of rarity. In H. Synge (ed.), *The Biological Aspects of Rare Plant Conservation*. Wiley, London, pp. 205–17.

Rabinowitz, D., S. Cairns, and T. Dillon. 1986. Seven forms of rarity and their frequency in the flora of British Isles. In M. E. Soulé (ed.), *Conservation Biology: Science of Scarcity and Diversity*. Sinauer, Sunderland, Mass., pp. 182–204.

RADAMBRASIL. 1973–82. *Levantamento de Recursos Naturais.* Ministério das Minas e Energia, Departamento de Produção Mineral (DNPM), Rio de Janeiro. 27 Vols.

Rahm, U. 1972. Zur oekologie der muriden in regenwaldgebiet des östlichen Kongo (Zaire). Revue Suisse de Zoologie 79: 1121–30.

Rankin–de Mérona, J. M., and D. D. Ackerly. 1987. Estudos populacionais de árvores em florestas fragmentadas e as implicações para conservação "in situ" das mesmas na floresta tropical da Amazônia Central. Revista IPEF (Instituto de Pesquisas e Estudos Florestais, Escola Superior de Agrocultura "Luiz de Queiroz," Universidade de São Paulo, Piracicaba, S.P., Brazil) Edição Especial 35: 47–59.

Rankin–de Mérona, J. M., R. W. Hutchings, and T. E. Lovejoy. 1990. Tree mortality and recruitment over a five-year period in undisturbed upland rain forest of the central Amazon. In A. Gentry (ed.), *Four Neotropical Rainforests.* Yale University Press, New Haven, Conn., pp. 573–84.

Rankin–de Mérona, J. M., G. T. Prance, R. W. Hutchings, M. F. da Silva, W. A. Rodrigues, and M. E. Uehling. 1992. Preliminary results of large-scale tree inventory of upland rain forest in the Central Amazon. Acta Amazónica 22: 493–534.

Ranzani, G. 1980. Identificação e caracterização de alguns solos da Estação Experimental de Silvicultura Tropical do INPA. Acta Amazónica 10: 7–41.

Rasmussen, D. R. 1979. Correlates of patterns of range use of a troop of yellow baboons (*Papio cynocephalus*). I. Sleeping sites, impregnable females, births, and male emigrations and immigrations. Anim. Behav. 27: 1098–1112.

Rathcke, B. J., and E. S. Jules. 1993. Habitat fragmentation and plant-pollinator interactions. Current Science 65: 273–77.

Ratsirarson, J., and J. A. Silander. 1996. Reproductive biology of the threatened Madagascar triangle-palm *Neodypsis decaryi* Jumelle. Biotropica 28: 737–45.

Raulls, K., J. D. Ballou, and A. Templeton. 1988. Estimates of lethal equivalents and the cost of inbreeding in mammals. Conserv. Biol. 2: 185–93.

Raw, A. 1989. The dispersal of euglossine bees between isolated patches of eastern Brazilian wet forest (Hymenoptera, Apidae). Rev. bras. Ent. 33: 103–7.

Rebelo, J. M. M., and C. A. Garófalo. 1991. Diversidade e sazonalidade de machos de Euglossini (Hymenoptera, Apidae) e preferência por iscas odores em um fragmento de floresta no sudeste do Brasil. Rev. bras. Biol. 51: 787–99.

Rebertus, A. J., G. B. Williamson, and E. B. Moser. 1989. Longleaf pine pyrogenicity and turkey oak mortality in Florida xeric sandhills. Ecology 70: 60–70.

Rebertus, A. J., G. B. Williamson, and W. J. Platt. 1993. Impacts of temporal variation in fire regime on savanna oaks and pines. In *The Longleaf Pine Ecosystem: Ecology, Restoration and Management.* Proceedings 18th Tall Timbers Fire Ecology Conference, Tallahassee, Florida pp. 215–25.

Reboud, X., and C. Zeyl. 1994. Organelle inheritance in plants. Heredity 72: 132–40.

Redford, K. 1992. The empty forest. Bioscience 42: 412–22.

Régis, M. 1989. "IBGE e Embrapa divergem sobre melhor ocupação para Amazônia." Jornal do Brasil [Rio de Janeiro] 9 July 1989, Section 1, p. 17.

Reis, M. S. 1996. Distribuição e dinâmica da variabilidade genética em populações naturais de palmiteiro (*Euterpe edulis* Martius). Ph.D. diss., Escola Superior de Agricultura "Luiz de Queiroz," ESALQ/USP, Piracicaba, SP, Brazil.

Remsen, J. V., Jr. and D. A. Good. 1996. Misuse of data from mist-net captures to assess relative abundance in bird populations. Auk 113: 381–98.

Renner, S. S. 1997. Effects of habitat fragmentation on plant-pollinator interactions in the tropics. In D. M. Newbery, N. Brown, and H. H. T. Prins (eds.), *Dynamics of Trop-*

ical Communities. Blackwell Science, Oxford, U.K., pp. 339–60.

Ribeiro, J. E. L., M. J. G. Hopkins, A. Vicentini, C. A. Sothers, M. A. Costa, J. M. Brito, M. A. D. Souza, L. H. Martins, L. G. Lohmann, P. A. Assumção, E. Pereira, C. M. Silva, M. R. Mesquita, and L. C. Procópio. 1999. *Flora da Reserva Ducke: Guia de identifição das plantas vasculares de uma floresta de terra-firme na Amazônia central.* National Institute for Research in the Amazon (INPA), Manaus, Brazil.

Richardson, D. R. 1988. Sand pine: An annotated bibliography. Florida Scientist 52: 65–93.

Richter, D. D., and L. I. Babbar. 1991. Soil diversity in the tropics. Advances in Ecological Research 21: 315–89.

Rímoli, J., and S. F. Ferrari. 1997. Comportamento e ecologia de macacos-pregos (*Cebus apella nigritus*), Goldfuss, 1809) na Estação Ecológica de Caratinga (MG.). Livros de Resumos do VIII Congresso Brasileiro de Primatologia, Sociedade Brasileira de Primatologia, João Pessoa, Paraíba, Brasil. Resumo nº 213.

Ripley, B. D. 1981. *Spatial Statistics.* Wiley, New York.

Risch, S. J., and C. R. Carrol. 1982. The ecological role of ants in two Mexican agroecosystems. Oecologia 55: 114–19.

Riswan, S., and L. Hartanti. 1995. Human impacts on tropical forest dynamics. Vegetatio 121: 41–52.

Ritland, K. 1983. Estimation of mating systems. In S. D. Tanksley and T. J. Orton (eds.), *Isozymes in Plant Genetics and Breeding. Part A.* Elsevier, Amsterdam, pp. 289–302.

Robinson, G. R., and S. N. Handel. 1993. Forest restoration on a closed landfill: Rapid addition of new species by bird dispersal. Conserv. Biol. 7: 271–78.

Robinson, J. G. 1986. Seasonal variation in use of time and space by the wedge-capped monkey, *Cebus olivaceous:* Implications for foraging theory. Smithsonian Contributions to Zoology, Smithsonian Institution, Washington, D.C.

Robinson, J. G., and C. H. Janson 1987. Capuchins, squirrel monkeys, and atelines: Socioecological convergence with old world primates. In B. B. Smuts, D. L. Cheney, R. M. Seyfarth, R. W. Wrangham, and T. T. Struhsaker (eds.), *Primates Societies.* University of Chicago Press, Chicago, pp. 69–82.

Robinson, M. H., and B. Robinson 1970. Prey caught by a sample population of the spider *Argiope argentata* in Panama. Zoological J. Linnean Society 49: 345–57.

Robinson, S. K., F. R. Thompson, III, T. M. Donovan, D. R. Whitehead, and J. Faaborg. 1995. Regional forest fragmentation and the nesting success of migratory birds. Science 267: 1987–90.

Rolstad, J. 1991. Consequences of forest fragmentation for the dynamics of bird populations: Conceptual issues and the evidence. Biol. J. Linnean Soc. 42: 149–63.

Root, R. B. 1967. The niche exploitation pattern of blue-gray gnatcatcher. Ecol. Monogr. 37: 317–50.

Rosenzweig, M. L. 1995. *Species Diversity in Space and Time.* Cambridge University Press, Cambridge, U.K.

Roubik, D. W. 1978. Competitive interactions between Neotropical pollinators and Africanized honey bees. Science 201: 1030–32.

Roubik, D. W. 1980. Foraging behavior of competing Africanized honey bees and stingless bees. Ecology 61: 836–45.

Roubik, D. W. 1983. Nests and colony characteristics of stingless bees from Panama (Hymenoptera, Apidae). J. Kans. Entomol. Soc. 56: 327–55.

Roubik, D. W. 1989. *Ecology and Natural History of Tropical Bees.* Cambridge University Press, Cambridge, England.

Roubik, D. W. 1996. African bees as exotic pollinators in French Guiana. In A. Matheson, S. L. Buchmann, C. O'Toole, P. Westrich, and I. Williams (eds.), *The Conservation of Bees.* Academic Press, London, pp. 173–82.

Roubik, D. W., and J. D. Ackerman. 1987. Long-term ecology of euglossine orchid-bees (Apidae, Euglossini) in Panama. Oecologia 73: 321–33.

Roubik, D. W., J. E. Moreno, C. Vergara and D. Wittmann. 1986. Sporadic food competition with the African honey bee: Projected impact on Neotropical social bees. J. Trop. Ecol. 2: 97–111.

Roubik, D. W., and H. Wolda. In press. A test of invading honey bee impact on native bees of Barro Colorado Island, Panama: Long-term light trap data. Environmental Entomology.

Rowe, N. 1996. *A Pictorial Guide to the Living Primates*. Pogonias, East Hampton, N.Y.

Royama, T. 1992. *Analytical Populations Dynamics*. Chapman and Hall, New York.

Rudran, R. 1978. Socioecology of blue monkeys (*Cercopithecus mitis stuhlmanni*) of the Kibale forest, Uganda. Smithsonian Contributions to Zoology 249: 1–88.

Ruokolainen, K., A. Linna, and H. Tuomisto. 1997. Use of Melastomaceae and pteridophytes for revealing phytogeographical patterns in Amazonian rain forests. J. Trop. Ecol. 13: 243–56.

Rylands, A. B., and A. Keuroghlian 1988. Primate populations in continuous forest fragments in central Amazonia. Acta Amazónica 18: 291–307.

Sabatier, D. 1985. Saisonnalité et determinisme du pic de fructification en foret guyanaise. Review Ecology (Terre Vie) 40: 298–320.

Sabatier, D., M. Grimaldi, M.–F. Prévost, J. L., Guillaumet, M. Godron, M. Dosso, and P. Curmi. 1997. The influence of soil cover organization on the floristic and structural heterogeneity of a Guianan rain forest. Plant Ecol. 131: 81–108.

Sabatier, D., and M. F. Prévost. 1990. Variations du peuplement forestier é l'échelle stationnelle: Le cas de la station des Nouragues en Guyane française. Actes de l'Atelier sur l'amènagement et la conservation de l'écosystème forestier tropical humide. MAB-UNESCO, pp. 169–87.

St. John, T. V. 1980. Uma lista de espécies de plantas tropicais brasileiras naturalmente infectadas com micorriza vesicular-arbuscular. Acta Amazónica 10: 229–34.

St. John, T. V. 1985. Mycorrhizae. In G. T. Prance and T. E. Lovejoy (eds.), *Key Environments: Amazonia*. Pergamon, Oxford, U.K., pp. 277–83.

Salati, E., and P. B. Vose. 1984. Amazon Basin: A system in equilibrium. Science 224: 129–38.

Saldarriaga, J. G. 1985. Forest succession in the upper Rio Negro of Colombia and Venezuela. Ph.D. diss., University of Tennessee, Knoxville.

Saldarriaga, J. G., and C. Uhl. 1991. Recovery of forest vegetation following slash-and-burn agriculture in the upper Rio Negro. In A. Gómez –Pompa, T. C. Whitmore, and M. Hadley (eds.), *Rain Forest Regeneration and Management. Man and the Biosphere Series*. Parthenon, Paris, pp. 303–12.

Saldarriaga, J. G., and D. C. West. 1986. Holocene fires in the northern Amazon Basin. Quaternary Research 26: 358–66.

Saldarriaga, J. G., D. C. West, M. L. Tharp, and C. Uhl. 1988. Long-term chronosequence of forest succession in the upper Rio Negro of Colombia and Venezuela. J. Ecology 76: 938–58.

Salmah, S., T. Inoue, and S. F. Sakagami. 1990. An analysis of apid bee richness (Apidae) in Central Sumatra. In S. F. Sakagami, R. Ohgushi and D. W. Roubik (eds.), *Natural History of Social Wasps and Bees in Equatorial Sumatra*. Hokkaido University Press, Sapporo, Japan, pp. 139–74.

Sanchez, P. A. 1976. *Properties and Management of Soils in the Tropics*. Wiley-Interscience, New York.

Sanchez, P. A. 1977. Advances in the management of OXISOLS and ULTISOLS in tropical South America. In *Proceedings of the International Seminar on Soil Environment and Fertility Management in Intensive Agriculture, Tokyo, Japan,* Society of the Science of Soil and Manure, Tokyo, pp. 535–66.

Sanchez, P. A., D. E. Bandy, J. H. Villachica, and J. J. Nicholaides III. 1982. Amazon basin soils: Management for continuous crop production. Science 216: 821–27.

Sanford, R. L., Jr., J. Aldarriaga, K. E. Clark, C. Uhl, and M. Herrera. 1985. Amazon rain-forest fires. Science 227: 53–55.

Santos, E. M. G. 1994. Ecologia da polinização, fluxo de pólen e taxa de cruzamento em *Bauhinia forficata* Link. (Caesalpinaceae). M.Sc. Thesis. ESALQ/USP, Piracicaba, SP, Brasil.

Santos, T., and J. L. Tellería. 1992. Edge effects on nest predation in Mediterranean fragmented forests. Biological Conservation 60: 1–5.

Santos, T., and J. L. Tellería. 1994. Influence of forest fragmentation on seed consumption and dispersal of Spanish juniper, *Juniper thurifera*. Biol. Conserv. 70: 129–34.

Sarre, S., G. T. Smith, and J. A. Meyers. 1995. Persistence of two species of gecko (*Oedura reticulata* and *Gehyra variegata*) in remnant habitat. Biological Conservation 71: 25–33.

Sarukhán, J. 1980. Demographic problems in tropical systems. In O. Solbrig (ed.), *Demography and Evolution in Plant Populations*. University of California Press, Berkeley, pp. 161–88.

SAS Institute. 1989. *SAS/STAT Users Guide*. Version 6, 4th edition, Volume 1, SAS Institute, Cary, N.C.

Saunders, D. A., and C. P. de Rebeira. 1991. Values of corridors to avian populations in a fragmented landscape. In D. A. Saunders and R. J. Hobbs (eds.) *Nature Conservation 2, The Role of Corridors*. Surrey Beatty and Sons, Norton, U.K., pp. 221–40.

Saunders, D. A., R. J. Hobbs and C. R. Margules. 1991. Biological consequences of ecosystem fragmentation: A review. Conserv. Biol. 5: 18–32.

Savill, P. S. 1983. Silviculture in windy climates. Forestry Abstracts 44: 473–88.

Scariot, A. O. 1996. The effects of rain forest fragmentation on the palm community in central Amazonia. Ph.D. diss., University of California, Santa Barbara.

Scariot, A. 1998. Forest fragmentation: the effects on palm diversity in central Amazonia. J. Ecology 87: 66–76.

Scariot, A. 1999. Forest fragmentation effects on diversity of the palm community in central Amazonia. J. Ecology 87: 66–76.

Schafer, D. E., and D. D. Chilcote. 1970. Factors influencing persistence and depletion in buried seed populations. II. The effects of soil temperature and moisture. Crop Science 10: 342–45.

Schelhas, J., and R. Greenberg (eds.). 1996. *Forest Patches in Tropical Landscapes*. Island, Washington, D.C.

Schemske, D. W., B. C. Husband, M. H. Ruckelhaus, C. Goodwillie, I. M. Parker, and J. G. Bishop. 1994. Evaluating approaches to the conservation of rare and endangered plants. Ecology 75: 584–606.

Schluter, D., and R. E Ricklefs. 1993. Species diversity: An introduction to the problem. In R. E. Ricklefs and D. Schluter (eds.), *Species Diversity in Ecological Communities*. University of Chicago Press, Chicago, pp. 1–10.

Schmidt, R. C. 1991. Tropical Rainforest Management: A Status Report. In A. Gomez-Pompa, T. C. Whitmore, and M. Hadlely (eds.), *Rainforest Regeneration and Management*. UNESCO, vol. 6, pp. 181–203.

Schoener, T. W. 1987. The geographical distribution of rarity. Oecologia 74: 161–73.

Schoener, T. W. 1991. Extinction and the nature of the metapopulation: A case system. Acta Oecologica 12: 53–75.

Schoener, T. W., and D. A. Spiller. 1987. High population persistence in a system with high turnover. Nature 330: 474–77.

Schwarzkopf, L., and A. B. Rylands. 1989. Primate species richness in relation to habitat structure in Amazonian rainforest fragments. Biological Conservation 48: 1–12.

Seastedt, T. R., and D. A. Crossley. 1984. The influence of arthropods on ecosystems. Bioscience 34: 157–61.

Sene, F. M., and F. C. Val. 1977. Ocorrência de *Drosophila malerkotliana* na América do Sul. Ciência e Cultura 29: 716.

Serrão, E. A. S., I. C. Falesi, J. B. da Viega, and J. F. Teixeira Neto. 1979. Productivity of cultivated pastures on low fertility soils in the Amazon of Brazil. In P. A. Sanchez and L. E. Tergas (eds.), *Pasture Production in Acid Soils of the Tropics: Proceedings of a Seminar held at CIAT, Cali, Colombia, April 17–21, 1978. CIAT Series 03 EG–05.* Centro Internacional de Agricultura Tropical (CIAT), Cali, Colombia, pp. 195–225.

Serrão, E. A. S., and A. K. O. Homma. 1993. Brazil. In R. R. Harwood (ed.), *Sustainable Agriculture and the Environment in the Humid Tropics.* National Academy Press, Washington, D.C., pp. 265–351.

Serrão, E. A. S., D. Nepstad, and R. Walker. 1996. Upland agricultural and forestry development in the Amazon: Sustainability, criticality, and resilience. Ecological Economics 18: 3–13.

Setz, E. Z. F. 1992. Comportamento de alimentação de *Pithecia pithecia* (Cebidae, Primates) em um fragmento florestal. Primatologia no Brasil, 3: 327–30.

Setz, E. Z. F. 1993. Ecologia alimentar de um grupo de parauacus (*Pithecia pithecia chrysocephala*) em um fragmento florestal na Amazônia Central. Ph.D. diss., Universidade Estadual de Campinas, São Paulo.

Setz, E. Z. F. 1994. Feeding ecology of golden-faced sakis. Neotropical Primates 2: 13–14.

Setzer, A. W., and M. C. Pereira. 1991. Amazonia biomass burnings in 1987 and an estimate of their tropospheric emission. Ambio 20: 19–22.

Sevenster, J. G. 1992. The community ecology of frugivorous *Drosophila* in a Neotropical forest. Ph.D. thesis, Leiden University, Netherlands.

Sevenster, J. G., and J. J. M. van Alphen. 1993a. Coexistence in stochastic environments through a life history trade off in Drosophila. In J. Yoshimura and C. W. Clark (eds.), *Adaptation in Stochastic Environ-*ments. Lectures Notes in Biomathematics 98: 155–72.

Sevenster, J. G., and J. J. M. van Alphen. 1993b. A life history trade-off in *Drosophila* species and community structure in variable environments. J. Animal Ecology 62: 720–36.

Shaffer, M. L. 1981. Minimum population sizes for species conservation. BioScience 31: 131–34.

Shelly, T. E. 1988. Relative abundance of day-flying insects in treefall gaps vs shaded understorey in a Neotropical forest. Biotropica 20: 114–19.

Shorrocks, B. 1982. The breeding sites of temperate woodland *Drosophila*. In M. Ashburner, H. L. Carson, and J. N. Thompson (eds.), *The Genetics and Biology of Drosophila*. Academic Press, London.

Shorrocks, B., and M. Bingley. 1994. Priority effects and species coexistence: Experiments with fungal-breeding Drosophila. J. Animal Ecology 63: 799–806.

Shukla, J., C. Nobre, and P. Sellers. 1990. Amazon deforestation and climate change. Science 247: 1322–25.

Siegel, S. 1975. *Estatística não-paramétrica*. McGraw-Hill, São Paulo.

Siegel, S., and N. J. Castellan, Jr. 1988. *Nonparametric Statistics: For The Behavioral Sciences*. McGraw-Hill International Series, Library of Congress CIP data, Singapore.

Sieving, K. E., and J. R. Karr. 1997. Avian extinction and persistence mechanisms in lowland Panama. In W. F. Laurance and R. O. Bierregaard, Jr. (eds.), *Tropical Forest Remnants: Ecology, Management, and Conservation of Fragmented Communities*. University of Chicago Press, Chicago.

Silva, J. M. C. da, C. Uhl, and G. Murray. 1996. Plant succession, landscape management, and the ecology of frugivorous birds in abandoned Amazonian pastures. Conserv. Biol. 10: 491–503.

Simberloff, D. S. 1986. Are we on the verge of a mass extinction in tropical rain forests? In D. K. Elliott (ed.), *Dynamics of Extinction*. Wiley, New York, pp. 165–80.

Simberloff, D. S. 1988. The contribution of population and community biology to conservation science. Ann. Rev. Ecol.Syst. 19: 473–511.

Simberloff, D. S. 1994. The ecology of extinction. Acta Palaeontologica Polonica 38: 159–74.

Simberloff, D. S., and L. G. Abele. 1982. Refuge design and island geographic theory: Effects of fragmentation. Am. Nat. 120: 41–45.

Simberloff, D. S., and J. Cox. 1987. Consequences and costs of conservation corridors. Conserv. Biol. 1: 63–71.

Simberloff, D. S., and T. Dayan. 1991. The guild concept and the structure of ecological communities. Ann. Rev. Ecol. Syst. 22: 115–43.

Simberloff, D. S., J. A. Farr, J. Cox, and D. W. Mehlman. 1992. Movement corridors: Conservation bargains or poor investments? Conserv. Biol. 6: 493–505.

Sinsch, U. 1990. Migration and orientation in anuran amphibians. Ethol. Ecolo. Evol. 2: 65–79.

Sizer, N. 1992. The impact of edge formation on regeneration and litterfall in a tropical rain forest fragment in Amazonia. Ph.D. thesis, Cambridge University, Cambridge, U.K.

Sizer, N., and E. V. J. Tanner. 1999. Responses of woody plant seedlings to edge formation in a lowland rainforest, Amazonia. Biological Conservation 91: 135–42.

Sjogren, P. 1991a. Extinction and isolation gradients in metapopulations: The case of pool frog (*Rana lessonae*). Biol. J. Linnean Soc. 42: 135–47.

Sjogren, P. 1991b. Genetic variation in relation to demography of peripheral pool frog populations (*Rana lessonae*). Evolutionary Ecology 5: 248–71.

Skole, D., and C. J. Tucker. 1993. Tropical Deforestation and Habitat Fragmentation in the Amazon: Satellite Data from 1978 to 1988. Science 206: 1905–10.

Smith, A. P. 1997. Deforestation, fragmentation, and reserve design in western Madagascar. In W. F. Laurance and R. O. Bierregaard, Jr. (eds.), *Tropical Forest Remnants: Ecology, Management, and Conservation of Fragmented Communities*. University of Chicago Press, Chicago, pp. 138–55.

Smith, A. T. 1980. Temporal changes in insular populations of the pika (*Ochotona princeps*). Ecology 61: 8–13.

Smith, N. J. H., E. A. S. Serrão, P. de T. Alvim, and C. Falesi. 1995. *Amazonia: Resiliency and Dynamism of the Land and its People*. United Nations University Press, Tokyo.

Smyth, A. J. 1966. *The Selection Of Soils For Cacao*. FAO Soils Bulletin No. 5. Food and Agriculture Organization of the United Nations (FAO), Rome.

Snook, L. K. 1996. Catastrophic disturbance, logging, and the ecology of mahogany (*Swietenia macrophylla* King): Grounds for listing a major tropical timber species in CITES. Bot. J. Linnean Soc. 122: 35–46.

Sokal, R. R., and F. J. Rohlf. 1981. *Biometry*. W. H. Freeman, New York.

Sokal, R.R., P. E. Smouse, and J. V. Neel. 1986. The genetic structure of a tribal population, the Yanomama Indians. XV. An effort to find patterns by autocorrelation analysis. Genetics 114: 259–87.

Sollins, P. 1998. Factors influencing species composition in tropical lowland rain forest: Does soil matter? Ecology 79: 23–30.

Soltis, D. E., and P. S. Soltis. 1990. *Isozymes in Plant Biology*. Chapman and Hall, London.

Sombroek, W. G. 1966. *Amazon Soils: A Reconnaissance of the Soils of the Brazilian Amazon Region*. Centre for Agricultural Publications and Documentation, Wageningen, Netherlands.

Sombroek, W. G. 1984. Soils of the Amazon region. In H. Sioli (ed.), *The Amazon: Limnology and Landscape Ecology of a Mighty Tropical River and its Basin*. Dr. W. Junk Publishers, Dordrecht, Netherlands, pp. 521–35.

Sombroek, W. G., P. M. Fearnside and M.

Cravo. 2000. Geographic assessment of carbon stored in Amazonian terrestrial ecosystems and their soils in particular. In R. Lal, J. M. Kimble, and B. A. Stewart (eds.)., *Global Climate Change and Tropical Ecosystems: Advances in Soil Science.* CRC Press, Boca Raton, Fla., pp. 375–89.

Soulé, M. E. 1983. What do we really know about extinctions? In C. M. Schoenwald-Cox, S. M. Chambers, B. MacBryde, and L. Thomas (eds.), *Genetics and Conservation: A Reference for Managing Wild Animal and Plant Populations.* Benjamin/Cummings, Menlo Park, Calif., pp. 111–24.

Soulé, M. E., A. C. Alberts, and D. T. Bolger. 1992. The effects of habitat fragmentation on chaparral plants and vertebrates. Oikos 63: 39–47.

Soulé, M. E., and K. A. Kohm. 1989. *Research Priorities For Conservation Biology.* Island, Washington, D.C., Critical Issues Series.

Soulé, M. E., and J. Terborgh. 1999. *Continental Conservation.* Island Press, Washington, D.C.

Souza, O. F. F. d., and V. K. B. Brown 1994. Effects of habitat fragmentation on Amazonian termite communities. J. Trop. Ecol. 10: 197–206.

Spironelo, W. R., 1987. Range size of a group of *Cebus apella* in Central Amazonia. Am. J. Primatol. 8: 522.

Spironelo, W. R. 1991. Importância dos frutos de palmeiras (Palmae) na dieta de um grupo de *Cebus apella* (Cebidae: Primates) na Amazônia central. Primatologia no Brasil 3: 285–96.

Spironello, W. R. 1999. The Sapotaceae community ecology in a central Amazonian forest: Effects of seed dispersal and seed predation. Ph.D. diss., University of Cambridge, Cambridge, U.K.

Stacey, P. B., and M. Taper. 1992. Environmental variation and persistence of small populations. Ecological Applications 2: 18–29.

Stacy, E. A., J. L. Hamrick, J. D. Nason, S. P. Hubbell, R. B. Foster, and R. Condit. 1996.

Pollen dispersal in low-density populations of three Neotropical tree species. Am. Nat. 148: 275–98.

Stockard, J., B. Nicholson, and G. Williams. 1985. An assessment of a rainforest regeneration program at Wingham Brush, New South Wales. Victorian Naturalist 103: 85–91.

Stone, T. A., P. Schlesinger, G. M. Woodwell, and R. A. Houghton. 1994. A map of the vegetation of South America based on satellite imagery. Photogrammetric Engineering and Remote Sensing. 60: 541–51.

Stork, N. E. 1988. Insect diversity: Facts, fiction, and speculation. Biol. J. Linnean Soc. 35: 321–37.

Stork, N. E., R. K. Didham, and J. Adis. 1997. Canopy arthropod studies for the future. in N. E. Stork, J. Adis, and R. K. Didham (eds.), *Canopy Arthropods.* Chapman and Hall, London, pp. 551–61.

Stotz, D. F. 1993. Geographic variation in species composition of mixed-species flocks in lowland humid forests in Brazil. Papéis Avulsos de Zoologia 38: 61–75.

Stotz, D. F., and R. O. Bierregaard, Jr. 1989. The birds of the fazendas Porto Alegre, Dimona, and Esteio north of Manaus, Amazonas, Brazil. Rev. Bras. Biol. 49: 861–72.

Stouffer, P. C. 1997. Interspecific territoriality in *Formicarius* antthrushes? The view from central Amazonian Brazil. Auk 114: 780–85.

Stouffer, P. C., and R. O. Bierregaard, Jr. 1995a. Use of Amazonian forest fragments by understory insectivorous birds. Ecology 76: 2429–45.

Stouffer, P. C., and R. O. Bierregaard, Jr. 1995b. Effects of forest fragmentation on understory hummingbirds in Amazonian Brazil. Conserv. Biol. 9: 1085–94.

Stouffer, P. C., and R. O. Bierregaard, Jr. 1996. Forest fragmentation and seasonal patterns of hummingbird abundance in Amazonian Brazil. Ararajuba 4: 9–14.

Stratford, J. A., and P. C. Stouffer. 1999. Local extinctions of terrestrial insectivorous birds in a fragmented landscape near

Manaus, Brazil. Conserv. Biol. 13: 1416–23.

Strong, D. R. 1977. Insect species richness: Hispine beetles of *Heliconia latispatha*. Ecology 58: 573–82.

Struhsaker, T. T. 1997. *Ecology of an African Forest: Logging in Kibale and the Conflict Between Conservation and Exploitation*. University Presses of Florida, Gainesville.

Stuart, M. D., L. L. Greenspan, K. E. Glander, and M. R. Clarke. 1990. A coprological survey of parasites of wild mantled howling monkeys, *Alouatta palliata palliata*. J. Wildlife Diseases 26: 547–49.

Stuart, M. D., K. B. Strier, and S. M. Pierberg. 1993. A coprological survey of parasites of wild muriquis, *Brachyteles arachnoides,* and brown howling monkeys, *Alouatta fusca*. J. Helminthological Society of Washington 60: 111–15.

Stuiver, M., and B. Becker. 1986. High-precision decadal calibration of the radiocarbon time scale, AD 1950–2500 BC. Radiocarbon 28: 863–910.

Sutton, S. L., C. P. Ash, and A. Grundy. 1983. The vertical distribution of flying insects in lowland rain-forest of Panama, Papua, New Guinea and Brunei. Zoological J. Linnean Society 78: 287–97.

Swofford, D. L., and R. B. Selander. 1989. Byosys–1. *A computer program for the analysis of allelic variation in population genetics and biochemical systematics. Release 1,7,* Illinois Natural History Survey, Illinois.

Tanaka, A., T. Sakuma, N. Okagawa, H. Imai, and S. Ogata. 1984. *Agro-Ecological Condition of the Oxisol-Ultisol Area of the Amazon River System*. Faculty of Agriculture, Hokkaido University, Sapporo, Japan.

Tanner, E. V. J., P. M. Vitousek, and E. Cuevas. 1998. Experimental investigation of nutrient limitation of forest growth on wet tropical mountains. Ecology 79: 10–22.

Taylor, O. 1977. The past and possible future spread of Africanized honeybees in the Americas. Bee World 55: 19–30.

Taylor, R. E., A. Long, and R. S. Kra. 1992. *Radiocarbon After Four Decades: An Interdisciplinary Perspective*. Springer-Verlag, Tucson, Ariz.

Taylor, R. J. 1987a. The geometry of colonization: 1. Islands. Oikos 48: 225–31.

Taylor, R. J. 1987b. The geometry of colonization: 2. Peninsulas. Oikos 48: 232–37.

Tellería, J. L., and T. Santos. 1995. Effects of forest fragmentation on a guild of wintering passerines: The role of habitat selection. Biological Conservation 71: 61–67.

Templeton, A. R., K. Shaw, E. Routhman, and S. K. Davis. 1990. The genetic consequences of habitat fragmentation. Ann. Missouri Bot. Gard. 77: 13–27.

Terayama, M., and K. Murata. 1990. Effects of area and fragmentation of forests for nature conservation: Analysis by ant communities. Bull. Biogeographical Society Japan 45: 11–18.

Terborgh, J. 1974. Preservation of natural diversity: The problem of extinction prone species. Bioscience 24: 715–22.

Terborgh, J. 1983. *Five New World Primates: A Study In Comparative Ecology*. Princeton University Press, Princeton, N.J.

Terborgh, J. 1986. Keystone plant resources in the tropical forest. In M. E. Soulé (ed.), *Conservation Biology: The Science of Scarcity and Diversity*. Sinauer, Sunderland, Mass., pp. 330–44.

Terborgh, J. 1992. Maintenance of diversity in tropical forest. Biotropica 24: 283–92.

Terborgh, J., and C. H. Janson. 1983. Ecology of primates in southeastern Peru. National Geographic Society Research Reports 15: 655–62.

Terborgh, J., L. Lopez, J. Tello, D. Yu, and A. R. Bruni. 1997. Transitory states in relaxing ecosystems of land bridge islands. In W. F. Laurance and R. O. Bierregaard, Jr. (eds.), *Tropical Forest Remnants: Ecology, Management, and Conservation of Fragmented Communities*. University of Chicago Press, Chicago, pp. 256–74.

Terborgh, J., S. K. Robinson, T. A. Parker, III, C. A. Munn, and N. Pierpont. 1990. Structure and organization of an Amazonian forest bird community. Ecol. Monogr. 60: 213–38.

Terborgh, J., and B. Winter. 1980. Some causes of extinction. In. M. E. Soulé and B. A. Wilcox (eds.) *Conservation Biology: An Evolutionary-Ecological Perspective.* Sinauer, Sunderland, Mass.

Thébaud, C., and D. Strasberg. 1997. Plant dispersal in fragmented landscapes: A field study of woody colonization in rainforest remnants of the Mascarene Archipelago. In W. F. Laurance and R. O. Bierregaard, Jr. (eds.), *Tropical Forest Remnants: Ecology, Management, and Conservation of Fragmented Communities.* University of Chicago Press, Chicago, pp. 321–32.

Thiollay, J.–M. 1989. Area requirements for the conservation of rainforest raptors and game birds in French Guiana. Conserv. Biol. 3: 128–37.

Thiollay, J.–M. 1992. Influence of selective logging on bird species diversity in a Guianan rain forest. Conserv. Biol. 6: 47–63.

Thiollay, J.–M. 1994. Structure, density, and rarity in an Amazonian rainforest bird community. J. Trop. Ecol. 10: 449–81.

Thoisy, B. de and C. Richard-Hansen. 1996. Diet and social behaviour changes in a red howler monkey (*Alouatta seniculus*) troop in a highly degraded rain forest. Folia Primatologia 68: 357–61.

Thomas, C. D., and T. M. Jones. 1993. Partial recovery of a skipper butterfly (*Hesperia comma*) from populations refuge: Lessons for conservation in a fragmented landscape. J. Anim. Ecol. 62: 472–81.

Thompson, K. 1992. The functional ecology of seed banks. In M. Fenner (ed.), *Seeds: The Ecology of Regeneration in Plant Communities.* C. A. B. International, Wallingford, U.K., pp. 230–58.

Tian, H., J. M. Mellilo, D. W. Kicklighter, A. D. McGuire, J. V. K. Helfrich III, B. Moore III and C. Vrsmarty. 1998. Effect of interannual climate variability on carbon stor-

age in Amazonian ecosystems. *Nature* 396: 664–67.

Tilman, D. 1996. Biodiversity: Population versus ecosystem stability. Ecology 77: 350–63.

Tilman, D. 1997. Community invasibility, recruitment limitation, and grassland biodiversity. Ecology 78: 81–92.

Tocher, M. D. 1996. The effects of forest disturbance on habitat use and abundance of frogs in central Amazonia. Ph.D. diss., University of Canterbury, New Zealand.

Tocher, M. D. 1998. A comunidade de anfíbios da Amazônia central: Diferenças na composição específica entre a mata primária e pastagens. In C. Gascon and P. Moutinho (eds.), *Floresta Amazônica: Dinâmica, Regeneração e Manejo.* Instituto Nacional de Pesquisas da Amazônia (INPA), Manaus, Amazonas, Brazil, pp. 219–32.

Tocher, M. D., C. Gascon, and B. L. Zimmerman. 1997. Fragmentation effects on a central Amazonian frog community: A ten-year study. In W. F. Laurance and R. O. Bierregaard, Jr. (eds.), *Tropical Forest Remnants: Ecology, Management, and Conservation Of Fragmented Communities.* University of Chicago Press, pp. 124–37.

Toda, M. J. 1992. Three-dimensional dispersion of drosophilid flies in a cool temperate forest of northern Japan. Ecological Research 7: 283–95.

Toft, C. A., and T. W. Schoener. 1983. Abundance and diversity of orb spiders on 106 Bahamanian islands: Insular biogeography at an intermediate trophic level. Oikos 41: 359–71.

Townshend, J. R. G., C. O. Justic, and V. Kalb. 1987. Characterization and classification of South American land cover types using satellite data. International J. Remote Sensing 8: 1189–1207.

Tucker, C. J., B. N. Holben, and T. E. Goff. 1984. Intensive forest clearing in Rondônia, Brazil, as detected by satellite remote sensing. Remote Sensing of the Environment 13: 255–61.

Tucker, G. F., and J. R. Powell. 1991. An improved canopy access technique. Northern J. Applied Forestry 8: 29–32.

Tuomisto, H., K. Ruokolainen, R. Kalliola, A. Linna, W. Danjoy, and Z. Rodriguez. 1995. Dissecting Amazonian biodiversity. Science 269: 63–66.

Tuomisto, H., K. Ruokolainen, and J. Salo. 1992. Lago Amazonas: Fact or fancy? Acta Amazónica 22: 353–61.

Turner, I. M., K. S. Chua, J. Ong, B. Soong, and H. Tan. 1996. A century of plant species loss from an isolated fragment of lowland tropical rain forest. Conserv. Biol. 10: 1229–44.

Turner, I. M., and R. T. Corlett. 1996. The conservation value of small, isolated fragments of lowland tropical rain forest. Trends Ecol. Evol. 11: 330–33.

Turner, I. M., H. T. W. Tan, Y. C. Wee, A. bin Ibrahim, P. T. Chew, and R. T. Corlett. 1994. A study of plant species extinction in Singapore: Lessons for the conservation of tropical biodiversity. Conserv. Biol. 8: 705–12.

Turner, M. G. 1989. Landscape ecology: The effect of pattern on process. Ann. Rev. Ecol. Syst. 20: 171–97.

Turner, M. G., R. V. O'Neill, Robert H. Gardner, and B. T. Milne. 1989. Effects of changing spatial scale on the analysis of landscape pattern. Landscape Ecology 3: 153–62.

Turton, S. M., and H. J. Freiburger. 1997. Edge and aspect effects on the microclimate of a small tropical forest remnant on the Atherton Tableland, northeastern Australia. In W. F. Laurance and R. O. Bierregaard, Jr. (eds.), Tropical Forest Remnants: Ecology, Management, and Conservation of Fragmented Communities. University of Chicago Press, Chicago, pp. 45–54.

Uhl, C. 1987. Factors controlling succession following slash-and-burn agriculture in Amazonia. J. Ecology 75: 377–407.

Uhl, C., P. Barreto, A. Veríssimo, A. C. Barros, P. Amaral, E. Vidal, and C. Souza Jr. 1998. Uma abordagem integrada de pesquisa sobre o manejo dos recursos florestais na Amazônia brasileira. In C. Gascon and P. Moutinho (eds.), Floresta Amazônica: Dinâmica, Regeneração e Manejo. Instituto Nacional de Pesquisas da Amazônia (INPA), Manaus, Amazonas, Brazil, pp. 313–32.

Uhl, C., and R. Buschbacher. 1985. A disturbing synergism between cattle-ranch burning practices and selective tree harvesting in the eastern Amazon. Biotropica 17: 265–68.

Uhl, C., R. Buschbacher, and E. A. S. Serrão. 1988. Abandoned pastures in eastern Amazonia. I. Patterns of plant secession. J. Ecology 76: 663–81.

Uhl, C., and J. B. Kaufman. 1990. Deforestation, fire susceptibility, and potential tree responses to fire in the eastern Amazon. Ecology 71: 437–49.

Uhl, C., J. B. Kaufman, and D. L. Cummings. 1988. Fire in the Venezuelan Amazon. 2: Environmental conditions necessary for forest fires in evergreen rain forest of Venezuela. Oikos 53: 176–84.

United States Department of Agriculture, Soil Survey Staff. 1975. Soil Taxonomy: A Basic System of Soil Classification for Making and Interpreting Soil Surveys. Soil Conservation Service, Washington, D.C.

Usher, M. B. 1987. Effects of fragmentation on communities and populations: A review with applications to wildlife conservation. Nature Conservation 103–21.

Valencia, R., H. Balslev, and G. Paz y Miño. 1994. High tree alpha-diversity in Amazonian Ecuador. Biodiversity and Conservation 3: 21–28.

van der Linde, M. 1995. Linking land use and biodiversity Drosophila species as an indicator system. Environmental Management Plan for the Secondary Forest of Sierre Madre Mountains of Region II. Philippines. Report of the Cagayan Valley Program for Development and Environment, Leiden University, Netherlands.

Vandermeer, J. H., M. A. Mallona, D. Boucher, I. Perfecto, and K. Yih. 1995.

Three years of in growth following catastrophic hurricane damage on the Caribbean Coast of Nicaragua: Evidence in
support of the direct regeneration hypothesis. J. Trop. Ecol. 11: 465–71.

Van Roosmalen, M. G. M. 1985. Habitat preferences, diet, feeding strategy, and social
organisation of the black spider monkey
(*Ateles paniscus paniscus,* Linnaeus 1758)
in Suriname. Acta Amazónica 15: 1–238.

Van Roosmalen, M. G. M., and L. L. Klein.
1988. The spider monkeys, genus *Ateles.*
In R. A. Mittermeier, A. B. Ryland, A. F.
Coimbra-Filho, and G. A. B. da Fonseca
(eds.), *Ecology and Behavior of Neotropical Primates,* vol 2. World Wildlife Fund,
Washington, D.C., pp. 455–58.

Van Roosmalen, M. G. M., R. A. Mittermeier,
and K. Milton. 1981. The bearded sakis,
genus Chiropotes. In A. F. Coimbra-Filho
and R. A. Mittermeier (eds.), *Ecology and
Behavior of Neotropical Primates,* vol. 1.
Academia Brasileira de Ciéncias, Rio de
Janeiro, Brazil, pp. 419–41.

Van Shaik, C. P., and E. Mirmanto. 1985.
Spatial variation in the structure and litterfall of Sumatran rainforest. Biotropica
17: 196–205.

Van Valen, L. 1975. Life, death, and energy
of a tree. Biotropica 7: 260–69.

Van Wambeke, A. 1978. Properties and po-
tentials of soils in the Amazon Basin. Interciencia 3: 233–41.

Váquez-Yanes, C., A. Orozco, G. François,
and L. Trejo. 1975. Observations on seed
dispersal by bats in a tropical humid region in Veracruz, Mexico. Biotropica 7:
73–76.

Varela, V. P., and J. D. Santos. 1992. Influence
of shading on the production of *Dinizia
excelsa* (Ducke) seedlings. Acta Amazónica 22: 407–11.

Vasconcelos, H. L. 1988. Distribution of *Atta*
(Hymenoptera–Formicidae) in "terra-firme"
rain forest of Central Amazonia: Density,
species composition and preliminary
results on effects of forest fragmentation.
Acta Amazónica 18: 309–15.

Vasconcelos, H. L. 1991. Mutualism between *Maieta guianensis,* a myrmecophytic melastome, and one of its ant inhabitants: Ant protection against insect
herbivores. Oecologia 81: 295–98.

Vasconcelos, H. L. 1999. Levels of leaf herbivory in Amazonian trees from different
stages in forest regeneration. Acta Amazónica 29: 615–23.

Vasconcelos, H. L. 1999. Effects of forest disturbance on the structure of ground-
foraging ant communities in central Amazonia. Biodiversity and Conservation 8:
409–20.

Vasconcelos, H. L., and J. M. Cherrett. 1995.
Changes in leaf-cutting ant populations
(Formicidae: Attini) after the clearing of
mature forest in Brazilian Amazonia.
Studies on Neotropical Fauna and Envi-
ronment 30: 107–13.

Vasconcelos, H. L., and J. M. Cherrett. 1997.
Leaf-cutting ants and early forest regeneration in central Amazonia: Effects of herbivory on tree seedling establishment.
J. Trop. Ecol. 13: 357–70.

Vázquez-Yánes, C., and A. Orozco-Segovia.
1984. Ecophysiology of seed germination
in the tropical humid forest of the world:
A review. In E. Medina, H. A. Mooney,
and C. Vázquez-Yánes (eds.), *Physiological Ecology of Plants in the Wet Tropics.*
Junk Publishers, The Hague, Netherlands,
pp. 37–50.

Venticinque, E. M. 1995. Dinâmica populacional de Anelosimus eximius (Simom
1891) (Araneae: Theridiidae) em mosaicos
ambientais na Amazônia Central. Tese de
Mestrado apresentado no I. B.–UNESP–
Campus de Botucatu.

Venticinque, E. M., and H. G. Fowler. 1998.
Sheet-web regularity: Fixed allometric relationship in the social spider Anelosimus eximius (Araneae: Theridiidae).
Ciência e Cultura 50: 371–73.

Venticinque, E. M., H. G. Fowler, and C. A.
Silva. 1993. Modes and frequencies of colonization and its relation to extinctions,
habitat, and seasonality in the social spi-

der *Anelosimus eximius* in the Amazon (Araneidae: Theridiidae). Psyche: 35–41.

Veríssimo, A., P. Barreto, M. Matos, R. Tarifa, and C. Uhl. 1992. Logging impacts and prospects for sustainable forest management in an old Amazonian frontier: The case of Paragominas. For. Ecol. & Manag. 55: 169–99.

Veríssimo, A., P. Barreto, R. Tarifa, and C. Uhl. 1995. Extraction of a high-value natural resource in Amazonia: The case of mahogany. For. Ecol. & Manag. 72: 39–60.

Vetter, R. 1995. Untersuchungen Über Zuwachsrhythmen an Tropischen Bäumen in Amazonien. Ph.D. diss., Albert-Ludwigs-Universität, Freiburg, Germany.

Viana, G. 1997. Asian Timber Company Activities in Brazil. External Commission from the Federal House, Brasilia, Brazil.

Viana, V., A. Tabanez, and J. Batista. 1997. Dynamics and restoration of forest fragments in the Brazilian Atlantic moist forest. In W. F. Laurance and R. O. Bierregaard, Jr. (eds.), *Tropical Forest Remnants: Ecology, Management, and Conservation of Fragmented Communities.* University of Chicago Press, pp. 351–65.

Vieira, I. C. G., C. Uhl, and D. Nepstad. 1994. The role of the shrub *Cordia multispicata* Cham. as a "succession-facilitator" in an abandoned pasture, Paragominas, Amazônia. Vegetatio 115: 91–99.

Vieira, L. S. 1975. *Manual da Ciência do Solo.* Editora Agronômica Ceres, São Paulo.

Vieira, L. S., and M. N. F. Vieira. 1983. *Manual de Morfologia e Classificação de Solos.* Editora Agronômica Ceres, São Paulo.

Villanueva, R. 1994. Nectar sources of European and Africanized honey bees (*Apis mellifera* L.) in the Yucatan Peninsula, Mexico. J. Apicultural Research 33: 44–58.

Villard, M. A., G. Merriam, and B. A. Maurer. 1995. Dynamics in subdivided populations of Neotropical migratory birds in a fragmented temperate forest. Ecology 76: 27–40.

Vitousek, P. M. 1984. Litterfall, nutrient

cycling, and nutrient limitation in tropical forests. Ecology 65: 285–98.

Vitousek, P. M. 1994. Beyond global warming: Ecology and global change. Ecology 75: 1861–76.

Vitousek, P. M., and R. L. J. Stanford. 1986. Nutrient cycling in moist tropical forest. Ann. Rev. Ecol. Syst. 17: 137–67.

Vivian, J. M., and D. Ghai (eds.). 1992. *Grassroots Environmental Action: People's Participation In Sustainable Development.* Routledge, London.

Vollrath, F. 1982. Colony foundation in a social spider. Z. Tierpsychology 60: 313–24.

Vollrath, F. 1986a. Eusociality and extraordinarily sex ratios in the spider *Anelosimus eximius* (Araneae: Theridiidae). Behaviour Ecology and Sociobiology 18: 283–87.

Vollrath, F. 1986b. Environment, reproduction, and the sex ratio of the social spider *Anelosimus eximius.* J. Arachnology 14: 267–81.

Walker, B. H., P. Lavelle, and W. Weischet. 1987. Yurimaguas technology. BioScience 37: 638–40.

Walkley, A., and I. A. Black. 1934. An examination of the Degtjareff method for determining soil organic matter and a proposed modification of the chromic acid titration method. Soil Science 37: 29–38.

Wallace, H. M. and S. J. Trueman. 1995. Dispersal of Eucalyptus torelliana seeds by the resin-collecting bee Trigona carbonaria. Oecologia 104: 12–16.

Wandelli, E. 1991. Eco-physiological response of the understory palm *Astrocaryum sociale* to environmental changes resulting from the forest edge effect. M. Sc. Thesis, Cambridge University, U.K.

Waterman, P. G. 1983. Distribution of secondary metabolites in rain forest plants: toward an understanding of cause and effect. In S. L. Sutton, T. C. Whitmore, and A. C. Chadwick (eds.), *Tropical Rain Forest: Ecology and Management.* Blackwell Scientific, Oxford, U.K., pp. 167–79.

Webster C. C., and P. N. Wilson. 1980. *Agri-*

culture in the Tropics, 2nd ed. Longman, London.

White, J. 1980. Demographic factors in plant populations. In O. Solbrig (ed.), *Demography and Evolution in Plant Populations.* University of California Press, Berkeley, pp. 21–48.

White, L. J. T. 1994a. The effects of commercial mechanised selective logging on a transect in lowland rainforest in the Lopé Reserve, Gabon. J. Trop. Ecol. 10: 313–22.

White, L. J. T. 1994b. Biomass of rain forest mammals in the Lope Reserve, Gabon. J. Anim. Ecol. 63: 499–512.

Whitmore, T. C. 1989. Canopy gaps and the two major groups of forest trees. Ecology 70: 536–38.

Whitmore, T. C. 1997. Tropical forest disturbance, disappearance, and species loss. In W. F. Laurance and R. O. Bierregaard, Jr. (eds.), *Tropical Forest Remnants: Ecology, Management, and Conservation of Fragmented Communities.* University of Chicago Press, Chicago, pp. 3–28.

Whittaker, R. H. 1965. Dominance and diversity in land plant communities. Science 147: 250–60.

Wiens, J. A. 1976. Population responses to patchy environments. Ann. Rev. Ecol. Syst. 7: 81–120.

Wiens, J. A. 1994. Habitat fragmentation: Island v. landscape perspectives on bird conservation. Ibis 137: S97–S104.

Wiens, J. A. 1995. Landscape mosaics and ecological theory. In L. Hansson, L. Fahrig and G. Merriam (eds.), *Mosaic Landscapes and Ecological Processes.* Chapman and Hall, London, pp. 1–26.

Wiens, J. A., and M. R. Moss (eds.). 1999. *Issues in Landscape Ecology.* International Association of Landscape Ecology, Guelph, Canada.

Wilkinson, L. 1990. *Systat: System for Statistics.* SYSTAT, Evanston, Ill.

Williams, C. B. 1943. Area and number of species. Nature 152: 264–67.

Williams, N. H., and C. H. Dodson. 1972. Selective attraction of male euglossine bees to orchid floral fragrances and its importance in long distance pollen flow. Evolution 26: 84–95.

Williams, N. H., and W. M. Whitten. 1983. Orchid floral fragrances and male euglossine bees: Methods and advances in the last sesquidecade. Biol. Bull. 164: 355–95.

Williams-Linera, G. 1990a. Origin and early development of forest edge vegetation in Panama. Biotropica 22: 235–41.

Williams-Linera, G. 1990b. Vegetation structure and environmental conditions of forest edges in Panama. J. Ecology 78: 356–573.

Williamson, G. B., R. de C. G. Mesquita, K. Ickes, and G. Ganade. 1998. Estratégias de árvores pioneiras nos Neotrópicos. In C. Gascon and P. Moutinho (eds.), *Floresta Amazônica: Dinâmica, Regeneração e Manejo.* Instituto Nacional de Pesquisas da Amazônia (INPA), Manaus, Amazonas, Brazil, pp. 131–44.

Williamson, G. B., G. E. Schatz, A. Alvarado, C. S. Redhead, A. C. Stam, and R. W. Sterner. 1986. Effects of repeated fires on tropical paramo vegetation. J. Trop. Ecol. 27: 62–69.

Williamson, M. 1981. *Island Populations.* Oxford University Press, Oxford, U.K.

Willis, E. O. 1979. The composition of avian communities in remanescent woodlots in southern Brazil. Papéis Avulos de Zoologia 33: 1–25.

Willis, E. O., and Y. Oniki. 1978. Birds and army ants. Ann. Rev. Ecol. Syst. 9: 243–63.

Willis, E. O., D. Wechsler, and Y. Oniki. 1978. On behavior and nesting of McConnell's flycatcher (*Pipromorpha macconnelli*): Does female rejection lead to male promiscuity? Auk 95: 1–8.

Willson, M. F., and F. H. J. Crome. 1989. Patterns of seed rain at the edge of a tropical Queensland rain forest. J. Trop. Ecol. 5: 301–8.

Wilms, W., V. L. Imperatriz-Fonseca, and W. Engels. 1996. Resource partitioning between highly eusocial bees and possible impact of the introduced Africanized honey bee on native stingless bees in the Brazilian Atlantic forest. Studies on Neotropical Fauna and Environment 31: 137–51.

Wilson, C. C., and W. L. Wilson. 1975. The influence of selective logging on primates and some other animals in East Kalimantan. Folia Primatologia 23: 245–74.

Wilson, E. O. 1958. Patchy distribution of ant species in New Guinea rain forests. Psyche 65: 26–38.

Wilson, E. O. 1971. *Insect Societies*. Harvard University Press, Cambridge, Mass.

Wilson, E. O. 1987a. The little things that run the world (the importance and conservation of invertebrates). Conserv. Biol. 1: 344–46.

Wilson, E. O. 1987b. The arboreal ant fauna of Peruvian Amazon forests: A first assessment. Biotropica 19: 245–51.

Wilson, E. O. 1988. The current state of biological diversity. In E. O. Wilson and F. M. Peter (eds.), *Biodiversity*. National Academy Press, Washington, D.C., pp. 3–18.

Wilson, E. O., and E. O. Willis. 1975. Applied Biogeography. In M. L. Cody and J. M. Diamond (eds.), *Ecology and Evolution of Communities*. Belknap/Harvard, Cambridge, Mass., pp. 522–36.

Wilson, W. L., and A. D. Johns. 1982. Diversity and abundance of selected animal species in undisturbed forest, selectively logged forest and plantations in East Kalimantan, Indonesia. Biological Conservation 24: 205–18.

Winston, M. L. 1992. The biology and management of Africanized honey bees. Ann. Rev. Entomol. 37: 173–93.

Wittmann, D., M. Hoffmann, and E. Scholz. 1988. Southern distributional limits of euglossine bees in Brazil linked to habitats of the Atlantic and Subtropical Rain Forest (Hymenoptera, Apidae, Euglossini). Entomol. Gener. 14: 53–60.

Wolda, H. 1977. Fluctuation in abundance of some Homoptera in a Neotropical forest. Geo-Eco-Tropica 31: 229–57.

Wolda, H. 1978. Seasonal fluctuations in rainfall, food, and abundance of tropical insects. J. Anim. Ecol. 47: 369–81.

Wolda, H. 1992. Trends in abundance of tropical forest insects. Oecologia 89: 47–52.

Wolda, H., and D. W. Roubik. 1986. Nocturnal bee abundance and seasonal bee activity in a Panamanian forest. Ecology 67: 426–33.

Woodroffe, R., and J. R. Ginsberg. 1998. Edge effects and the extinction of populations inside protected areas. Science 280: 2126–28.

Woods, P. 1989. Effects of logging, drought, and fire on structure and composition of tropical forests in Sabah, Malaysia. Biotropica 21: 290–98.

Worbes, M., and W. J. Junk. 1989. Dating tropical trees by means of C-14 from bomb tests. Ecology 70: 503–7.

Wright, I., J. Gash, H. da Rocha, W. Shuttleworth, C. Nobre, G. Maitelli, C. Zamparoni, and P. Carvalho. 1992. Dry season micrometeorology of central Amazonian ranchland. Quarterly J. Royal Meteorological Soc. 118: 1083–99.

Wright, S. 1965. The interpretation of population structure by F-statistics with special regard to systems of mating. Evolution 19: 395–420.

Young, A. 1976. *Tropical Soils and Soil Survey*. Cambridge University Press, Cambridge, U.K.

Young, A., T. Boyke and T. Brown. 1996. The population genetic consequences of habitat fragmentation for plants. Trends Ecol. Evol. 11: 413–18.

Zagatto, E. A. C., A. C. Jacintho, B. F. Reis, F. J. Krug, F. H. Bergamin, L. C. R. Pessenda, S. Moratti and M. F. Gine. 1981. *Manual de Análises de Plantas e Águas Empregando Sistemas de Injeção de Fluxo*. Centro de Energia Nuclear na Agricultura/ Universidade de São Paulo (CENA/USP), Piracicaba, São Paulo.

Zar, J. H. 1984. *Biostatistical Analysis*. Prentice-Hall, Englewood Cliffs, N.J.

Zhang, S-Y. 1994. Utilisation de l'espace, stratégies alimentaires et rôle dans la dissémination des graines du singe capucin *Cebus apella* (Cebidae, Primates) en Guyane française. Ph.D. diss., Université de Paris VI Pierre et Marie Curie, France.

Zhang, S-Y. 1995a. Activity and ranging patterns in relation to fruit utilisation by brown capuchins (*Cebus apella*) in French Guiana. International J. Primatology 16: 489–507.

Zhang, S-Y. 1995b. Sleeping habits of brown capuchin monkeys (*Cebus apella*) in French Guiana. American J. Primatology 36: 327–35.

Zimmerman, B. L. 1991. Distribution and abundances of forest frogs at a site in central Amazonia. Ph.D. diss., Florida State University, Tallahassee.

Zimmerman, B. L., and R. O. Bierregaard, Jr. 1986. Relevance of the equilibrium theory of island biogeography and species-area relations to conservation with a case from Amazonia. J. Biogeography 13: 133–43.

Zimmerman, B. L., and M. T. Rodrigues. 1990. Frogs, snakes, and lizards of the INPA-WWF reserves near Manaus, Brazil. In A. H. Gentry (ed.), *Four Neotropical Rainforests*. Yale University Press, New Haven, Conn., pp. 426–54.

Contributors

John B. Adams
Geological Sciences
University of Washington
P.O. Box 351310
Seattle, WA 98103-1310

Peter Becker
Biology Department
Universiti Brunei Darussalam
Gadong
Brunei 3186

Julieta Benitez-Malvido
Universidad Nacional Autónoma de
    México
Departamento de Ecología de los Recursos
    Naturales
Unidad Académica Morelia
Antigua Carretera a Pátzcuaro No. 8701
Col. Ex-Hacienda de San Miguel de la
    Huerta
Morelia, Michoacán
México
jbenitez@ate.oikos.unam.mx

Richard O. Bierregaard, Jr.
Biology Deptartment
University of North Carolina at Charlotte
9201 University City Blvd.
Charlotte, NC 28223
rbierreg@email.uncc.edu

Sérgio H. Borges
Museu Paraense Emílio Goeldi
Departamento de Zoologia
Av. Perimetral 1901/1907—Terra Firme
66077-530 Belém, PA
Brazil

Valerie K. Brown
Department of Agriculture
University of Reading
Earley Gate
P.O. Box 236
Reading RG6 6AT
United Kingdom

Karine S. Carvalho
Departamento de Ciéncias Biológicas
Universidade do Sudoeste da Bahia
45200-000 Jequié BA
Brazil

Jeffrey Q. Chambers
University of California
Ecology, Evolution, and Marine Biology
Santa Barbara, CA 93106
chambersjq@yahoo.com

Jacques H. C. Delabie
Laboratório de Mirmecologia
CEPEC/CEPLAC
C. Postal 07
45600 Itabuna, BA
Brazil

Og DeSouza
Departamento de Biologia Animal
Universidade Federal de Viçosa
36.571-000 Viçosa MG

Brazil
og.souza@mail.ufv.br

Christopher Dick
Harvard University Herbaria
22 Divinity Ave.
Cambridge, MA 02138
dickc@naos.si.edu

Raphael K. Didham
Zoology Department
University of Canterbury
Private Bag 4800
Christchurch
New Zealand
r.didham@zool.canterbury.ac.nz

Tim van Eldik
Mil Madeireira Itacoatiara Ltda.
Itacoatiara, AM
Brazil

Philip M. Fearnside
National Institute for Research in the
    Amazon (INPA)
C. Postal 478
69011-970 Manaus, AM
Brazil
pmfearn@inpa.gov.br

Carlos Roberto Fonseca
Departamento de Zoologia, IB, CP 6109
UNICAMP—Universidade Estadual de
    Campinas
Campinas, SP 13081-970
Brazil

Harold G. Fowler
UNESP—Universidade Estadual Paulista
Campus de Rio Claro
Departamento de Ecologia
Rio Claro, SP 13500
Brazil

Gislene Ganade
Departamento de Zoologia, IB
C. Postal 6109
UNICAMP—Universidade Estadual de
    Campinas

Campinas, SP 13081-970
Brazil
gganade@obelix.unicamp.br

Claude Gascon
Vice-President for Field Support Programs
Conservation International
1919 M St., Suite 600
Washington, DC 20036
c.gascon@conservation.org

Kellen A. Gilbert
Department of Sociology and Criminal
    Justice
Southeastern Louisiana University
Hammond, LA 70402-0736
kgilbert@selu.edu

Niro Higuchi
National Institute for Research in the
    Amazon (INPA)
C. Postal 478
69011-970 Manaus, AM
Brazil
niro@inpa.gov.br

Roger W. Hutchings H.
CPEC-INPA
C. Postal 478
69011-970 Manaus-AM
Brazil

Paulo Y. Kageyama
Escola Superior de Agricultura "Luiz de
    Queiroz"
Universidade de São Paulo/ESALQ/USP
C. Postal 09, Depto de Ciências
Florestais, 13418-970
Piracicaba, SP
Brazil

Valerie Kapos
World Conservation Monitoring Centre
219 Huntingdon Road
Cambridge CB3 0DL
United Kingdom

Dwight Kincaid
Department of Biological Sciences
Lehman College

Bedford Park Boulevard W.
Bronx, NY 10468

William F. Laurance
Smithsonian
Tropical Research Institute
Apartado 2072, Balboa
Panama
wfl@inpa.gov.br

Niwton Leal Filho
Instituto Nacional de Pesquisas da
    Amazônia (INPA)
C. Postal 478
69011-970 Manaus, AM
Brazil

Nadja Lepsch-Cunha
Projeto Dinémica Biológica dos Fragmentos
    Florestais (PDBFF)
Ecologia
Instituto Nacional de Pesquisas da
    Amazônia (INPA)
C. Postal 478
69011-970 Manaus, AM
Brazil
nadja@inpa.gov.br

Miles Logsdon
School of Oceanography
POB 357940
University of Washington
Seattle WA 98195-7940
mlog@u.washington.edu

Thomas E. Lovejoy
SI-230
Smithsonian Institution
1000 Jefferson Drive, S.W.
Washington, D.C. 20560

Jay R. Malcolm
Faculty of Forestry
University of Toronto
Toronto, Ontario M5S 3B3
Canada
jay.malcolm@utoronto.ca

Marlúcia B. Martins
Depto de Zoologia
Museu Paraense Emílio Goeldi/CNPq/MCT
C. Postal 399
66000 Belém, PA
Brazil
marlucia@museu-goeldi.br

Rita C. G. Mesquita
Departmento de Ecologia
Instituto Nacional de Pesquisas da
    Amazônia (INPA)
C. Postal 478
69011-970 Manaus, AM
Brazil

Joel Meyer
Biological Dynamics of Forest Fragments
    Project
Instituto Nacional de Pesquisas da
    Amazônia (INPA)
C. Postal 478
69011-970 Manaus, AM
Brazil

Scott A. Mori
Institute of Systematic Botany
New York Botanical Garden
Bronx, NY 10458-5126
smori@nybg.org

Márcio Luiz de Oliveira
Universidade Federal do Acre
Departamento de Ciéncias da Natureza
BR 364, Km 4 69915.900
Rio Branco AC
Brazil
mlolivei@ufac.br

Judy M. Rankin–de Mérona
25 Domaine du Montabo
97300 Cayenne
France
rankin-de-merona@mdi-guyane.fr

Aldicir Scariot
CENARGEN/EMBRAPA
C. Postal 02373
70.770-900 Brasília, DF

Brazil
scariot@cenargen.embrapa.br

José H. Schoereder
Departamento de Biologia Geral
Universidade Federal de Viçosa
36.571-000 Viçosa, MG
Brazil

Eleonore Z. F. Setz
Zoologia—I.C.B.
UNICAMP—Universidade Estadual de
    Campinas
Campinas, São Paulo
Brazil

John Southon
Center for AMS, L-397
Lawrence Livermore National Laboratory
Livermore, CA 94551-9900

Wilson R. Spironello
Biological Dynamics of Forest Fragments
    Project
Instituto Nacional de Pesquisas da
    Amazônia (INPA)
C. Postal 478
69011-970 Manaus, AM
Brazil
wilson@inpa.gov.br

Philip C. Stouffer
Department of Biological Science
Southeastern Louisiana University
Hammond, LA 70402-0736
stouffer@selu.edu

Mandy D. Tocher
Science and Research Unit
Science, Technology, and Information
    Services
Department of Conservation
P.O. Box 5244
77 Stuart Street
Dunedin
New Zealand
mtocher@doc.govt.nz

Heraldo L. Vasconcelos
Coordenação de Pesquisas em Ecologia
Instituto Nacional de Pesquisas da
    Amazônia (INPA)
C. Postal 478
69011-970 Manaus, AM
Brazil
heraldo@inpa.gov.br

Eduardo Martins Venticinque
Instituto Nacional de Pesquisas da
    Amazônia (INPA)
C. Postal 478
69011-970 Manaus, AM
Brazil
edmventi@inpa.gov.br

G. Bruce Williamson
Dept. of Biology
Louisiana State University
Baton Rouge, LA 70803
btwill@inpa.gov.br

Changes in forest dynamics, 159–160

Checkerboard landscapes, 346–347

Chloroplast DNA genome (cpDNA), 92

Clay. *See* Soils

Clear-cutting, 5, 29, 119, 344

Clearings, avoidance by animals, 374

CNPq. *See* Brazilian National Research Council

Coarse-grained landscape, 346–57

Co-evolved species associations, 52, 99, 373, 374

Colonization: federally sponsored, 26–27; shape effects in, 20

Community-based conservation, 384

Connectivity and faunal corridors, 166. *See also* Corridors

Conservation, 6–7, 10, 379–84; community based, 384; education and, 384; employment opportunities and, 384; endangered species orientation, 6–7; *ex situ* efforts, 45; focus of efforts, 381–82; forest fragment, minimum size for, 379–81; land-use change and, 30; people's role, 384; principles of, 371–85; protected area size debate, 7; research projects and, 384; understory birds in fragments and second-growth areas, 248–61; urgency around Manaus, 382

Conservation Biology, 6–7

Conservation lessons: abandoned pastures and forest restoration, 323–24; African honeybees and *Dinizia excelsa* (Fabaceae), 157; ants, 206–7; biological dynamics in fragmented ecosystem, 21; brown capuchin's ecology and home range requirements, 283; deforestation, 30, 119–20; Drosophilid fruit-fly guilds, 185–186; extending models to diverse landscape configurations, 357; fire effects on regeneration, 333–334; frogs in matrix and continuous forest, 246; genet-

ics of rare tropical trees, 92–95; hyperdiverse flora of central Amazon, 52–53; invertebrates, 228–29; landscape fragmentation effects on palm communities, 135; Lecythidaceae study plot, 66; Phytodemographic Project, 167; primates, 270; regeneration in tropical rainforest fragments, 145, 157; selective logging, 345; social spiders, 197–98; soil and development, 311–12; synthesis of habitat fragmentation and plant communities, 167; tree age structure, 77; understory birds, 260–61

Constitution of Brazil and zoning, 293–94

Continuous forest: fragments compared, 374–76; frogs in, 235–47; invertebrates in, 222

Coprophagus beetles, 373–74

Core-area model for rainforest fragments, 165, 381

Corners of fragments. *See* Forest fragmentation

Corridors: edge penetration effects, 380; faunal, 2, 23, 102–3, 106, 166, 286; forested, 270; land-use guidelines, 383–84; maximally effective, 383; retention of habitat, 383; use by flora and fauna, 2

Crossroads for species migration, 52–53, 382

Decomposition rates, changes in, 99, 375

Deer, vulnerability to hunting, 373

Defensive compounds, 48–49

Definitions: density compensation, 18; ecosystem fragmentation, 13; endogamy, 18; genetic drift, 18; guilds, 175; habitat fragmentation, 13, 97; immigration rate, 14; inbreeding depression, 18, 82; local extinction, 195; matrix habitat, 22; mesoscale processes, 16; metapopulations, 170; outbreeding depression, 18; point species richness, 205; sociotomy, 188; species density, 205; terra firme forest, 47

Deforestation, 22–30, 335–36;

acceleration of, 22–30, 371; amount of, 25; bees and, 146; Brazilian natural resource management, 25; buffer zones, 24; causes of, 2–3, 22; central Amazonia, 27–30; characterizing changing spatial structure of landscape, 358–68; conservation lessons, 30; current trends, 372; development and, 25–27, 372; drivers of, 97; eastern Amazonia, 25–26; edge effects, 107–20; farming and, 26; federally sponsored colonization, 26–27; fire and, 29; fishbone pattern, 27, 37, 120, 371–73; food web distortion, 162; forest fragmentation and, 22–30; frogs and, 235–47; habitat corridors, retention of, 383; human migration and, 371; IBAMA, 25, 342, 383; landscape grain, 346–57; "Legal" Amazon, 336; Manaus area, 382; pasture creation, 24; process of, 346; reduction through land-use guidelines, 383–84; remote sensing and spatial analysis of, 358–68; results of, 25, 26, 28, 371; road construction and, 24, 25, 28, 371–72, 382–83; settlement restrictions, 29–30; spatially nonrandom, 371–72; states affected, 25; timber extraction, 24, 28–29; tree mortality, damage, and recruitment, 107–20; urban populations and, 29–30; yearly gross rate of, 338. *See also* Brazilian Forest Code; Edge effects; Forest fragmentation; Habitat fragmentation; Logging; Road construction

Density compensation, 18–19

Developed areas, planning and management, 285–368; abandoned pastures and forest restoration, 313–24; fire's effect on forest regeneration, 325–34; model of edge effects from neotropics, 346–57; remote sensing to view landscape in entirety, 358–68; selective logging and tropical hardwood market, 335–45; soils, 291–312

Smithsonian's Museum of Natural History: BDFFP with,
10–11; Biodiversity Program,
10; BIOLAT Program, 10;
voucher specimens, 58
Snakes in primary and second
growth forests, 377
SNI (National Information Service), 294
Social spiders (*Anelosimus eximius*), 169–70, 187–98; conservation lessons, 197–98;
described, 188–89; edge effects, 196; environmental
stochasticity, 195–96; implications of metapopulation
dynamics, 197; life span,
169–70, 196; local extinction,
187–98; as metapopulation,
193–97; population dynamics,
189–93; precipitation and,
190–91, 193; predation and,
196; prey availability, 196;
rainfall, 193, 196; rescue
effect, 197; seasonal fluctuations of populations, 169–70;
sociotomy, 188; structural differences in vegetation, 196;
successional mosaic and, 197;
tramp species, 197
Sociotomy, 188
Soils, 291–312; agricultural potential, 305–8; aluminum toxicity, 285, 297–98, 306; in
BDFFP, 33, 39, 48–49; biomass and, 309–10; chrono-
sequences, 310; classification
of, 292–93; climatic changes
and, 311; conservation lessons,
311–12; development in Amazonia and, 285–86, 291–312;
distribution of natural vegetation and, 309–10; fertility, 48,
305–8, 372, 383–84; fertilizers, 307–8, 315; greenhouse
gases and, 310; importance of
soil properties, 294–99; lessons for development, 305–9;
phytoliths, 64; prospects for
resilience, 309; response to
forest fragmentation, 310–11;
survey of BDFFP, 39, 291–312;
tree species occurrence and,
309–310; variability, 293;
water capacity, 301
Solitary wasps, 373
Soybeans, 306, 371, 372
Species-area curves, 65

Species-area relationships,
14–15
Species richness: ants, 205;
birds, 232; in forest fragments,
13–21, 97–99, 381–82. *See
also specific species*
Spiders, social, 169–70, 187–98.
See also Social spiders
Steep slope forests, land-use
guidelines, 383–84
Stochastic effects, 18, 373; social
spiders and, 195–96
Stream buffer zones, 24–25
Students in BDFFP, 6, 9, 10, 384
Successional species, 159–60
Suppressed trees, 70
Sustainable forestry, 75–76
Swallow-tailed kites, 267
Synthesis of lessons from frag-
mented forest, 106, 167,
371–85; implications for land-
scape management, 158–67

Tamarins: golden-handed (*Saguinus midas*), 232, 266; Golden-
Lion project, 384; reintroduction of, 384. *See also* Primates
Tapirs, 373, 379
Tax-exempt status of Manaus, 29
Termites, vulnerability to fragmentation effects, 373
Terra firme forest defined, 47
Timber extraction, 24; clear-
cutting vs. low-impact logging,
29; concessions, 372; conserved tree cover requirement,
24; deforestation and, 28–29;
as driver of expansion, 373;
ecosystem conservation, 345;
faunal impact, 29; fire and, 29,
325–26; investment in, 28;
markets, 71; 100-meter buffer
zones, 24; production, 344;
scarce stocks, 25. *See also*
Logging
Topography surveys of BDFFP,
39
Tramp species, 197
Tree age study, 68–77; [14]C
dating, 68, 69–70, 76; carbon
cycling, 74–75, 77; *Cariniana
micrantha,* 70; central Amazon tropical forests, 68–77;
commercial silvicultural
strains, 74; conservation lessons from, 77; correlation of
size and age, 70; difficulty in
dating, 68; discussion, 70–71;

Frankfurt Biosphere Model,
74; genetic diversity, 71–74;
largest trees, 70; mahogany, 71;
mega-ENSO events, 71; methods for determining, 68–70;
Mil Madeireira logging operation, 68, 71; oldest trees, 70;
pathways to reach canopy, 70;
population dynamics and conservation, 70–71; radiocarbon
dates, 44–45, 69–70, 76; results of, 70; shade-tolerant,
slow-growing species, 70; suppressed trees, 70; sustainable
forestry and, 75–76; timber
markets, 71; tree rings in tropical forests, 68, 69, 76–77; tropical rainforest trees, 44–45
Tree families in BDFFP, 50
Tree islands, 331
Tree mortality, 107–20, 159–60,
162–64, 310–11, 376; burn to
clear felled trees and, 118;
causes of, 117–19; edge-core
comparisons, 109–14; location
of study, 108; methods of
study, 109; results of study,
109–17; undisturbed continuous forest and forest fragment
comparisons, 114–17; wind
damage, 117
Tree phenology, 141, 150,
152–54
Trees. *See specific types*
Tropical forests. *See* Deforestation; Forest fragmentation;
Logging; Timber extraction
Tropical hardwoods: dwindling
stocks outside of Amazon,
372; environmental impact of
land use, 344; evaluation of
forest management in Brazilian Amazon, 342–45; future
growth of timber extraction,
28; international market,
337–42; predicted shift in
market, 341–42; production of,
336–37; roundwood, 71; tree
age structure in tropical
forests, 68–77; value of,
287–88. *See also* Logging;
Timber extraction; Tropical
trees
Tropical trees: age of, 44–45,
68–77; deforestation effect at
edge of fragments, 107–20; *dinizia excelsa* (Fabaceae) case
study, 151–57; genetics of rare

Taxa Index